Lecture Notes of the Institute for Computer Sciences, Social Informatics and Telecommunications Engineering 13

Fabrizio Granelli Charalabos Skianis
Periklis Chatzimisios Yang Xiao
Simone Redana (Eds.)

Mobile Lightweight Wireless Systems

First International ICST Conference, MOBILIGHT 2009
Athens, Greece, May 18-20, 2009
Revised Selected Papers

 Springer

Volume Editors

Fabrizio Granelli
University of Trento
I-38050 Trento, Italy
E-mail: granelli@disi.unitn.it

Charalabos Skianis
University of the Aegean
83200 Karlovasse, Samos, Greece
E-mail: cskianis@aegean.gr

Periklis Chatzimisios
University of Macedonia
59200 Naousa, Greece
E-mail: pchatzimisios@ieee.org

Yang Xiao
University of Alabama
Tuscaloosa, AL 35487-0290 USA
E-mail: yangxiao@ieee.org

Simone Redana
Nokia Siemens Networks
81541 Munich, Germany
E-mail: simone.redana@nsn.com

Library of Congress Control Number: 2009934005

CR Subject Classification (1998): C.2, C.2.1, C.1.3, C.3, I.5.4, C.2.5

ISSN   1867-8211
ISBN-10   3-642-03818-2 Springer Berlin Heidelberg New York
ISBN-13   978-3-642-03818-1 Springer Berlin Heidelberg New York

springer.com

© ICST Institute for Computer Science, Social Informatics and Telecommunications Engineering 2009
Printed in Germany

Typesetting: Camera-ready by author, data conversion by Scientific Publishing Services, Chennai, India
Printed on acid-free paper   SPIN: 12733057   06/3180   5 4 3 2 1 0

Preface

The First International Conference on Mobile Lightweight Systems (MOBILIGHT) was held in Athens during May 18–20, 2009.

The decision to organize a scientific event on wireless communications, where competition is really enormous, was motivated by discussions with some colleagues about the current unprecedented request for lightweight, wireless communication devices with high usability and performance able to support added-value services in a highly mobile environment. Such devices follow the user everywhere he/she goes (at work, at home, while travelling, in a classroom, etc.), but also result in exciting research, development and business opportunities.

Such a scenario clearly demands significant upgrades to the existing communication paradigm in terms of infrastructure, devices and services to support the anytime, anywhere, any device philosophy, introducing novel and fast-evolving requirements and expectations on research and development in the field of information and communication technologies. The core issue is to support the desire of wireless users to have 24/7 network availability and transparent access to "their own" services.

In this context, we envisioned an international forum where practitioners and researchers coming from the many areas involved in lightweight wireless systems design and deployment would be able to interact and exchange experiences. For this reason, MOBILIGHT was targeted to information exchange and cross-fertilization among the different worlds of academia, research centers and industry through the organization of specific and interacting tracks related to: (a) technology, including wireless (WPAN, WLAN, WMAN/cellular) as well as architectures and design methodologies to support seamless access to the communication facility; (b) services, in the vision of "always on" requirement; (c) business models, opportunities and solutions.

The final technical program was organized in such a way as to provide keynote speeches, tutorials and panels in the morning, and technical sessions and workshops in the afternoon.

At MOBILIGHT 2009, keynote speakers covered up-to-date research and development topics, including future Internet services architecture and research funding opportunities within the European Space Agency. Tutorials focused on power-saving mechanisms in IEEE 802.16e/m and middleware for RFID network deployment and operation.

Panels were aimed at allowing discussion between the audience and top-level experts from research and industry. Selected topics this year included: "Anytime, Anywhere, Any Device: The (Wireless) Way Forward" and "Mobile Lightweight Wireless Systems - Rising to the Challenge!".

The technical program of the afternoon focused on 40 high-quality selected papers (both from an open call and by invitation), providing current advancements in the fields related to mobile lightweight services, networks and devices. Some special sessions included the latest results from research projects funded by the European

Commission in the framework of mobility and networking, including NEWCOM++, PASSENGER, LOOP.

The effort for organizing such an event was huge, and organizing it for the first time made it even more difficult, requiring hard teamwork and true dedication. For this reason, and for their outstanding contributions, we like to express our gratitude to all the members of the Organizing Committee, who really did a wonderful job. In particular, our gratitude goes to the Technical Co-chairs, Periklis, Simone and Yang, for their constant support and fruitful suggestions, to ICST and CREATE-NET for providing technical and financial sponsorship of the event and, last but not least, to Imrich Chlamtac, for his precious suggestions and his vision. But most of all, we would like to thank the authors and contributors for trusting the Organizing Committee and giving us the chance to set up a high-level technical program. Lastly, we would also like to thank all Technical Committee members and additional reviewers for the thorough review reports and constructive remarks that ensured a high technical quality level.

Fabrizio Granelli
Harry Skianis

Organization

Steering Committee Chair

Imrich Chlamtac   Create-Net, Italy

General Co-chairs

Fabrizio Granelli   University of Trento, Italy
Charalabos Skianis   University of Aegean, Samos, Greece

Technical Program Committee Co-chairs

Periklis Chatzimisios   University of Macedonia, Greece
Simone Redana   Nokia Siemens Networks, Munich, Germany
Yang Xiao   University of Alabama, USA

Tutorials / Workshops Chair

Alexey Vinel   Russian Academy of Sciences, Russia
Christos Verikoukis   CTTC – SPAIN

Publications Chair

Mauro Biagi   University of Rome, Sapienza, Italy

Publicity / Sponsorship Chair

Qiang Ni   Brunel University, Uxbridge, The West of London, UK

Panel Chair

Adlen Ksentini   IRISA University of Rennes, France

Local Arrangements Chair

Nikos Papaoulakis   National Technical University of Athens, Greece
Charalampos Z. Patrikakis   Agricultural University of Athens, Greece

Web Chair

Eirini Karapistoli   Aristotle University of Thessaloniki, Greece

Conference Coordinator

Gergely Nagy   ICST

Technical Program Committee

Ozgur B. Akan   Middle East Technical University, Turkey
Chadi Assi   Concordia University, Canada
Albert Banchs   Universidad Carlos III Madrid, Spain
Paolo Bellavista   University of Bologna, Italy
Christos Bouras   RACTI, University of Patras, Greece
Andrea F. Cattoni   DIBE - University of Genoa, Italy
Periklis Chatzimisios   University of Macedonia, Greece
Hsiao-Hwa Chen   National Cheng Kung University, Taiwan
Hui Chen   Virginia State University, USA
Tasos Dagiuklas   Tech. Institute of Messolonghi, Greece
Spyros Denazis   Hitachi Europe, France
Dimitris Dernikas   AIRCOM International, UK
Xiaojiang Du   North Dakota State University, USA
Chuan Heng Foh   Nanyang Tech. University, Singapore
Nelson Fonseca   State University of Campinas, Brazil
Alex Gluhak   Ericsson LMI, Ireland
Fabrizio Granelli   DISI - University of Trento, Italy
Jussi Haapola   CWC, Finland
Ibrahim Habib   City University of New York, USA
Laurent Herault   MINATEC, France
Fei Hu   Rochester Institute of Tech., USA
Athanasios Kanatas   University of Piraeus, Greece
Helen Karatza   Aristotle University of Thessaloniki, Greece
George Karetsos   Tech. Research Center Thessaly, Greece
Tom Karygiannis   National Institute of Standards and Tech., USA
Dzmitry Kliazovich   DISI - University of Trento, Italy
George Kormentzas   University of Aegean, Greece
Yevgeni Koucheryavy   Tampere University of Technology, Island
Fotis Lazarakis   NCSR "Demokritos", Greece
Haizhon Li Li   University of South Carolina, USA
Ming Li   California State University, USA
Michael D. Logothetis   DECE - University of Patras, Greece
Pascal Lorenz   University of Haute Alsace, France
Stefan Mangold   Swisscom, Switzerland
Athanassios Manikas   Imperial College London, UK

Table of Contents

Thorough Analysis of Downlink Capacity in a WCDMA Cell

Ioannis B. Daskalopoulos, Vassilios G. Vassilakis, and Michael D. Logothetis

WCL, Dept. of Electrical & Computer Engineering, University of Patras, Patras, Greece
{idaskalopou,vasilak,m-logo}@wcl.ee.upatras.gr

Abstract. In WCDMA networks, the Call Blocking Probability (CBP) assessment is necessary for proper cell capacity determination, in respect of traffic load in erlangs, and network dimensioning. This paper focuses on the downlink capacity estimation, through CBP calculation in a WCDMA cell. To this end, we study an analytical model for the WCDMA cell by taking into consideration the effects of the following: the multi-service environment, the soft blocking, the imperfect power control and multipath propagation. In this model, the maximum transmission power of a base station in the downlink is considered as the shared system resource. To analyze the system, we follow the methodology proposed by Mäder & Staehle, and describe the WCDMA cell by a Markov chain, where each system state represents a certain number of resources occupied by mobile users. We solve the Markov chain and provide an efficient recurrent formula for the system occupancy distribution, as well as the so-called local blocking probabilities. Based on them, we calculate the CBP of different service-classes accommodated in the cell, versus the total offered traffic load. We evaluate the analytical model through simulation. The results show that the accuracy of the model is very satisfactory. The main contribution of this paper is the improved determination of several parameters involved in the downlink capacity estimation, in comparison to the calculations that appear in the work of Mäder & Staehle. In addition, we show the effect of the intra-cell interference (due to orthogonality factor) on the erlang capacity of the cell.

Keywords: WCDMA, downlink capacity, soft blocking, Markov chain, call blocking probability.

1 Introduction

Each cell covering a geographical area of a mobile cellular network is controlled by a *Base Station* (BS) which is named NodeB in Wideband Code Division Multiple Access (WCDMA) networks. Most of Third Generation (3G) networks operate with WCDMA over the air interface. The system bandwidth in WCDMA is 5MHz with 3.84 Mcps system chiprate [1]. WCDMA networks support applications with different QoS requirements and rates, while offering wide range of voice and data services.

Second Generation (2G) systems were designed for symmetric traffic such as voice and SMS. The 3G systems have introduced services, such as multimedia, internet and video stream, which have asymmetric traffic. Given that the offered-traffic load is

F. Granelli et al. (Eds.): MOBILIGHT 2009, LNICST 13, pp. 1–14, 2009.

heavier in the downlink than in the uplink, the downlink plays more important role, rather than uplink, in the cell capacity determination [1]. Especially in WCDMA networks, the downlink capacity determination is complicated because of soft blocking, multipath propagation, and intra-/inter-cell interferences.

In WCDMA systems, two blocking considerations are possible. The first is the *hard blocking* while the second is the *soft blocking*. Hard blocking means that blocking of a call occurs with probability one in some system states (*hard capacity*), while no blocking occurs (the blocking probability is zero) in all other states. On the other hand, soft blocking means that blocking of a call may occur in every system state with some probability. Soft blocking is a result of inter-cell interference, pseudo-orthogonality, multipath propagation and thermal noise. Due to the soft blocking, in WCDMA systems the resultant capacity is not deterministic, but it is a stochastic value. Thus, we talk about *soft capacity*. We consider both *hard* and *soft capacity*.

For the analysis of traditional connection-oriented networks with Poisson arriving calls, the well-known Erlang Multirate Loss Model (EMLM) is used (also known as Kauffman and Roberts (KR) recursion) [2], [3]. This is a recurrent formula that achieves efficient and accurate calculation of Call Blocking Probabilities (CBP). This recursion has been extended for the CBP calculation in the uplink of WCDMA systems [4]-[7]. In [4] CBP are calculated for Poisson arriving calls, imperfect power control, user activity and inter-cell interference. In [5], the authors calculate CBP in the WCDMA uplink, using the cell load estimation method based on the wideband received power. This work was further extended in [6] by providing an explicit distinction between the new and the handoff calls. In [7], the authors calculate CBP in the WCDMA uplink, using a throughput-based cell load estimation method.

As far as CBP calculation in the downlink of a WCDMA system is concerned, little progress has been done in comparison to the uplink. In [8], an analytical method is proposed which results in a closed formula for CBP calculation, in the absence of multipath signal propagation. In the same paper, in the case of multipath signal propagation, a Chernoff bound is determined for CBP. In [9] the CBP calculation is based on an analytical model which takes into account multiple service-classes, user activity, imperfect power control and multipath propagation. In this model, the maximum transmission power of the base station in the downlink is considered as the shared system resource. Based on this model, Mäder & Staehle (the authors of [9]) have developed an algorithm for the CBP determination per service-class.

In this paper, to analyze a WCDMA system in the downlink, we follow the methodology proposed by Mäder & Staehle, and describe the WCDMA cell by a Markov chain, where each system state represents a certain number of resources occupied by Mobile users (MUs). We solve the Markov chain and provide an efficient recurrent formula for the system occupancy distribution, as well as the so-called local blocking probabilities. Based on them, we calculate the CBP of different service-classes accommodated in the cell, versus the total offered traffic load in the cell. Both hard and soft blocking is considered. Then, we set a CBP boundary for each service-class and according to these boundaries the WCDMA cell capacity in erlangs is determined for the downlink, as the maximum traffic load which satisfies all CBP boundaries. That is, the erlang cell-capacity is defined by the maximum traffic load for which the CBP of each service class is below than the corresponding CBP boundary. We evaluate the presented analytical model through simulation. The results show that the accuracy of the model is very satisfactory. The main contribution of this paper is the improved determination of several parameters involved in the downlink capacity estimation, in comparison to the calculations that appear in the work of Mäder & Staehle. In addition, we evaluate the effect of the intra-cell interference due to the orthogonality factor on the erlang capacity.

The paper is organized as follows. In section 2 we describe the system model. In section 3 we present an algorithm for the CBP calculation of each service-class. Section 4 shows the validation of the algorithm through comparison with simulation results, the effect of the intra-cell interference (orthogonality factor) on the erlang capacity and the accuracy of the model. We conclude in section 5.

2   System Model

We consider a WCDMA reference cell surrounded by a number of neighbor cells. We use the index x to denote the BS that controls the reference cell. MUs generate calls within the reference cell. A call may belong to one out of S independent service-classes. By M_x we denote the number of all MUs within the reference cell (i.e. MUs that are power-controlled by the BS x). At any time instant some of these MUs are *active*, i.e. have a call in progress, whereas the rest of them are *passive*. The number of active users is denoted by A_x ($A_x \le M_x$). The position of MUs within the reference cell is assumed an i.i.d. (independent and identically distributed) random variable.

By $S_{x,\max}$ we denote the maximum transmission power of BS x. By $S_{x,c}$ we denote the power that BS x transmits for common channels; this power is assumed to be constant. A part of the transmission power of BS x is devoted to satisfy the QoS requirements of all active MUs. More precisely, the signal power, $S_{k,x}$, transmitted by BS x to an active MU k ($k=1, \dots, A_k$) depends on the position and on the service-class of the MU k. The total signal transmission power from BS x (to all active MUs and for common channels) at a time instant is denoted by S_x:

$$S_x = S_{x,c} + \sum_{k=1}^{A_k} S_{k,x} \tag{1}$$

The maximum transmission power of BS x, $S_{x,\max}$, can be considered as shared system resource, whereas the power requirements, $S_{k,x}$, of MUs, as resource requirements. A neighbor BS y transmits with S_y. This power, similarly to [9], is modeled as a log-normal random variable with mean $E[S_y]$ and variance $VAR[S_y]$. Due to path loss, the signal power received at the MU k is less than the power $S_{x,k}$, transmitted by BS x towards the MU k. This attenuation is described by the attenuation factor $\hat{d}_{k,x}$ in dB [9]:

$$\hat{d}_{k,x} = -128.1 - 37.6 \log_{10}(\text{dist}(x,k)) \tag{2}$$

In the above equation, the distance, dist(x,k), from BS x to the MU k is in Km. In the following, the linear value of the attenuation factor is denoted by $d_{k,x}$.

For the purposes of our analysis, the coverage area of the whole network is divided into small square subareas. Every subarea may fully or partially belong to the coverage area, F_x, of BS x, as shown in Fig. 1. The total traffic-load offered by MUs of a subarea f is denoted by a_f. We assume that the size of each subarea is small enough in order for the distances of all MUs within the same subarea from BS x to be equal. The probability that a subarea f is within the coverage area of the BS x is given by [9]:

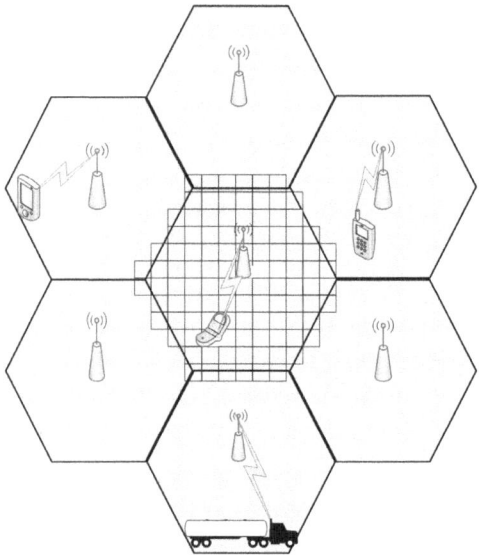

Fig. 1. Cellular concept and division of a cell into small subareas

$$p(f \in F_x) = P(d_{k.x} < \min\{d_{k.y}\}), \ y \in Y \tag{3}$$

where Y is the set of BSs that are neighbors for the BS x.

We consider that a service-class s ($s=1, ...,S$) is characterized by:

- R_s: Transmission bit rate.
- $(E_b/N_0)_s$: QoS parameter - Signal energy per bit divided by noise spectral density, required to meet a predefined Block Error Rate.
- v_s: The user activity factor at physical layer.

Furthermore, we take under consideration three kinds of interference, namely, the thermal noise, N_0, the inter-cell interference, I_{inter}, and the intra-cell interference, I_{intra}.

One of the advantages of the WCDMA technology is that it separates different signals in the cell by using orthogonal spreading codes [1], aiming at illuminating the intra-cell interference. In practice, however, due to multipath propagation, the complete illumination is not possible. For this reason, the orthogonality factor, a, is introduced in order to describe the fraction of power which is seen by a MU as interference from other

MUs that are power controlled by the same BS. On the other hand, inter-cell interference results from signals coming from neighbor cells (without enough attenuation).

In order for a MU to be serviced by a BS, the latter must satisfy the MU's E_b/N_0 requirements, which depend on the MU service-class and on the distance of the MU from the BS. The required (target) E_b/N_0 of the MU k that is power-controlled by the BS x is denoted by $\varepsilon_{k,x}$ and is given by the outer loop power control [9]:

$$\varepsilon_{k,x} = \frac{W}{R_k} \frac{S_{k,x} d_{k,x}}{WN_0 + \sum_{y \neq x} S_y d_{k,y} + a d_{k,x} (S_x - S_{k,x})} \tag{4}$$

where W is the WCDMA system chip rate.

We make a reasonable consideration that the power control is not perfect. In that case, the ratio E_b/N_0 fluctuates around the target E_b/N_0 which results from eq. (4). The ratio E_b/N_0 can be modeled by a lognormal random variable [9].

Each time a MU starts a new call, the Call Admission Control (CAC) estimates the increase caused to the transmitting power of the BS. If the total required power of the BS after a new call acceptance is going to exceed $S_{x,max}$, then the new call will be blocked; otherwise it will be accepted. The increase of BS power caused by a new call acceptance depends on the QoS parameter of the MU and its distance from the BS. Hence, in the downlink, the CAC is performed according to the following condition:

$$S_x < S_{max} \tag{5}$$

Repeating, the CAC needs to know the current BS transmission power and the increase in the transmission power that a new call will cause. From (4) we can calculate the required transmission power of BS x towards the MU k [9]:

$$S_{k,x} = \omega_k \left(WN_0 \delta_{k,x} + \sum_{y \in Y} S_y \Delta_{y,k} + aS_x \right) \tag{6}$$

where $\delta_{k,x} = \dfrac{1}{d_{k,x}}$, $\Delta_{k,y} = \dfrac{d_{k,y}}{d_{k,x}}$ and ω_k is the *service load factor* of the MU k [1]:

$$\omega_k = \frac{\varepsilon_{k,x} R_k}{W + a\varepsilon_{k,x} R_k} \tag{7}$$

The sum of the service load factors of all active MUs in the cell defines the *cell load* n_x of the BS x:

$$n_x = \sum_{k=1}^{A_x} \omega_k \tag{8}$$

The cell load n_x can be considered as the shared system resource and the service load factor ω_k as the resource requirement of the MU k.

It is also useful to define the *position and service load factor* $\omega_{k,y}$:

$$\omega_{k,y} = \begin{cases} \omega_k \delta_{k,x} & \text{if } y = 0 \\ \omega_k a & \text{if } y = x \\ \omega_k \Delta_{k,y} & \text{if } y \neq x \text{ and } y \neq 0 \end{cases} \tag{9}$$

n (9), $\omega_{k,y}$ is the cell load introduced either due to the thermal noise ($y=0$), either due to BS x ($y=x$), or due to a neighbor BS ($y \neq x$ and $y \neq 0$).

The sum of the above position and service load factors of all active MUs in the cell defines the combined cell load $n_{x,y}$:

$$n_{x,y} = \sum_{k=1}^{A_x} \omega_{k,y} \tag{10}$$

In order to calculate the transmission power, S_x, of BS x we must sum all $S_{k,x}$ of the active users in the cell and the power $S_{x,c}$. Hence, from (1) and (4) we derive [9]:

$$S_x = \frac{1}{1 - n_{x,x}} \left(n_{x,0} WN_0 + \sum_{y \in Y} n_{x,y} S_y + S_{x,c} \right) \tag{11}$$

From (8) - (11) we have [9]:

$$\sum_{k=1}^{A_x} \omega_k \left(WN_0 \delta_{k,x} + \sum_{y \in Y} \Delta_{k,y} S_y + a S_{x,\max} \right) < S_{x,\max} - S_{x,c} \tag{12}$$

In the above equation, we denote by Q_k the quantity in the brackets. It is called *positional load factor* since it depends only on the position of the user k in the cell (i.e. Q_k is independent on ω_k and depends only on $\delta_{k,x}$ and $\Delta_{k,x}$) [9]:

$$Q_k = WN_0 \delta_{k,x} + \sum_{y \in Y} \Delta_{k,y} S_y + a S_{x,\max} \tag{13}$$

Based on (12), (13) and taking into account the activity factor, v_k, we obtain [9]:

$$\sum_{k=1}^{M_x} v_k \omega_k Q_k < S_{x,\max} - S_{x,c} \tag{14}$$

3 Algorithm for Capacity Calculation

When the system supports S service-classes its state space is S-dimensional. An example of the state space for $S=2$ is shown in Fig. 2. In this figure, different system states are denoted by circles, while the transitions between the states are denoted by arrows. A state is defined as an S-dimensional vector whose elements are the numbers of in-service calls of different service-classes. Upon each arrow the transition rate

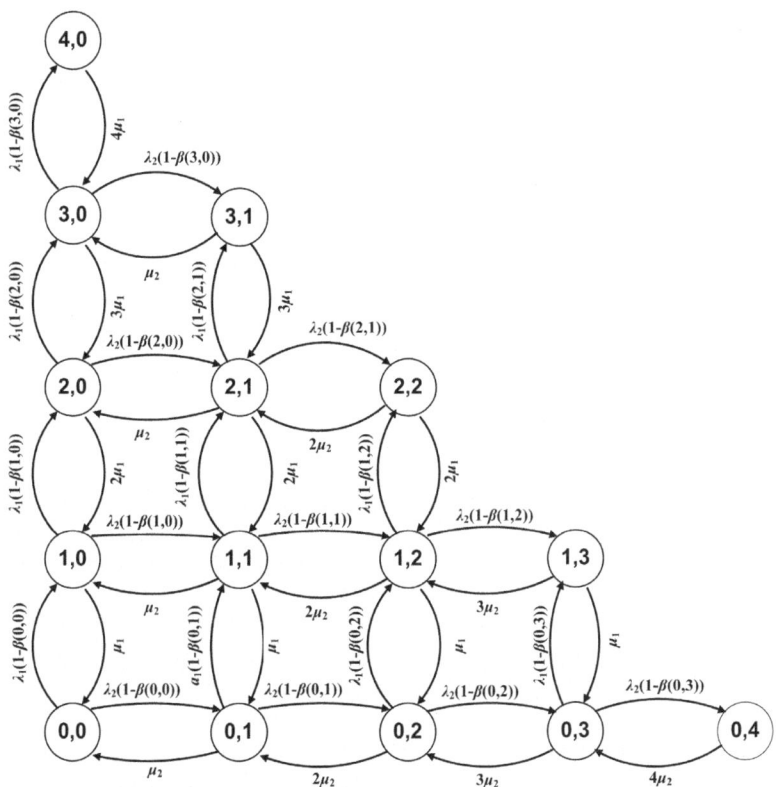

Fig. 2. Micro-state transition diagram for a system with $S=2$ service-classes

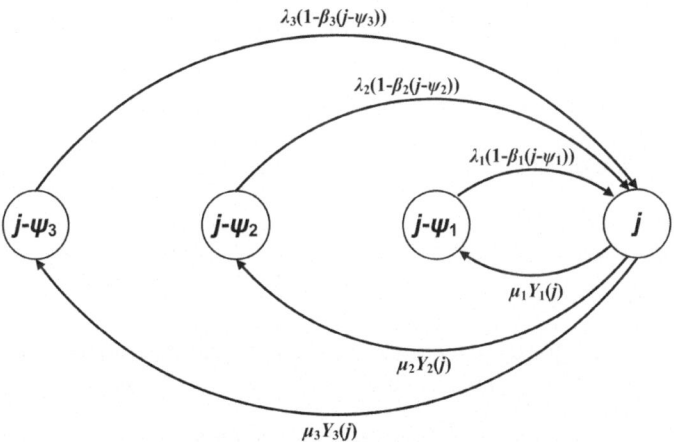

Fig. 3. Macro-state transition diagram (1-dimensional Markov chain)

from one state to another is shown. Even for $S=2$, solving the resultant Markov chain (in order to determine the state probabilities) is not an easy task. This problem is even more complicated for $S>2$. To simplify the problem, the S-dimensional state space is transformed into 1-dimensional. The idea is to combine all states (called micro-states) which have the same number of occupied resources into one state (macro-state).

The goal of the analysis presented below is to calculate the probability $q(j)$ of each (macro-)state j. Then, the CBP of each service-class can be determined as it is shown at the end of this section (see eq. (33)).

The first step is the discretization of ω_s with the aid of the *basic unit*, g [9]:

$$\psi_s = \left(\left\lfloor \frac{v_s \omega_s}{g} + \frac{1}{2} \right\rfloor \right) \tag{15}$$

From now, the discrete value ψ_s will be considered as the service-class s call resource requirement.

A segment of the 1-dimensional Markov chain for a system with three service-classes is shown in Fig. 3. In this figure, by λ_s and μ_s we denote the mean arrival rate and mean service rate of service-class s call, respectively. By $\beta_s(j)$ we denote the *local blocking probability*, defined as the probability that a new call of service-class s is blocked when the system is in state j. By $Y_s(j)$ we denote the mean number of service-class s calls in state j. Note that the transition rates from lower states ($j - \psi_s$) to higher (j) are reduced by the factors $1 - \beta_s(j - \psi_s)$, which denote the probability of non-blocking in state $j - \psi_s$.

For the calculation of the un-normalized state probabilities, $\tilde{q}(j)$, we use a modification of the KR Recursion while capturing the effect of soft blocking [9]:

$$\tilde{q}(j) = \begin{cases} 1, & \text{for } j = 0 \\ \dfrac{1}{j} \displaystyle\sum_{s=1}^{S} \left(1 - \beta_s\left(j - \psi_s\right)\right) a_s \psi_s \tilde{q}\left(j - \psi_s\right), & \text{for } j = 1, ..., j_{max} \end{cases} \tag{16}$$

where $a_s = \lambda_s / \mu_s$ is the offered traffic-load of service-class s and j_{max} is the maximum reachable system state.

Then, the normalized state probabilities are computed by:

$$q(j) = \frac{\tilde{q}(j)}{\displaystyle\sum_{j \leq j_{max}} \tilde{q}(j)} \tag{17}$$

Let us denote by $P_s(j)$, the conditional probability that the current state j has been reached from the state $j - \psi_s$, through the arrival of a service-class s call:

$$P_s(j) = \frac{(1 - \beta_s(j - \psi_s)) a_s q(j - \psi_s)}{\displaystyle\sum_{s=1}^{S} (1 - \beta_s(j - \psi_s)) a_s q(j - \psi_s)} \tag{18}$$

Since a MU position in the cell is i.i.d., the quantities $E[\delta_{k,x}]$, $E[\Delta_{k,y}]$ and $E[Q_k]$ are independent of k; therefore, hereinafter we use the notations $E[\delta_x]$, $E[\Delta_y]$ and $E[Q]$.

The first moment of $S_x(j)$ can be computed as follows [9]:

$$E[S_x(j)] = \begin{cases} 0 & \text{for } j = 0 \\ \sum_{s=1}^{S} P_s(j)\big(E[S_x(j-\psi_s)] + v_s E[\omega_s]E[Q]\big) & \text{for } 0 < j \leq j_{max} \end{cases} \tag{19}$$

While, for the second moment we have [9]:

$$E\big[S_x(j)^2\big] = \begin{cases} 0 & \text{for } j = 0 \\ \sum_{s=1}^{S} P_s(j)(E\big[S_x(j-\psi_s)^2\big] + v_s E\big[\omega_s^2\big]E\big[Q^2\big] + \\ \quad 2v_s E[\omega_s]E[Q]E[S_x(j-\psi_s)] & \text{for } 0 < j \leq j_{max} \end{cases} \tag{20}$$

From (20) with the introduction of mean cell load n_a, we obtain:

$$E\big[S_x(j)^2\big] = \begin{cases} 0 & \text{for } j = 0 \\ \sum_{s=1}^{S} P_s(j)(E\big[S_x(j-\psi_s)^2\big] + v_s E\big[\omega_s^2\big]E\big[Q^2\big] \\ \quad +2v_s E[\omega_s]E[QQ']E[n_a(j-\psi_s)]) & \text{for } 0 < j \leq j_{max} \end{cases} \tag{21}$$

where

$$E[n_a(j)] = \begin{cases} 0 & \text{for } j = 0 \\ \sum_{s=1}^{S} P_s(j)\big(E[n_a(j-\psi_s)] + v_s E[\omega_s]\big) & \text{for } 0 < j \leq j_{max} \end{cases} \tag{22}$$

Here, we must define the first, second and combined moment of the positional load factor. The first moment is obtained by the following equation [9]:

$$E[Q] = WN_0 E[\delta_x] + \sum_{y \in Y} E[\Delta_y]E[S_y] + aS_{max} \tag{23}$$

We calculate the second moment as follows:

$$E\big[Q^2\big] = (WN_0)^2 E\big[\delta_x^2\big] + 2WN_0 \sum_{y \in Y} E[S_y]E[\Delta_y \delta_x] + 2aS_{x,max}WN_0 E[\delta_x]$$

$$+ (aS_{x,max})^2 + \sum_{x \neq y}\sum E[S_{y1}]E[S_{y2}]E[\Delta_{y1}\Delta_{y2}] + \sum_{y \in Y} E\big[S_y^2\big]E\big[\Delta_y^2\big]$$

$$+ 2aS_{x,max} \sum_{y \in Y} E[S_y]E[\Delta_y] \tag{24}$$

We calculate the combined moment by the following equation:

$$E[QQ'] = (WN_0)^2 E[\delta_x]^2 + (aS_{x,\max})^2 + 2aS_{x,\max}WN_0 E[\delta_x]$$
$$+ 2WN_0 \sum_{y \in Y} E[S_y] E[\Delta_y] E[\delta_x] + \sum_{y \in Y} E[S_y^2] E[\Delta_y]^2 \qquad (25)$$
$$+ \sum_{x \neq y} \sum E[S_{y1}] E[S_{y2}] E[\Delta_{y1}] E[\Delta_{y2}] + 2aS_{x,\max} \sum_{y \in Y} E[S_y] E[\Delta_y]$$

The mean of $\delta_{k,x}$ and $\Delta_{k,x}$ are given by [9]:

$$E[\delta_x] = \sum_{f \in F_x} \frac{a_f p(f \in F_x)}{\sum_{s=1}^{S} a_{x,s}} E\left[\frac{1}{d_{f,x}} \middle| f \in F_x\right] \qquad (26)$$

$$E[\Delta_y] = \sum_{f \in F_x} \frac{a_f p(f \in F_x)}{\sum_{s=1}^{S} a_{x,s}} E\left[\frac{d_{f,y}}{d_{f,x}} \middle| f \in F_x\right] \qquad (27)$$

As we have already shown, one can compute the first and the second moment of the transmitting power $S_x(j)$ based on (19), (21) and (22). After a new call of service-class s is accepted in the system, the new transmitting power becomes $S_x(j)+S_s$, where $S_s = \omega_s Q$ is the additional power required for the new call. The first and the second moments of this new transmitting power are calculated according to [9]:

$$E[S_x(j) + S_s] = E[S_x(j)] + E[\omega_s] E[Q] \qquad (28)$$

$$E[(S_x(j) + S_s)^2] = E[S_x(j)^2] + 2E[\omega_s] E[QQ'] E[n_a(j)] + E[\omega_s^2] E[Q^2] \qquad (29)$$

Due to the fact that the new user is assumed to be active at the beginning of his call, the activity factor is neglected in (28), (29).

If now we assume that the random variable $S_x(j)+S_s$ is lognormally distributed, the local blocking probability $\beta_s(j)$ can be calculated by (based on (5), (28) and (29)) [9]:

$$\beta_s(j) = 1 - CDF_{\mu,\sigma}\left(S_{x,\max} - S_{x,c}\right) \qquad (30)$$

Where $CDF()$ is the Cumulative Distribution Function of the random variable $S_x(j)+S_s$.

We calculate the parameters μ and σ by:

$$\mu = \ln\left(E[S_x(j) + S_s]\right) - \frac{1}{2}\sigma^2 \qquad (31)$$

$$\sigma = \sqrt{\ln\left(CV^2 + 1\right)} \tag{32}$$

where CV is the coefficient of variation of the random variable $S_x(j)+S_s$.
Finally, the CBP of a service-class s can be calculated by [9]:

$$P_{block}(s) = \sum_{j < j_{max} - \psi_s} \beta_s(j)q(j) + \sum_{j_{max} - \psi_s < j \le j_{max}} q(j) \tag{33}$$

4 Application Example – Numerical Results

We evaluate the accuracy of the analytical model of Section 3, through comparison with simulation. The simulation model is based on the system model described in Section 2; it is developed by using the SIMSCRIPT II.5 simulation tool [10].

We consider the WCDMA network of Fig. 1 where three service-classes are accommodated. We calculate CBP by the analytical model and simulation. Then, we set a CBP boundary for each service-class and according to these boundaries the WCDMA cell capacity in erlangs is determined for the downlink, as the maximum traffic load which satisfies all CBP boundaries. Also, through the analytical model, we determine the erlang cell capacity for different orthogonality factors. The system parameters are given in Table 1, whereas the service-class parameters in Table 2. Regarding the user activity factors, we consider two scenarios. In the first scenario, the activity factors are $v_1=0.3$, $v_2=0.7$ and $v_3=1.0$. In the second scenario, the activity factors are $v_1=0.4$, $v_2=0.6$ and $v_3=0.8$.

Table 1. System Parameters

Distance between BSs	2 Km
Maximum transmission power of the BS	8 W
Transmission power required for common channels	2 W
Mean transmission power of neighbor BSs	3 W
St. dev. of transmission power of neighbor BSs	200 mW
Thermal noise power spectral density	-174 dBm/Hz
Bandwidth	5 MHz
Orthogonality factor	0.1

Table 2. Service-class Parameters

Service class	1	2	3
Transmission bit rate	12.2 Kbps	64 Kbps	144 Kbps
Target E_b/N_0	5.5	4	3.5
E_b/N_0 st.dev.	1.2	1.2	1.2
Traffic mix	96 %	3 %	1 %

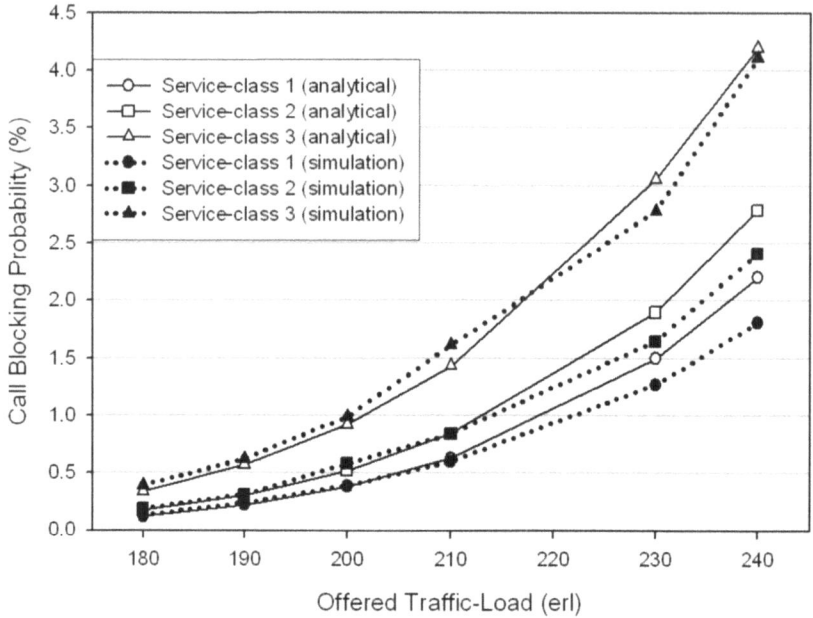

Fig. 4. CBP vs. Offered traffic-load for the 1st scenario

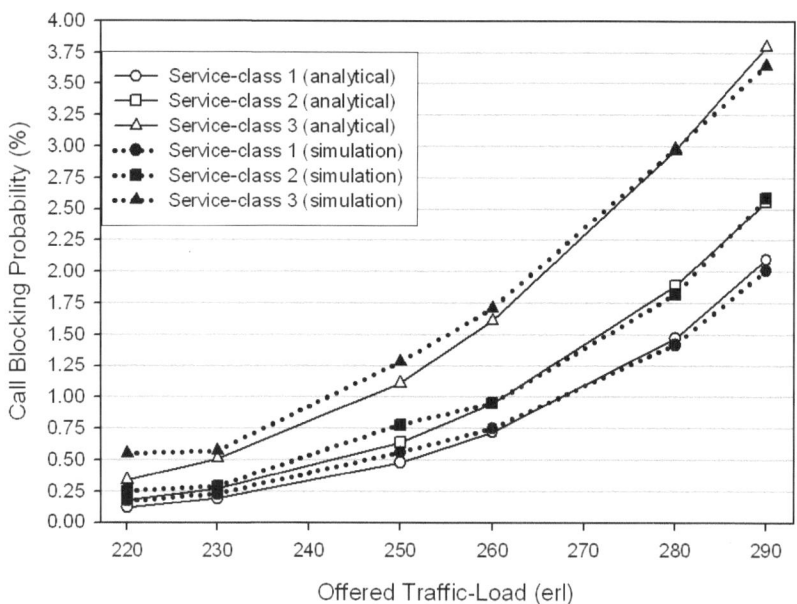

Fig. 5. CBP vs. Offered traffic-load for the 2nd scenario

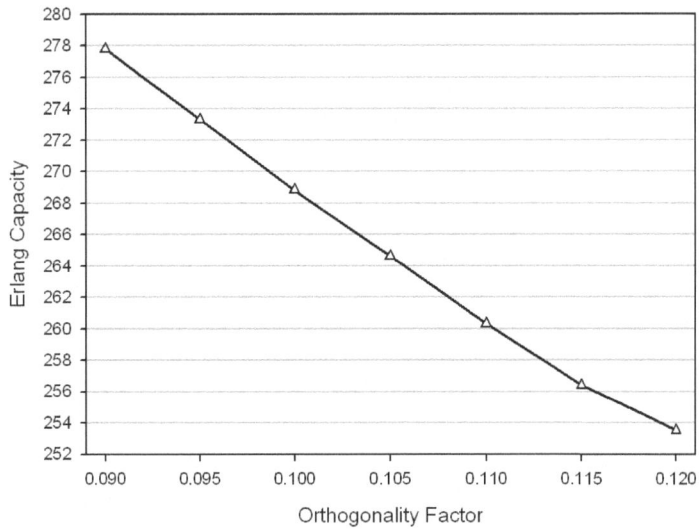

Fig. 6. Erlang capacity vs. Orthogonality factor

In Fig. 4 we present both analytical and simulation CBP results versus the offered traffic-load for the three service-classes, for the first scenario. The CBP results versus the offered traffic-load for the second scenario are presented in Fig. 5. In both Figs. 4 and 5 we observe that the accuracy of the calculations is satisfactory, especially for low, reasonable offered traffic-load.

In order to reveal the importance of the orthogonality factor for WCDMA systems, we determine the erlang capacity of the system for different orthogonality factors. To this end, first we calculate the CBP of each service-class (for each orthogonality factor) and then, based on the selected CBP boundaries per service-class, we determine the erlang capacity of the system. The CBP boundaries used in our example are 1%, 3% and 5% for the 1st, 2nd and 3rd service-class, respectively. In Fig. 6 we present the erlang capacity versus the orthogonality factor. We observe that even small improvements in the orthogonality factors have huge impact in the erlang capacity of a WCDMA system.

5 Conclusion

In this paper, we described the WCDMA cell by a 1-dimensional Markov chain and provided an efficient recurrent formula for the system occupancy distribution, as well as the so-called local blocking probabilities. Based on them, we calculated the CBP of different service-classes accommodated in the cell, versus the total offered traffic load. We also calculated the erlang capacity of the cell for different orthogonality factors. The analytical model was evaluated through simulation. The results showed that the accuracy of the model is very satisfactory.

Acknowledgment

This research project (PENED) is co-financed by E.U.-European Social Fund (80%) and the Greek Ministry of Development-GSRT (20%).

References

[1] Holma, H., Toskala, A. (eds.): WCDMA for UMTS. John Wiley & Sons Ltd., Chichester (2002)

[2] Kaufman, J.: Blocking in a shared resource environment. IEEE Trans. Commun. COM-29(10), 1474–1481 (1981)

[3] Roberts, J.W.: A service system with heterogeneous user requirements. In: Pujolle, G. (ed.) Performance of Data Communications systems and their applications, pp. 423–431. North Holland, Amsterdam (1981)

[4] Staehle, D., Mäder, A.: An analytic Approximation of the Uplink Capacity in a UMTS Network with Heterogeneous traffic. In: 18th International Teletraffic Congress (ITC18), Belrin (2003)

[5] Kallos, G.A., Vassilakis, V.G., Logothetis, M.D.: Call Blocking Probabilities in a W-CDMA Cell with Fixed Number of Channels and Finite Number of Traffic Sources. In: Proc. CSNDSP 2008, Graz, Austria (2008)

[6] Vassilakis, V.G., Kallos, G.A., Moscholios, I.D., Logothetis, M.D.: The Wireless Engset Multi-Rate Loss Model for the Call-level Analysis of W-CDMA Networks. In: Proc. 18th IEEE PIMRC 2007, Athens, Greece (2007)

[7] Vassilakis, V.G., Logothetis, M.D.: The Wireless Engset Multi-rate Loss Model for the Handoff Traffic Analysis in W-CDMA Network. In: Proc. 18th IEEE PIMRC 2008, Cannes, France (2008)

[8] Choi, W., Kim, J.: Forward-link Capacity of a DS/CDMA System with Mixed multirate Sources. IEEE Trans. On Veh. Tech. 50, 737–749 (2001)

[9] Mäder, A., Staehle, D.: Analytic Modeling of the WCDMA Downlink Capacity in Multi-Service Environments. In: ITC Specialist Seminar on Performance Evaluation of Wireless and Mobile Systems, Antwerp, Belgium, pp. 217–226 (2004)

[10] Russell, E.C.: Building Simulation Models with SIMSCRIPT II.5, CACI Products Company (Feburary 1999),
http://www.simscript.com/cust_center/ss3docs/zbuildin.pdf

Fast Randomized STDMA Link Scheduling

Sergio Gomez, Oriol Gras, and Vasilis Friderikos[*]

Department of Electronic Engineering
Division of Engineering
King's College London
Strand Campus
WC2R 2LS, London, UK
vasilis.friderikos@kcl.ac.uk

Abstract. In this paper a fast randomized parallel link swap based packing (RSP) algorithm for timeslot allocation in a spatial time division multiple access (STDMA) wireless mesh network is presented. The proposed randomized algorithm extends several greedy scheduling algorithms that utilize the physical interference model by applying a local search that leads to a substantial improvement in the spatial timeslot reuse. Numerical simulations reveal that compared to previously scheduling schemes the proposed randomized algorithm can achieve a performance gain of up to 11%. A significant benefit of the proposed scheme is that the computations can be parallelized and therefore can efficiently utilize commoditized and emerging multi-core and/or multi-CPU processors.

Keywords: Spatial-TDMA, Wireless Mesh Networks, Scheduling, Routing, Wireless Multi-Hop.

1 Introduction

Wireless Mesh Networks (WMNs) have recently emerged as a key technology to fulfil a diverse set of applications. The envisioned applications for WMNs range from being a viable alternative to wire line last mile broadband Internet service delivery at home or offices to backhaul support for wireless local area networks to different cellular networks such as for example LTE [1], [2]. One of the most important building blocks of wireless mesh networks is how to perform efficient scheduling so that high levels of throughput can be attained. For collision-free WMNs that support Spatial Time Division Multiple Access (STDMA) the critical aims is to increase the spectral efficiency by minimizing the frame length (i.e., number of timeslots) that a predefined number of transmitting and receiving pairs of nodes can successfully transmit [3]. Finding the optimal reuse of timeslots, i.e., the shortest frame length, has been shown to be an *NP*-complete optimization problem [4]. To provide a feasible STDMA timeslot allocation a number of sub-optimal algorithms with polynomial time complexity have been previously proposed [5], [6], [7].

---

[*] Corresponding author.

F. Granelli et al. (Eds.): MOBILIGHT 2009, LNICST 13, pp. 15–24, 2009.

In this paper, a very fast randomized link scheduling algorithm for STDMA wireless mesh networks that is build upon previously proposed greedy scheduling schemes is proposed. As will become evident in the sequel, in the numerical investigations (section 4), the proposed scheme can significantly decrease the frame length by up to 11%, providing in that respect better spatial reuse of timeslots in the mesh network compared to previous well known greedy scheduling algorithms. Another key benefit of the proposed scheduling scheme is that the computations can be parallelized. Clearly, among the applications that can significantly gain from multi-core and multi-CPU enabled network elements are the scheduling algorithms. To this end, the proposed fast scheduling algorithm falls within the family of the so-called "embarrassingly" parallel problems [17] since different iterations of the algorithm can be executed without requiring any communication between them.

The rest of the paper is organized as follows. In section 2, closely related previous research works are discussed and the main contributions of the paper are lined up. Section 3 specifies the system model that has been adopted in the analysis, describes the STDMA link scheduling problem and details the proposed randomized scheduling algorithm. Numerical investigations are reported in section 4 and finally the paper concludes in section 5.

2 Previous Work

The concept of Spatial-TDMA has first been presented in the seminal work of Kleinrock [10]. A significant part of previous research in the area of STDMA scheduling has been concentrating on graph based representation of the STDMA scheduling problem and associated graph theoretic tools; conceiving in that respect the STDMA scheduling as a graph colouring problem [11], [12], [13]. Despite their attractiveness, graph colouring based algorithms can resolve only the problems of *primary* and *secondary* conflicts between the links that need to be scheduled [6]. Hence, their drawback is that they do not consider the effect of aggregate interference, as reflected at the Signal to Interference Noise Ratio($SINR$) constraint for successful packet transmission and, therefore they may lead to schedules which are infeasible [6], [14]. To resolve this issue a number of previous works have explicitly taken into consideration the $SINR$ constraints (the so-called *physical interference* model) together with power control for constructing minimum frame length schedules [7], [8], [15]. In [16] a randomized distributed STDMA scheduling algorithm (DRAND) is presented. The difference with our proposed scheme is that DRAND does not take the $SINR$ constraints into account. Also the randomization has a different rational compared to the proposed RSP algorithm. In DRAND the randomization is on how neighbour nodes are selecting timeslots, whereas in RSP the randomization is on how to deviate from an already feasible allocation and search alternative feasible (hopefully better) solutions.

Recently, the problem of scheduling has also been considered jointly with the routing decisions. The rational being that due to the broadcast nature of the wireless transmission medium, it is possible that better spatial reuse of timeslots can be achieved by considering the problem of routing and scheduling jointly [9].

3   Problem Description and STDMA Link Scheduling

3.1   Preliminaries

We consider a WMN, which can be modelled by a network graph$G(V, E)$, where V is the set of nodes (mesh routers and clients) and E expresses the set of wireless links. Each node is equipped with one wireless interface card, and hereafter the terms radios and nodes are used interchangeably since they coincide. We further assume that all nodes in the mesh network operate at the same frequency band (frequency reuse factor is one) and we do not consider spurious or other inter channel interference. The packet length is normalized and occupies a single timeslot. For a single transmission bit-rate, each link $(i, j) \epsilon L$ needs to satisfy a signal to interference noise-ratio threshold (γ) for successful packet decoding; this constraint can be written as follows,

$$\frac{g_{ij}p_{ij}}{\sum_{(m,n)\epsilon L\{i,j\}} g_{mj}p_{mn}+W} \geq \gamma \tag{1}$$

where p_{ij} denotes the transmission power for link(i, j), g_{ij} is the link gain for link (i, j) and W expresses the lump sum power of background and thermal noise.

3.2   Greedy STDMA Link Scheduling

The strategy followed by several scheduling algorithms that utilize the physical interference model consists in firstly sorting the links based on a pre-defined criterion and then greedily packing the links into timeslots to generate feasible schedules. We detail in the sequel two well known heuristic scheduling algorithms, namely the Greedy Physical [8] and the Packing Heuristic [9] algorithms. These algorithms will be the basis upon where the proposed RSP algorithm is developed.

Note that these algorithms do not perform power control. As will be explained, each link transmits above the minimum power needed to transmit on its own, i.e., when there is no interference to allow concurrent transmissions to take place. Furthermore, both algorithms assume that is implicit that a node neither transmit and receive at the same timeslot nor transmit/receive to/from more than one node at the same timeslot. This can be accomplished binding the following two constraints: the *indegree* constraint ensures that only one node can send traffic to the same receiving node in each timeslot; the *outdegree* constraint ensures that a transmitting node can only send traffic to one receiving node per timeslot [18].

3.2.1   Greedy Physical (GP)

Greedy physical starts sorting the links to be scheduled according to the interference number, which is detailed next. The interference number of a link $E_i \epsilon E$ is the number of links $E_j \epsilon E \setminus \{E_i\}$ that cannot establish a communication at the same time such the set E_i and E_j does not share an endpoint and is infeasible. A set of two links is considered infeasible when the receiver nodes do not satisfy the SINR restriction described in (1). Thereafter, a list is created sorting the links with higher interference number first and then links are packed according to the scheduling algorithm stated in Table 1.

Table 1. Pseudo-code of the GP algorithm

Input	L	A list containing all links sorted by its interference number
Output	S	A feasible schedule
	TS	Frame length found for S

1: $TS \leftarrow 0$
2: for each Link in L **do**
3:   Schedule link L_i in the first available slot such that the resulting set of scheduled transmission is feasible with the physical interference model.
4:   **If** currently available slots are not sufficient to schedule L_i, add a new slot at the end of the schedule S and schedule link L_i in this slot.
5:   Let $TS \leftarrow TS + 1$
6:   **endif**
7: end

3.2.2 Packing Heuristic (PH)

The Packing Heuristic presented in this paper is the same algorithm used in [9] and it is also a variation of the heuristic used in [19] and [20], where different weights are utilized to sort the links. This algorithm tries to pack as many links as possible in each timeslot, having as a starting point a list where the links are sorted with the links that require higher transmitted power first. The pseudo-code of the algorithm is shown in Table 2 below.

Table 2. Pseudo-code of the PH algorithm

Input	A	A list containing all links sorted by its power levels (highest power first)
Output	B	A feasible schedule
	TS	Frame length found for S

1: $t \leftarrow 1$
2: $B \leftarrow$ Empty List
3: At timeslot t schedule the first link in list A for transmission and shift it from list A to list B.
4: repeat
5:   Proceed down the current list A scheduling links for transmission in timeslot t, if feasible, and shifting them to list B if they transmit.
6:   Let $t \leftarrow t + 1$
7: until A is empty
8: Let $TS \leftarrow t - 1$

It has to be noticed that there is only one difference between the Packing Heuristic and the Greedy Physical that results in different schedules. The difference is the way the links are sorted in the initial list. In the Packing Heuristic the first links to be scheduled are the ones that have the highest transmitted power, whereas in the Greedy Physical the priority is given to the links that cause more interference. There is another difference between the GP and the PH though, in this case, does not lead to any different schedule. This difference lies in the way the algorithm proceeds to pack the links. In the Packing Heuristic we fix a timeslot and we try to pack in it all the links that have not yet transmitted, whereas in the Greedy Physical we fix a link and we try to pack it in the first timeslot available.

3.3 Randomized Link Swap Packing (RSP) Algorithm

The RSP algorithm is based on altering the interference number list by swapping N_S (number of swaps) times the order of two elements selected randomly from the list. The number of swaps applied to the list characterizes the degree to which the original list is distorted. After the swapped list is generated, the links are scheduled according to the GP or PH algorithms as described in the previous sections. Hence, a new feasible schedule is obtained. Different criteria can be applied in order to determine the best schedule when schedules with the same frame length as the best one found so far are generated. For instance, to improve the interference robustness of the network, possible criteria are (i) to choose the schedule with the best averaged SINR or (ii) the schedule with the maximum average min-SINR across all timeslots. This process is repeated for a pre-defined number of iterations (M_{ITER}). The pseudo-code of the proposed RSP algorithm is shown in Table 3 below.

Table 3. Pseudo-code of the RSP algorithm

Input	L	A list containing all links sorted by its interference number, or power levels
	N_S	Number of swaps
	M_{ITER}	Maximum number of iterations
	P	Number of Processors
Output	$S_{BEST,p}$	A feasible schedule with the minimum frame length found so far at processor p
	$T_{BEST,p}$	The minimum frame length found so far at processor p

1: $S_{BEST}, T_{BEST} \leftarrow$ Schedule(L)
2: for each processor **do** in parallel
3:   **for** i=1: $\lceil M_{ITER}/P \rceil$ **do**
4: $L_{SWAP} \leftarrow L$
5:   **for** j=1: N_S **do**
6: $L_{SWAP} \leftarrow$ Swap two elements from L_{SWAP}
7:   **end**
8: $S, T_S \leftarrow$ Schedule(L_{SWAP})
9:   **If** $T_S \leq < T_{BEST}$ **then**
10: $T_{BEST,p} \leftarrow T_S$
11: $S_{BEST,p} \leftarrow$ BestSchedule $(S, S_{BEST,p})$

12:   endif
13: end
14: $T_{BEST} \leftarrow \min\{T_{BEST,p}\}$
15: $S_{BEST} \leftarrow$ BestSchedule $\{S_{BEST,\forall p}\}$
16: end

As can be observed from the Table 2 above, the proposed RSP algorithm can be easily parallelized and run in P processors. In fact, the RSP algorithm can run without requiring any communication between the different processors, therefore there is no communication cost or delay for exchanging information between the different processors. Hence, the RSP algorithm enables embarrassingly parallel computations since different schedules can be calculated independently, offering a convenient way to use multiple processors concurrently to solve the problem. We note that a brute force enumeration of all possible ways to pack the N − 1 links in a predefined number of timeslots would be (N − 1)!. But since both the GP and the PH heuristics provide good initial feasible solutions, few iterations of the above proposed approach can provide significant benefits. As shown in the next section, the gains with the number of iterations follow a concave like function, which means that the net benefit of performing higher number of iterations diminishes with the number of iterations.

4 Numerical Investigations

4.1 Setting of the Simulation

The wireless mesh network is deployed in a square area AxA Km^2 containing N wireless nodes that are random uniformly distributed. Two nodes in the mesh network can establish a link if the receiving node satisfies the Signal to Interference Noise Ratio ($SINR$) threshold criterion. A special node in the topology acts as the gateway node for providing Internetworking; throughout the numerical investigations and without loss of generality a single gateway node is considered. Based on all feasible links that can be constructed when no co-channel interference is considered, a shortest path spanning tree is constructed rooted at the gateway node to all other nodes in the network. The spanning tree is based on the minimum power routing (MPR) scheme, as described and analyzed in [9]. The MPR scheme is based on Dijkstra's algorithm and uses the required transmitted power to combat the path loss as the cost of the link. To calculate the required transmission power level for link (i, j) the following simple path loss has been considered hereafter,

$$PL(d(i,j)) = PL(d_0) + 10\eta log\left(\frac{d(i,j)}{d_0}\right) \tag{2}$$

where $d(i,j)$ express the Euclidean distance of link (i, j), $PL(d_0)$ is the close-in reference distance loss, which is assumed to be equal to 78 dB for distance d_0 equal to 50 meters, and η denotes the path loss exponent, which can in general take values between 2 to 5 depending on the environment. Finally, it should be noted that only unidirectional links in the downlink scenario (from the gateway to the nodes) are considered. Similar results are expected to hold also for the uplink scenario but are not considered in this paper. Since a shortest path spanning tree is created that rooted at the designated gateway node, the links that need to be scheduled are always $N - 1$. The complete set of the simulation parameters used in the numerical investigations are summarized in Table 4 below.

Table 4. Simulation Parameters Used

Notation	Explanation	Values
A	Length of the Square Area	850 meters
N	Number of Nodes	20 - 120
L	Number of Links	19 - 119
d_0	Close-in reference distance	50 meters
γ	SINR threshold	8dB
$\eta \ P_{max}$	Path loss exponent	3.5
f_C	Maximum transmitted power	20 Watt
W	Carrier frequency	3.8GHz
	Thermal & background Noise	-132dBW

4.2 Results

We evaluate the performance of our proposed scheduling algorithm by comparing it with two well known and tested greedy STDMA scheduling schemes that utilize the physical interference model, namely the Packing Heuristic [9] and the Greedy Physical [8] algorithms as have been explained in detail in section 3.2. Notice that all results have been averaged over 200 WMNs topologies with randomly distributed nodes.

The quality of the solution provided by the RSP algorithm scheme (T_{RSP}) is compared to the corresponding solutions from the GP (T_{GP}) and the improvement (**I**) is measured as follows, $\mathbf{I}(\%) = (T_{GP} - T_{RSP})/T_{GP}$. The same measure is used to compare the solution of the RSP with the PH (T_{PH}) algorithm.

Fig. 1 shows the performance gains on the minimum frame length using the proposed randomized scheduling scheme compared to the Greedy Physical algorithm with respect to the number of iterations. Observe that substantial improvements can be achieved with a reduced number of iterations, for instance, with just 15 iterations the schedule allocation is ameliorated above 5 % for topologies with 40 and 60 nodes. This improvement is even better as the number of iterations increases. However, it is becoming less significant as the number of iterations augments. Note that the same behaviour holds when the RSP is applied to the Packing Heuristic, as Fig. 2 shows. In this case, the gain obtained is slightly higher and, in consequence, with less than 10 iterations we achieve an improvement above 5%.

Figure 3 describes the performance improvement on the minimum frame length using RSP (with different number of link swaps) compared to the GP and PH for different number of nodes in the network. As has been mentioned above, the number of swaps applied to the list influences the degree to which the original list is distorted. Observe from figure 3 that after a small number of swaps the performance stops increasing.

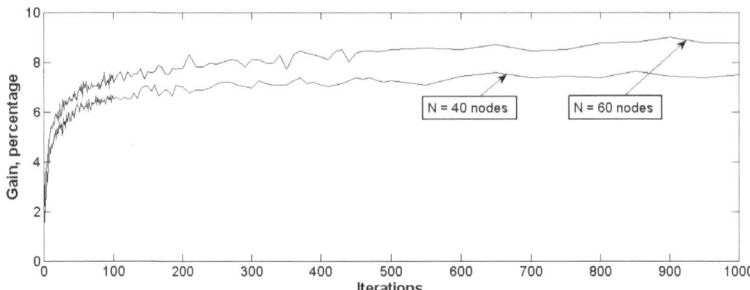

Fig. 1. Performance gains on the minimum frame length using the RSP algorithm compared to the GP algorithm with respect to the number of iterations for topologies with 40 and 60 nodes. These results have been calculated using 3 swaps.

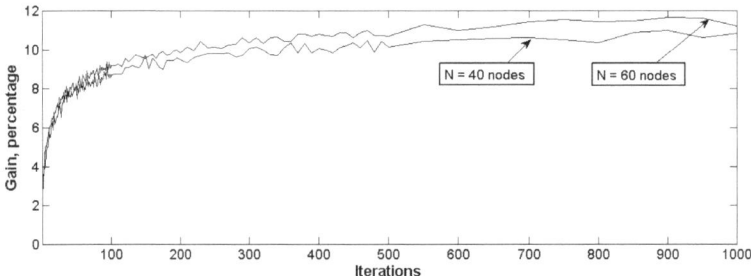

Fig. 2. Performance gains on the minimum frame length using the RSP algorithm compared to the PH algorithm for topologies with 40 and 60 nodes. These results have been calculated using 3 swaps.

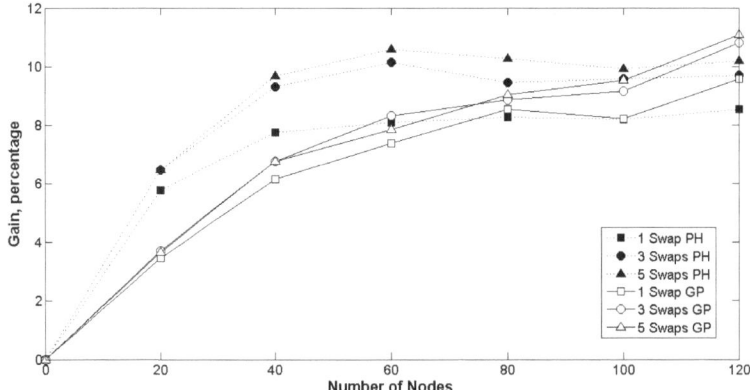

Fig. 3. Performance gains on the minimum frame length using the RSP algorithm (with different number of link swaps) compared to the GP and PH

5 Conclusions

In this paper, a fast randomized link scheduling algorithm for Spatial-TDMA enabled wireless mesh networks is detailed. The randomization is based on swapping links on a list that is created by well known greedy scheduling algorithms such as the Greedy Physical and the Packing Heuristic. In that way, the order of the scheduling is affected and by varying the number of swaps that are performed a larger set of feasible solutions space can be explored. Extensive numerical investigations reveal that the proposed fast scheduling scheme can improve by more than 10% the timeslot reuse compared to the previous mentioned link scheduling algorithms. Another important characteristic of the proposed scheme is that its structure is amenable for parallel processing and therefore, emerging multi-core and multi-CPU enabled network elements can be fully utilized. The simplicity of the algorithm, the achieved gains and the potential of parallel computation clearly demonstrate the potential benefits of the proposed scheme.

References

1. Akyildiz, I.F., Wang, X., Wang, W.: Wireless mesh networks: a survey. Computer Networks 47(4), 445–487 (2005)
2. Bruno, R., Conti, M., Gregori, E.: Mesh Networks: Commodity Multihop Ad Hoc Networks. IEEE Communications, 123–131 (2005)
3. Nelson, R., Kleinrock, L.: Spatial-TDMA: A collision-free multihop channel access protocol. IEEE Transactions on Communications 33, 934–944 (1985)
4. Ephremides, A., Truong, T.V.: Scheduling broadcasts in multihop radio networks. IEEE Transactions on Communications 38(4), 456–460 (1990)
5. Krumke, S., Marathe, M., Ravi, S.: Models and Approximation Algorithms for Channel Assignment in Radio Networks. ACM Wireless Networks 7, 575–584 (2001)
6. Gronkvist, J., Hansson, A.: Comparison Between Graph-Based and Interference-Based STDMA Scheduling. IEEE MobiHoc, 255–258 (2001)
7. Jain, K., Padhye, J., Padmanabhan, V., Qiu, L.: Impact of Interference on Multi-Hop Wireless Network Performance. In: ACM Mobicom, pp. 66–80 (2003)
8. Brar, G., Blough, D.M., Santi, P.: Computationally efficient scheduling with the physical interference model for throughput improvement in wireless mesh networks. In: 12th annual international conference on Mobile computing and networking (MOBICOM), Los Angeles, CA, USA, September 23-29 (2006)
9. Friderikos, V., Papadaki, K.: Interference Aware Routing for Minimum Frame Length Schedules in Wireless Mesh Networks. EURASIP Journal on Wireless Communications and Networking (2008)
10. Nelson, R., Kleinrock, L.: Spatial-TDMA: A collision-free multihop channel access protocol. IEEE Transactions on Communications 33(9), 934–944 (1985)
11. Hajek, B., Sasaki, G.: Link scheduling in polynomial time. IEEE Transactions on Information Theory 34, 910–917 (1988)
12. Prohazka, C.G.: Decoupling link scheduling constraints in multihop packet radio networks. IEEE Transactions on Computers (March 1989)
13. Chou, A.M., Li, V.O.: Slot allocation strategies for TDMA protocols in multihop packet radio networks. In: IEEE INFOCOM 1992 (1992)

14. Behzad, A., Rubin, I.: On the Performance of Graph-based Scheduling Algorithms for Packet Radio Networks. In: IEEE GLOBECOM, San Francisco, CA (December 2003)
15. Das, A., Marks, R., Arabshahi, P., Gray, A.: Power Controlled Minimum Frame Lenght Sheduling in TDMA Wireless Networks with Sectored Antennas. In: IEEE INFOCOM 2005, Miami (March 2005)
16. Rhee, I., Warrier, A., Min, J., Xu, L.: DRAND: Distributed Randomized TDMA Scheduling For Wireless Adhoc Networks. In: ACM MobiHOC (2006)
17. Foster, I.: Designing and Building Parallel Programs: Concepts and Tools for Parallel Software Engineering. Addison Wesley, Reading (1995)
18. Friderikos, V., Papadaki, K., Wisely, D., Aghvami, H.: Multi-rate power-controlled link scheduling for mesh broadband wireless access networks. IET communications 1(5), 909–914 (2007)
19. Papadaki, K., Friderikos, V.: Approximate dynamic programming for link scheduling in wireless mesh networks. Computers & Operations Research 35(12), 3848–3859 (2008)
20. Gronkvist, J.: Traffic controlled spatial reuse TDMA in multi-hop radio networks. In: 9th IEEE International Symposium on Personal, Indoor and Mobile Radio Communications (PIMRC 1998), Boston, Mass, USA, September 1999, vol. 3, pp. 1203–1207 (1999)

Efficient QoS-Driven Resource Allocation in Integrated CDMA/WLAN Networks - An Autonomic Architecture

Georgios Aristomenopoulos, Timotheos Kastrinogiannis,
and Symeon Papavassiliou

Institute of Communications and Computer Systems (ICCS)
School of Electrical and Computer Engineering, National Technical University of Athens,
Zografou, 15780, Greece
{aristome,timothe}@netmode.ntua.gr, papavass@mail.ntua.gr

Abstract. In this paper the problem of proficient joint radio resource management in an integrated WLAN/CDMA-cellular heterogeneous environment is considered. Nodes'/Networks' autonomicity is envisioned as the enabler for devising a proficient QoS-aware service orientated wireless interworking architecture founded on a common utility based framework that provides enhanced flexibility in reflecting different access networks' type of resources and diverse QoS prerequisites under common optimization problems. A decentralized node-networks assignment mechanism is introduced, aiming at QoS provisioning and efficient resource utilization. Numerical results are presented that validate the efficacy of the proposed architecture.

Keywords: Integrated WLAN/CDMA-cellular networks, autonomic networking, QoS provisioning, load balancing, self-optimization.

1 Introduction

The complementary characteristics of broadband Wireless Local Area Networks (WLANs) and Code Division Multiple Access (CDMA) cellular networks have recently attained much interest towards realizing an integrated system that efficiently enables seamless broadband Internet access for mobile users with multimode access capabilities [1]. However, when aiming at satisfying various Quality of Service (QoS) constraints within the integrated system, separate and independent studies on optimal resource allocation and QoS provisioning in either network, may prove inadequate. Heading towards optimal utilization of resources over an integrated CDMA/WLAN network, current research efforts are targeting on QoS traffic class mapping among different access networks [2] and on proficient call admission control mechanisms aiming at services' seamless continuity [3], [4], or load balancing [5].

Due to the heterogeneity of the wireless environment, in most cases only the mobile node has the complete view of its own environment, in terms of available access networks in its locality, the corresponding available resources and QoS support mechanisms. This becomes even more critical when the available networks belong to different operators. Therefore, contrary to traditional architectures where network/nodes' performance is controlled in a centralized way, future wireless networking [10]

F. Granelli et al. (Eds.): MOBILIGHT 2009, LNICST 13, pp. 25–34, 2009.
© ICST Institute for Computer Sciences, Social-Informatics and Telecommunication Engineering 2009

envisions as its foundation element an autonomic self-optimized wireless node with enhanced capabilities in terms of acting/re-acting to mobility, connectivity or even QoS- performance related events.

Such a vision and evolution, makes the design of a flexible autonomic QoS-aware joint network selection mechanism, a promising alternative service oriented paradigm that allows to fully exploit the proliferation of wireless networks, as opposed to the more conventional existing access oriented designs. In this paper, we describe an autonomic QoS-aware joint resource allocation architecture for integrated WLAN/CDMA-cellular systems that aims at maximizing the overall integrated network's revenue, while enabling users to efficiently self-adapt at QoS-triggered occurrences towards self-optimizing their services' performance.

The rest of the paper is organized as follows. In Section 2, we present the key features of the introduced autonomic joint WLAN/CDMA architecture. In Section 3, intra-cell autonomic resource allocation and QoS provisioning mechanisms are presented, followed by the introduction of an autonomic QoS and service oriented joint network selection mechanism. In Section 4, some quantitative comparative results are presented that demonstrate the efficacy of the proposed approach, while Section 5 concludes the paper.

2   Towards Autonomic Integrated WLAN/CDMA Networks – Motivation and Goals

Our approach in this paper is motivated by the fact that future autonomic networking aims at implementing self-* functions for self-optimization and self-adaptation to context or situation driven behavior changes in systems, services or applications [6]. In order to provide the needed flexibility and functional scalability in the joint resource management process in an integrated WLAN/CDMA network, we introduce autonomicity as the vehicle allowing the design of a novel autonomic framework that maximizes overall integrated network's revenue, enabling self-adaptation and self-optimization functionalities in both mobile nodes and base stations or access points.

The fundamental concept of an autonomic system is a control loop(s). Inputs to the control loop consist of various status signals, information and views continuously exposed from the system, component(s) or resource(s) being controlled (e.g. protocols, nodes, functionalities, etc.), along with (usually policy-driven) management rules that orchestrate the behavior of the system or component. Outputs are commands to the system or component(s) to adjust its operation, along with status to other autonomic systems or components.

Henceforth, future autonomics envisions the aggregation of node-scoped control loops, i.e. within a single node, in terms of interacting intra/inter-node control loops or triggered/managed low level control loops by higher level control loops within the node or the network as a system. Intuitively, the above view leads to a hierarchal control loops paradigm that enables the efficient design of autonomic nodes, systems and architectures. In this paper, node/network's atomicity is employed as an enabler en route for devising a flexible and proficient QoS-aware service orientated wireless interworking architecture.

The distinct features of the proposed architecture are summarized as follows:

- Optimal utility-based resource allocation and QoS provisioning within each system's cell (WLAN or CDMA). Thus, an Autonomic Radio Recourse Management (ARRM) mechanism is introduced to achieve the above goal.
- Efficient joint resource management via a flexible network selection mechanism, which determines whether or not to admit and to which network cell (WLAN or CDMA network) a new or vertical/horizontal handoff service arrival. Towards enabling that, an Autonomic JOint Network Selection (AJONS) decentralized mechanism is introduced.

3 Autonomic QoS-Aware Joint Resource Control

In the proposed integrated CDMA/WLAN architecture a proper Autonomic Radio Resource Management mechanism (ARRM) residing at the base station of each cell in the network is responsible for optimally and independently allocating cell's available radio resources among all active users already attached to the specific network. Moreover, a new user entering the network or an already attached user willing to perform vertical or horizontal handoff due to connectivity, mobility or QoS-triggered events, is accountable for selecting the most appropriate access network type to be attached to, as well as the corresponding base station (cell) from the ones available in his locality using only locally available information. Such a Self-optimization behavior is enabled via a new novel scheme Autonomic JOint Network Selection (AJONS).

3.1 Towards Optimal Resource Allocation

Due to the different wireless access technologies type of resources, as well as the users' services diverse expectations the concept of utilities from the field of economics has been adopted for developing QoS-aware resource allocation mechanisms. A utility function reflects a user's degree of satisfaction with respect to his service performance, and therefore services with assorted QoS prerequisites can be represented in a normalized way. The presented architecture is founded on a common utility-based framework in order to reflect users' QoS requirements in both network types in a unified way, towards achieving seamless and efficient integration.

We consider a set S_{CDMA} (S_{WLAN}) of N_{CDMA} (N_{WLAN}) continuously backlogged users attached to a cell of the CDMA (WLAN) network. Each mobile user is associated with a proper utility function U_i which represents his degree of satisfaction in accordance to his expected actual downlink transmission rate R_i. We assume that U_i has the following properties.

1. U_i is an increasing, twice continuous differentiable function of R_i.
2. $U_i(0) = 0$ and also upper bounded.
3. U_i is a sigmoidal-like or strictly concave or convex function of its rate allocation.

Typically, most utility functions that have been used in wired or wireless networks can be represented by the latter three types of functions illustrated in Fig. 1 [7].

In current CDMA cellular systems QoS-aware resource allocation is commonly performed via power control. Periodically (T_s), resource allocation utility-based optimization problems are set and solved by each cell's ARRM to acquire optimal user's resource assignment. To obtain users' power vector $\bar{P}$ that maximizes total

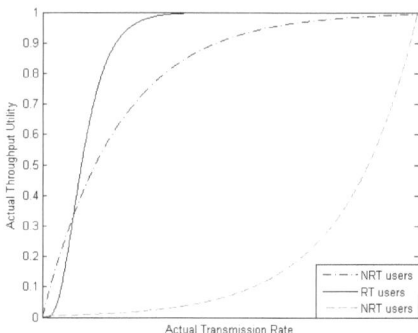

Fig. 1. Basic utilities types as a function of a user's achieved goodput

system utility, the solution of the following non-convex maximization problem must must be derived [7]:

$$\max_{\bar{P}} \sum_{i=1}^{N_{CDMA}} U_i\left(R_i(\bar{P})\right) \quad s.t. \quad \sum_{i=1}^{N_{CDMA}} P_i \leq P_{\max}, \ 0 \leq P_i \leq P_{\max} \tag{1}$$

where P_{max} denotes the maximum transmission power of a CDMA base station.

To support QoS in WLANs, IEEE 802.11e [8] has been introduced which allows specific parameters that affect a user's j performance (e.g. its maximum (minimum) contention window) to be altered by the access point. The corresponding non-convex utility based optimization problem can be formally defined as:

$$\max_{\bar{R}} \sum_{j=1}^{N_{WLAN}} U_j(R_j) \quad s.t. \quad \sum_{j=1}^{N_{WLAN}} R_j \leq C_{\max}, \ 0 \leq R_j \leq C_{\max} \tag{2}$$

where C_{max} is system's maximum effective capacity. Problem (2) is set and solved within short-term time periods denoted as WLAN's time-frame (T_f). Mapping the derived optimal rate vector $\bar{R}^*$ to appropriate users' contention windows can be easily achieved as described in [8].

Towards solving (1) and (2), each user already attached to a cell $b \in \{CDMA \text{ or } WLAN\}$ solely computes (the Lagrange) $\lambda_{i,b}^{\max} = \min\left\{\lambda \geq 0 \mid \max_{0 \leq R \leq R_{\max}} \{U_{i,b}(R_i) - \lambda R_i\} = 0\right\}$ towards maximizing its net utility (i.e. the utility minus a corresponding cost), which represents user's i maximum willingness to pay per unit resource [7]. It is shown that each mobile has a unique $\lambda_i^{\max}$ which can be calculated as follows:

$$\lambda_i^{\max} = \begin{cases} \left.\dfrac{\partial U_{i,b}(R_i)}{\partial R_i}\right|_{R=0} & \text{if } U_{i,b} \text{ is concave} \\[2ex] \left.\dfrac{\partial U_{i,b}(R_i)}{\partial R_i}\right|_{R=R^*} & \begin{array}{l} \text{if } U_{i,b} \text{is a sigmoidal} \\ \text{function and } R^* \text{ exists} \end{array} \\[2ex] \dfrac{U_{i,b}(R_{\max})}{R_{\max}} & \text{otherwise} \end{cases} \tag{3}$$

where R_i^* is the unique positive solution of:

$$U_{i,b}(R_i) - R_i \frac{\partial U_{i,b}(R_i)}{\partial R_i} = 0 \quad for \; 0 \le R_i \le R_{max}.$$

(4)

Then, the base station obtains a unique equilibrium price per unit of resource λ_b^* that optimizes cell's resource allocation and broadcasts it. Finally, for those users that $\lambda_{i,b}^{max} \ge \lambda^*$, their allocated resources can be easily derived, following the approaches provided in [7] and [8]. It is noted that the value of a user's willingness to pay $\lambda_{i,b}^{max}$ defines his superiority against others (i.e. higher values of $\lambda_{i,b}^{max}$ declare higher possibility in QoS requirement fulfillment when attached to cell b), while at the same time cell's equilibrium price per unit of resource λ_b^* indicates its congestion level (i.e. lower values of λ_b^* dictate higher availability in resources).

3.2 Autonomic Intra-cell QoS-Aware Radio Resource Allocation in WLAN and CDMA Cellular Networks

In order to enable intra-cell mobile nodes'/network autonomicity we introduce two control loops residing at mobile nodes and base stations. The first manages a node's QoS performance and the second one manages a cell's resource control mechanism, while their collaboration realizes autonomic QoS radio resource management within the cells of an integrated WLAN/CDMA system (ARRM), as depicted in Fig.2.

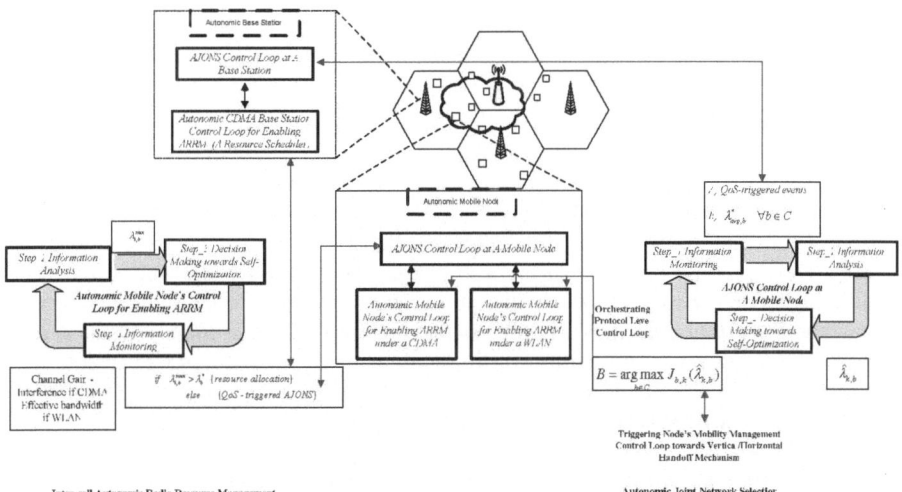

Fig. 2. Autonomic intra-cell QoS-aware radio resource management & Autonomic Joint Network Selection Mechanism (AJONS)

Autonomic Base Station Control Loop for Enabling ARRM

Periodically (i.e. on a time-slot basis regarding a CDMA cell (T_s) and every time-frame concerning a WLAN (T_f)), a control loop residing at a base station performs the following steps:

Step_1. Monitors its environment and gathers QoS related information concerning: a) active mobile users' services' QoS requirements (i.e. users' utilities) b) active mobile nodes' channel conditions and overall interference (CDMA cell) or current cell's maximum effective capacity (WLAN).

Step_2. Sets the corresponding constrained non-convex utility-based optimization problem, as defined in (1) or (2) and obtains its solution.

Step_3. Disseminates the acquired optimal resource allocation vectors to the cell's active autonomic nodes.

Autonomic Mobile Node's Control Loop for Enabling ARRM

Mobile node i is already attached to cell $b \in \{CDMA, WLAN\}$

Step_1 Information Monitoring: Constantly monitors a user's service performance and networking environment conditions (i.e. user's channel quality and overall attached cell interference or effective capacity).

Step_2 Information Analysis: Analyzes its current status with respect to QoS requirements and computes its current willingness to pay $\lambda_{i,b}^{max}$.

Step_3 Decision Making towards Self-Optimization: Interacts with the cell's b base station towards determining cell's equilibrium price per unit of resource λ_b^*. Thus:

If $\lambda_{i,b}^{max} < \lambda_b^*$, and the user is currently not selected to access system's resources and thus triggers user's network selection algorithm towards performing a QoS-triggered handoff (as detailed in the following section). Otherwise, determines its allocated resources.

Establishing control loops that steer node's and base stations' QoS resource allocation mechanisms allows us to further enhance them with self-optimization attributes and manageability attributes. The necessity of the latter emerges by the heterogeneity of the wireless environment where multiple access networks' mechanisms regarding a node's service QoS functionalities must simultaneously coexist and/or collaborate. To effectively accomplish that, an orchestrator is required. Such an orchestrator should also be a control loop, superior in a hierarchy of control loops, with advanced accountabilities that manages inferior in the hierarchy control loops within a mobile node. The operation, the liabilities, the goals and the algorithms that enable such a superior control loop, and thus an autonomic mobile node, to make QoS-aware self-optimization decisions are studied in the following.

3.3 Autonomic Joint Network Selection Mechanism (AJONS)

The goal of Autonomic JOint Network Selection mechanism (AJONS) is to enable autonomic mobile nodes to exploit locally available information from the corresponding base stations of the existing cells in their locality in order to dynamically determine whether or not, and to which network to be attached to, either when entering the system or at the event of a QoS-triggered handoff. Such a procedure mainly aims at guarantying the services' QoS constraints in both WLANs and CDMA networks as well as maximizing the average network revenue via endorsing cell's load balancing. The short-time intervals that both CDMA (T_s) and WLAN's (T_f) ARRM mechanisms set and solve the corresponding resource allocation problems, towards exploiting multi-users' diversity, makes the use of the instantaneous short-term pricing values λ^* as a cell's congestion indicator insufficient, since it would be undesirably sensitive to short-term cell's status variation. Moreover, this would lead users to constantly alter access network or cell preference; thus triggering QoS-driven ping-pong effects.

To overcome such a drawback, AJONS mechanism requires the setup and solution, of both CDMA and WLAN QoS and resource allocation problems, as defined in (1) and (2) in such time intervals that allows to the derived equilibrium price per unit of resource λ^*, denoted as λ^*_{avg}, to efficiently reflect long-term cells' load and environmental variations, in parallel to cell's ARRM mechanism. We refer to the above long-term time period as AJONS time frame (T_{AJONS}) and is defined as $T_{AJONS} = M \cdot \max(T_s, T_f)$ $M \in \aleph$. The smaller the value of M, the more sensitive AJONS mechanism is in short-term variations of the interworking environment.

As the system evolves, periodically, every T_{AJONS}, each base station solves problem (1) in the case of a CDMA cell or problem (2) in the case of a WLAN cell, regarding its already attached users and considering exponentially averaged values for nodes' and cells' characteristics respectively. Subsequently, each cell's b averaged equilibrium price per unit of resource, $\lambda^*_{avg,b}$ is disseminated via broadcasting to the mobile nodes. Each autonomous node k, either entering the integrated system or reacting to QoS-triggered events computes its maximum willingness to pay per resource unit $\lambda^{max}_{k,b}$ that he would acquire if he selected cell b to attach to, for each of the corresponding existing cells in his locality. In the following we assume a set C of N_C network cells, belonging to either of the considered access technologies, to be available for the user to receive service from. In this way the user possesses all the necessary required information to compute for each cell $b \in C$ the normalized indicator $\hat{\lambda}_{k,b}$, defined as follows:

$$\hat{\lambda}_{k,b} = \begin{cases} \dfrac{\lambda^{max}_{k,b} - \lambda^*_{avg,b}}{\lambda^*_{avg,b}} & if \ \lambda^{max}_{k,b} \geq \lambda^*_{avg,b} \\ 0 & otherwise \end{cases}. \tag{5}$$

Since $\lambda^{max}_{k,b}$ can be interpreted as the maximum value of resource unit of user k at cell b and $\lambda^*_{avg,b}$ as the long-term price of resource unit at cell b, then $\hat{\lambda}_{k,b}$ can be interpreted as the normalized profit per resource unit that user k can acquire once selecting cell b to attach to. Afterwards, the node selects the cell $B \in C$ at which he will be finally attached to, in accordance to the following policy:

$$B = \arg \max_{b \in C} J_{b,k}(\hat{\lambda}_{k,b}) \tag{6}$$

where $J_{b,k}$, is an non-negative, increasing, concave function of $\hat{\lambda}_{k,b}$ and is employed to either reflect network type related parameters or to allow network's operator to impose specific policies regarding billing, access priorities, and congestion avoidance.

The intuition behind the proposed network selection strategy is twofold. From network's perspective, the higher the congestion of a cell, the higher its equilibrium price per unit $\lambda^*_{avg,b}$ will be in the event that a user selects to attach to the cell, thus discouraging or preventing the user from being attached to that cell. Such an approach will eventually lead towards a load balanced integrated network. From user's perspective, the higher his utility-based satisfaction from being attached to a cell is (i.e. his service QoS-aware performance), the higher his maximum willingness to pay will be for the specific offered service quality, steering him to select the most profitable cell and network type. In the rest of this section we outline two control loops (illustrated in Fig.2) for accomplishing the above described distributed asynchronous QoS-triggered joint network selection.

AJONS Control Loop at a Base Station

Step_1. Periodically, every T_{AJONS}, sets and obtains the solution of cell's b constrained non-convex utility-based resource allocation optimization problem defined in (1) and (2) using exponentially averaging, within a T_{AJONS} time interval, for the parameters: users' channel quality regarding all users already attached to cell b when b is a CDMA cell, or effective capacity when b is a WLAN, respectively.

Step_2. Disseminates the acquired equilibrium price per unit of resource $\lambda^*_{avg,b}$ to the autonomous mobile nodes/users in the cell.

AJONS Control Loop at a Mobile Node

Step_1. Constantly monitors user's services performance and reacts to QoS-triggered events (i.e. Step_3 of Autonomic Mobile Node's Control Loop for Enabling ARRM) or mobility triggered events.

Step_2. Obtains locally available networks' average equilibrium price per unit of resource $\lambda^*_{avg,b}$, disseminated from all network's cells in his locality (i.e. $b \in C$).

Step_3. Computes the normalized profit per resource unit $\hat{\lambda}_{k,b}$ for each $b \in C$ and selects the most profitable network to handover/attach (i.e. cell B in accordance to (4)).

Step_4. Disseminates this decision to lower level control loops that execute the attachment/handoff.

4 Numerical Results and Discussions

In this section we present some indicative numerical results considering an integrated CDMA/WLAN (IEEE 802.11e) system with one CDMA cell, and one WLAN network overlapping with the CDMA cell. We assume that the CDMA network's base station is located at the cell's center and that its maximum transmission power is $P_{max}=10$. Moreover, we assume that CDMA system's spreading bandwidth is $W = 10^8$ and all users' maximum downlink rate is $R_i^{max} = 2 \cdot 10^3 \, kbps$. Regarding the

WLAN, the system's access point is also located at the center of its coverage area and operates in *5GHz* band with maximum network data rate of *54Mbit/s*. WLAN effective capacity, C_{max} is dynamically calculated using a simulator that incorporates the IEEE 802.11e scheme. We model the path gain from the CDMA base station to user *i* as $G_i = K_i / s_i^n$ where s_i is the distance of user *i* from the base station and *n* is the path loss exponent (*n=4*) and K_i is a log-normal distributed random variable with mean *0* and variance $\sigma^2 = 8(dB)$. New users periodically enter the system (i.e. every T_{AJONS}) requesting Real Time (RT) and Non-Real Time (NRT) services in a random manner while moving in arbitrary patterns. We use the following sigmoidal function to represent real-time users' $U_i(R_i)$, i.e $U_i(R_i) = m\{1/1 + e^{-a(R_i - p)} - d\}$, where we set $m = (1 + e^{ap})/e^{ap}$ and $d = 1/(1 + e^{ap})$ for normalization purposes (i.e. $U(0) = 0$ and $U(\infty) = 1$), while regarding non-real-time services a concave function $U_i(R_i) = 1 - exp(-gR_i)$ is applied, with *g=0.8*. For demonstration only purposes we set *a=3* and *p=3* [7].

In order to better illustrate the efficacy of the proposed autonomic joint network selection and QoS-triggered handoff mechanism in terms of achieved overall integrated network utility-based performance, we compare the performance of ARRM/AJONS architecture against three other network selection schemes. The first one makes use of Radio Signal Strength quality for determining the cell that a user should be attached to (referred as RSS) [9]. The second approach applies a Service Differentiation scheme (SDiff), where RT users are served by the CDMA cellular network while NRT by the WLAN [3]. Finally, INS scheme performs only Initial Network Selection at the time of a new user's arrival adopting AJONS mechanism, while vertical handovers are not permitted over the duration of its service. Let us underline that under all examined schemes optimal intra-cells' radio resource management is achieved by ARRM scheme. Finally, for demonstration purposes we set $J_{b,k}(\hat{\lambda}_{k,b}) = \hat{\lambda}_{k,b}$ and *M=1*.

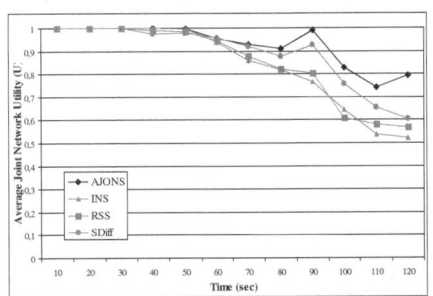 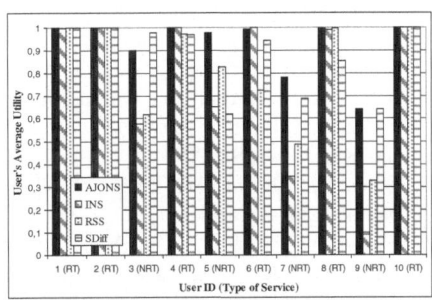

Fig. 3. a) Overall system utility and **b)** Users' average utility based performance

Fig.3a illustrates average joint integrated network's utility performance achieved under ARRM/AJONS, INS, RSS and SDiff schemes. The results reveal the superiority of the proposed autonomic scheme in terms of overall system performance, especially as the system evolves and the overall load increases. Moreover, the normalized profit per resource unit ($\hat{\lambda}$) exploited by ARRM/AJONS scheme is able to reflect not only performance parameters regarding both types of networks (i.e. congestion level, available resources and channel conditions), but more importantly user's service QoS-aware metrics, thus steering users towards making appropriate attachment decisions. On the other hand, myopic network selection criteria (i.e. RSS, SDiff) or even static network attachment schemes where no vertical handoffs are allowed (i.e. INS) are not capable of responding to networking environment variation (e.g. network cells' load and or users' channels time-varying nature) resulting to low overall system's performance, and thus to users' service QoS degradation. The latter behaviour is revealed in Fig.3b where users' average utility based performance is illustrated as a function of their ID and requested type of service.

5   Conclusions

In this paper, a novel QoS-centric joint resource allocation architecture for a WLAN/CDMA integrated network, founded on nodes/networks' autonomicity is discussed and evaluated. Autonomicity is deployed to facilitate the realization of multiple self-optimization functionalities towards integrated system's proficient utilization and efficient support of QoS provisioning and services' continuity.

Acknowledgments. This work has been partially supported by EC EFIPSANS project (INFSO-ICT-215549).

References

1. Luo, L., Mukerjee, R., Dillinger, M., Mohyeldin, E., Schulz, E.: Investigation of Radio Resource Scheduling in WLANs Coupled with 3G Cellular Network. In: IEEE Comm. Magazine, June 2003, vol. 41, pp. 108–115 (2003)
2. Xiao, Y., Leung, K.K., Pan, Y., Du, X.: Architecture, mobility management, and quality of service for integrated 3G and WLAN networks. Wireless Communications & Mobile Computing 5(7), 805–823 (2005)
3. Song, W., Jiang, H., Zhuang, W., Shen, X.: Resource management for QoS support in cellular/WLAN interworking. IEEE Network Mag. 19(5), 12–18 (2005)
4. Yu, F., Krishnamurthy, V.: Optimal Joint Session Admission Control in Integrated WLAN and CDMA Cellular Networks with Vertical Handoff. IEEE Trans. on Mobile Computing 6(1), 126–139 (2007)
5. Song, W., Zhuang, W., Cheng, Y.: Load balancing for cellular/WLAN integrated networks. IEEE Networks 21(1), 27–33 (2007)
6. Cheng, Y., Farha, R., Kim, M.S., Leon-Garcia, A., Hong, J.W.: A generic architecture for autonomic service and network management. Computer Communications 29(18), 3691–3709 (2006)
7. Lee, J.W., Mazumdar, R.R., Shroff, N.B.: Downlink power allocation for multi-class wireless systems. IEEE/ACM Trans. on Net. 13(4), 854–867 (2005)
8. Banchs, A., Pérez-Costa, X., Qiao, D.: Providing throughput guarantees in IEEE 802.11e wireless LANs. In: Proc. of the 18th International Teletraffic Congress (ITC18), Berlin, Germany (September 2003)
9. Liang, L., Wang, H., Zhang, P.: Net Utility-Based Network Selection Scheme in CDMA Cellular/WLAN Integrated Networks. In: Proc of WCNC 2007, March 2007, pp. 3313–3317 (2007)
10. [TS23402] 3GPP TSG SA, 3GPP TS 23.402 V8.1.1. Architecture enhancements for non-3GPP accesses (Release 8)

Efficient Anonymous Authentication Protocol Using Key-Insulated Signature Scheme for Secure VANET

Youngho Park, Chul Sur, Chae Duk Jung, and Kyung-Hyune Rhee

Division of Electronic Computer and Telecommunication Engineering, Pukyong
National University, 599-1 Daeyon 3Dong Nam-Gu, Busan, Republic of Korea
{pyhoya,kahlil,jcd0205,khrhee}@pknu.ac.kr

Abstract. In this paper, we propose an efficient authentication protocol with conditional privacy preservation for secure vehicular communications. The proposed protocol follows the system model to issue on-the-fly anonymous public key certificates to vehicles by road-side units. In order to design an efficient message authentication protocol, we consider a key-insulated signature scheme for certifying anonymous public keys of vehicles to such a system model. We demonstrate experimental results to confirm that the proposed protocol has better performance than other protocols based on group signature schemes.

Keywords: vehicular network, security, anonymous, authentication, key-insulated signature.

1 Introduction

As vehicular communications bring the promise of improved road safety and optimized road traffic through cooperative systems applications, vehicular ad hoc networks(VANET) have received a great deal of attention from both academia and industry. Considering the useful applications in VANET, a prerequisite for the successful deployment of VANET is to make vehicular communications secure first of all [6][8].

For example, it is essential to make sure that life-critical information cannot be illegally inserted or modified by an attacker in safety applications, and it should also protect the privacy of the drivers and passengers as far as possible. Therefore, it becomes fundamental requirement to provide anonymous message authentication for secure vehicular communications. Moreover, there is a common need for a security infrastructure for establishing mutual trust and enabling cryptographic schemes. The security infrastructure includes all technical and organizational measures and facilities needed to provide for the security goals.

Raya et al. [7] proposed some building blocks for secure vehicular communication. As a straightforward solution in their protocol, each vehicle possesses a set of anonymous keys to sign a message and these keys are periodically changed to avoid being tracked. However, it has some critical disadvantages; it requires

F. Granelli et al. (Eds.): MOBILIGHT 2009, LNICST 13, pp. 35–44, 2009.

a large number of anonymous public key certificates, and hence less efficient in storage costs. Moreover, it requires a long revocation list and a long time to update the certificate revocation list due to the large number of public keys.

Lin et al. [3] proposed a secure and privacy preservation protocol using group signature scheme, named GSIS, to resolve the requirement of a large number of public key certificates. In their work, vehicles possess only their own group signing key issued by a trusted group manager, and each vehicle signs a message by using group signature scheme to be authenticated as a legitimate sender of the message. However, although it does not require a large storage space, the time for message verification accompanied with revocation check grows linearly with the number of revoked vehicles, and hence less efficient in computational cost.

Lu et al. [4] proposed a system model for efficient privacy preservation protocol, named ECPP, which also uses a group signature scheme. Compared with the GSIS, instead of using group signature scheme for anonymous message authentication, each RSU(Road Side Unit), on vehicle's request, issues on-the-fly short-time anonymous public key certificate to the requesting vehicle by using group signature scheme. Since the RSU checks the validity of the requesting vehicles during the short-time anonymous public key certificate issuance protocol, such revocation check by vehicle itself of GSIS is not required. Therefore, message verification is more efficient than GSIS.

It is evident that Lu et al., in ECPP, introduced a somewhat reasonable system model for implementing a practical short-time anonymous public key certificates management in VANET. However, although efficient group signature schemes have been proposed in cryptographic literatures, group signature itself is still a rather much time consuming operation. Hence, in our opinion, key-insulated signature(KIS) [2] scheme may be an alternative solution suitable for this network architecture.

Based on these observation, in this paper, we propose an efficient anonymous authentication protocol(EA^{2}P) in VANET. Our system model and roles of each entity on VANET are similar to ECPP's. However, we consider the KIS scheme to our system model as our cryptographic building blocks to issue on-the-fly short-time anonymous public key certificates by RSUs. We demonstrate experimental results to confirm that our protocol has better performance than other protocols based on group signature schemes.

The rest of this paper is organized as follows: We describe our system architecture in Section 2, and we propose our protocols in Section 3. We analyze the performance of our protocol as comparing with ECPP in Section 4, and conclude in Section 5.

2   System Model

Because safety applications on VANET are in the beginnings and the primary VANET's goal is to increase road safety, we also consider a simple public safety

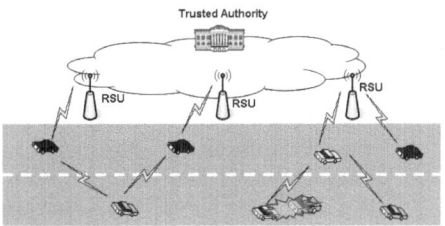

Fig. 1. Vehicular network model

message application using IEEE 802.11p incorporated with DSRC [9]. As shown in Figure 1, vehicular network consists of three entities and each entity has the following roles.

TA: TA(Trusted Authority), such as Governmental Transportation Authority, is in charge of the registration of RSUs deployed on the road side and OBUs equipped on the vehicles. The TA can reveal the real identity of a message originator by incorporating with its subordinate RSUs when a disputed situation is occurred.

RSU: RSUs are controlled by the TA and responsible for issuing short-time anonymous public key certificates to OBUs by using KIS scheme. RSUs do not disclose any inner information without the authorization of the TA.

OBU: The OBUs are installed on the moving vehicles. They mainly communicate with each other for sharing local traffic information, and with RSUs for requesting the short-time anonymous public key certificate.

3   Proposed Protocol: EA²P

We apply the key-insulated signature scheme [5] and key agreement scheme based on bilinear pairings [1] to our short-time anonymous public key certificate issuance protocol. Table 1 shows the notations used in our EA²P.

3.1   System Initialization

TA chooses random numbers $s_0, x, x' \in Z_q^*$ and sets s_0 and $x_0 = x - x'$ as the master secrets for ID-based private key extraction and key-insulated signing key extraction, respectively. TA calculates $y_0 = g_2^{s_0}$, $y_1 = g_1^{x_0}$ and $y_1' = g_1^{x'}$, and then publishes system parameters $\langle \mathbb{G}_1, \mathbb{G}_2, \mathbb{G}_T, q, g_1, g_2, \hat{e}, y_0, y_1, y_1', H_1, H_2, H_3 \rangle$. Here, $\langle y_1, y_1' \rangle$ is the public KIS verification key to be used for checking short-time anonymous public key certificate. TA issues ID-based private keys and KIS signing keys according to the initial registration process of Figure 2. We assume that those keys are distributed through out-of-band channel.

Table 1. Notations for EA^2P

Notation	Description
$\mathbb{G}_1, \mathbb{G}_2, \mathbb{G}_T$	cyclic groups of the same prime order q.
$\hat{e} : \mathbb{G}_1 \times \mathbb{G}_2 \to \mathbb{G}_T$	bilinear map from $\mathbb{G}_1 \times \mathbb{G}_2$ to $\mathbb{G}_T$.
$g_1 \in \mathbb{G}_1, g_2 \in \mathbb{G}_2$	generators of $\mathbb{G}_1$ and $\mathbb{G}_2$.
K_{TA}	TA's secret key for message encryption.
PID_i	pseudo-id for a real vehicle identity VID_i.
RSU_j	identity of an RSU.
OBU_i	on-board unit of a vehicle VID_i.
$ok_i, rk_j \in \mathbb{G}_1$	ID-based private keys for OBU_i and RSU_j respectively.
$kk_j \in \mathbb{G}_1$	RSU_j's secret KIS signing key.
sk_i, pk_i	OBU_i's short-time private/public key pair.
$Cert_i$	short-time anonymous public key certificate for pk_i.
$Enc_K(), Dec_K()$	encryption and decryption under the key K.
MAC_K	message authentication code using the key K.
$H_1 : \{0,1\}^* \to \mathbb{G}_1$	cryptographic one-way hash functions.
$H_2 : \mathbb{G}_1 \times \{0,1\}^* \to Z_q^*$	
$H_3 : \mathbb{G}_1^3 \times \{0,1\}^* \to Z_q^*$	

---

1. for OBU_i:
   (a) compute $PID_i = Enc_{K_{TA}}(VID_i)$.
   (b) set $ok_i = H_1(PID_i)^{s_0}$ as VID_i's ID-based private key for PID_i.
   (c) issue $\langle ok_i \rangle$ to OBU_i.
2. for RSU_j:
   (a) set $rk_j = H_1(RSU_j)^{s_0}$ as RSU_j's ID-based private key.
   (b) choose $r_j \in Z_q^*$ and compute $v_j = g_1^{r_j}$ and $c_j = H_2(v_j, T)$, where T is time period.
   (c) calculate $x_j = c_j r_j + x_0 \pmod{q}$.
   (d) set $kk_j = x_j + x' \pmod{q}$ as secret signing key.
   (e) store $\langle RSU_j, v_j \rangle$.
   (f) issue $\langle rk_j, kk_j, v_j \rangle$ to RSU_i.

Fig. 2. Initial registration and key issuance of the TA

3.2   Short-Time Anonymous Public Key Certificate Issuance

Instead of having a large number of pre-issued short-time anonymous public key certificates, each OBU_i can request a $Cert_i$ to the RSU_j within OBU_i's communication range when the OBU_i is necessary to renew its anonymous public key. Figure 3 shows the certificate issuance protocol.

The detailed protocol steps are described as follows.

1. When OBU_i with pseudo-id PID_i requests a $Cert_i$ to RSU_j, they should authenticate each other to determine whether the OBU_i can provide the RSU_j with its PID_i, and to convince the given RID_j and PID_i are valid.

OBU_i	RSU_j
1. $a \in Z_q^*$, g_2^a, $\phi_i = H_1(PID_i)$.	

$$\xrightarrow{\quad \text{req1}: \ g_2^a, \phi_i \quad}$$

	2. $b \in Z_q^*$, g_2^b, $\phi_j = H_1(RSU_j)$.
	$k = \hat{e}(\phi_i^b, y_0) \cdot \hat{e}(rk_j, g_2^a)$.
	$\pi_j = MAC_k(RSU_j, \phi_i, \phi_j, g_2^a, g_2^b)$.

$$\xleftarrow{\quad \text{res1}: \ \phi_j, g_2^b, \pi_j \quad}$$

3. $k = \hat{e}(ok_i, g_2^b) \cdot \hat{e}(\phi_j^a, y_0)$.

  check $\pi_j \overset{?}{=} MAC_k(RSU_j, \phi_i, \phi_j, g_2^a, g_2^b)$.

  choose sk_i, pk_i, t_i.

  $C_i = Enc_k(PID_i, pk_i, t_i)$.

  $\pi_i = MAC_k(PID_i, RSU_j, \phi_i, \phi_j, g_2^a, g_2^b, pk_i, t_i)$.

$$\xrightarrow{\quad \text{req2}: \ C_i, \pi_i \quad}$$

	4. $\langle PID_i, pk_i, t_i \rangle = Dec_k(C_i)$.
	check $\pi_i \overset{?}{=} MAC_k(PID_i, RSU_j, \phi_i, \phi_j, g_2^a, g_2^b, pk_i, t_i)$.
	$u_j \in Z_q^*$, $w_j = g_1^{u_j}$.
	$z_j = H_3(v_j, w_j, pk_i, t_i)$.
	$\sigma_j = u_j z_j + kk_j \pmod{q}$.
	$Cert_i = \langle pk_i, t_i, \sigma_j, z_j, v_j \rangle$.
	store $\langle PID_i, Cert_i \rangle$.

$$\xleftarrow{\quad \text{res2}: \ Cert_i \quad}$$

5. $c_j = H_2(v_j, T)$.

  check

  $z_j \overset{?}{=} H_3(v_j, (g_1^{\sigma_j}(v_j^{c_j} y_1 y_1')^{-1})^{1/z_j}, pk_i, t_i)$.

Fig. 3. Short-time anonymous public key certificate issuance protocol between OBU_i and RSU_j

OBU_i chooses a random value $a \in Z_q^*$ to compute g_2^a and $\phi_i = H_1(PID_i)$, and then sends a request with $\langle g_2^a, \phi_i \rangle$ to RSU_j.

2. Upon receiving the request, RSU_j chooses a random value $b \in Z_q^*$ and sets g_2^b and $\phi_j = H_1(RSU_j)^b$. RSU_j calculates $k = \hat{e}(\phi_i^b, y_0) \cdot \hat{e}(rk_j, g_2^a)$ to computes $\pi_j = MAC_k(RSU_j, \phi_i, \phi_j, g_2^a, g_2^b)$, and then sends $\langle g_2^b, \phi_j, \pi_j \rangle$ to the OBU_i as a response.

3. The OBU_i computes $k = \hat{e}(ok_i, g_2^b) \cdot \hat{e}(\phi_j^a, y_0)$, and checks $\pi_j \overset{?}{=} MAC_k(RSU_j, \phi_i, \phi_j, g_2^a, g_2^b)$ to authenticate the RSU_j. If it holds, the OBU_i selects its anonymous private/public key pair $\langle sk_i, pk_i \rangle$ and short-time period t_i ($t_i < T$). Then OBU_i requests a $Cert_i$ for the public key pk_i to be used for the time period t_i by providing $C_i = Enc_k(PID_i, pk_i, t_i)$ and $\pi_i = MAC_k(PID_i, RSU_j, \phi_i, \phi_j, g_2^a, g_2^b, pk_i, t_i)$.

4. When receiving certificate request, RSU_j first decrypts C_i to get OBU_i's pseudo-id PID_i, public key pk_i and t_i, and then looks up the up-to-date revocation list retrieved from the TA to check the validity of the given PID_i. If the PID_i is revoked one, the RSU_j refuses to issue the short-time public key certificate. Otherwise, RSU_j verifies $\pi_i \overset{?}{=} MAC_k(PID_i, RSU_j, \phi_i, \phi_j, g_2^a, g_2^b, pk_i, t_i)$. If it holds, the OBU_i is ultimately authenticated, and then RSU_j generates a $Cert_i = \langle pk_i, t_i, \sigma_j, z_j, v_j \rangle$ by using RSU_j's KIS signing key kk_j. In fact, $\langle \sigma_j, z_j, v_j \rangle$ is RSU_j's digital signature for certifying the given public key pk_i. In the end, RUS_j issues a $Cert_i$ to OBU_i and stores $\langle Cert_i, PID_i \rangle$ in its local certificate list for assisting TA by way of provision against a liability investigation. Note, in certificate generation, that no identity-related information is included in the $Cert_i$.

5. To verify the validity of the $Cert_i$, OBU_i computes $c_j = H_2(v_j, T)$ for the current date T and checks $z_j \overset{?}{=} H_3(v_j, (g_1^{\sigma_j}(v_j^{c_j} y_1 y_1')^{-1})^{1/z_j}, pk_i, t_i)$ by using TA's KIS public key $\langle y_1, y_1' \rangle$. If it holds, the OBU_i comes to possess the private key sk_i and the corresponding anonymous public key certificate $Cert_i$. Then, OBU_i can use this key for the purpose of anonymous message authentication for the short-time period t_i in VANET.

3.3 Anonymous Message Authentication

Once obtaining a $Cert_i$, the OBU_i can send safety messages in authenticated manner during the short-time period t_i. With the proposed protocol, an OBU_i which intends to send a safety message msg composed of traffic-related information without OBU_i's identity can run the following steps.

1. OBU_i signs the msg with its short-time signing key sk_i for signature $sig_i = Sig(sk_i, msg)$, where $Sig()$ is ordinary digital signature algorithm such as ECDSA, and forms the message $Msg = [msg \mid sig_i \mid Cert_i]$, and then broadcasts Msg.

2. Upon receiving a safety message, each receiving OBU first checks the validity of the signature $\langle \sigma_j, z_j, v_j \rangle$ in the $Cert_i$ for the current date T by using TA's KIS public key $\langle y_1, y_1' \rangle$. Here, the same verification procedure in step 5 of Figure 3 is used. If the $Cert_i$ is valid, then the receiver retrieves the public key pk_i from the $Cert_i$ and verifies the signature sig_i using the pk_i. If sig_i is verified as valid, the safety message can be accepted, otherwise discarded.

3.4 Vehicle Tracing

When we deploy vehicular safety applications, liability requirement should be considered in addition to privacy preservation requirement. Hence, anonymity should be conditional depending on scenarios such as law enforcement. In our EA^{2}P, if a disputed circumstance occurs to a safety message $Msg = [msg \mid sig_i \mid Cert_i]$, TA is involved in tracing the originator of this message.

1. TA first retrieves the partial public key v_j from the $Cert_i$ and searches its trace list to find the RSU_j for the v_j, then requests the pseudo-id of the $Cert_i$ holder.
2. On TA's demand, the RSU_j retrieves the pseudo-id corresponding to the $Cert_i$ by searching its local certificate list and returns the pseudo-id PID_i to the TA.
3. Then, the TA can ultimately recover the real identity from the returned pseudo-id by $VID_i = Dec_{K_{TA}}(PID_i)$.

4   Performance Evaluation

In order to show the efficiency of our protocol in terms of RSU valid serving ratio and efficient message verification, we compare our EA^{2}P with ECPP of Lu et al.'s in this section. For fairness in comparisons, we selected the same security measures of Lu et al.'s; We assumed an MNT curve of embedding degree $k = 6$ and $|q| = 160$bits for bilinear pairing implemented on Pentium IV 3.0 GHz. We do not put restriction to any digital signature scheme, but we assume the ECDSA adopted by IEEE 1609.2 standard [10] for message authentication. Table 2 and Table 3 show the measures to estimate and to compare our EA^{2}P with ECPP, respectively.

Table 2. Cryptographic operation time

	Description	time
T_{pair}	bilinear pairing operation	4.5 ms
T_{mul}	point multiplication	0.6 ms
T_{ecdsa}	ECDSA signature verification	1.28 ms

Table 3. Protocol execution time and message size of EA^{2}P and ECPP

	Description	ECPP	EA^{2}P
T_{gen}	time for certificate issuance protocol	34.8 ms	20.4 ms
T_{cert}	time for certificate verification	18.9 ms	2.4 ms
T_{sig}	time for signature verification	1.2 ms	1.28 ms
$|Sig|$	signature size for a safety message	40 bytes	40 bytes
$|pk| + |Cert|$	public key certificate size	147 bytes	84 bytes

4.1   RSU Serving Ratio

RSU's main operation is to issue anonymous public key certificates to OBUs on requests within RSU's valid coverage (R_{rng}), so RSU's performance always depends on vehicles density(d) and speed(v) on the road. To measure RSU valid serving ratio, we followed Lu et al.'s analysis method. Then, the valid serving ratio S_{ratio}, which is the fraction of the number of actually processed certificates to the number of requests, can be defined by

$$S_{ratio} = \begin{cases} 1, & \text{if } \frac{R_{rng}}{T_{gen} \cdot v} \cdot \frac{1}{d \cdot \rho} \geq 1; \\ \frac{R_{rng}}{T_{gen} \cdot v} \cdot \frac{1}{d \cdot \rho}, & \text{otherwise.} \end{cases}$$

where ρ is the probability for each OBU to issue a certificate request.

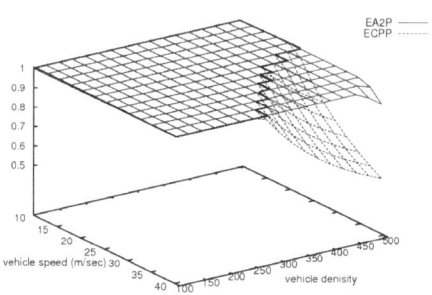

Fig. 4. RSU valid serving ratio of EA^{2}P and ECPP

Figure 4 shows RSU valid serving ratio of EA^{2}P and ECPP with different vehicle density and different vehicle speed for $R_{rng} = 300$m and $\rho = 0.8$. In this estimation, as shown in Table 3, $T_{gen} = 20.4$ms of EA^{2}P short-time anonymous public key certificate issuance protocol in Figure 3 was measured by $4T_{pair} + 4T_{mul}$, and 34.8ms of ECPP by $6T_{pair} + 13T_{mul}$ with respective. From these results, we can observe that RSU in our EA^{2}P can efficiently process OBUs' short-time anonymous public key certificate requests in most scenarios. On the other hand, ECPP cannot effectively process OBUs requests in some cases. Therefore, our EA^{2}P has the advantage in scalability for RSUs than ECPP.

4.2   Efficient Message Verification

When we authenticate a safety message, it requires to verify the certificate and signature of the safety message. Therefore, the required time cost of EA^{2}P is $T_{EA^2P} = T_{sig} + T_{cert} = 4.68$ms, and that of ECPP is $T_{ECPP} = T_{sig} + T_{cert} = 20.1$ms. The computational gains of EA^{2}P against ECPP is due to the certificate verification cost because our protocol considered key-insulated signature scheme to generate short-time anonymous public key certificate while ECPP applied group signature scheme which requires relatively much computations.

In actual vehicular communications, each vehicle is supposed to receive a lot of messages from many other vehicles within the same communication range. Therefore, it is required to measure the throughput of received messages. Suppose that there are n vehicles sending k messages every second within the same

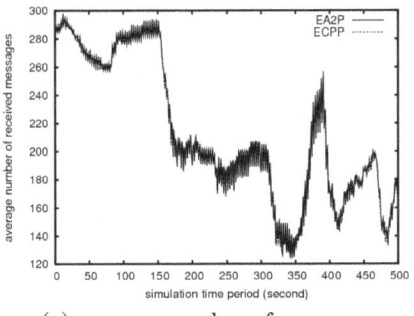 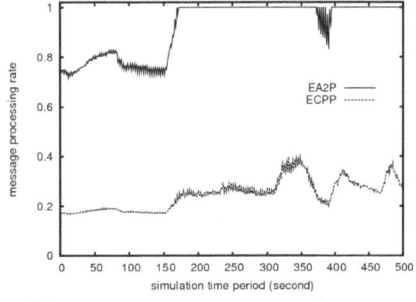

(a) average number of messages   (b) average message processing rate

Fig. 5. Average message processing rate of 400 vehicles for the received messages during 500 seconds simulation

communication range and the processing time per message is T_p. In the worst case, where all vehicles contend for the channel, $n_{msg} = n \times k$ messages are received per second, then the message processing rate per second is numerically calculated as $1/(T_p \times n_{msg})$.

In order to consider some actual vehicular communication on the road, we simulated message transmission on VANET by using network simulator, and then we traced the number of received messages and estimated the message processing rates. Figure 5 shows these results. We used TraNS [13] with ns2-2.33 [12] and IEEE 802.11p parameters for ns2 [11]. We put 400 vehicles on a grid-shape road of 600m×700m rectangular size. Each vehicle moves with a maximum speed of 16.7m/s (i.e., 60km/hr) and sends out a message every 300ms within 100m nominal radio range. The simulation was run for 500 seconds and we measured the received messages every second.

From Figure 5, we can observe that EA²P and ECPP received similar number of messages during the simulation, but our EA²P shows about minimum 72% processing rate while ECPP processes about minimum 17% and maximum 40% of received messages, which is much less than EA²P's. As a result, we can conclude that the proposed EA²P is more practical.

5   Conclusion

It is a fundamental security requirement to provide anonymous message authentication mechanism for secure vehicular communications. In this paper, we have proposed an efficient and effective anonymous authentication protocol based on the system model which on-the-fly short-time anonymous public key certificate is issued by an RSU on OBU's request when it needed. To implement a concrete protocol, we considered a key-insulated signature scheme to issue anonymous public key certificate by RSUs. By doing so, our protocol is more efficient and effective in RSU valid serving capability and message verification than those of group signature-based protocols. We have demonstrated, through the performance evaluation, that the proposed protocol can achieve much better performance than ECPP based on group signature scheme. As a result, our protocol can be implemented for practical secure vehicular communications.

Acknowledgement

This study was financially supported by Pukyong National University in the 2008 Post-Doc. program. This work was partially supported by the Korea Research Foundation Grant funded by the Korean Government (MOEHRD, Basic Research Promotion Fund) (KRF-2008-521-D00454).

References

1. Chen, L., Cheng, Z., Smart, N.P.: Identity-based key agreement protocols from pairings. International Journal of Information Security 6(4), 213–241 (2007)
2. Dodis, Y., Katz, J., Xu, S., Yung, M.: Key-insulated public key cryptosystems. In: Knudsen, L.R. (ed.) EUROCRYPT 2002. LNCS, vol. 2332, pp. 65–82. Springer, Heidelberg (2002)
3. Lin, X., Sun, X., Shen, X.: GSIS: a secure and privacy preserving protocol for vehicular communications. IEEE Transaction on Vechicular Technology 56(6), 3442–3456 (2007)
4. Lu, R., Lin, X., Zhu, H., Ho, P.-H., Shen, X.: ECPP: Efficient conditional privacy preservation protocol for secure vehilce communications. In: Proceedings of The IEEE INFOCOM 2008, pp. 1229–1237 (2008)
5. Ohtake, G., Hanaoka, G., Ogawa, K.: An efficient strong key-insulated signature scheme and its application. In: Mjølsnes, S.F., Mauw, S., Katsikas, S.K. (eds.) EuroPKI 2008. LNCS, vol. 5057, pp. 150–165. Springer, Heidelberg (2008)
6. Parno, B., Perrig, A.: Challenges in securing vehicular networks. In: Proceedings of the Fourth Workshop on Hot Topics in Networks (HotNets-IV) (2005)
7. Raya, M., Hubaux, J.-P.: Securing vehicular ad hoc networks. Journal of Computer Security 15(1), 39–68 (2007)
8. Zarki, M.E., Mehrotra, S., Tsudik, G.: Security Issues in a Future Vehicular Network. In: European Wireless 2002 (2002)
9. Dedicated Short Range Communications (DSRC),
   http://www.leearmstrong.com/dsrc/dsrchomeset.htm
10. IEEE Standard 1609.2 - IEEE Trial-Use Standard for Wireless Access in Vehicular Environments - Security Services for Applications and Management Messages (July 2006)
11. IEEE 802.11p parameters for NS2,
    http://dsn.tm.uni-karlsruhe.de/Overhaul_NS-2.php
12. Network Simulator-NS2, http://www.isi.edu/nsnam/ns/
13. TraNS - Realistic Simulator for VANET, http://trans.epfl.ch/

Application-Aware Dynamic Retransmission Control in Mobile Cellular Networks

Nadhir Ben Halima, Dzmitry Kliazovich, and Fabrizio Granelli

DISI - University of Trento
Via Sommarive 14, I-38050 Trento, Italy
{nadhir,kliazovich,granelli}@disi.unitn.it

Abstract. This paper proposes an application-aware cross-layer approach between application/transport layers on the mobile terminal and link layer at the wireless base station to enable dynamic control on the strength of per-packet error protection for multimedia and data transfers. Specifically, in the context of cellular networks, the proposed scheme allows to control the desired level of Hybrid ARQ (HARQ) protection by using an in-band control feedback channel. Such protection is dynamically adapted on a per-packet basis and depends on the perceptual importance of different packets as well as on the reception history of the flow[1].

Keywords: Hybrid ARQ (HARQ), cellular networks, service aware protocols.

1 Introduction

Nowadays, networking services are evolving to a "triple play" vision, implying delivery of data, voice and video to the end user using the same IP transport facility.

While no solution for end-to-end quality of service (QoS) assurance over heterogeneous networks is available, still several approaches exist for improving data transfer performance on the wireless access trunk [6].

In the specific framework of multimedia (e.g. voice and video), several works are available based on the Unequal Error Protection (UEP) paradigm [1-5]. The goal of UEP is to provide higher protection to the most perceptually relevant data, where protection can be achieved through means of adaptive power levels, forward error correction codes, retransmission control, etc. Nevertheless, since UEP is usually performed or managed at source level and thus without specific knowledge of the contingent operating scenario, such solutions (while increasing the complexity of multimedia codecs) can lead to non-optimal performance due to waste of available capacity in case network / channel conditions are good (and no packet drops are experienced) or for time-varying performance oscillations of the transport infrastructure (particularly true in the case of wireless networks).

---

[1] This work has been supported by the Italian National Project: Wireless multiplatfOrm mimo active access netwoRks for QoS-demanding muLtimedia Delivery (WORLD), under grant number 2007R989S, and by the PAT grant RObust Streaming Environment (ROSE).

F. Granelli et al. (Eds.): MOBILIGHT 2009, LNICST 13, pp. 45–52, 2009.

The proposed scheme represents a novel paradigm of dynamic and "link-level" UEP, focused on the access network and the actual "reception history" at the receiver. The core idea is to adaptively tune the level of HARQ protection based on the relative importance on the overall user experience of the packet being transmitted by the base station. Such approach enables to differentiate protection on the basis of the actual content of the packet: for voice and video flows, the impact of losing the current packet is estimated in terms of audio or visual quality as measurable by MOS or PSNR, respectively, for data transfer TCP throughput is chosen as the main quality feedback metric.

The structure of the paper is as follows: Section 2 describes in details the proposed framework, while performance evaluation is presented in Section 3. Finally, Section 4 concludes the paper with final remarks and outlines about future work on the topic.

2   Proposed Approach

The main idea of the proposed approach, called Application-Aware Dynamic Re-transmission Control (A^2DRC), is to allow the mobile terminal receiver to control the level of HARQ protection applied by the base station for every frame transmitted on the radio link. The decision of the mobile terminal is based on the potential benefit in correctly receiving the next packet given the current reception history and the actual perceptual relevance of the packet itself.

As an evolution of Automatic Repeat Request (ARQ) approach used in UMTS, the Hybrid ARQ (HARQ) scheme employed in HSDPA allows incorrectly received data blocks to be soft-combined in the effort to correct propagated link errors [8].

In this paper, the level of HARQ protection (also indicated as "HARQ Strength" in the following) is considered in terms of the maximum number of stop-and-wait re-transmission attempts taken for a packet delivery in case of failure.

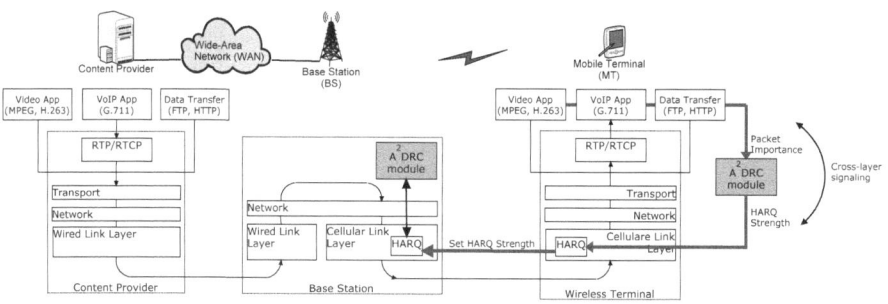

Fig. 1. Architectural principles of A^2DRC in a 3G cellular network. Blocks and links high-lighted in "blue" underline the modules and signaling links employed by A^2DRC approach.

Fig. 1 illustrates architectural principles of the proposed A^2DRC approach. As out-lined in the previous sections, A^2DRC operates on the wireless 3G link. At the mobile terminal side, whenever a packet is received by the application, the latter can specify packet importance for subsequent incoming packets for a given flow. The packet importance is then transferred into corresponding values of HARQ protection by the A^{2}DRC module (implemented within the protocol stack at the mobile terminal) and delivered to the HARQ entity at the link layer of the Base Station using cross-layer signaling. At the link layer, the specified HARQ protection parameter is sent along with HARQ acknowledgement, which is generated for every frame received according to stop-and-wait HARQ type.

The A^{2}DRC module implemented at the BS analyses incoming traffic and specifies the HARQ entity to use the requested HARQ protection on a per-packet basis.

In this paper, we address three different classes of services – voice, video, and data transfer – improving delivery performance by adapting network response to the relevance of packet being delivered over the radio link.

The level of HARQ protection in the proposed approach varies on the basis of a packet importance metric, which consists of two components:

- *Initial packet importance* corresponds to the level of quality reduction for a given flow in case the packet is lost during transmission or corrupted at the receiver [9]. The quality of the flow is determined by end-to-end application requirements and user demands. For example, commonly used metric for VoIP is Mean Opinion Score (MOS), for video is Peak Signal-to-Noise Radio (PSNR), and for TCP-based data is transfer throughput level.

- *Dynamic packet importance* component accounts for the "reception history" of the flow and adjusts initial packet importance. For example, the importance of frame i in a video sequence can be dynamically adjusted in case its decoding depends on the neighboring frames $i-1$ and $i+1$ and frame $i-1$ is not correctly received.

Packet Importance Metrics in Video Streams. For sake of a clear explanation, we consider a scenario with a mobile node receiving MPEG-4 video flows from a streaming server located in the wired Wide-Area Network (WAN). However, similar reasoning can be applicable to H.263 and H.264 encoded video streams, as well as embedded video streams.

An MPEG-4 video is composed of Groups of Pictures (GOPs), consisting of video frames of three types: I-Frames (Intra coded frames) which are encoded without reference to any other frame in the sequence, P-Frames (Predicted frames) which are encoded as differences from the last I- or P-frame, and B-Frames (Bidirectional frames) which are encoded as the difference from the previous or following I- or P-frames.

Due to the correlation property of P- and B-frames, the effective impact deriving from the loss of an I-frame can be clearly considered much higher than that of P- or B-frame. In addition, the loss of one I- or P-packet may generate error propagation: while the loss of a B-frame does not affect the quality of the consecutive frames, the loss of an I-frame may disable correct decoding of subsequent P- and B- frames. This leads to the conclusion that I-frames are more important than P-frames, which are more important than B-frames.

Fig. 2 shows the quality reduction of a real video flow transmitted using VideoLan software [12] in terms of PSNR measured at the receiver versus the loss of different types of packets within a GOP. The horizontal scale indicates which frame within the GOP was lost, while the first value (obtained with no losses) serves as a reference point.

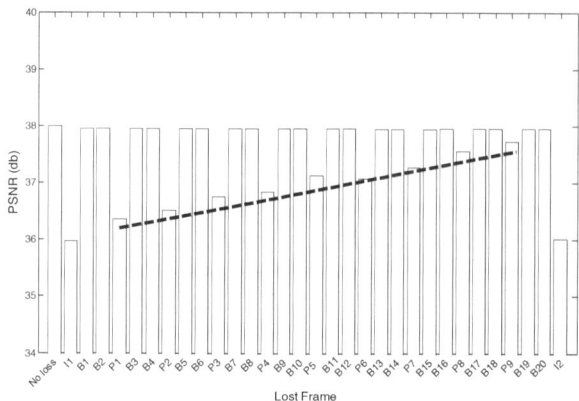

Fig. 2. Quality of the received video flow for different frames lost

Following such observation, the importance of P-frames P_{imp} is defined ranging linearly from I_{imp} to B_{imp}, where I_{imp} is the importance level of I-frames and B_{imp} is the importance level of B-frames with $I_{imp} \geq P_{imp} \geq B_{imp}$.

Packet importance metrics in VoIP flows. At the receiver side, speech frames are de-multiplexed and inserted into a playout buffer. The playout buffer plays an important role in perceived speech quality since it enforces speech frames delivery at the same interval at which they are generated by the encoder. This is done through re-ordering, delaying or even dropping the frames which arrive later than their expected playback time. However, whenever the frame is dropped it causes a relevant decrease of the quality of the voice stream.

Based on the above, initially, equal packet importance (i.e., "initial packet importance") is associated to all transmitted speech frames. However, in case the receiver detects frame losses after out-of-order frame reception, it increases importance (and error redundancy) for the subsequent packets of the stream (i.e. increases the "dynamic packet importance"). Summarizing, A²DRC aims at avoiding bulk frame losses, which are critical for the quality of the speech stream, while single frame losses can be easily compensated or concealed by the decoder.

Packet importance metric in file transfer. Packet losses can severely decrease the data transfer performance of TCP which is the most widely used protocol in Internet.

The proposed A²DRC scheme dynamically adapts the level of HARQ protection used on the radio link based on the value of the TCP congestion window computed at the receiver node.

The core idea is to provide higher protection on the radio link (and more retransmission attempts) when congestion window is small and lower protection for high window values. Indeed, when congestion window is small, any link error will trigger window reduction to its half – unnecessarily reducing the throughput of the TCP flow. In the opposite case, the impact of link errors becomes less significant, since the window will be possibly reduced due to congestion-related losses.

Fig. 3 presents congestion window evolution in TCP New Reno and the corresponding proposed variation of the packet importance metric. Specifically, the proposed approach assigns the highest importance ("High Imp") to TCP segments produced

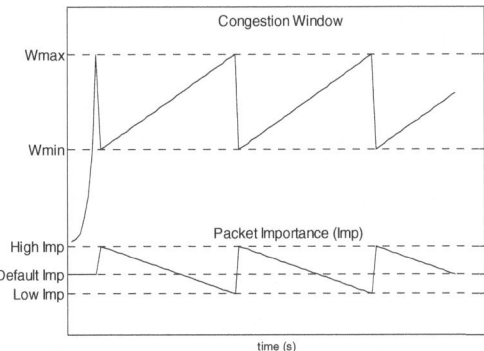

Fig. 3. Variation of HARQ strength used for TCP packet transmission based on the congestion window parameter estimated at the receiver

right after each window reduction and decreases it down to the "Low Imp" threshold following linear or any other monotonically decreasing function.

Summarizing, A^2DRC provides higher protection for low congestion window values or flow sending rates. This reduces the probability of packet losses due to link errors on the wireless channel, which is a well-known reason for TCP performance degradation [7].

3   Performance Evaluation

The proposed scheme is evaluated in the context of an UMTS/HSDPA cellular network. Network Simulator 2 (NS-2) [13] with the additional Enhanced UMTS Radio Access Network Extensions (EURANE) module [14] was used for the experiments. Figure 4 illustrates the reference scenario and the main parameters employed in the experiments.

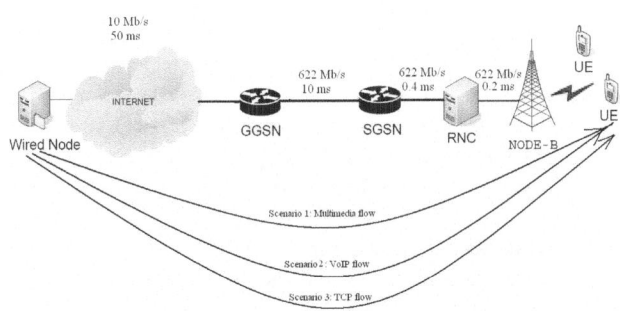

Fig. 4. Simulation scenario used for ns-2 experiments

In video transfer scenario, the FH is a video server which transmits video streams to the video receiver located at UE. Results are presented for the "Foreman" video sequence, using MPEG-4 (open-source ffmpeg [15]) video coding. The video format is Quarter Common Intermediate Format (QCIF, 176 * 144). The GOP structure is IBBPBBPBBPBBPB. The "Foreman" video trace is composed of 300 frames (10 I-frames, 102 P-frames and 188 B frames). An integrated environment methodology proposed and developed within the framework of EvalVid [11] was for experiments, enhanced as in [17] for including NS-2.

The crucial portion of the multimedia stream is retransmitted by A^2DRC with a higher HARQ strength, i.e. packets belonging to an I-frame are retransmitted with HARQ Strength = 8, while packets belonging to B-frames are retransmitted with a HARQ strength = 2. P-frames are retransmitted with a variable HARQ strength ranging from 8 to 3 depending on the position of the frame in the GOP. Default value of HARQ strength is set to 4 for all packets in the legacy scenario (i.e. without A^2DRC). Achieved results are illustrated in Fig.5, where A^2DRC increases the range of packet error tolerance to 10^{-2}-10^{-1}.

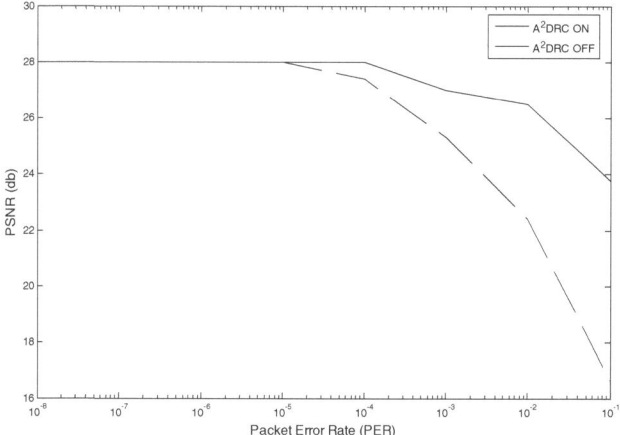

Fig. 5. Quality of "Foreman" video clip for different error rates

VoIP Transfer Performance. Experiments on VoIP flows are performed using the simulation model presented in [10]. Initially, equal HARQ strength equal to 3 is associated to all transmitted speech frames. However, in case the receiver detects frame losses after out-of-order frame reception, it increases HARQ strength linearly for the subsequent packets of the stream (with HARQ strength max equal to 8) in order to avoid bulk frame losses. Once no loss is detected, A^2DRC decreases the HARQ strength to the initial value.

Achieved results (Fig. 6) demonstrate that A^2DRC is able to provide a relevant improvement in terms of MOS both for G.711 and GSM AMR speech flows. In average, application of A^2DRC scheme enables the codec to deliver the same speech quality for error rate of 5% higher if compared with the case when A^2DRC is not enabled.

In File Transfer Performance scenario, FTP/TCP flow sent by FH is received by the UE. For the entire duration of the flow the receiver maintains up-to-date value of the congestion window (_cwnd_) computed by counting the number of packets received for the last RTT. Whenever the loss detection signal (three duplicate acknowledgements) is sent to the sender packet importance is increased according the function

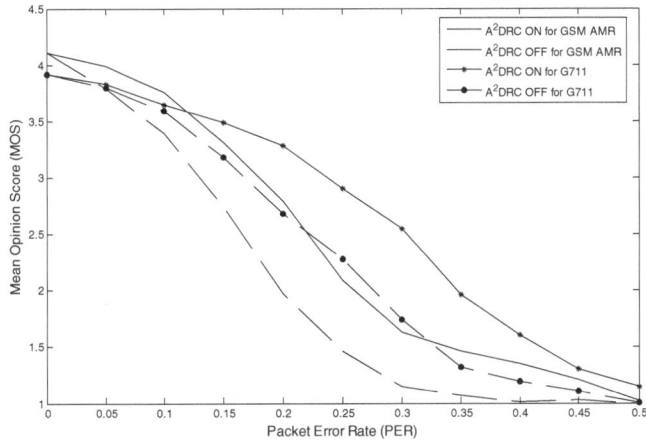

Fig. 6. G.711 and GSM AMR Voice MOS against Packet Error Rate

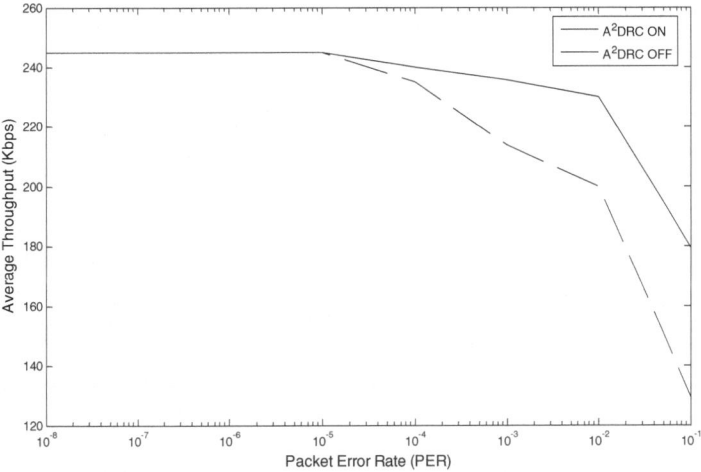

Fig. 7. Average TCP Throughput as a function of Packet Error Rate on the wireless channel

presented in Section 2 causing higher strength of HARQ process and, as a result, producing higher resistance to the link errors. Figure 7 presents TCP throughput achieved by the flows for different PERs of the wireless link. As expected, higher protection against the link errors for low congestion values of the congestion window brings evident performance improvement and underlines advantages of dynamic error protection techniques based on application awareness introduced by A^{2}DRC.

4   Conclusions

In order to enable dynamic control on the level of per-packet HARQ protection, the paper proposed a cross-layer solution between application/transport layers on a mobile terminal and link layer at the base station. Protection level is dynamically set on a per-packet basis and depends on the importance of different packets as well as on the reception history of the flow. Experimental results show that our scheme improves audio and video flows as well as TCP-based data transfers.

References

1. Horn, U., Stuhlmüller, K., Link, M., Girod, B.: Robust Internet video transmission based on scalable coding and unequal error protection. Image Communication 15(1-2), 77–94 (1999)
2. Wu, Z., Bilgin, A., Marcellin, M.W.: Unequal error protection for transmission of JPEG2000 codestreams over noisy channels. In: ICIP (2002)
3. van der Schaar, M., Radha, H.: Unequal packet loss resilience for Fine-Granular-Scalability video. IEEE Trans. on Multimedia 3, 381–394 (2001)
4. Yang, X.K., Zhu, C., Li, Z.G., Feng, G.N., Wu, S., Ling, N.: A degressive error protection algorithm for MPEG-4 FGS video streaming. In: Proc. of IEEE ICIP, Rochester, USA (September 2002)
5. Zhang, Q., Zhu, W., Zhang, Y.-Q.: Network-adaptive scalable video streaming over 3G wireless network. In: IEEE ICIP 2001, Thessaloniki, Greece (2001)
6. Granelli, F., Kliazovich, D., da Fonseca, N.L.S.: Performance Limitations of IEEE 802.11 Networks and Potential Enhancements. In: Xiao, Y., Pan, Y. (eds.) Wireless LANs and Bluetooth. Nova Science Publishers, Hardbound (2005)
7. Lim, H., Xu, K., Gerla, M.: TCP performance over multipath routing in mobile ad hoc networks. In: IEEE International Conference on Communications, vol. 2, pp. 1064–1068 (2003)
8. Chase, D.: Code combining: A Maximum-Likelihood Decoding App. for Comb. an Arbitrary Number of Noisy Packets. IEEE Trans. on Comm. (May 1985)
9. Hoene, C., Rathke, B., Wolisz, A.: On the Importance of a VoIP Packet. In: Proc. of ISCA Tutorial and Research Workshop on th Auditory Quality of Systems, Herne, Germany (April 2003)
10. Bacioccola, A., Cicconetti, C., Stea, G.: User-level Performance Evaluation of VoIP Using ns-2. In: Workshop on Network Simulation Tools (NSTools), Nantes, France (October 2007)
11. Klaue, J., Rathke, B., Wolisz, A.: EvalVid – A Framework for Video Transmission and Quality Evaluation. In: 13th International Conference on Modeling, Techniques and Tools for Computer Performance Evaluation, Urbana, Illinois (2003),
    http://www.tkn.tu-berlin.de/research/evalvid
12. VLC media player home page, http://www.videolan.org/vlc/
13. NS-2 simulator tool home page, http://www.isi.edu/nsnam/ns/
14. Enhanced UMTS Radio Access Extensions, http://www.ti-wmc.nl/eurane/
15. FFmpeg multimedia system, http://ffmpeg.mplayerhq.hu/

Congestion Avoidance Control through Non-cooperative Games between Customers and Service Providers

Dimitris E. Charilas, Athanasios D. Panagopoulos, Panagiotis Vlacheas,
Ourania I. Markaki, and Philip Constantinou

National Technical University of Athens, Department of Electrical & Computer Engineering,
Heroon Polytechneiou 9, Zographou, 15773, Athens, Greece
dcharilas@mobile.ntua.gr, thpanag@ece.ntua.gr,
panvlah@telecom.ntua.gr, omarkaki@epu.ntua.gr,
fkonst@mobile.ntua.gr

Abstract. Congestion avoidance control refers to controlling the load of the network by restricting the admission of new user's sessions and resolving the unwanted overload situations. Admission control and Load control constitute key mechanisms regarding Radio Resource Management. As the wireless world is moving towards heterogeneous wireless networks, these types of control are facing more challenges, since efficiency and fairness are required. Game theory provides an appropriate framework for formulating fair and efficient congestion avoidance control problems. In this paper we formulate a non-cooperative game between service providers and customers. On the one hand, the service providers wish to maximize their revenue, but on the other hand, the users wish to maximize the quality of service received, keeping at the same time the expenses as low as possible. Therefore a balance has to be established among these contradictory demands. Our effort also concentrates in the proper modeling of the user's level of satisfaction, so as to provide a logical decision-taking framework. The proposed scheme is then tested using the ns2 simulator. Results show that both parties can benefit from this mechanism.

Keywords: Congestion Avoidance, Admission Control, Load Control, Quality of Service, Game theory, Payoff, Non-cooperative games, Nash Equilibrium.

1 Introduction

Wireless networks have limited radio resources and should be managed very carefully so that they operate under normal conditions, thus assuring that the users will receive the requested Quality of Service (QoS) for their requested applications/services. To retain its customer base, the service provider must make sure that customers are satisfied with the level of QoS they receive, taking into account the premium they pay. The level of customer satisfaction received can be represented by *utility*-based functions, due to the fact that each customer spends his/her disposable income in the way that yields him/her the greatest amount of satisfaction. This leads to the maximization of the utility functions [1]-[5]. Because of the limited radio resources, there have been defined Radio Resource strategies in order not only to assure the QoS guarantees to

F. Granelli et al. (Eds.): MOBILIGHT 2009, LNICST 13, pp. 53–62, 2009.

the users, but also to assure that the users won't violate these agreements from their side. When, for example, there are too many users admitted and they don't receive the agreed QoS, the network is overloaded. This state of the network constitutes the congestion situation. Congestion situations are very harmful for the network because they cause many problems, such as increased interference, loss of packets, low bandwidth availability and from the user's point of view it causes decreased QoS reception, which leads to the user's disappointment.

The **Congestion Avoidance Control** mechanism [2] consists of Admission control (AC) and Load Control (LC). **Admission Control** is one of the key Radio Resource Management (RRM) mechanisms that ensure the proper operation of a network, by admitting or rejecting new user requests based on criteria such as the load of the network. In general, the Admission Control mechanism ensures that the admittance, of a new flow into a resource-constrained network, does not violate the QoS commitments already made by the network to the admitted flows. **Load Control** is also one of the key RRM mechanisms that serve for the effective performance of a wireless network by keeping the load of the network at normal boundaries. It performs traffic balancing between nodes or cells of the same mode preventing congestion situations. Reactive load control is employed to encounter overload situations of the network, when the users' QoS is at high risk. In these cases, the load control performs several actions to decrease the amount of traffic in the congested cell. These unwanted congested states my be prevented by the load control mechanism that monitors continuously the system.

On the other hand, **Game theory** is a mathematical tool developed to understand competitive situations in which rational decision makers interact to achieve their objectives. Game theory techniques have recently been applied to various engineering design problems in which the action of one component impacts (and perhaps conflicts with) that of any other component. In [6] the authors review popular game theory techniques, with regard to, the wireless networks' resource management problem and propose a game theoretic framework for optimizing bandwidth allocation and admission control issues in wireless networks. Existing game theory-based approaches to wireless networks' resource management consider that the game is played either among users, competing for network resources, or among networks, which try to maximize their efficiency by serving the largest possible number of available service requests. A different approach is adopted in [7],[8], where the AC problem is formulated as a non cooperative, non zero-sum game between the service provider and the customers, so as to increase the provider's revenue and offer differentiated QoS to the users. The authors of [7] also provide the required framework for n-player games. Finally, in [4], [9], the network selection problem is modeled by defining a game between the access networks involved in 4G converged environments. Decisive factor for the admission of service requests in all afore-mentioned works is the maximization of the payoff. The utility of the players is taken as the combination of strategies chosen in the game [10]. In this paper we intend to expand the methodology presented in [8], in order to include the efficient load control and guaranteed congestion avoidance mechanisms. The proposed scheme is then tested through extended simulations.

The paper is organized as follows. Section 1 is the introduction to the problem that we are dealing with. Sections 2 and 3 model the AC and LC games respectively. Section 4 shows the results deriving from simulations conducted on ns2. Finally, Section 5 summarizes the work.

2 The Admission Control Game

In this section we analyze the competitive customer vs. provider scenario as a non-cooperative two-player game. In the proposed scheme we consider that each customer has a contract with a specific service provider, thus him being the default network choice ("home" provider); nevertheless, in case of insufficient resources, the customer is free to pursue higher QoS at another provider, given that there is some kind of federation agreement between the visited and the home provider as in the roaming scenario (possibly under a small monetary penalty). Suppose that there are N users and M service providers, which means that each user at any time can choose any provider, giving a total of M^N possible states. Also, let $n_i(t)$ be the number of users subscribed to provider i at time t, $1 \leq i \leq M$. Furthermore, we assume that the user is not allowed to be subscribed simultaneously to multiple providers, meaning $\sum_{i=1}^{M} n_i(t) = N$. Each user-provider combination is considered as a two-player game G_j, $1 \leq j \leq M$.

Admission control takes place each time a new session request is received and decides whether it should be allocated resources or be rejected due to lack of resources. The decision is based on measurements extracted from on-going sessions of the same service type. Therefore, each time a new request is made, an instance of the game is played, as depicted in Figure 1. We assume that the service provider has two choices: either admit (S_1) or reject (S_2) the request. The customer also possesses two strategies: either leave (C_1) or stay (C_2) with the service provider, leaving us with four possible strategy combinations. The payoffs of the two players are expressed by the matrices A = $[a_{ij}]_{2 \times 2}$ and B = $[b_{ij}]_{2 \times 2}$. Table 1 presents schematically the relationships between payoffs and player strategies.

Table 1. Relationships between payoffs and player strategies

	Customer Leaves (C_1)	Customer Stays (C_2)
Provider Admits (S_1)	a_{11}, b_{11}	a_{12}, b_{12}
Provider Rejects (S_2)	a_{21}, b_{21}	a_{22}, b_{22}

Assume that a customer requests a session admission. We define his payoff matrix B = $[b_{ij}]_{2 \times 2}$ as follows:

$$B = \begin{pmatrix} b_{11} & b_{12} \\ b_{21} & b_{22} \end{pmatrix} = \begin{pmatrix} w_1 R - w_2 L_c & R \\ w_1 R_o - w_2 L_c & R_o \end{pmatrix} \tag{1}$$

Let us explain these payoff values.

- The term b_{21} denotes the case in which the user decides to leave, while the provider chooses not to admit access (strategy S_2C_1). In this case the user has a certain revenue R_o and is obliged to pay a penalty L_c for early termination, given that he has a contract with the service provider. Both these terms are multiplied by weights (w_1 and w_2 respectively), which reflect the user's preference to save money and satisfaction respectively. The values of these weights may be specified through the user profile and indicate whether the user is risk neutral, risk seeking or risk adverse.

- Similarly, b_{22} describes the state where the service provider denies admission but the user chooses to stay, therefore having only the R_o revenue. Note that R_o is not calculated similarly to R, since no service is in progress. Its value may be fixed.
- The term b_{12} is defined as the user's revenue, in case he chooses to stay and he is granted admission.
- Finally, b_{11} is simply defined as the revenue minus the penalty, both multiplied by the weights w_1 and w_2 respectively.

The user's revenue expresses in monetary value the quality of service offered to him, taking into account the cost, and is modeled as

$$R = QoS\,(\%) \cdot q - C_{cust} \tag{2}$$

where q is a constant factor mapping the $QoS(\%)$ value to monetary value (specified by the provider), C_{cust} is the cost of the service from the customer's point of view and $QoS(\%)$ is given by (3). This difference between the monetary value of the QoS offered to the user and the actual price charged is known in microeconomic terms as the Consumer Surplus.

$$QoS(\%) = \sum_{i=1}^{4} w_i \cdot Parameter_i\% \tag{3}$$

QoS(%) expresses a percentage of user satisfaction, taking into considetation the normalized mean values of QoS parameters such as Delay, Jitter, Throughput and Packet Loss. The weights w_i, $1 < i < 4$, vary with respect to the service and can be calculated based on the network's performance, as explained in [11][12]. The reason for this choice is because user satisfaction is subjective and therefore difficult to characterize mathematically. As a result, for the purposes of this work, we will use (3) to estimate the level of each customer's satisfaction or dissatisfaction, taking into account, four QoS parameters instead of only Call Blocking Probability, as in [7]. We consider this approach more concrete since it reflects the network's current status and is not dependent on constant variables, whose optimal values need to be specified.

Whenever the AC Game is performed, statistics are extracted from the data sent or received during a specified time interval. The duration of this interval is referred to as a "**window**". Therefore, the QoS assigned to a session or service will indicate only the most recent state information. Note that in case of requests during the first window, the AC Game is not performed since the window is not yet completed, meaning that these requests will always be granted admission. Now we define the provider's payoff matrix $A = [a_{ij}]_{2x2}$ as:

$$A = \begin{pmatrix} a_{11} & a_{12} \\ a_{21} & a_{22} \end{pmatrix} = \begin{pmatrix} R_t + L_c + C - F - L & R_t + C - F \\ R_t + L_c - L & R_t \end{pmatrix} \tag{4}$$

Let us explain these payoff values as well.

- The term a_{22} represents the total revenue of the provider, deriving from all ongoing sessions. In other words, it shows what the provider is gaining in strategy S_2C_2.
- Similarly, the term a_{21} shows the provider's gain if the customer chooses to leave, therefore we need to subtract the loss L from the total revenue and add the penalty L_c.

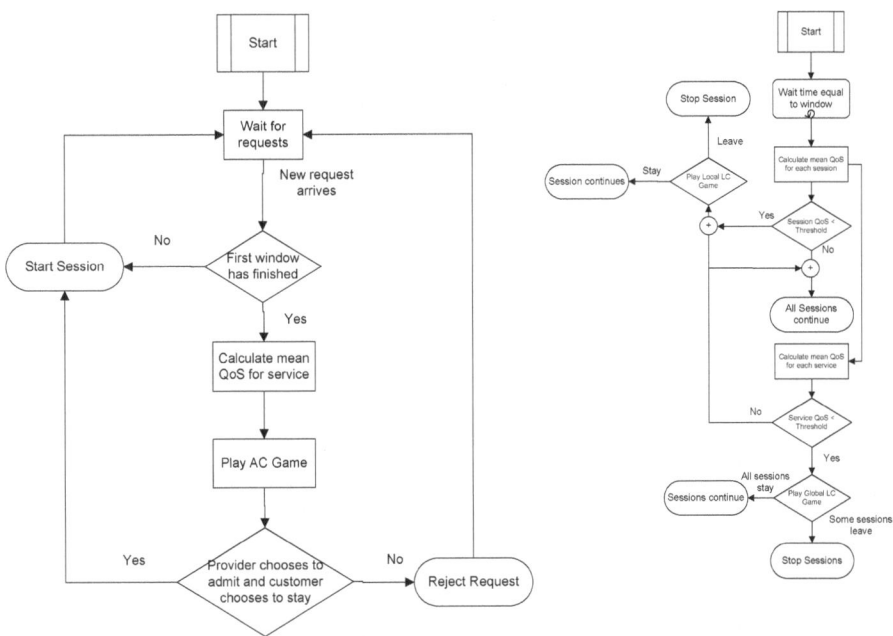

Fig. 1. Admission Control (AC) Game   **Fig. 2.** Load Control (LC) Game

- The term a_{12} denotes the provider's payoff in strategy S_1C_2. This seems to yield the highest payoff for the provider. However, in a fully loaded system admitting a new customer may reduce the resources offered to on-going sessions (which means QoS degradation), thus causing other customers to leave and consequently leading to revenue loss. This degradation can be specified through real-time monitoring in simulations. Therefore, the total revenue equals the revenue gained from on-going sessions C plus the expected revenue from the new session minus the potential loss due to the dissatisfaction of other customers. The latter is expressed by the term F, in which its calculation will be discussed later.

- In a similar way, the term a_{11} shows the provider's revenue in case the customer chooses to **leave**. The term L corresponds to the revenue loss, while L_c corresponds to the penalty the customer has to pay for leaving.

Assuming the fact that the more the user is satisfied, the less likely he is going to leave and since $QoS(\%)$ expresses the user's satisfaction, we can estimate the probability that one customer leaves his current provider through the following equation

$$p_{leave} = 1 - \left(QoS\left(\%\right)/100 \right) \tag{5}$$

This allows us to express F in monetary value as the sum of all possible losses from all current customers, as

$$F = \sum_{i=1}^{n(t)} p_{leave\_i} L_i \tag{6}$$

where p_{leave_i} is the probability that customer i chooses to leave and L_i is the corresponding revenue loss of the network provider.

As far as the solution of the game is concerned, two cases are distinguished. Assuming the case where the system is not full, the user request will be accepted and the probability that a customer leaves p_{leave_i} is near to 0. In this case, there is a Nash equilibrium at strategy pair S_1C_2, that means the service provider accepts the request while the user remains with the provider. Assuming now the case where the system is loaded to a certain extent or even overloaded, the user request may be not accepted and the probability that a customer leaves p_{leave_i} is non zero. Even in this case, there is also a pure strategy Nash equilibrium at the pair S_iC_j, which depends on the relation between some terms in the payoffs. The new request is accepted if the revenue C generated from admitting the request is greater than the possible revenue loss F is the user leaves. Otherwise, the provider is better to reject the request. Based on these two proofs, a pure strategy Nash equilibrium is guaranteed and a solution to the admission control game is available at any point.

$$i = \begin{cases} 1 & if\ a_{11} \ge a_{21} \\ 2 & if\ a_{11} < a_{21} \end{cases} \Rightarrow i = \begin{cases} 1 & if\ C \ge F \\ 2 & otherwise \end{cases} (7a) \quad j = \begin{cases} 1 & if\ \{i=1\ and\ b_{11} \ge b_{12}\}\ or\ \{i=2\ and\ b_{21} \ge b_{22}\} \\ 2 & if\ \{i=1\ and\ b_{11} < b_{12}\}\ or\ \{i=2\ and\ b_{21} < b_{22}\} \end{cases} (7b)$$

3   The Load Control Game

The **Load Control (LC) Game** is played similarly to the AC Game. The main difference is that the LC Game is played periodically while the sessions are running. Through this process we intend to terminate sessions that greedily consume the system's resources, causing this way degradation to the QoS offered to the rest of the customers and thus reducing the provider's total revenue. Moreover, unsatisfied customers are granted the opportunity to seek more efficient networks, based on their preferences. Risk adverse customers for example will tolerate low levels of QoS and prefer to stay with the same provider in order to minimize the total cost.

Two types of LC Games are distinguished: the Local LC Game and the Global LC Game. Figure 2 shows the Load Control process. As mentioned before, the LC process is repeated periodically, independently from the AC process. We refer to the time interval between two successive rounds as a "**window**", since the LC Game will consider only the data received or sent during the last window. Therefore, the QoS assigned to a session or service will indicate only the most recent condition, enabling the LC scheme to react quickly in case of QoS degradation. In the beginning of every load control round, the QoS level for each session is extracted; then the average QoS level is specified for each service type. The QoS threshold is the lowest level of QoS that can be tolerated.

If the QoS of at least one service type is found below the acceptance threshold, then the **Global LC Game** is triggered, during which LC Games are played between the provider and all running sessions. This game may result in either disappointed customers leaving the provider or the provider terminating unprofitable customers. If either one decides that a connection should be terminated, then the session ends and the customer is prompted to another service provider. Penalty is submitted only if the customer chooses to leave willingly. On the other hand, if the Global LC Game is not triggered and at least one session presents a QoS below the acceptance threshold, then the **Local LC Game** is triggered. In this case, an LC Game is played between the provider and each session that triggered the game, leaving the other sessions unaffected. This type of game may also result in some sessions being terminated.

4 Simulation Results

The proposed scheme was tested on the ns2 platform, using both CBR and TCP traffic. More specifically, we consider the voice service as a CBR/UDP service, defined by the G.711 codec, setting packets of 120 bytes and interval equal to 15 ms. Secondly, we consider the FTP/TCP service with packets of 512 bytes. Whenever a game is called, the trace files produced by ns2 are processed with the help of an awk script and the mean values for QoS parameters are extracted.

For the purposes of this study we have adopted the parameter values shown in Table 2. Parameters C_{cust}, L_c, c, L and R_o express monetary value (for example euros or American dollars), referring to the total duration of a session. These values may be derived from the provider's statistics. On the other hand, q refers only to the duration of a single time window and its value should be specified properly. In other words, q indicates the monetary value of a session with 100% QoS for a time equal to the duration of a window. Note that based on the selected values for C_{cust}, L_c and q, the QoS(%) threshold is specified as 40% for the voice service and 70% for the FTP service. The QoS percentage is estimated through the normalization of current parameter values in the interval specified by a minimum and maximum value, indicating the system's worst and optimal performance respectively, as shown in Table 3. This should be taken into account during the explanation of results, since different normalization intervals differentiate the same QoS percentage in each service.

Furthermore, in the frame of ensuring the best possible QoS for CBR connections, a new FTP request is allowed to play the AC game only if the mean QoS for all ongoing voice sessions is over 60%. The service mix was set to 2:1, while the simulation duration was set to 120 seconds. This time has been proven to be enough for the system to be congested at least once. Also note that, for simplicity reasons, the values of weights w_1 and w_2 were considered the same for all requests. In reality, those values should be different for each user in order to reflect his actual preferences.

Table 2. Simulation parameters

Q	L_c	C_{cust}	C	L	R_0	W_1	W_2	window
0.03	0.3	1.32 for Voice 2.4 for FTP	3	1	0.1	0.6	0.4	10 sec

Table 3. Normalization values

	Voice			FTP		
	Min	**Max**	**w**	**Min**	**Max**	**w**
Delay(ms)	10	60	0.4	10	50	0.2
Jitter(ms)	0	10	0.4	0	20	0.1
Thr/put(kbps)	10	65	0.1	10	100	0.7
P. Loss (%)	0	1	0.1	0	1	0.0

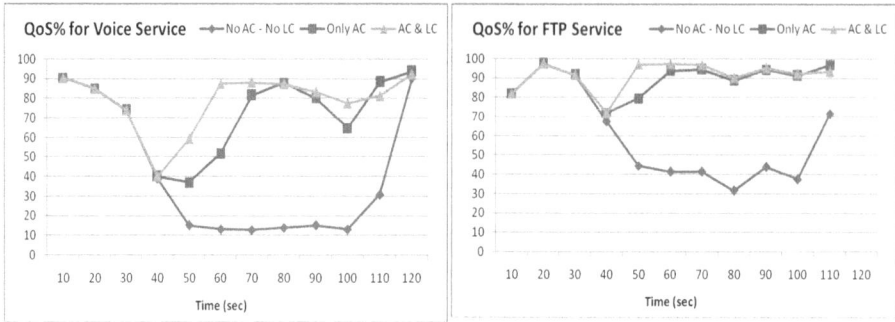

Fig. 3. Impact of AC and LC Games to QoS(%) a) Voice b) FTP

In order to monitor the improvement to offered QoS, we have implemented three tests. Firstly, we do apply neither AC nor LC mechanisms, meaning that all requests are accepted and no sessions are interrupted. Secondly we apply only the AC scheme and finally we apply both AC and LC schemes. In all cases, QoS is recorded for both service types at the end of each time window. Simulation results are depicted in Figure 3, where it can be seen that the third test provided the most satisfactory results. As expected, the absence of all kinds of congestion avoidance mechanisms resulted in major congestion, leaving the system unable to serve all sessions and thus resulting in unacceptable QoS levels. It can be easily observed that approximately after 40 seconds the system is congested. During the following time window, incoming requests are rejected, while load control terminates certain sessions as well. The system then recovers until a minor congestion takes place, approximately at $t=100$ seconds.

The main goal of this paper is to provide a scheme that maximizes not only the QoS offered to customers, but also the provider's gain. So far we have proven that the first part of this goal is indeed achieved. Therefore it is essential to examine whether the proposed scheme is actually in the provider's best interest as well. It is assumed that the provider's billing scheme takes into account the QoS percentage offered to customers, meaning that the customer pays an amount proportional to the level of QoS he receives. In this way, the provider has interest in optimizing the balance between the number of handled requests and QoS offered, instead of only maximizing the number of accepted requests. In conclusion, the total revenue deriving from all ongoing sessions during a time window is shown by (8)

$$\mathrm{R}_{total} = \sum_{i=0}^{i=service\_num} \sum_{k_i=0}^{k_i} \left(QoS_{k_i} (\%)/100 \right) \cdot w_{cost} \tag{8}$$

where *service_num* indicates the number of service types, k the number of on-going sessions for the specific service type and w_{cost} the per window cost of a 100% QoS served session. In other words, w_{cost} constitutes the maximum amount of money the customer may pay during a single time window. As the QoS received decreases, so does the amount of money the customer is obliged to pay. Figure 4 shows the total revenue that derives from the three tests described in the previous section. Similarly

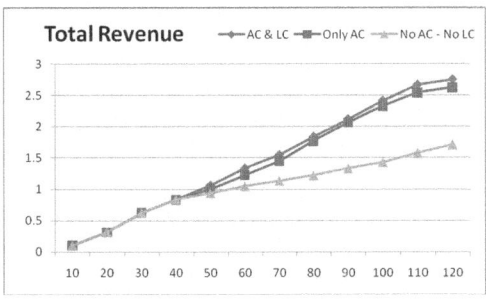

Fig. 4. Total Revenue for the provider

to QoS, the combination of AC and LC indeed offers the best revenue to the provider, thus this mechanism protects the interests of both parties.

The choice of window duration is considered of high importance, since an extremely high value may not allow the system to react quickly in case of congestion, while an extremely low value may add significant computational load and thus reduce the system's performance. When the window is set to a high duration, the system's reaction to the congestion is delayed, which lowers the quality of voice sessions. This occurs due to the fact that the system is unable to instantly detect congestions; therefore more requests are admitted in comparison with a smaller window. On the other hand, when the time window is set to a small duration, the system becomes extremely "strict" and results in rejecting voice requests which after a few seconds could be appropriately served. The window size should be chosen according to the expected average number of on-going sessions, since an increasing number of sessions also increases complexity and the algorithm becomes more time-consuming. Ideally, the window size should be given an initial value and dynamically adapt based on the system's status.

5 Conclusions

In this paper a methodology for integrating game theory in congestion avoidance mechanisms has been presented. The problem is seen as follows: the goal of service providers is to maximize their revenue without congesting their networks, though the goal of the customers is to get the maximum QoS with the minimum paying. So a congestion avoidance mechanism should take into account these factors in its decisions in order to maximize the satisfaction of both groups and simultaneously not congest the wireless networks. In this paper, the previously described problem has been modelled as a non-cooperative game between service providers and customers. Emphasis has also been given, on how we perceive and estimate the user's satisfaction based on network measurements. This approach provides a more subjective framework for the estimation of QoS as it is perceived by the end user. Finally, the simulations of the proposed scheme are presented and some useful conclusions are drawn.

References

[1] Das, S.K., Lin, H., Chatterjee, M.: An Econometric Model for Resource Management in Competitive Wireless Data Networks. IEEE Network (November 2004)

[2] Holma, H., Toscala, A.: WCDMA for UMTS. J. Wiley & Sons, Chichester (2001) Revised Edition

[3] Perez-Romero, J., Sallent, O., Ruiz, D., Agusti, R.: An Admission Control Algorithm to Manage High Bit Rate Static Users in W-CDMA. In: 13th IST Mobile & Wireless Communications Summit, Lyon, France (June 2004)

[4] Josephina, A., Andreas, P.: 4G Converged Environment: Modeling Network Selection as a Game. In: 16th IST Mobile and Wireless Communications Summit (2007)

[5] Ormond, O., Murphy, J., Muntean, G.-M.: Utility-based Intelligent Network Selection in Beyond 3G Systems. In: 2006 IEEE International Conference on Communications, June 2006, vol. 4, pp. 1831–1836 (2006)

[6] Niyato, D., Hossain, E.: Radio resource management games in wireless networks: an approach to bandwidth allocation and admission control for polling service in IEEE 802.16. IEEE Wireless Communications 14(1), 27–35 (2007)

[7] Lin, H., et al.: ARC: An Integrated Admission and Rate Control Framework for Competitive Wireless CDMA Data Networks Using Noncooperative Games. IEEE Trans. Mobile Comp. 4(3), 243–258 (2005)

[8] Vlacheas, P., Charilas, D., Tragos, E., Markaki, O.: Maximizing Quality of Service for Customers and Revenue for Service Providers through a Noncooperative Admission Control Game. In: ICT Mobile Summit 2008, Stockholm (June 2008)

[9] Charilas, D., Markaki, O., Tragos, E.: A Theoretical Scheme for applying game theory and network selection mechanisms in access admission control. In: International Symposium on Wireless Pervasive Computing (ISWPC) (May 2008)

[10] de Sousa Jr., V.A., de, R.A., Neto, O., de, F., Chaves, S., da Silva, A.P., Cavalcanti, F.R.P.: Conception and Evaluation of Access Selection Algorithms for Cooperative Beyond 3G Systems. In: VI International Telecommunications Symposium (ITS 2006), Fortaleza-CE, Brazil, September 3-6 (2006)

[11] Charilas, D., Markaki, O., Nikitopoulos, D., Theologou, M.: Packet-Switched Network Selection with the Highest QoS in 4G Networks. Elsevier Computer Networks 52(1), 248–258 (2008)

[12] Markaki, O., Charilas, D., Nikitopoulos, D.: Enhancing Quality of Experience in Next Generation Networks through Network Selection Mechanisms. In: Mobile Terminal Assisted Enhanced Services Provisioning in a B3G Environment Workshop, PIMRC (September 2007)

Application of Fuzzy AHP and ELECTRE to Network Selection

Dimitris E. Charilas, Ourania I. Markaki, John Psarras, and Philip Constantinou

National Technical University of Athens, Department of Electrical & Computer Engineering,
Heroon Polytechneiou 9, Zographou, 15773, Athens, Greece
dcharilas@mobile.ntua.gr, omarkaki@epu.ntua.gr,
john@epu.ntua.gr, fkonst@mobile.ntua.gr

Abstract. In a heterogeneous wireless network environment services are ubiquitously delivered over multiple wireless access technologies. Ranking of the alternatives and selection of the most efficient and suitable access network to meet the QoS requirements of a specific service, as these are defined by the user, constitutes thus an important issue. Decisions on which network to connect to are however difficult to be reached, since multiple factors of different relative importance have to be taken into consideration. This paper addresses this difficulty by adopting Multi Attribute Decision Making (MADM) methods. Fuzzy AHP, a MADM method, is initially applied to determine the weights of certain Quality of Service indicators that act as the criteria impacting the decision process. The fuzzy extension of the method, and consequently the use of fuzzy numbers, is adopted in order to incorporate the existence of fuzziness as a result of subjective evaluations. Afterwards, ELECTRE, a ranking MADM method, is applied to rank the alternatives, in this case wireless networks, based on their overall performance.

Keywords: Wireless Networks, Fuzzy Logic, Fuzzy Triangular Numbers, Multi Attribute Decision Making, Analytic Hierarchy Process, ELECTRE.

1 Introduction

The heterogeneous wireless networks integrate different access networks, such as IEEE 802.15 WPAN, IEEE 802.11 WLAN, IEEE 802.16 WMAN, GPRS/ EDGE, cdma2000, WCDMA and satellite network, etc. The proliferation of wireless access technologies, along with the evolution of the end-user terminals (smart phones, PDAs, etc. . .) are leading fast towards a ubiquitous, pervasive and rich connectivity offer, such that the end users won't just be always connected, but also always covered by multiple access networks / technologies. Selection of the most efficient and suitable access network to meet a specific application's QoS requirements has thus recently become a significant topic, the actual focus of which is maximizing the QoS experienced by the user. The end-users can potentially take wise decisions on which access network to connect to on the basis of several merit functions including the current load of the network and the cost-for-connectivity. However, since multiple factors have to be taken into account, it is no longer easy to rank the candidate networks according to

F. Granelli et al. (Eds.): MOBILIGHT 2009, LNICST 13, pp. 63–73, 2009.
© ICST Institute for Computer Sciences, Social-Informatics and Telecommunication Engineering 2009

preference on a single criterion. In such cases, multiple criteria have to be combined and scaled in a meaningful way. In addition, various criteria in the decision process may oppose to each other, e.g., a desirable increase in QoS may require an undesirable increase of price. Thus, trade-offs are sometimes required.

Current approaches for network selection may involve either Fuzzy Logic-based schemes or Multi Attribute Decision Making (MADM) methods. In the first one, a group of fuzzy logic rules in the form of linguistic IF-THEN expressions have to be defined to model network selection [1] [2]. However, such rules have to be configured by the user manually prior to selection and their complexity becomes overwhelming high as the number of attributes increases. Thus, the scalability of the fuzzy logic-based schemes is extremely low, which limits their usage in wireless networks selection.

MADM on the other hand involves the selection of a series of criteria that impact the decision process and the comparison of the alternatives taking into account the relative importance of these criteria. More specifically, a weight is assigned to each criterion to reflect its importance in the final decision and then alternatives are evaluated based on their overall performance. MADM provides a solid framework for network selection, since the latter constitutes a multi-criteria problem. Several researchers have considered the use of MADM algorithms to rank candidate networks in a preference order [3]. Numerous types of MADM algorithms exist [4]. Several of them may be suitable for solving a decision problem so that the decision maker may encounter the task of selecting amongst a number of feasible methods the most appropriate one. Common QoS parameters, such as Delay, Jitter, Packet Loss, Throughput etc can be used as criteria in such schemes. Furthermore, additional criteria may involve Availability, Reputation, Access Delay, Cost etc. MADM requires also an initial indication on each parameter's relative importance with regard to other parameters. As far as network selection is concerned, such an indication may be acquired from questionnaires or measurements [5] [6]. Criteria such as user satisfaction or cost are rather subjective and it is preferable that they are expressed in linguistic terms that correspond to fuzzy numbers. Wenhui Zhang in [7] addresses the issue of imprecise criteria, whose values cannot be easily obtained.

D. Charilas et al. in [5] [6], as well as Abbas Jamalipour et al. in [8] [9], demonstrate an example of how network selection can be modeled using two MADM methods, AHP and GRA. Farooq Bari and Victor Leung demonstrate in [10] how network selection can be approached using another MADM method, called ELECTRE. This method performs pair-wise comparisons among the alternatives for each one of the criteria separately to establish outranking relationships between them.

In this paper network selection is modeled using two MADM methods: Fuzzy AHP (FAHP), which is an extended version of AHP to support fuzzy numbers, and ELECTRE. The relative importance of selection criteria is modeled as a fuzzy number since objective judgment cannot be guaranteed. Using vague comparisons, FAHP is applied to provide a fuzzy weight for each criterion involved in the selection process. Through the process of defuzzification crisp weights are finally obtained, and are used as input data for ELECTRE. The latter provides a ranking of all alternatives.

The paper is structured as follows. Section 1 is an introduction to the problem addressed. Section 2 provides an insight to fuzzy numbers, while Section 3 presents the MADM methods, FAHP and ELECTRE, to be used in the following analysis. In section 4 a numerical example is provided, to illustrate the proposed scheme. Finally, Section 5 concludes the work.

2 Fuzzy Numbers

The fuzzy sets theory, introduced by Zadeh (1965) to deal with vague, imprecise and uncertain problems, has been used as a modeling tool for complex systems that can be controlled by humans but are hard to define precisely. The main characteristic of fuzziness is the grouping of individuals into classes that do not have sharply defined boundaries. An uncertain comparison can be represented by a fuzzy number. A fuzzy set is one that assigns grades of membership between 0 and 1 to objects using a particular membership function $\mu_A(x)$. A triangular fuzzy number is a special type of fuzzy number whose membership is defined by three real numbers, expressed as (l, m, u), where l is the lower limit value, m is the most promising value and u is the upper limit value. Particularly, when $l = m = u$, fuzzy numbers become crisp numbers. The triangular fuzzy numbers are represented as shown in Figure 1.

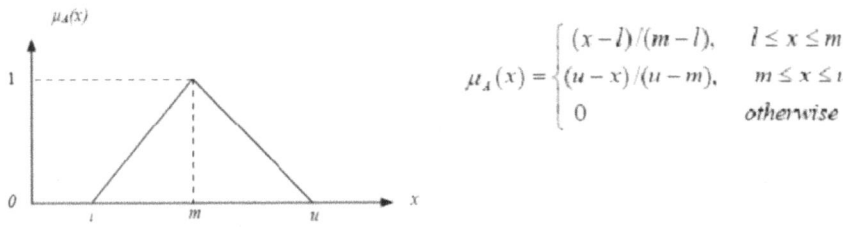

$$\mu_A(x) = \begin{cases} (x-l)/(m-l), & l \le x \le m \\ (u-x)/(u-m), & m \le x \le u \\ 0 & otherwise \end{cases}$$

Fig. 1. Triangular Fuzzy Number

3 Multi-attribute Decision Making Methods

3.1 Analytic Hierarchy Process (AHP)

The Analytic Hierarchy Process (AHP) is one of the extensively used multi-attribute decision-making methods. Since 1977 that Saaty [11] proposed AHP as a decision aid to help solve unstructured problems in economics, social and management sciences, the method has been applied in a variety of contexts, including telecommunications [12]. One of the main advantages of this method is the relative ease with which it handles multiple criteria. In addition to this, AHP is easy to understand and it can effectively handle both qualitative and quantitative data. Finally, the use of AHP does not involve complex mathematics.

The AHP decision problem is structured hierarchically at different levels, each level consisting of a finite number of decision elements. The top level of the hierarchy represents the overall goal, while the lowest level is composed of all possible alternatives. One or more intermediate levels embody the decision criteria and sub-criteria. The relative importance of the decision elements (weights of criteria and scores of alternatives) is assessed indirectly through a series of comparative judgments during the second step of the decision process. The decision-maker is required to provide his/her preferences by comparing all criteria, sub-criteria and alternatives with respect to upper level decision elements. The standard preference scale used for AHP ranges from 1 that indicates "equal importance" to 9 representing "extreme importance"; however sometimes different evaluation scales can be used, such as 1 to 5. The values of the weights and scores are elicited from these comparisons and represented in a decision table.

3.2 Fuzzy AHP

In spite of its popularity the conventional AHP method is often criticized for its inability to adequately handle the inherent uncertainty and imprecision associated with the mapping of the decision-maker's perception to exact numbers, i.e. for the fact that it does not reflect human thinking. A natural way to cope with uncertainty in judgments is to express the comparison ratios as fuzzy sets or fuzzy numbers, which reflect the vagueness of human thinking. In this frame, a vague evaluative judgment resulting from the comparison of any two elements at the same level of the decision hierarchy can be represented by a fuzzy number. Therefore, fuzzy AHP (FAHP), a fuzzy extension of AHP, has been developed to solve hierarchical fuzzy problems [13]. In the fuzzy-AHP procedure, the pair wise comparisons in the judgment matrix are fuzzy numbers that are modified according to the designer's focus. Based on a set of standardized answers (linguistic variables) provided through appropriate question forms, the corresponding triangular fuzzy values are defined and the pair wise comparisons matrix $\tilde{A}$ is constructed as

$$
\tilde{A} = (\tilde{a}_{ij})_{n \times n} = \begin{bmatrix} (1,1,1) & (l_{12}, m_{12}, u_{12}) & \cdots & (l_{1n}, m_{1n}, u_{1n}) \\ (l_{21}, m_{21}, u_{21}) & (1,1,1) & \cdots & (l_{2n}, m_{2n}, u_{2n}) \\ \vdots & \vdots & \vdots & \vdots \\ (l_{n1}, m_{n1}, u_{n1}) & (l_{n2}, m_{n2}, u_{n2}) & \cdots & (1,1,1) \end{bmatrix}
$$

where $\tilde{a}_{ij}$ denotes a triangular fuzzy number depicting the relative strength of two elements. Note that the decision matrix is symmetric, meaning that

$$
\tilde{a}_{ji} = \left[\tilde{a}_{ij} \right]^{-1} = (l_{ij}, m_{ij}, u_{ij})^{-1} = (\frac{1}{u_{ij}}, \frac{1}{m_{ij}}, \frac{1}{l_{ij}})
$$

Final weights of alternatives can be acquired from different methods that have been proposed in the literature. One of the most popular is the Fuzzy Extent Analysis, proposed by Chang (1996) [14]. The value of fuzzy synthetic extent with respect to the i_{th} object is defined with the help of fuzzy arithmetic operations as:

$$
\tilde{S}_i = \sum_{j=1}^{n} \tilde{a}_{ij} \otimes \left[\sum_{i=1}^{n} \sum_{j=1}^{n} \tilde{a}_{kj} \right]^{-1} \tag{1}
$$

The fuzzy addition operation of n extent analysis values as well as the inverse of the vector are given by equations (2) and (3) respectively.

$$\sum_{j=1}^{n} \tilde{a}_{ij} = (\sum_{j=1}^{n} l_{ij}, \sum_{j=1}^{n} m_{ij}, \sum_{j=1}^{n} u_{ij}) \tag{2}$$

$$\left[\sum_{i=1}^{n} \sum_{j=1}^{n} \tilde{a}_{ij} \right]^{-1} = \left(\frac{1}{\sum_{i=1}^{n}\sum_{j=1}^{n} u_{ij}}, \frac{1}{\sum_{i=1}^{n}\sum_{j=1}^{n} m_{ij}}, \frac{1}{\sum_{i=1}^{n}\sum_{j=1}^{n} l_{ij}} \right) \tag{3}$$

The possibility of $\tilde{S}_1 \geq \tilde{S}_2$ is defined as $V(\tilde{S}_1 \geq \tilde{S}_2) = SUP_{x \geq y}[\min(\tilde{S}_1(x), \tilde{S}_2(y))]$, x and y being the values on the axis of the membership function of each criterion. This expression can be equivalently written as:

$$V(\tilde{S}_1 \geq \tilde{S}_2) = \begin{cases} 1, & m_1 \geq m_2 \\ 0, & l_2 \geq u_1 \\ \dfrac{l_2 - u_1}{(m_1 - u_1) - (m_2 - l_2)}, & otherwise \end{cases} \tag{4}$$

To compare $\tilde{S}_1$ and $\tilde{S}_2$, both the values of V ($\tilde{S}_1 \geq \tilde{S}_2$) and V ($\tilde{S}_2 \geq \tilde{S}_1$) are needed. The possibility for a convex fuzzy number to be greater than k convex fuzzy numbers $\tilde{S}_i$ ($i = 1,2,...,k$) is defined by:

$$V(\tilde{S} \geq \tilde{S}_1, \tilde{S}_2, ..., \tilde{S}_k) = V\left[(\tilde{S} \geq \tilde{S}_1) and (\tilde{S} \geq \tilde{S}_2) and...and (\tilde{S} \geq \tilde{S}_k) \right] = \min V(\tilde{S} \geq \tilde{S}_i), i = 1,2,3,...,k \tag{5}$$

Assuming that $d'_i = \min V(\tilde{S}_i \geq \tilde{S}_k)$, the weight vector is given by $w' = (d'_1, d'_2, ..., d'_n)^T$ Via normalization, the normalized (non-fuzzy) weight vector is

$$W = (d_1, d_2, ..., d_n)^T \tag{6}$$

3.3 ELECTRE

In this section we present ELECTRE, another type of MADM algorithm, which performs pair wise comparisons among the alternatives, in order to establish outranking relationships between them. The method was first developed by Bernard Roy and its acronym stands for ELimination Et Choix Traduisant la REalité (ELimination and Choice Expressing REality). The problem of network selection can be modeled through the use of ELECTRE as P = (A, C, w), where

- $A = \{1, \dots, N\}$ denotes the set of alternatives, in this case candidate networks
- $C = \{1, \dots, M\}$ denotes the set of criteria impacting the decision process of network selection,
- $w = \{1, \dots, M\}$ denotes the set of weights assigned to the selected criteria depending on the information about the specific service requested or the user QoS profile, so that $\sum_{i=1}^{M} w_i = 1$.

Based on the scores achieved for each one of the selected criteria (attributes), the i_{th} candidate network can be represented by a vector as follows:

$$NW_i = [D_i \quad J_i \quad B_i \quad T_i \quad C_i]$$

For N alternative networks to be considered in the selection process, a matrix NW is formulated as follows:

$$NW = \begin{bmatrix} D_1 & J_1 & B_1 & T_1 & C_1 \\ D_2 & J_2 & B_2 & T_2 & C_2 \\ \vdots & \vdots & \vdots & \vdots & \vdots \\ D_N & J_N & B_N & T_N & C_N \end{bmatrix}$$

At this point attention must be drawn to the fact that the utility associated with each alternative is a monotonically decreasing function of Delay, Jitter, BER and Cost and a monotonically increasing function of Throughput. However, since ELECTRE presupposes a monotonically increasing or decreasing level of importance for all criteria considered, the modification of the method presented in [10] is adopted here as well in order to provide a complete ranking of all the alternatives. The main idea of this modification lies in the concept of a reference network, which is considered as an access network that demonstrates the desired performance with regard to all criteria in question as this is perceived by the user. The reference access network is represented as

$$NW_{ref} = [D_{ref} \quad J_{ref} \quad B_{ref} \quad T_{ref} \quad C_{ref}]$$

and is used to calculate the absolute difference between the value of each one of the attributes of matrix NW and the corresponding reference attribute value, allowing to assume that all criteria have a monotonically decreasing utility: the larger the attribute value, the farther it is from the desired or reference value. The application of ELECTRE also presupposes that the values assigned to all criteria are measured on a common scale. In order thus to remove the impact of the use of diverse measurement units, each one of the adjusted attribute values in row i of a specific column j is normalized using the equation:

$$\overline{V_i} = Value_{i,norm} = \frac{max_{j=1\dots N}\{V_{j,adj}\} - V_{i,adj}}{max_{j=1\dots N}\{V_{j,adj}\} - min_{j=1\dots N}\{V_{j,adj}\}}$$

where $V_{i,adj} = V_i - V_{i,ref}$ and $max_{j=1\dots N}\{V_{j,adj}\}$, $min_{j=1\dots N}\{V_{j,adj}\}$ indicate respectively the maximum and minimum measurement achieved for the specific criterion. The normalized matrix is represented as follows:

$$NW_{norm} = \begin{bmatrix} \overline{D_1} & \overline{J_1} & \overline{B_1} & \overline{T_1} & \overline{C_1} \\ \overline{D_2} & \overline{J_2} & \overline{B_2} & \overline{T_2} & \overline{C_2} \\ \vdots & \vdots & \vdots & \vdots & \vdots \\ \overline{D_N} & \overline{J_N} & \overline{B_N} & \overline{T_N} & \overline{C_N} \end{bmatrix}$$

The next step involves the utilization of the relative importance of each one of the criteria impacting network selection, which has to be taken into consideration before performing pair-wise comparisons among the alternatives. The impact of the relative weights is integrated in the decision process through the calculation of a new matrix NW_w as follows:

$$NW_w = \begin{bmatrix} w_D\overline{D_1} & w_J\overline{J_1} & w_B\overline{B_1} & w_T\overline{T_1} & w_C\overline{C_1} \\ w_D\overline{D_2} & w_J\overline{J_2} & w_B\overline{B_2} & w_T\overline{T_2} & w_C\overline{C_2} \\ \vdots & \vdots & \vdots & \vdots & \vdots \\ w_D\overline{D_N} & w_J\overline{J_N} & w_B\overline{B_N} & w_T\overline{T_N} & w_C\overline{C_N} \end{bmatrix}$$

The main part of ELECTRE lies in the construction of the concordance and discordance matrices, which provide measurements of satisfaction and dissatisfaction of the decision maker when one alternative is compared to another. The values of the aforementioned matrices are calculated using concordance and discordance sets, where a concordance set constitutes a list of the attributes, for which network X is superior to network Y and a discordance set is its complementary set, comprising of the list of attributes for which network X is worse than the compared alternative Y. More specifically, the elements of the concordance and discordance matrices are calculated by equations (7) and (8) respectively.

$$c_{XY} = \sum_{j \in CS_{XY}} W_j \, , \quad Concordance \ Set = \{ j : NW_{w_{X,j}} \geq NW_{w_{Y,j}} \} \tag{7}$$

The concordance and discordance matrices are finally represented as

$$d_{XY} = \frac{\sum_{j \in DS_{XY}} \left| \left(NW_{w_{X,j}} \right) - \left(NW_{w_{Y,j}} \right) \right|}{\sum_j \left| \left(NW_{w_{X,j}} \right) - \left(NW_{w_{Y,j}} \right) \right|}, \quad Discordance \ Set = \{ j : NW_{w_{X,j}} < NW_{w_{Y,j}} \} \tag{8}$$

$$C = \begin{bmatrix} - & C_{12} & \cdots & C_{1N} \\ C_{21} & - & \cdots & C_{2N} \\ \vdots & \vdots & \vdots & \vdots \\ C_{N1} & C_{N2} & \cdots & - \end{bmatrix} \quad D = \begin{bmatrix} - & D_{12} & \cdots & D_{1N} \\ D_{21} & - & \cdots & D_{2N} \\ \vdots & \vdots & \vdots & \vdots \\ D_{N1} & D_{N2} & \cdots & - \end{bmatrix}$$

Elements of the diagonal in both matrices are not defined. The application of the method proceeds with the calculation of the net concordance and discordance indexes, where the net concordance index constitutes a measure of relative dominance of an alternative i over other alternatives when compared with a measure of dominance of other alternatives over the alternative i and a net discordance index provides a measure of relative weakness of alternative i over other alternatives when compared with a measure of weakness of other alternatives over alternative i. The net concordance and discordance indexes are calculated by equations (9) and (10). Alternatives are finally ranked based on the concordance and discordance indices as well by taking the average of these two rankings. The network with the highest average rank-ing is considered to be best alternative.

$$C_i = \sum_{\substack{j=1 \\ j \neq i}}^{N} C_{ij} - \sum_{\substack{j=1 \\ j \neq i}}^{N} C_{ji} \tag{9}$$

$$D_i = \sum_{\substack{j=1 \\ j \neq i}}^{N} D_{ij} - \sum_{\substack{j=1 \\ j \neq i}}^{N} D_{ji} \tag{10}$$

4   Numerical Example

In this section we intend to demonstrate how the above methods can be used to model the problem of network selection. The following scenario is considered: a user wishes to access the Voice service in a cell, where four networks are available. All networks may serve the user; however, in order to optimize the QoS offered to the latter, the most efficient network has to be selected. The criteria to be taken into account in the decision process are Delay, Jitter, BER, Throughput and Cost. Of course, Delay and Jitter are of greater importance compared to other parameters, while Throughput holds minimum importance. Cost yields a variety of importance, depending on whether the user is characterized as risk seeking, risk neutral or risk adverse. Since the relative importance cannot be accurately determined, we consider the use of fuzzy logic as a valuable asset in our approach. Uncertainty in comparisons can be represented by fuzzy numbers, so that a set of possible values rather than a single value is acquired. The corresponding fuzzy numbers may derive from questionnaires or network measurements [6] [7], where the worst and best scores may form respec-tively the lower and upper bound of fuzzy numbers.

The first step of the network selection process is to determine the relative impor-tance of each criterion. Since this issue escapes the scope of this work, it is assumed that individual criteria importance has already been obtained. Table 1 collects the input data. Note that the decision matrix is symmetric, meaning that $a_{ij}=1/a_{ji}$. This allows to fill in only the cells where criterion i is more important than criterion j. The rest of the cells are filled in by reversing the corresponding fuzzy numbers. Application of FAHP (section 3.2) provides the weights for each parame-ter. From (1) we obtain S_d=(0.1551,0.2486,0.38669), S_j=(0.1674,0.2557,0.42141), S_b=(0.0818,0.1243,0.18757),S_t=(0.0756,0.1227,0.17559),S_c=(0.1658,0.2486,0.37755). The normalized weights of Table 2 deriving from (6) can be now used in the applica-tion of ELECTRE.

Table 3 presents the attribute values for all alternatives, as well as the reference values used for normalization. Tables 4 and 5 depict the normalized and weighted values respectively, as described in section 3.3.

Afterwards, the concordance and disconcordance matrices are formed according to (7) and (8), as shown in Table 6. The corresponding indexes are finally calculated through equations (9) and (10), providing the final ranking of the networks. Table 7 shows that both indexes point out Network 3 as the best option.

Table 1. Fuzzy Relative criteria importance

	Delay	Jitter	BER	Throughput	Cost
Delay	(1,1,1)	(0.667,1,1.429)	(1.5,2,2.4)	(1.6,2,2.8)	(0.8,1,1.333)
Jitter	(0.7,1,1.5)	(1,1,1)	(1.6,2,2.6)	(1.8,2.2,3)	(0.909,1,1.667)
BER	(0.417,0.5,0.667)	(0.385,0.5,0.625)	(1,1,1)	(0.75,1,1.5)	(0.385,0.5,0.556)
T/put	(0.357,0.5,0.625)	(0.333,0.455,0.556)	(0.667,1,1.333)	(1,1,1)	(0.357,0.5,0.556)
Cost	(0.75,1,1.25)	(0.6,1,1.1)	(1.8,2,2.6)	(1.8,2,2.8)	(1,1,1)

Table 2. Normalized weights

	Delay	Jitter	BER	Throughput	Cost
Normalized weights	0.3148	0.3249	0.036302	0.009555	0.3144

Table 3. Attribute values

	Delay(msec)	Jitter(msec)	BER($\cdot 10^6$)	Throughput(kbps)	Cost(units)
Network 1	155.5	4	0.085	18.8	25
Network 2	151.9	2.54	0.09	21.5	31
Network 3	121.3	2.79	0.06	19.6	31
Network 4	117.4	3.58	0.08	16.4	35
Reference	100	1	0.01	25	22

Table 4. Normalized values

	Delay	Jitter	BER	Throughput	Cost
Network 1	0	0	0.167	0.471	1
Network 2	0.095	1	0	1	0.4
Network 3	0.898	0.833	1	0.628	0.4
Network 4	1	0.305	0.333	0	0

Table 5. Normalized and weighted values

	Delay	Jitter	BER	Throughput	Cost
Network 1	0	0	0.001	0.004	0.314
Network 2	0.03	0.325	0	0.01	0.126
Network 3	0.283	0.271	0.036	0.006	0.126
Network 4	0.315	0.099	0.012	0	0

Table 6. Concordance and disconcordance matrices

$$C = \begin{bmatrix} - & 0.351 & 0.314 & 0.323 \\ 0.649 & - & 0.649 & 0.649 \\ 0.686 & 0.665 & - & 0.685 \\ 0.676 & 0.351 & 0.315 & - \end{bmatrix} \qquad D = \begin{bmatrix} - & 0.649 & 0.756 & 0.568 \\ 0.351 & - & 0.833 & 0.451 \\ 0.243 & 0.167 & - & 0.089 \\ 0.432 & 0.549 & 0.91 & - \end{bmatrix}$$

Table 7. Concordance and disconcordance indexes

	Rank C	Rank D
Network 1	4	4
Network 2	2	2
Network 3	1	1
Network 4	3	3

5 Conclusions

This paper addressed the problem of network selection that characterizes service delivery over a heterogeneous mix of wireless access technologies. In this frame, application of Fuzzy AHP and ELECTRE, i.e. two MADM algorithms, for ranking network alternatives was described. Since the relative importance of each criterion over another cannot be precisely defined, fuzzy numbers were adopted in order to integrate the uncertainty of subjective judgment into the problem analysis. The proposed methodology was finally tested through a numerical example, which pointed out how the most efficient alternative, in this case network, is selected.

References

[1] Dang, M.S., Prakash, A., Anvekar, D.K., et al.: Fuzzy logic based handoff in wireless networks. In: 2000 IEEE 51st VTC 2000-Spring, Vehicular Technology Conference Proceedings, Tokyo, May 2000, vol. 3, pp. 2375–2379 (2000)

[2] Kher, S., Somani, A.K., Gupta, R.: Network selection using fuzzy logic. Broadband Networks

[3] Yu, Y., Yong, B., Lan, C.: Utility dependent network selection using MADM in heterogeneous wireless networks. In: 18th Annual IEEE International Symposium on Personal, Indoor and Mobile Radio Communications (PIMRC 2007) (2007)

[4] Triantaphyllou, E.: Multi-Criteria Decision Making Methods: A Comparative Study. Kluwer Academic Publishers, Dordrecht (2002)

[5] Markaki, O., Charilas, D., Nikitopoulos, D.: Enhancing Quality of Experience in Next Generation Networks through Network Selection Mechanisms. In: Mobile Terminal Assisted Enhanced Services Provisioning in a B3G Environment Workshop, PIMRC (September 2007)

[6] Charilas, D., Markaki, O., Nikitopoulos, D., Theologou, M.: Packet-Switched Network Selection with the Highest QoS in 4G Networks. Elsevier Computer Networks 52(1), 248–258 (2008)

[7] Zhang, W.: Handover Decision Using Fuzzy MADM in Heterogeneous Networks. In: IEEE Wireless Communications and Networking Conference (WCNC 2004) (March 2004)

[8] Song, Q., Jamalipour, A.: Network Selection in an integrated Wireless LAN and UMTS Environment Using Mathematical Modeling and Computing Techniques. IEEE Personal Communications 12(3), 42–48 (2005)

[9] Song, Q., Jamalipour, A.: A Network Selection Mechanism for Next Generation Networks. In: 2005 IEEE International Conference on Communications. ICC 2005, May 16-20, vol. 2, pp. 1418–1422 (2005)

[10] Bari, F., Leung, V.: Application of ELECTRE to Network Selection in a Hetereogeneous Wireless Network Environment. In: Wireless Communications and Networking Conference (2007)
[11] Saaty, T.L.: Relative Measurement and its Generalization in Decision Making: Why Pairwise Comparisons are Central in Mathematics for the Measurement of Intangible Factors - The Analytic Hierarchy/Network Process. Review of the Royal Spanish Academy of Sciences, Series A, Mathematics 102(2), 251–318 (2008)
[12] Douligeris, C., Pereira, I.: An Analytical Hierarchy Process Approach to the Analysis of Quality in Telecommunication Systems (1992)
[13] Mikhailov, L., Tsvetinov, P.: Evaluation of services using a fuzzy analytic hierarchy process. In: Applied Soft Computing 2004, vol. 5, pp. 23–33 (2004)
[14] Mahmoodzadeh, S., Shahrabi, J., Pariazar, M., Zaeri, M.S.: Project Selection by Using Fuzzy AHP and TOPSIS Technique. In: Proceedings of World Academy of Science, Engineering and Technology, October 2007, vol. 24 (2007)

An Adaptive QoS Routing Solution for MANET Based Multimedia Communications in Emergency Cases

Tipu Arvind Ramrekha and Christos Politis

Kingston University, London, KT1 2EE, UK
{a.ramrekha,c.politis}@kingston.ac.uk

Abstract. The Mobile Ad hoc Networks (MANET) is a wireless network deprived of any fixed central authoritative routing entity. It relies entirely on collaborating nodes forwarding packets from source to destination. This paper describes the design, implementation and performance evaluation of CHAMELEON, an adaptive Quality of Service (QoS) routing solution, with improved delay and jitter performances, enabling multimedia communication for MANETs in extreme emergency situations such as forest fire and terrorist attacks as defined in the PEACE[1] project. CHAMELEON is designed to adapt its routing behaviour according to the size of a MANET. The reactive Ad Hoc on-Demand Distance Vector Routing (AODV) and proactive Optimized Link State Routing (OLSR) protocols are deemed appropriate for CHAMELEON through their performance evaluation in terms of delay and jitter for different MANET sizes in a building fire emergency scenario. CHAMELEON is then implemented in NS-2 and evaluated similarly. The paper concludes with a summary of findings so far and intended future work.

Keywords: MANET, emergency communication, quality of service, wireless networks, NS-2, adaptive routing.

1 Introduction

There is a definite rise in popularity for portable wireless devices which are equipped for multimedia communication. This trend is likely to mould the next generation network architecture as described by [1] and illustrated in fig. 1, where such highly mobile wireless multimedia devices will be preponderant in society. Therefore, in such a context, the IP Multimedia Subsystem (IMS) as proposed by the 3rd Generation Partnership Project (3GPP) [2] has added importance. The IMS plans to provide facilities for possible high quality IP based multimedia communication in case of emergency scenarios. In many emergency scenarios, such as natural or manmade disasters, the rescuers will have difficulty using traditional legacy networks due to destruction or collapse of infrastructure in such events or in case of remote disaster locations, the non-existence of any cost effective means of communication.

The MANET [4] is defined as a wireless network consisting of a group of at least two wireless mobile nodes which can either communicate directly to each other

---

[1] PEACE is a partly funded EU project. For more info visit: http://www.ict-peace.eu/

F. Granelli et al. (Eds.): MOBILIGHT 2009, LNICST 13, pp. 74–84, 2009.
© ICST Institute for Computer Sciences, Social-Informatics and Telecommunication Engineering 2009

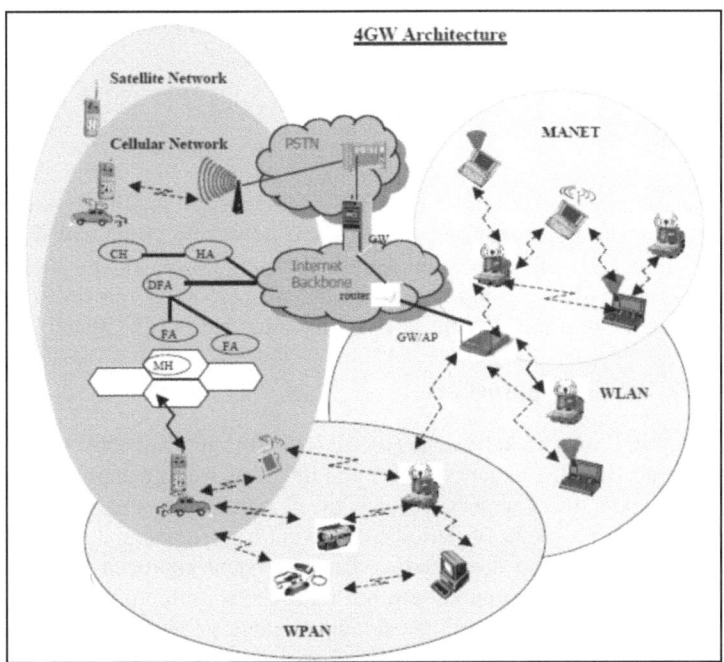

Fig. 1. Next generation network architecture illustrating the integration of *MANET* [3]

through a wireless interface or communicate through the forwarding of packets through other mobile nodes within the same group. Hence, a node in the MANET should essentially collaborate as a router of packets while also having the possibility of acting as a sender or receiver. The mobility of the nodes is a very useful feature for users but result in a very dynamic topology where routing of packets can be complex.

The nature of MANET makes it suitable to be utilised in the context of an Emergency for all rescue teams. The network layer Quality of Service (QoS) is critical for multimedia communication and encompasses several issues which need consideration [5] due to the topological dynamism of the MANET. The QoS of IP based multimedia communications can therefore be decisive in the effectiveness and efficiency of rescue missions in extreme emergency cases as described in PEACE project.

The project considers the provisioning of both day-to –day and extreme emergency communications in next generation all-IP networks. In the case of extreme emergency, it investigates the possibility for rescuers to use mobile devices such as PDAs for ad-hoc multimedia communication because in such scenarios, including natural and manmade disasters, the rescuers will have difficulty using traditional legacy networks due to destruction or collapse of infrastructure or in case of remote disaster locations, the non-existence of any cost effective means of communication.

The paper is organised as follows. In section 2, a summary of the various types of routing protocols for MANET is provided along with a small discussion on their merits. A more detailed explanation of AODV and OLSR routing protocols is provided in Section 3 together with their performance evaluation in NS-2 for a fire emergency scenario. Section 4 describes the CHAMELEON routing protocol and compares its performance against both OLSR and AODV. Finally section 5 contains conclusions from the findings in this paper and future work.

2 Background

The Internet Engineering Task Force (IETF) is responsible for the research and development of various Internet protocols and it has a working group called manet WG [6] which has the responsibility to standardise IP routing protocols for the purpose of static and dynamic MANET topologies. There are numerous categories of routing protocols which are designed for the MANET as described in [7], [3] and [4].

2.1 MANET Routing Approaches

The main MANET routing approaches are described in this section. A proactive approach is derived from the distance vector and link state routing approaches for wired internet. Each node stores a possible path to all possible destination nodes at any given time through a mixture of periodic and event triggered signalling mechanisms. A reactive approach is also described as the on-demand approach where routes are updated in MANET when required and not throughout its lifetime thus diminishing the overhead due to frequent message flooding. Then, geographical approaches use the proactive or reactive techniques described above adding geographical information to help the routing in the form of actual geographic coordinates (GPS) or relative to a fixed coordinate system. Also, the hybrid approach is an approach usually comprising of a mixture of proactivity and reactivity displayed by the routing protocol depending on certain situations.

2.2 Comparing Routing Approaches

The MANET, as described in the above sections, is a dynamic network. The approaches described above apart from the hybrid approach offer a specific solution to a possibly changing problem of routing in a dynamic topology of mobile nodes. A hybrid approach on the other hand can propose a more adaptive solution.

A reliable hybrid protocol could be constructed by using protocols which are being considered by the IETF. The Dynamic Source Routing (DSR), AODV, OLSR, and Topology Broadcast Based on Reverse-Path Forwarding (TBRPF) protocols, summarised in table 1 below, have evolved into a Request for Comment (RFC) through the IETF. However, research [8] has shown that AODV and OLSR are the two protocols which are most attractive for a hybrid solution in the case of a QoS routing solution for multimedia transmission.

Table 1. A Summary of RFC routing protocols being considered by IETF

Type	Reactive	Proactive
Source routing	DSR	OLSR
Hop-by-hop routing	TORA, AODV	DSDV, TBRF

3   Comparing AODV and OLSR

This section will investigate the performance of two routing protocols in MANET. The protocols are in the RFC stage at the IETF and therefore have the potential of developing into standardised protocols.

3.1 Ad Hoc on-Demand Distance Vector (AODV) Routing

The AODV Routing Protocol [9] provides a reactive approach towards routing. Route discovery is carried out by first sending a MANET wide broadcast request to search for the destination node and then receiving a unicast reply with the required path to the destination node. AODV nodes maintain a route table containing next hop routing information for a given destination and per hope sequence numbers nodes is stored. Each cached value in the table has an associated lifetime before expiration if unused.

3.2 Optimized Link State Routing (OLSR)

The OLSR protocol [10] is a variant of traditional link state routing for MANET. OLSR has multipoint relays (MPR) which reduces flooding and link state update overhead. Each node computes its own MPR from set of neighbouring nodes, the MPR Selectors. The MPR Selectors are the minimum set of single hop bidirectional linked neighbours that can connect the source to the maximum number of two hop bidirectional linked neighbours. They are the only nodes that rebroadcast a broadcast message from the sender. The sender also only exchange link state messages with its MPR selectors. Topology control (TC) messages are used to propagate topology information in MANET. The TC message are destined for MPR selectors only which is the only set of node advertised in the network. Nodes then accordingly rectify their routing tables through shortest path first algorithms such as refined Dijkstra's algorithms.

3.3 Emergency Scenario

The scenario considered is a fire situation in the Faculty of Computing and Information Systems (Sopwith building) at Kingston University, London and will be used for evaluating the performances of AODV, OLSR and CHAMELEON. There will be rescuers equipped with wireless multimedia transmission enabled portable devices, represented as mobile nodes, in the building for the emergency situation. The building has an area of 30m x 50m. The initial positions of the nodes are randomly distributed over the building. The movement of each node is random, but with a pre-configured speed of 1.5m/s and a maximum of 2.0m/s. They are assumed to have a pause time of 30s at random to take actions on the fire situation. The nodes start moving in random directions again after the pause and this movement pattern continues until simulation time ends. The simulations will be run for five scenario sets which include 2, 5, 10, 20, 50 and 90 who transmit both TCP and UDP packets to their colleagues i.e. a varying number of rescuers to demonstrate the reaction of routing protocols with respect to overall MANET size including traffic increase. The parameters used for simulating such an environment in NS-2 is summarised in table 2 below.

Table 2. Simulation parameters used for performance evaluation of AODV, OLSR and CHAMELEON protocols

Parameter	Value	Parameter	Value
TCP Packet Size	572 bytes	Simulation area	30m x 50m
TCP data rate	128 kbps	Data Traffic	FTP and CBR
TCP window	20	MAC trans. range	NS default for 802.11 option
TCP max. packets	10000	Simulation time	300s for 2 nodes
			360s for 5, 10, 20 nodes
			300s for 50, 70, 90 nodes
UDP Packet Size	512 bytes	OLSR setting	Default[11]
UDP data rate	64kbits/s	AODV setting	NS Default
UDP max. packets	10000	Mobility Model	Random
TCP Packet Size	572 bytes	Simulation area	30m x 50m
TCP data rate	128 kbps	Data Traffic	FTP and CBR

3.4   Comparative Results and Discussion

Delay performance comparison for UDP traffic link. Delay was considered on an end to end basis. The graph in fig. 2 shows that delay increases with an increase in number of nodes in the network. The OLSR protocol can be observed to have similar delay properties for the first ten nodes while its performance degrades for larger networks where the AODV protocol displays lower delay time. Also, the delay for AODV UDP packets itself decreases for number of nodes greater than 20 most probably because of non-expiry of route timeouts as described in section 2.

Jitter performance comparison for UDP traffic link. The graph in fig. 3 clearly shows that, for UDP traffic, OLSR has lower jitter magnitude than AODV for MANET sizes of less than 10 nodes while AODV is a better routing protocol for MANET sizes of more than 10 nodes. Consequently, it can be deduced from earlier results in this section that the proactive behaviour of OLSR is better than the reactive behaviour of AODV for MANET sizes of less than 10. Nevertheless, the performance of AODV is much better, for delay and jitter, for MANET sizes of more than 10. The resulting routing protocol will then be able to sustain a lower delay and jitter value irrespective of the number of nodes in such a MANET of size range between 2 and 90.

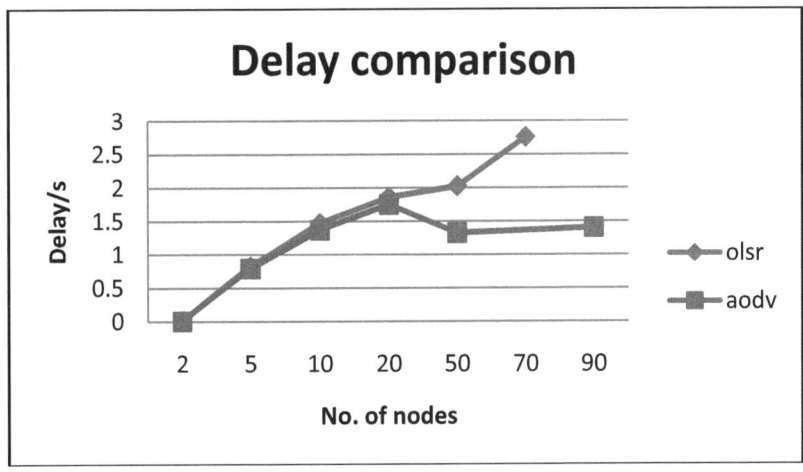

Fig. 2. Delay comparison between OLSR and AODV for varying MANET size for UDP links

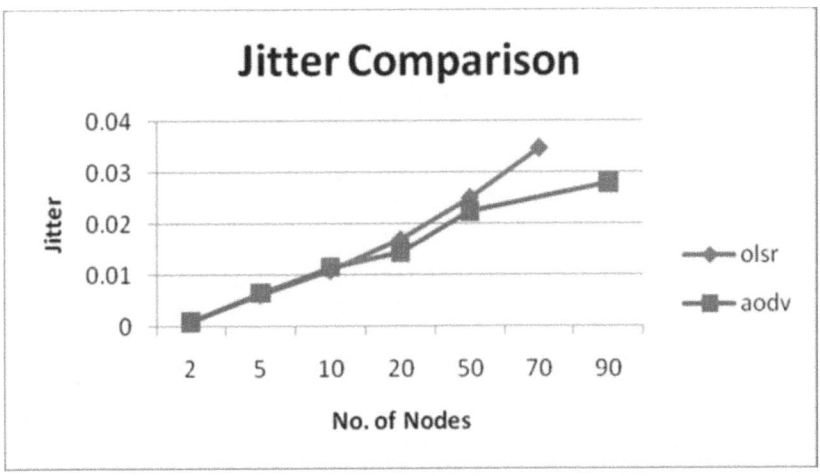

Fig. 3. Jitter comparison between OLSR and AODV for varying MANET size for UDP links

4 CHAMELEON

The CHAMELEON is a QoS routing protocol for variable size MANET in Emergency cases such as the fire situation in this paper.

4.1 CHAMELEON Algorithm

The hybrid protocol will have two modes of operations which are the proactive mode and the reactive mode. The proactive mode of operation will operate for a MANET size of up to 10 nodes and the reactive mode will operate for bigger sized MANETs

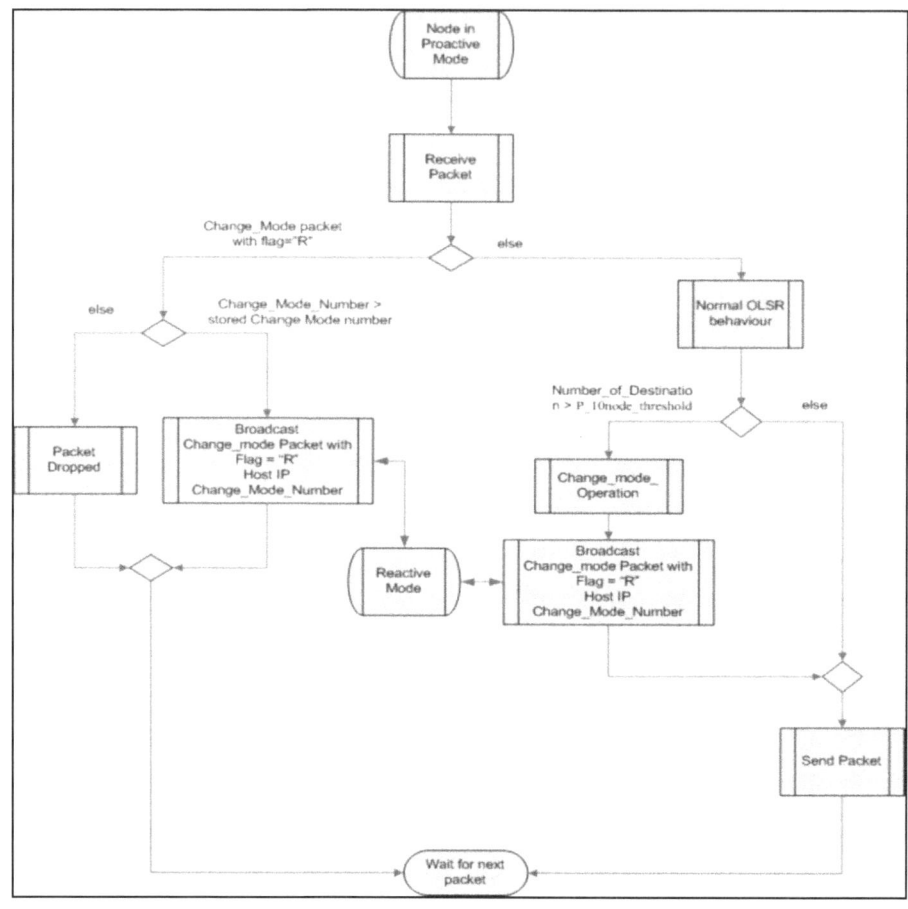

Fig. 4. CHAMELEON algorithm flowchart

of size greater than 10. The reactive mode operates in a similar fashion to OLSR with minor modifications on some features while the reactive mode operates in a similar fashion to AODV. A Change_Mode packet is introduced in the protocol. It is responsible for signalling neighbours through broadcasts that a threshold has been exceeded and behaviour has to be changed accordingly. Here, the presence of more than 10 nodes in the MANET is detected as exceeding the threshold and the protocol needs to shift from a proactive to a reactive routing behaviour as shown in fig. 4. The 10 node threshold is detected by using the P_10node_threshold value, which is equal to the value of the number of host stored in the OLSR routing tables.

A node can change to a reactive behaviour either by receiving a non-looped Change_mode packet with a flag 'R' or through detecting an exceeded P_10 node_threshold value from its routing table after computations when information from Hello and TC packets are received. The node which thus detects more than 10 nodes in the network first, broadcasts the Change_Mode_Packets which are valid for one-hop only. The Change_Mode packet is only considered in the case where the mode of operation is still proactive. The change in behaviour mode is carried out before the next packet to be sent is created to maximise efficiency. The algorithm for CHAMELEON requires that both the reactive and proactive structures are maintained simultaneously.

4.2 Results

The simulation results of CHAMELEON implemented through NS-2 and run for the same scenario as in table 2 are shown in fig. 5 for delay results and fig. 6 for jitter results respectively. It can be observed from fig. 5 below that the delay results for

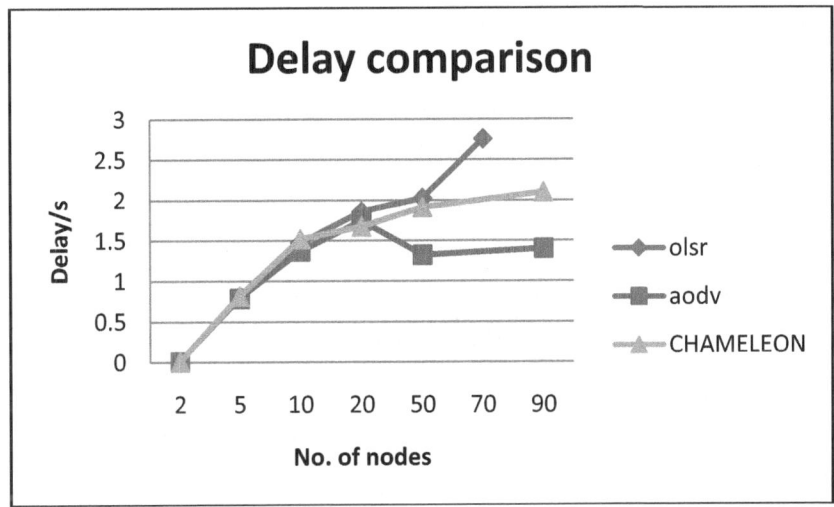

Fig. 5. Delay comparison results for CHAMELEON in UDP links for varying MANET sizes

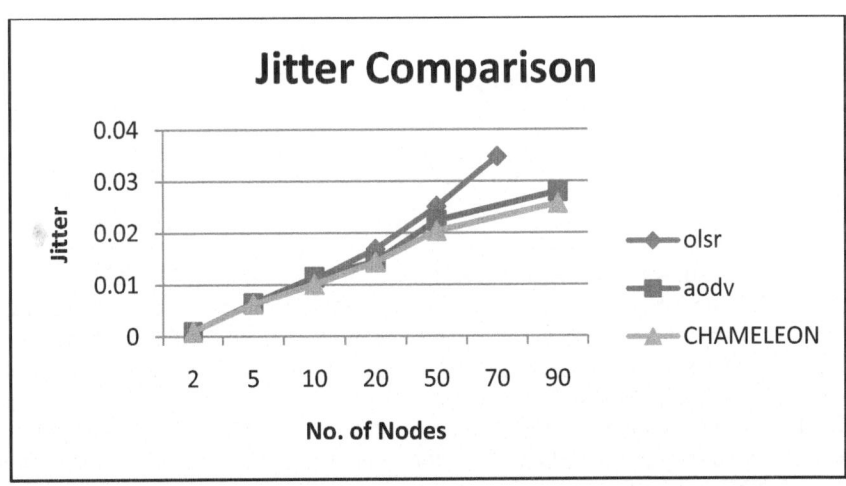

Fig. 6. Jitter comparison results for CHAMELEON in UDP links for varying MANET sizes

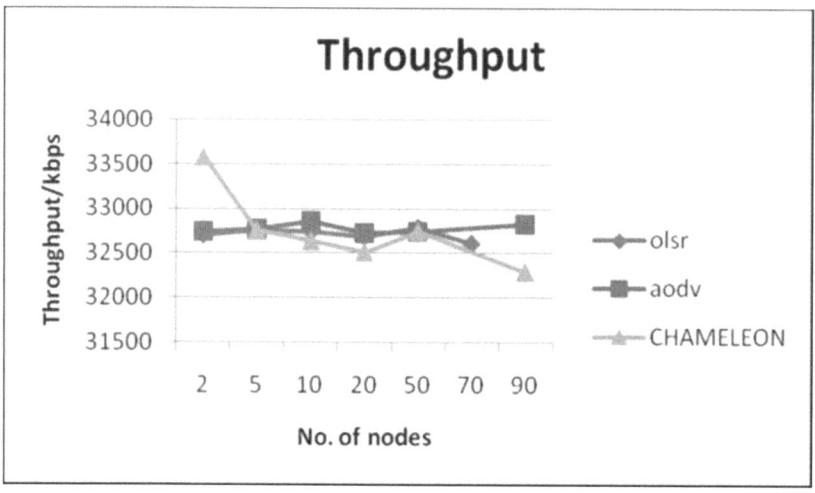

Fig. 7. Throughput comparison results for CHAMELEON in UDP links for varying MANET sizes

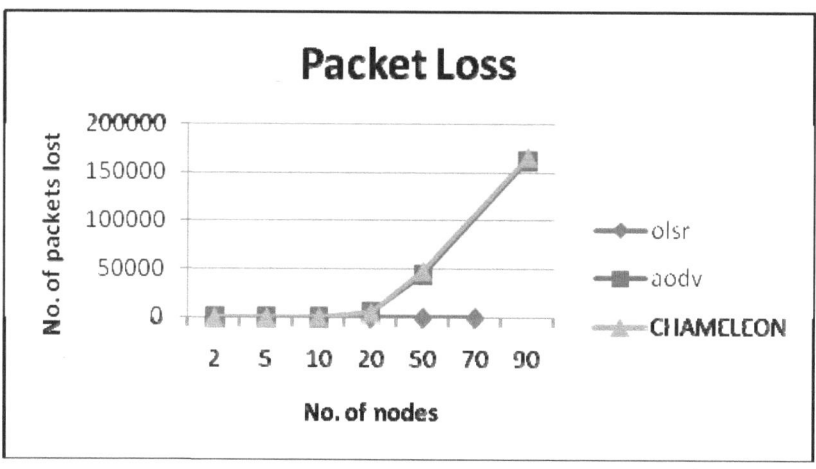

Fig. 8. Packet loss comparison results for CHAMELEON in UDP links for varying MANET sizes

CHAMELEON shows that its delay is similar to AODV and OLSR for MANET sizes of up to 10 nodes. For a MANET size of more than 10 nodes, the CHAMELEON delay is less than OLSR delay and slightly more than AODV delay. However, as shown in fig. 6 below, the jitter in CHAMELEON is less than both AODV and OLSR for the whole range of MANET size investigated.

Further results, as shown in fig. 7 demonstrate that CHAMELEON has an approximate average throughput of 32kbps. This throughput is comparable to the throughputs of both AODV and OLSR protocols and has a low variance irrespective of the number of nodes in the network. Therefore, voice over IP (VOIP) communication could be considered in such a network. Nevertheless, a codec such as 'G.729 (A)' would have to be used for good quality VOIP communication.

However, CHAMELEON has the disadvantage of displaying a similar trend to the AODV protocol in terms of packet losses for large networks in the case of the simulated building fire emergency scenario, as shown in fig. 8. OLSR has a relatively good performance with respect to packet losses in the same scenario. Applications enabling file transfers and using CHAMELEON or AODV for routing would therefore have bad performance as these applications require data transmission with minimal data loss.

5 Conclusion and Future Work

The findings in this paper show that a hybrid approach for providing a QoS solution for MANET routing can be beneficial. The results in section 4 show that CHAMELEON has an overall improved delay and jitter performance over both AODV and OLSR routing protocols over the range of investigated MANET sizes. The slightly greater average delay as compared to that of the AODV protocol can be explained by the fact that the CHAMELEON algorithm comprises of additional *Change_mode* packet processing time and behaviour shifting time delays. Since the performance results show improved overall delay and jitter performances, CHAMELEON can provide better QoS for multimedia communications over a larger MANET size range in case of such emergency scenarios as compared to AODV and OLSR respectively.

Future work includes the need to secure the CHAMELEON routing protocol against the risk of attacks in MANET used for emergency situations as proposed in [12]. The performance of the protocol should also be investigated for a larger set of scenarios. The algorithm for CHAMELEON can be refined to minimise processing delay so that its overall routing delay time can be decreased as compared to that of both AODV and OLSR. Also, further study on CHAMELEON could be carried out to identify the cause of packet losses and consequently measures could be derived to minimise packet loss to improve the QoS further.

Acknowledgments. The authors wish to acknowledge the support of the ICT European Research Programme and all the partners in PEACE: PDMF&C, Instituto de Telecomunicaes, FhG Fokus, University of Patra, Thales, Telefonica, CeBit.

References

1. Basagni, S., et al.: Mobile Ad Hoc Networking. Wiley-IEEE press (2004) ISBN: 978-0-471-37313-1
2. 3GPP Specification: IP Multimedia Subsystem (IMS) emergency sessions (2008), http://www.3gpp.org/FTP/Specs/html-info/23167.htm (accessed on July 31, 2008)
3. Mishra, A.: Security and Quality of Service in Ad Hoc Netwroks, Johns Hopkins University, United States of America. Cambridge University Press, New York (2008)
4. Conti, M., Giordano, S.: Multihop Ad Hoc Networking: The Theory. IEEE Communications Magazine 45(4), 78–86 (2007)

5. Liu, J., Chlamtac, I.: Mobile Ad-Hoc Networking with a View of 4G Wireless: Imperatives and Challenges. In: Basagni, S., Conti, M., Giordano, S. (eds.) Mobile Ad Hoc Networking, pp. 1–47. Wiley-IEEE Press, USA (2004)
6. IETF Secretariat: Mobile Ad-hoc Networks (manet) (2008), http://www.ietf.org/html.charters/manet-charter.html (accessed on July 1, 2008)
7. Belding-Royer, E.M.: Routing Approaches in Mobile Ad Hoc Networks. In: Basagni, S., Conti, M., Giordano, S. (eds.) Mobile Ad Hoc Networking, pp. 275–301. Wiley-IEEE press, USA (2004)
8. Huhtonen, A.: Comparing AODV and OLSR Routing Protocols. Seminar on Internetworking, Sjökulla, April 26-27 (2004)
9. Perkins, C.E., Royer, E.M.: The Ad Hoc On-Demand Distance Vector Protocol. In: Perkins, C.E. (ed.) Ad Hoc Networking, pp. 173–219. Addison-Wesley, Reading (2000)
10. Clausen, T., Jacquet, P., Laouiti, A., Muhlethaler, P., Qayyum, A., Viennot, L.: Optimized Link State Routing Protocol. In: Proceedings of IEEE INMI, Lahore, Pakistan (December 2001)
11. MASIMUM: Documentation and information related to UM-OLSR software (2008), http://masimum.dif.um.es/?Software:UM-OLSR (accessed on July 31, 2008)
12. Panaousis, E.A., Politis, C.: Towards Secure Ad-Hoc Communications in Extreme Emergency Case. In: WWRF/22nd meeting, May 2009, Paris (to be published) (2009)

A Distributed Energy-Aware Trust Management System for Secure Routing in Wireless Sensor Networks

Yannis Stelios, Nikos Papayanoulas, Panagiotis Trakadas,
Sotiris Maniatis, Helen C. Leligou, and Theodore Zahariadis

TEI of Chalkis, Dept. of Electrical Engineering, Psahna, Greece
Tel.: +30-2228099550
{jstellios,nicolappj,trakadasp,smaniatis,leligou,
zahariad}@teihal.gr

Abstract. Wireless sensor networks are inherently vulnerable to security attacks, due to their wireless operation. The situation is further aggravated because they operate in an infrastructure-less environment, which mandates the cooperation among nodes for all networking tasks, including routing, i.e. all nodes act as "routers", forwarding the packets generated by their neighbours in their way to the sink node. This implies that malicious nodes (denying their cooperation) can significantly affect the network operation. Trust management schemes provide a powerful tool for the detection of unexpected node behaviours (either faulty or malicious). Once misbehaving nodes are detected, their neighbours can use this information to avoid cooperating with them either for data forwarding, data aggregation or any other cooperative function. We propose a secure routing solution based on a novel distributed trust management system, which allows for fast detection of a wide set of attacks and also incorporates energy awareness.

Keywords: Wireless sensor networks, trust management, secure routing.

1 Introduction

Wireless Sensor Networks (WSN) offer efficient solutions in a great variety of application domains such as military fields, healthcare, homeland security, industry control, intelligent green aircrafts and smart roads. Although security is a key user requirement, which can be specified in a list of detailed security requirements, (see [1] - [3]), which include node verification, user authorization, data confidentiality, data integrity and freshness, privacy, secure localization and trusted resource allocation, its satisfaction proves to be a difficult task mainly due to the limited node and network resources. Well established security solutions designed for infrastructure based networks, cannot be applied in wireless sensor networks due to the limited memory space, processing power and energy (battery powered) as well as due to the absence of a trusted third party. The end result is that new solutions to defend against security attacks are needed [3].

In this infrastructure-less environment, nodes rely on the cooperation among each other for forwarding their packets to the sink node. A wide set of attacks addresses the

F. Granelli et al. (Eds.): MOBILIGHT 2009, LNICST 13, pp. 85–92, 2009.

routing procedure specifically. For example, in the blackhole attack, a node exhibits selfish behaviour and refuses to forward its neighbours traffic [3]. The situation can be further aggravated if it additionally advertises routes passing through it, alluring traffic. Another set of attacks is based on the modification of packets (either data or routing), which can disrupt the routing procedure driving nodes to route traffic incorrectly or falsifying the data that will finally reach the sink node. Other types of attacks include the Sybil attack, where a node pretends to possess certain characteristics which it does not really possess, and wormhole attack where more than one nodes collude in order to get the transmitted data or just to disrupt the routing procedure.

To combat such behaviours, an approach borrowed from human societies has been proposed (see [2]): nodes establish trust relationships between each other and base their routing decisions not only on geographical or pure routing information, but also on their expectation (trust) that their neighbours will sincerely cooperate. In other words, a trust management system is implemented. While key-based techniques can be used to provide data integrity, a trust model is mostly used for higher layer decisions such as routing [4], [5] and data aggregation [6], but also cluster head election [7] and, more surprisingly, for key distribution [8]. We propose a novel trust model which defends against routing attacks including black-hole, grey-hole, integrity-modification and confidentiality-authentication. Trust is then combined with energy awareness and location information to perform secure routing decisions.

In the rest of the paper, we first briefly report the related work found in the literature on trust models, while in section 3 we detail our innovative trust model and the secure routing protocol. In section 4, we evaluate the efficiency of our secure routing algorithm as a function of malicious nodes. Conclusions are drawn in the final section 5.

2 Trust Models for Sensor Networks - Related Work

Trust is the confidence of a node s_i that a node s_j will perform as expected i.e. on the node's s_j cooperation. To evaluate the trustworthiness of its neighbours, a node monitors their behaviour (direct observations) but may also communicate with other nodes to exchange their opinions. The methods for obtaining trust information and defining each node's trustworthiness are referred to as trust models. The aim is to improve security and thus increase the throughput, the lifetime and the resilience of a sensor network even in the presence of adversaries.

Depending on the distribution of the trust establishment functionality in the network, the trust models can be distinguished in centralized and distributed. In the centralized case, (an example of which can be found in [10]), a head node, which is assumed to be trusted, undertakes the responsibility to decide the nodes' trustworthiness, based either on trust data it has collected on its own, or on trust data received by other nodes. The advantage of this approach is that the head node can be selected to be the most powerful node in order to be able to monitor the behaviour of all others, alleviating the need for monitoring from the rest network nodes. However, it represents a single point of failure. Alternative trust architecture can be formed organizing the network in clusters and assigning the monitoring functionality to the cluster heads ([6], [7], [11]). In this case, the trustworthiness of each node is taken into account for the selection of the cluster head. In the purely distributed case, (like the one presented in [12]), each node monitors the behaviour of its neighbours and based on the collected measurements, it calculates their trustworthiness, which is then taken into account when routing decisions are made. In this case, the trust establishment functionality is uniformly distributed all over the network, and so does the implementation cost.

Different behaviour aspects can be monitored in a wireless sensor networks. Monitoring a certain behaviour aspect enables the detection of different security attacks. For example, each time node s_1 (see fig. 1) selects node s_3 for forwarding its packet, it enters the promiscuous mode in order to check whether node s_3 successfully forwarded it. After a number of cooperations, comparing the successfully forwarded packets to the number of packet s_1 sent to s_3, the source node (node s_1) can assess the sincere execution of the routing protocol while a systematic failure reveals a selfish and/or malicious node acting as a blackhole. Temporary failures due to channel errors will slightly affect the success over failure ratio when enough interactions have taken place. For each behaviour, based on the collected measurements either a trust value can be derived (in many cases a ratio of success over failure e.g. [10]), or distinct trust levels can be distinguished [13].

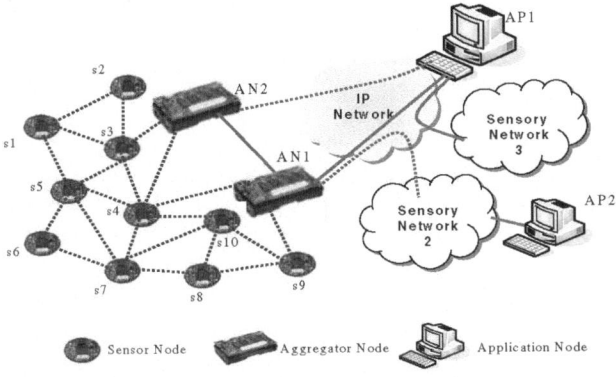

Fig. 1. Aggregator Nodes (ANs) collect data from the sensor nodes (si) and communicate with application nodes (AP) which provide the desired services

The detection of an unexpected behaviour based only on direct measurements with an adequate confidence requires the collection of an important number of evidences (samples). This procedure can be accelerated taking advantage of the neighbours' experiences. In other words, each node (say s_1 in Figure 1) may calculate its neighbour's (for example, node s_3) trust value either based on its own observations (direct evidence) [12] or combine it with information obtained from other nodes (for example nodes s_2, s_5). The information provided by s_2 and s_5 is called reputation or indirect evidence [13]. In this concept, every node can build a relation with its neighbours, based on the collection of actions (events) performed by other nodes in the neighbourhood. It is worth stressing that the trust information exchange can be exploited by adversaries to ruin the routing functionality of the network (as supported in [14]). Sharing information makes the system vulnerable to false reports (bad-mouthing attack), i.e. there are specific attacks targeting the reputation exchange protocol.

3   A Novel Energy-Aware Trust Model

In this section, we propose a novel distributed trust model suitable for the demanding and highly unreliable environments of wireless sensor networks (WSN). This trust model brings two important innovations: first, it defends against a wide set of attacks by monitoring multiple behaviour aspects (and not just a few as most trust models in the literature do) second, it incorporates energy-awareness which allows for better load balancing and higher resilience against the attacks. Moreover, a routing cost function incorporating trust, energy and location information is derived to guide routing decisions. The proposed trust model is a decentralized trust scheme, suitable for typical sensor network architecture (as the one shown in Figure 1).

3.1   Trust Metrics

One of the most important aspects of trust management schemes is the process of data collection. The direct trust value of a neighboring node can be determined by its multi-attribute, time-varying trust value depending on a set of events. We have selected a set of metrics that reveal the cooperation willingness of the nodes as regards routing. In more detail, each sensor monitors its neighbours as regards:

- Packet forwarding: To protect against black-hole and grey-hole attacks, every node should be evaluated regarding its willingness and sincerity in forwarding the received packets, cooperating in the routing procedure. This can be checked either through overhearing, or based on link layer acknowledgements, i.e. the source node checks whether its neighbour has forwarded the message.
- Network layer ACK: We also suggest that for each transmitted packet, the source node evaluates its next hop neighbour based on the reception (or not) of the relevant network layer ACK from the Base Station. The reception of the Net-ACK is evidence that the next hop node or any other node in the path is not colluding with another adversary in order to disrupt the network operation. In other words, the correct reception of the network layer ack ascertains that the message has reached a higher layer node in the proposed architecture, providing trust info for the whole path.
- Integrity: For the proper operation of the WSN, it is important that the nodes do not intentionally falsify both the data and the control messages. To avoid such malicious behaviours, each node overhears the wireless medium so that it receives the forwarded message. Then it processes it to check its integrity, i.e. that it is not altered violating the communication protocol rules.
- Authentication – Confidentiality. A node can collect trust information about neighbouring nodes during interactions regarding the proper use of the applied security measures. For example, a node might use a mechanism to authenticate the message of a neighbouring node or the base station. Furthermore, integrity measures and confidentiality measures (e.g. elliptic curve cryptography) can be applied for the communication between neighbouring nodes. Consequently, the proper use of these security mechanisms is considered as input for trust value computation.
- Remaining energy. To avoid the node with high trust value die out early, (which would possibly result in low connectivity), the node's remaining energy is considered as a trust value component. For the collection of this information, each node sends its

remaining energy value piggybacked in the periodically exchanged HELLO message and/or in every interaction (exchanged packet). This way energy awareness becomes an inherent feature of the trust model.

Although other application-aware trust metrics have been proposed in the literature, the implementation of such sophisticated functionality would exceed the sensor node capabilities. In this view, to enhance the flexibility of the model, since its implementation is based on embedded system software, we strongly propose the development of a security toolbox, which incorporates the trust model. The full set of metrics can be implemented in this software component and each time it is used in a specific sensor type, the configuration can change so that the trust model takes into account fewer metrics than the complete design. The level of security achieved depends on the number and type of behaviours monitored while it should also be stressed that the desired level of security strongly depends on the application.

3.2 Trust Quantification

Coming to the quantification of trust, for each trust metric except the remaining energy, node A calculates a trust value regarding node B based on the following equation:

$$T_i^{A,B} = \frac{S_i^{A,B}}{S_i^{A,B} + F_i^{A,B}} \, , \qquad (1)$$

where S_i and F_i stand for the number of successful and failed co-operations respectively. (Each node is responsible for computing its own trust value per neighbour in the network, collecting events from direct interactions. The S_i and F_i values for all neighbours and all the above described events are maintained in a trust repository.) As regards the remaining energy, this is calculated as

$$T_{RE} = V_{now}/V_{initial} \, , \qquad (2)$$

where V_{now} and $V_{initial}$ stand for the remaining energy levels reported at the last and initial message received.

Furthermore, if the trust model is used to perform routing decisions, we propose to incorporate in the total node trust value, its distance to the base station, in order to form a unique routing cost function guiding the node selection and thus address trust and routing information together. In this case, the value of the "trust" metric related to the distance to the base station can be computed as follows:

$$T_d = 1-(d_i/\Sigma d_i) \, , \qquad (3)$$

where d_i is the distance of neighbour i to the base station while Σdi stands for the sum of the distances of all its neighbours to the base station. To calculate the total direct trust value, all the trust values are summed up in a weighted manner (W_i represents the significance of i-th trust metric) based on the following equation:

$$DT^{A,B} = (\sum W_i * T_i^{A,B}) \, . \qquad (4)$$

The node that is assigned the highest total trust value will be selected for forwarding. If no malicious nodes exist in the network, the node closest to the base station will be chosen.

4 Performance Evaluation

The performance of the proposed trust management system has been evaluated through computer simulations. The JSim platform [15] has been used to model our approach. The simulated network topology includes 25 sensor nodes (n_0 to n_{24}) placed on a 5x5 grid. The adopted transport protocol operates in a request-response manner and implements retransmissions at the application level with a timeout set to 0.5s. The initial trust value for all neighbors has been set equal to 1. The modeled trust metrics include forwarding, network ACK, remaining energy, and distance to the Base Station. In all the tested scenarios, 4 connections are active.

To evaluate the efficiency of our model, we ran three scenario sets. In the first scenario set, the node distance is kept fixed (equal to 100m) while weights of the trust metrics were equal to: W_d (weight of distance to the sink node) = 0.5, W_e (weight of the remaining energy) = 0.1, W_f (weight of the forwarding metric) = 0.2, W_a (weight of the ack metric) = 0.2. Nodes performing the black-hole/grey hole attack drop all/randomly the received traffic. The obtained results are shown in Fig. 2.

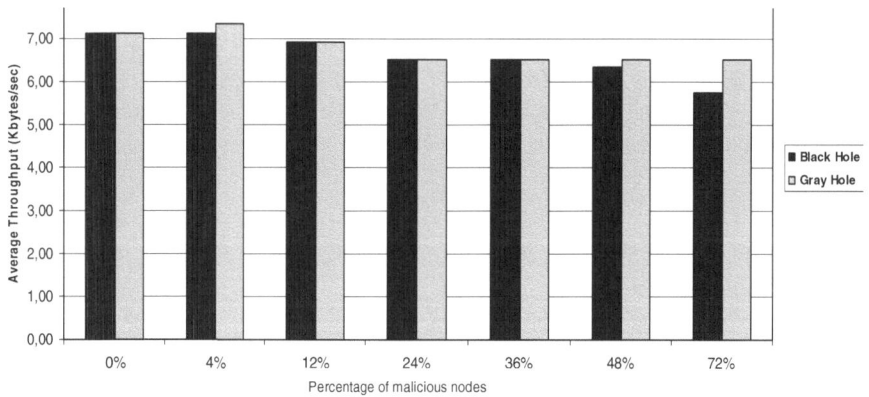

Fig. 2. Performance for different number of malicious nodes

The results show that the attackers are successfully detected and thus the throughput degradation compared to the case where no malicious node exists in the network is limited, even when malicious nodes represent the 72% of the network nodes. The detection of these attacks is based on the forwarding and net-ack trust metrics. Comparing black-hole to grey hole attacks, the impact of grey hole attack is less heavy and this is more evident when the malicious node reach high percentages.

We have run a second scenario set, where we modified the weights assigned to the trust metrics as follows: W_d = 0.2, W_e= 0.2, W_f = 0.5, W_a = 0.1. The results have shown that the impact of the modification of weights is almost negligible, since the network is dense enough so that a trust-full neighbour is always found towards the destination. To further investigate the impact of the density of the network we have performed the third scenario set, where malicious nodes perform only black-hole attacks and the transmission range of each node is 260m. The average throughput for different distance values are shown in figure 3.

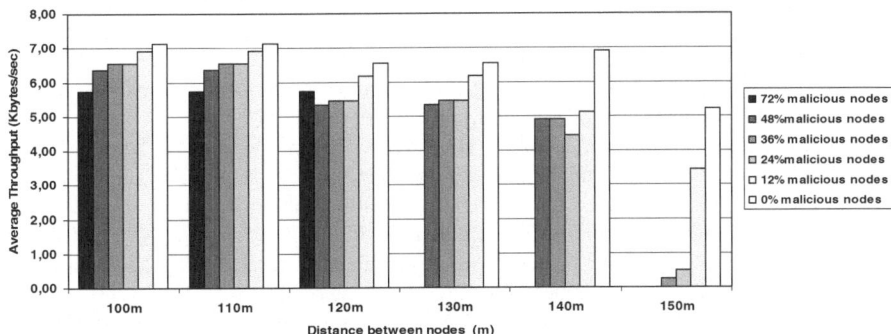

Fig. 3. Performance for different distance between neighbours

It is obvious that when the network is dense (i.e. low distance among nodes compared to the transmission range), this reflects to a high number of neighbours per node. As a result, a trusted neighbour can be found even when 72% of nodes are acting maliciously. The network operation becomes more vulnerable when the distance exceeds 150m. In this case less nodes exist in the neighbour list limiting the choices and thus the performance strongly depends on the number of attackers.

5 Conclusions

In the unmanaged environment of WSNs, the list of security attacks addressing the routing procedure is very long. Although cryptography and strong authentication schemes are powerful tools to safeguard packet integrity and node authenticity, they do not detect the (large set of) routing attacks such as selfish behaviours, blackholes, and bad mouthing. The establishment of trust relationships among nodes, exactly as happens in human societies, based on behaviour monitoring, is a usefull and effective tool. The choice of the behaviours to monitor is directly associated with the attacks that can be detected and against which protection is aimed. We proposed an efficient and scalable trust model incorporating energy awareness (which is a key issue for long network lifetime). Our model was shown to efficiently detect the malicious nodes and retain connectivity even when malicious nodes represent the 72% of the existing nodes. The performance of our scheme for different network densities has also been investigated and reported. Our future work includes the extension of our model to also incorporate indirect trust information.

Acknowledgments. The work presented in this paper was partially supported by the EU-funded FP7 211998 AWISSENET project.

References

1. Giruka, V.C., Singhal, M., Royalty, J., Varanasi, S.: Security in wireless sensor networks. Wireless Comm. Mob. Comput. 8, 1–24 (2008)
2. Kannhavong, B., Nakayama, H., Nemoto, Y., Kato, N., Jamalipour, A.: A survey of routing attacks in mobile ad hoc networks. IEEE Wireless Comm., 85–91 (2007)
3. Karlof, C., Wagner, D.: Secure Routing in Wireless Sensor Networks: Attacks and Countermeasures. In: IEEE Int. Workshop on Sensor Network Protocols and Applications, pp. 113–127 (2003)
4. Li, H., Singhal, M.: A Secure Routing Protocol for Wireless ad hoc Networks. In: 39th Hawaii International Conf. on system Sciences (2006)
5. Rezgui, A., Eltoweissy, M.: TARP: A Trust-Aware Routing Protocol for Sensor-Actuator Networks. In: IEEE Int. Conf. on Mobile Adhoc and Sensor Systems (2007)
6. Hur, J., Lee, Y., Yoon, H., Choi, D., Jin, S.: Trust evaluation model for wireless sensor networks. In: Advanced Comm. Tech. Conference, ICACT 2005, pp. 491–496 (2005)
7. Crosby, G.V., Pissinou, N.: Cluster-based Reputation and Trust for Wireless Sensor Networks. In: Consumer Communications and Networking Conference, CCNC (2007)
8. Lewis, N., Foukia, N.: Using Trust for Key Distribution and Route Selection in Wireless Sensor Networks. In: IEEE Globecom 2007 (2007)
9. Mahoney, G., Myrvold, W., Shoja, G.C.: Generic Reliability Trust Model. In: 3rd Annual Conference on Privacy, Security and Trust (2005)
10. Tanachaiwiwat, S., Dave, P., Bhindwale, R., Helmy, A.: Location-centric Isolation of Misbehavior and Trust Routing in Energy-constrained Sensor Networks. In: IEEE Int. Conf. on Performance, Computing, and Communications (2004)
11. Ghazaleh, N., Kang, K.D., Liu, K.: Towards Resilient Geographic Routing in Wireless Sensor Networks. In: 1st ACM Workshop on QoS and Security for Wireless and Mobile Networks (2005)
12. Pirzada, A., McDonald, C.: Trust Establishment In Pure Ad-hoc Networks. Wireless Personal Comm. 37, 139–163 (2006)
13. Marias, G., Tsetsos, V., Sekkas, O., Georgiadis, P.: Performance evaluation of a self-evolving trust building framework. In: 1st International Conference on Security and Privacy for Emerging Areas in Communication Networks (2005)
14. Sun, Y.L., Han, Z., Ray Liu, K.J.: Defense of Trust Management Vulnerabilities in Distributed Networks. IEEE Communications Magazine 25, 112–119 (2008)
15. http://www.j-sim.org/
16. Karp, K., Kung, H.T.: GPSR: Greedy Perimeter Stateless Routing for WirelessNetworks. In: MobiCom 2000 (2000)

A Mobile Multi-hop Relay Base Station (MRBS) – Relay Station (RS) Link Level Performance of Coding/Modulation Schemes, on the Basis of the REWIND Research Program

Ioannis P. Chochliouros[1], Avishay Mor[2], Konstantinos N. Voudouris[3], O. Amrani[2], and George Agapiou[1]

[1] Hellenic Telecommunications Organization (O.T.E.) S.A., Research Programs Section, 99, Kifissias Avenue, 15126 Maroussi, Athens, Greece
ichochliouros@oteresearch.gr
[2] DesignArt Networks Ltd.,
Ha'Haroshet Street, P.O. Box 2278, Ra'anana, Israel
avishaym@designartnetworks.com
[3] Technological Educational Institution (T.E.I.) of Athens, Dept. of Electronics
Ag. Spyridonos & Milou 1 Street, 12210 Egaleo, Athens, Greece
kvoud@ee.teiath.gr

Abstract. Among the essential aims of the European REWIND Research Program is to proceed to the algorithmic research and technology development of appropriate Mobile Multi-hop Relay networks based on the WiMAX technology, so that to increase coverage and throughput issues. The present work provides a description of the related link-level algorithms and simulations. The study performed evaluates the link-level performance of various coding and modulation schemes with different antenna configurations over several links. In particular, it studies the "backhaul channel" which is required to support the aggregate cell traffic. High-rate convolution turbo codes combined with high-order modulation schemes are employed. The performance and gains associated with multiple-antenna deployments such as MISO and MIMO techniques are evaluated.

Keywords: CTC (Convolutional Turbo Coding), IEEE 802.16j standard, MIMO (Multiple-Input, Multiple-Output) techniques, MISO (Multiple-Input, Single-Output) techniques, Mobile Multi-hop Relay (MMR) specification, MMR Base Station (MRBS), Relay Station (RS), WiMAX.

1 Introduction

There are various choices available for operators when deploying Base Stations (BSs) to improve indoor or outdoor coverage or to increase network capacity. These can include macro-cells, micro-cells, or pico-cells in an outdoor environment; pico-cells in public indoor locations (or within enterprise buildings), and; femto-cells for residential use ([1], [2]). The primary difference between these cells (performance-wise)

F. Granelli et al. (Eds.): MOBILIGHT 2009, LNICST 13, pp. 93–102, 2009.

is the size of coverage. Macro-cells are these BSs with the longest range, but are also the most expensive to purchase, deploy and maintain. Micro-, pico- and femto-BSs are used to fill in coverage gaps and establish coverage in buildings where macro-cell signals can hardly penetrate. A significant side-effect of placing a large number of BSs in a region is that each one needs a dedicated broadband backhaul connection. Thus, micro-, pico-, and femto-cells can use either wireline or wireless links for their backhaul, depending on the cost, availability and scalability of different solutions. In particular, they can support in-band backhaul to enable operators to use their spectrum holdings to carry backhaul traffic to the nearest macro-BS or to the nearest micro-cell or pico-cell with wireline backhaul.

Under these circumstances, the *IEEE 802.16j Mobile Multi-hop Relay (MMR) specifications* [3] are aimed to extend BS reach and coverage, while minimizing wireline backhaul requirement. A relay-based architecture will allow operators to use in-band wireless backhaul while retaining all the standard WiMAX (Worldwide Interoperability for Microwave Access) functionality and performance ([4], [5]). For example, as shown in Fig.1 the MMR Base Station (MRBS) provides the primary area of coverage. It also has a backhaul connection, such as leased copper, fiber optics, or microwave radio link. The Relay Station (RS) extends the BS coverage. A mobile subscriber station (SS) can connect to a BS (i.e. a MRBS or a RS). Some "combined" BS/RS deployments can also "reduce" (or "eliminate") Network coverage "holes"; or can be used to serve temporary network deployments for disaster/emergency situations and for special events [6].

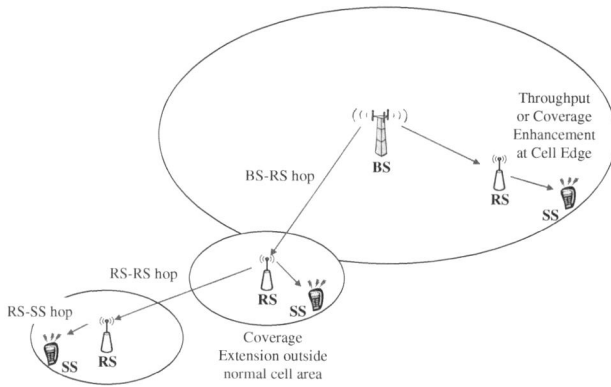

Fig. 1. Enhancement of "cell" edge performance through BS coverage extension by using RSs

In the above usage cases/scenarios it is clear that the backhaul channel has to support the aggregate cell traffic and is therefore crucial to study and optimize. We assume an in-band wireless backhaul connection, utilizing the same channel in the 2.5GHz frequency band as it is used for the WiMAX network deployment. The channel models assumed throughout are WiMAX-related, in 2.5GHz frequency band, using the WiMAX System Evaluation Methodology.

As the in-band backhaul link utilizes the WiMAX channel bandwidth, which is used for access, and in order to minimize the overhead of the relay backhaul link on the overall access capacity, it is essential that the spectral efficiency of the backhaul links will be as highest as possible. In a network of macro-BSs and multiple relays around each BS, the overall network performance is highly dependant on the backhaul links capacity performance [7]. The backhaul link's performance is subject to typical link budget and SNR (signal to noise ratio) performance, taking into account the overall interference pattern, in the network level.

The present work originates from the "core" context of the EU-funded "*REWIND*" Research Project (*ICT-FP7, Grant Agreement No.216751*), which intends to develop a specific Relay Station (RS) implementation on the basis of the WiMAX technology; in particular, *REWIND* will proceed to the algorithmic research and technology development of the corresponding MMR-based networks in order to increase coverage and throughput issues. An essential part of the Project is responsible for the design of the novelty software and hardware functional areas of the corresponding RS product, mainly including: algorithmic research and simulations; system architecture and requirements specifications, and; DSP (Digital Signal Processing) and MAC software code development and integration.

2   Essential Assumptions

The present work provides a description of the RS-MRBS link level algorithms and simulations. Thus, we evaluate the link-level performance of various coding and modulation schemes with different antenna configurations over several links. In particular, we study the backhaul channel which shall then be used as a "building block" for the network-level simulation. High-rate convolution turbo-codes combined with high-order modulation schemes are employed.

The performance and gains associated with multiple-antenna deployments such as MISO (Multiple-Input, Single-Output) [8] and MIMO (Multiple-Input, Multiple-Output) ([9], [10]) techniques are also evaluated. Obtained results can then be utilized by the system level simulation procedures.

In particular, a MATLAB environment[1] was developed for running Link Level BER (bit error rate) / BLER (block error rate) (PER) performance tests, focusing on: (i) WiMAX-compliant modem configurations and reference channel models, and; (ii) Enhanced backhaul configurations.

This can support laboratory debugging / testing as follows:

- On the transmitter (Tx) by exporting waveforms to signal generator;
- On the receiver (Rx) by importing recordings from VSA (vector signal analysis) and a logic analyzer;
- By generating board configuration scripts according to system design and settings.

The simulated Relay Station contains 2 Transmit (Tx) antennas and up to 6 (i.e. 2, 4, or 6) Receive (Rx) antennas, operating in MIMO receiver techniques ([11], [12]), in order to increase the spectral efficiency of the link between the MR-BS (Multi-hop

---

[1] MATLAB is a high-level language and interactive environment that enables you to perform computationally intensive tasks faster than with traditional programming languages such as C, C++, and FORTRAN.

Relay Base Station) and the RS. The setting is based on the premise that no more than 2 Downlink (DL) streams "share" the same time and frequency resource in the MRBS-RS link. Therefore, the MRBS-RS link is a two-stream backhaul utilizing 2 to 6 Relay receive antennas. For the simulation environment, we have utilized several statistical path-loss, shadowing and indoor loss models ([13], [14]). The simulation Environment is built as a MATLAB project. It can be compiled as a stand-alone application (currently on Linux).

3   MRBS-RS Link Level Simulations

Mobile radio channels can be narrowband (i.e., flat fading channels) or broadband (i.e., frequency selective fading channel). So, different channel models have to be developed and examined. In mobile radio channels, the high mobility causes rapid variations across the time-dimension, the large multi-path delay spread causes severe frequency-selective fading, and the large multi-path angular spread causes significant variations in the spatial channel responses. For best performance, the transmitter and receiver algorithms must accurately track all dimensions of channel responses (space, time, and frequency) [15].

This section describes the main building blocks employed for the link-level simulation. These are divided into: Transmitter, Channel, and Receiver. The reason for using these "blocks" is to provide realistic results (rather than theoretic ones) able to capture the effect of various practical issues that can be generally modelled as "implementation loss". Each specific "block" is described as follows:

The *Transmitter block* contains the following modules:

- A burst modulator (which consists of a payload FEC (Forward Error Correction) encoder and a randomizer);
- A symbol builder (which includes a frequency domain 2 OFDM (Orthogonal Frequency Division Multiplex) streams constructor according to zone / burst permutation type [16], MCS (modulation/coding schemes), and proper MIMO settings);
- A beam-former (considered for the mapping of the 2 streams onto N Tx antennas);
- A Decision Feedback Equalization-DFE (which includes IFFT (inverse Fast Fourier Transform), CP (Cyclic Prefix), windowing, sampling rate conversion and timing correction, carrier frequency correction and digital up-conversion, I/Q precompensation [17]).

The *Channel simulator block* contains the following modules:

- A MIMO fading channel simulator [18];
- Carrier frequency offset;
- Phase noise, and;
- Timing offset.

The *Receiver block* contains the following modules:

- A DFE scheme able to perform I/Q post-compensation, carrier frequency correction and digital down-conversion, sampling rate conversion and timing correction, CP removal, FFT (Fast Fourier Transform);

- Automatic Gain Control (AGC);
- Preamble sync;
- A pilots tracking "block" possessing CFO (Clock Frequency Offset) and STO (Symbol Timing Offset), as well as several specific "estimators", i.e.: channel estimators; noise variance estimator; MMSE (minimum mean square error) equalizer taps estimator; post-MMSE CINR (carrier-to-interference and noise ratio) estimator);
- A symbol decomposer (containing a frequency domain 2 OFDM streams spatial equalizer);
- A LLR (log-likelihood ratio) computation "block" [19];
- A Burst demodulator (able to perform a physical-to-logical reorder of LLRs, and including a payload FEC decoder and de-randomizer).

3.1 Simulation Parameters

The following set of specific parameters has been assumed for the simulations:

The *channel models* used have considered the following cases:

- AWGN (Additive White Gaussian noise) [20];
- ITU Pedestrian B [21];
- Backhaul Type A (derivative of SUI-1[2]) [22].

The *coding scheme* has been based both on convolution turbo coding and on a specific encoding scheme (as defined in IEEE 802.16e specifications).

The *modulation schemes* have comprised square QAM (Quadrature Amplitude Modulation) constellations from QPSK (Quadrature Phase Shift Keying) to 256QAM depending on link conditions and SNR operation region.

The following simulations have been realized either by using Matrix A (STC: Space-Time Coding [23]) (i.e. the case of scenarios 1 and 2) or Matrix B (MIMO-SM (Spatial Multiplexing)) (i.e. the case of scenario 3) smart antenna methods, as described in the following sections.

3.2 Simulation Results

Scenario 1: Here, the objective was to realize a reference performance measure in Single-Input, Single-Output (SISO) antenna system [24]. To this aim, we have considered an AWGN channel, with various combinations of coding rates and modulation schemes.

Fig.2 presents the BER performance in various modulation and coding schemes. The steepness of the performance curves is due to the so-called water-fall region (WFR), associated with the turbo codes used.

Scenario 2: In the above usage cases/scenarios it is clear that the backhaul channel has to support the aggregate cell traffic and is therefore crucial to study and optimize. We assume an in-band wireless backhaul connection, utilizing the same

---

[2] Stanford University Interim (or "SUI") models were used for evaluation of suggested 802.16 physical layer modifications.

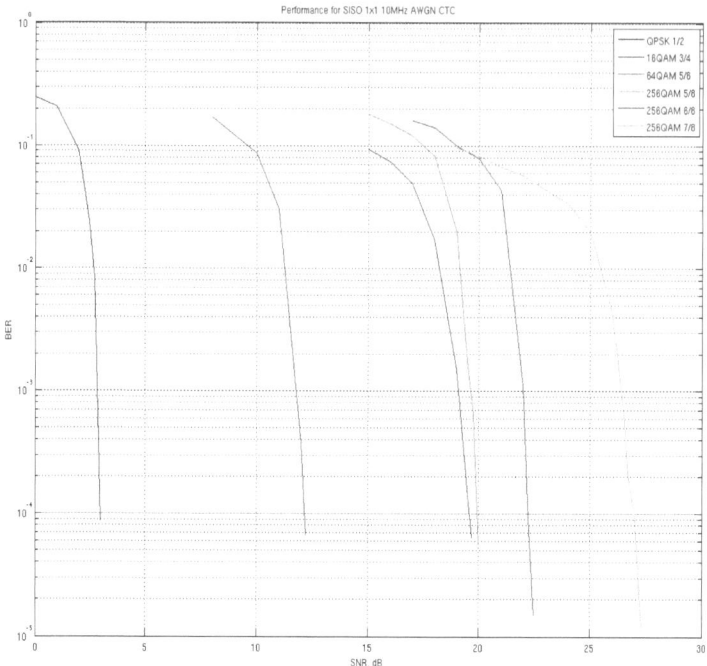

Fig. 2. BER performance for SISO channel with CTC (Convolutional Turbo Code)

channel in the 2.5GHz frequency band as it is used for the WiMAX network deployment. The channel models assumed throughout are WiMAX-related, in 2.5GHz frequency band, using the WiMAX System Evaluation Methodology.

In this scenario, the objective was to measure the obtained diversity gains by using multiple Receive antennas (from 2, 4 or 6 antennas); a single transmit antenna has been used throughout. For this scenario the channel was the backhaul channel Type A.

Fig.3 presents the BER performance in different modulation and coding schemes, with 2, 4 or 6 receive antennas. The results show significant gains in performance (about 8-10 dB) due to receive diversity obtained by multiple receive antennas.

Scenario 3: In this scenario, the essential objective was to determine the SNR regions required for optimized backhaul link performance, with high-level modulation schemes (16, 64 and 256QAM modulations) combined with turbo coding and various antenna configurations.

Two transmit antennas have been assumed for transmitting two independent streams.

The set of curves depicted in Fig.4, presents the BER performance in different modulation and coding schemes, and summarizes the extensive simulation results collected for the backhaul channel.

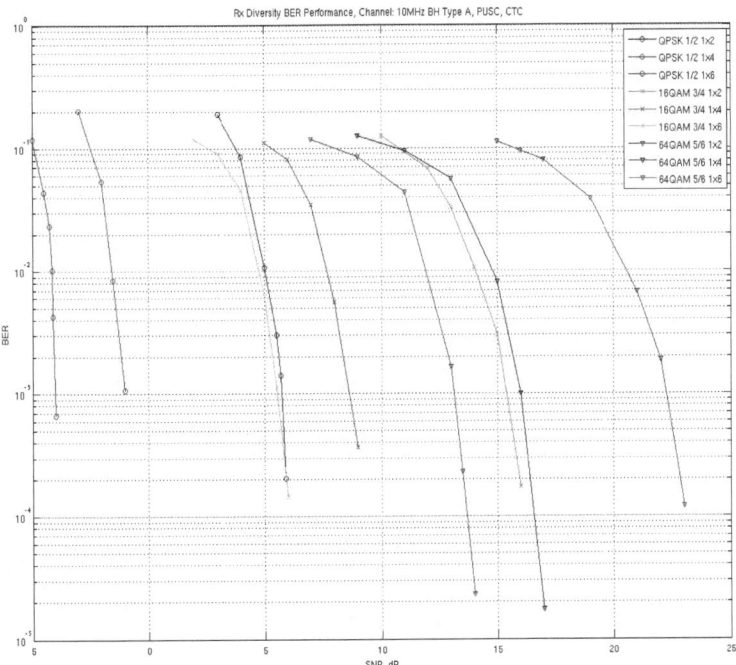

Fig. 3. BER performance for Receive-diversity (SIMO channel) with CTC

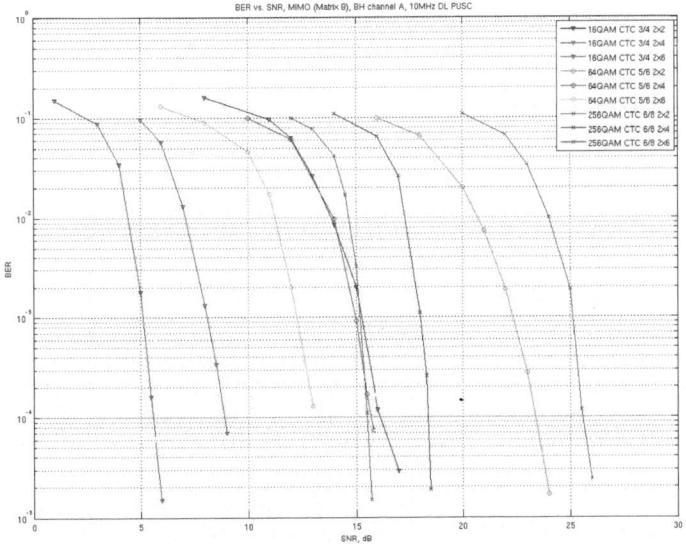

Fig. 4. Matrix-B, MIMO 2x2, BH Channel

The following Table 1 summarizes the obtained results for BER=10^{-4}.

Table 1. Several Results for different parameter combinations

QAM constellation	Coding rate	Spectral efficiency [bps/Hz]	Antenna configuration Tx/Rx	Required SNR [dB]
16	3/4	6	2x2	16.5
			2x4	9
			2x6	6
64	5/6	10	2x2	23
			2x4	16
			2x6	13
256	6/8	12	2x2	25.5
			2x4	18
			2x6	15.5

Note that very high operating rates (spectral efficiencies) can be obtained in the backhaul link using smart antenna array configurations. It is apparent that increasing the number of receive antennas from 2 to 4 provides a gain of about 7dB, while increasing the number to 6 receive antennas provides additional gain of about 3dB.

4 Conclusion

The present work evaluates, by simulation, the link-level performance for various coding and modulation schemes with different antenna configurations, with the aim to develop an innovative RS product on the basis of the IEEE 802.16j MMR specifications. Several practical links have been taken into account ([25], [26]), focusing on: WiMAX-compliant modem configurations and reference channel models; enhanced backhaul configurations and channel models.

Multiple receive antennas employing maximum ratio combining (MRC) for a single transmit stream offer substantial diversity gain compared to a single receive antenna when using the ITU pedestrian B channel model. When using a 2x2 antenna configuration over the ITU pedestrian B channel, Matrix B offers no SNR gains as compared to Matrix A when the spectral efficiency and BER/BLER conditions are the same for both schemes.

Considering the backhaul link, denoted backhaul Type A (SUI-1), using 2 streams and 2 transmit antennas, spectral efficiencies of up to 12 bps/Hz are achievable with as little as 15.5dB SNR when the receiver employs 6 receiving antennas (25.5dB SNR is required to achieve the same spectral efficiency with only 2 receive antennas).

Acknowledgments. The present work has been performed in the scope of the RE-WIND ("RElay based WIreless Network and StandarD") European Research Project and has been supported by the Commission of the European Communities - *Information Society and Media Directorate General* (FP7, Collaborative Project, *ICT-The Network of the Future*, Grant Agreement No.216751).

References

1. Tse, D., Viswanath, P.: Fundamentals of Wireless Communication. Cambridge Press (2005)
2. Rappaport, T.S.: Wireless Communications: Principles and Practice, 2nd edn. Prentice Hall, Upper Saddle River (2002)
3. IEEE 802.16j-06/026r4, IEEE Standard for Local and metropolitan area networks. Part 16: Air Interface for Fixed and Mobile Broadband Wireless Access Systems (June 2007), http://www.ieee802.org/16/relay/docs/80216j-06_026r4.zip
4. Pabst, R., Walke, B., Schultz, D.C., et al.: Relay-Based Deployment Concepts for Wireless and Mobile Broadband Radio. IEEE Communications Magazine 42(9), 80–89 (2004)
5. Hoymann, C., Klagges, K., Schinnenburg, M.: Multihop Communication in Relay Enhanced IEEE 802.16 Networks. In: 17th Annual IEEE International Symposium on Personal, Indoor and Mobile Radio Communications, Helsinki, Finland (2006)
6. Yu, Y., Murphy, S., Murphy, L.: Planning Base Station and Relay Station Locations, in IEEE 802.16j Multi-hop Relay Networks. In: IEEE International Conference on Communications (ICC), May 19-23, pp. 2586–2591 (2008)
7. Soldani, D., Dixit, S.: Wireless Relays for Broadband Access. IEEE Communications Magazine, 58–66 (2008)
8. Ertel, R.B., Cardieri, P., Sowerby, K.W., Rappaport, T.S., Reed, J.H.: Overview of Spatial Channel Models for Antenna Array Communication Systems. IEEE Personal Communications 5(1), 10–22 (1998)
9. Yu, K., Ottersten, B.: Models for MIMO Propagation Channels, a Review - Special Issue on Adaptive Antennas and MIMO Systems. Wiley Journal on Wireless Communications and Mobile Computing 2(7), 553–666 (2002)
10. Fleury, B., Yin, X., Kocian, A.: Impact of the Propagation Conditions on the Properties of MIMO Channels. In: International Conference on Electromagnetics in Advanced Applications 2003 (ICEAA 2003), Turin, Italy, September 8-12 (2003)
11. Gesbert, D., Bolcskei, H., Gore, D.A., Paulraj, A.J.: Outdoor MIMO Wireless Channels: Models and Performance Prediction. IEEE Trans. on Commun. 50(12), 1926–1934 (2002)
12. Xu, H., Chizhik, D., Huang, H., Valenzuela, R.: A Generalized Space-Time Multiple-Input Multiple-Output (MIMO) Channel Model. IEEE Trans. on Wireless Commun. 3(3), 96–975 (2004)
13. Meinilä, J., Jämsä, T., Kyösti, P., Laselva, D., El-Sallabi, H., Salo, J., Schneider, C., Baum, D.: IST-2003-507581 WINNER - Deliverable D5.2: Determination of Propagation Scenarios (July 2004), http://www.ist-winner.org/Documents/Deliverables/D5-2.pdf
14. Abaii, M., Auer, G., Cho, W., Cosovitch, I., et al.: IST-4-027756-WINNER II, Deliverable D6.13.7 v1.00: Test Scenarios and Calibration Cases Issue 2 (December 2006), http://www.signal.uu.se/Publications/WINNER/WIN2D6137.pdf
15. Brüninghaus, K., Astely, D., Sälzer, T., et al.: Link Performance Models for System Level Simulations for Broadband Radio Access Systems. In: 16th IEEE International Symposium on Personal Indoor and Mobile Radio Communications (PIMRC 2005), Berlin, September 11-14 (2005)
16. Hanzo, L., Munster, M., Choi, B.J., Keller, T.: OFDM and MC-CDMA for Broadband Multi-User Communications. In: WLANs and Broadcasting. John Wiley & Sons, Chichester (2008)

17. Valkama, M., Renfors, M., Koivunen, V.: Advanced Methods for I/Q Imbalance Compensation in Communication Receivers. IEEE Trans. on Signal Processing 49(10), 2335–2344 (2001)
18. Foschini, G.J., Gans, M.J.: On Limits of Wireless Communications in a Fading Environment. Wireless Personal Communications 6(3), 311–335 (1998)
19. Sayana, K., Zhuang, J.: Link Performance Abstraction based on Mean Mutual Information per bit (MMIB) of the LLR Channel. IEEE 802.16 BWA WG, C802.16m-07/097 (2007)
20. Pauluzzi, D.R., Beaulieu, N.C.: A comparison of SNR Estimation Techniques for the AWGN Channel. IEEE Trans. on Commun. 48(10), 1681–1691 (2000)
21. International Telecommunications Union - Radiocommunication Sector (ITU-R): Recommendation ITU-R M.1225 - Guidelines for evaluation of radio transmission technologies for IMT-2000 (February 1997), http://www.itu.int/rec/R-REC-M.1225/en
22. Erceg, V., Hari, K.V.S., Smith, M.S., Baum, D.S., et al.: IEEE 802.16.3c-01/29r4: Channel Models for Fixed Wireless Applications. IEEE 802.16 Broadband Wireless Access Working Group (January 2001)
23. Raleigh, G.G., Cioffi, J.M.: Spatio-Temporal Coding. IEEE Trans. on Commun. 46(3), 357–366 (1998)
24. Rappaport, T.S., Seidel, S.Y., Takamizawa, K.: Statistical Channel Impulse Response Models for Factory and Open Plan Building Radio Communication System Design. IEEE Trans. on Commun 39(5), 794–807 (1991)
25. Andrews, J.G., Ghosh, A., Muhamed, R.: Fundamental of WiMAX - Understanding Broadband Wireless Networking. Prentice-Hall, Englewood Cliffs (2007)
26. Okuda, M., Zhu, C., Viorel, D.: Multihop Relay Extension for WiMAX Networks - Overview and Benefits of IEEE 802.16j standard. Fujitsu Sci. Tech. J. 44(3), 292–302 (2008)

A Face Centered Cubic Key Agreement Mechanism for Mobile Ad Hoc Networks[*]

Ioannis G. Askoxylakis[1], Konstantinos Markantonakis[2], Theo Tryfonas[3],
John May[3], and Apostolos Traganitis[1]

[1] Foundation for Reserach and Technology-Hellas – Institute of Computer Science,
N. Plastira 100, 70013 Heraklion, Greece
{asko,tragani}@ics.forth.gr
[2] Royal Holloway University of London,
Information Security Group, UK
K.Markantonakis@rhul.ac.uk
[3] University of Bristol, Faculty of Engineering,
Queen's Building
University Walk, Clifton, Bristol, BS8 1TR
{t.tryfonas,j.may}@bristol.ac.uk

Abstract. Mobile ad hoc networking is an operating mode for rapid mobile node networking. Each node relies on adjacent nodes in order to achieve and maintain connectivity and functionality. Security is considered among the main issues for the successful deployment of mobile ad hoc networks (MANETs). In this paper we introduce a weak to strong authentication mechanism associated with a multiparty contributory key establishment method. The latter is designed for MANETs with dynamic changing topologies, due to continuous flow of incoming and departing nodes. We introduce a new cube algorithm based on the face-centered cubic (FCC) structure. The proposed architecture employs elliptic curve cryptography, which is considered more efficient for thin clients where processing power and energy consumption are significant constraints.

Keywords: MANET security, password authentication, elliptic curve cryptography, face-centered cubic (FCC) structure.

1 Introduction

Security is a primary concern for providing protected communications to mobile nodes that operate in hostile environments. Unlike the wireline networks or the mobile networks with hierarchical architecture like cellular networks, the unique nature and characteristics of Mobile Ad-hoc Networks (MANETs) pose a number of nontrivial challenges to security design, architecture and services. In MANETs nodes rely on each other in order to achieve and maintain connectivity and functionality.

---

[*] This work was supported in part by the European Commission in the 7th Framework Programme through project EU-MESH (Enhanced, Ubiquitous, and Dependable Broadband Access using MESH Networks), ICT-215320, http://www.eu-mesh.eu

F. Granelli et al. (Eds.): MOBILIGHT 2009, LNICST 13, pp. 103–113, 2009.

A MANET is a type of network, which is typically composed of equal mobile hosts that we call nodes. When the nodes are located within the same radio range, they can communicate directly with each other using wireless links. This direct communication is employed without hierarchical control. The absence of central control, such as base stations, introduces several problems, such as configuration advertising, discovery, maintenance, as well as ad hoc addressing, self-routing and security [1].

In this environment, trust cannot be provided among the nodes of the network without the existence of initial specific prior known information. This special kind of information is necessary in order to build trust between all participating nodes. An ad hoc network is established among the existing nodes, if from preexisting, commonly known information, we reach a state where a common Session Key is agreed among the nodes. Securing ad hoc networks is not trivial, mainly due to their dynamic topology and the vulnerability of the wireless links, which can be the medium for passive and active attacks.

In emergency situations or military operations the need for establishing a wireless network quickly and securely is crucial. The objective is to interconnect all computing and communication devices in a way that they will be able to share all necessary information securely, since nobody can guarantee that the "high tech" enemies will not try to disrupt or intercept the operation efforts.

The technical goal is to make sure that no other entity outside the *group* (*we define all the legitimate members of the established wireless network as group, e.g., soldiers of a military unit*) should be able to gain access within the new network. However, since neither a certification authority nor a secure communication channel exists, the enemy has the ability to eavesdrop and modify exchanged messages transmitted over the air. Additionally, since no central identification authority is present, group member impersonation is easy, jeopardizing the security of the whole system.

Considering all these issues, the main challenge that arises is the setting up of a wireless network where the legitimate members of a group will be able to establish a protected wireless network. Moreover, in the case where a new node arrives at place, desiring to become a member in an already established group, joining, without delaying or even intercepting the existing group, is also challenging. The case where a group member is captured by the enemy and therefore the group key is compromised is also part of the considered scenario.

2 Security Requirements

It is broadly known that security mechanisms cannot create trust [2]. The members of a team that wish to establish a MANET know and trust one another physically. Otherwise, they would never be able to achieve mutual trust regardless of the authentication mechanism used. Our goal is to exploit the existing physical mutual trust in order to secure the ad hoc network.

A password authentication mechanism seems to be a rational approach that can deliver a proper solution without adding new requirements like the use of dedicated hardware (i.e smart cards). In a password based authentication scheme the use of a sufficiently large and randomly generated data string that can be used as a password would be an obvious approach. This way all nodes could agree on a password and, by using a trivial authentication protocol, achieve mutual authentication.

In such a scenario, the underlying security depends on the size and the randomness of the chosen password. However, the larger the password gets the more difficult it is to memorize and use. Moreover, since the response time is vital during emergency operations, the use of large passwords can be proved inconvenient. Therefore the use of short, user-friendly passwords is an essential requirement.

The use of short passwords provides weak authentication since the password selection set is quite limited and thus the corresponding authentication procedure is vulnerable to dictionary attacks [3]. Therefore, we need an authentication protocol that will lead to a reasonable degree of security even if the authentication procedure has been initiated from a small, weak password.

Security threats can classified into two broad categories depending on their origin: external and internal attacks. External attacks originate outside the group while internal originate from already authenticated nodes belonging to the group. For instance, consider a group of soldiers operating in a hostile environment, trying to keep their presence and mission unknown to the enemy, and the case where a soldier, member of the mission group, is captured by the enemy who is now in a position to attack from inside. Another example, less extreme, is an ad hoc network formed in a classroom during a test exam between the laptops or PDAs of the students and the teacher's workstation. According to this scenario, not only we must secure the network from an external intruder but also from a student who temporarily exits the classroom in order to retrieve the solutions and then returns. In all cases, the misbehaving nodes must definitely be expelled from the established network.

At this stage it makes sense to outline the main security requirements of the proposed architecture:

Weak-to-strong password-based authentication. Use of an authentication scheme that will lead to a reasonable degree of security although the authentication procedure has been initiated from a small, weak password.

Secure authentication. Only the entities that hold the correct password will eventually become members of the MANET.

Forward authentication. Even if a malicious partner manages to compromise a network entity in a later phase, he will still be unable to participate in the already existing network.

Contributory key establishment. The MANET is established when a session key is generated and agreed among all network nodes. The session key should be generated throughout in a contributory manner, by all participating entities.

Security architecture for thin clients. A MANET is typically composed of mobile devices with limited processing power and energy consumption. The cryptographic algorithms used for authentication and key agreement should have minimal impact in terms of computational overhead.

The rest of the paper is organized as follows: In Section 3, we start with a review of the previous work concerning two-party and multiparty key agreements and we give a brief introduction on weak to strong authentication and the elliptic curve theory. We describe the state of the art in multiparty key agreement protocols and particularly the d-cube and the body centered cubic algorithms and examine their properties. In Section 4, we propose a modification of the body-centered cubic algorithm, called face centered cubic algorithm designed for the dynamic changing topologies and we compare them. A discussion concerning the implementation issues and the problems that arise is presented in Section 5. Finally, in Section 6, we provide our concluding remarks along with suggestions for future work.

3 Related Work

3.1 Key Exchange and Elliptic Curve Cryptography

Common cryptographic protocols based on keys chosen by the users are weak to dictionary attacks. Bellowin and Merrit [4] proposed a protocol called *encrypted key exchange (EKE)* where a strong shared key is derived from a weak one. However, this protocol has a disadvantage. The creation of the common session key takes place with unilateral prospective, that is, only by the entity that first initiated the whole procedure. Thus the key agreement scheme is not contributory. In [5], Asokan and Ginzboorg proposed a contributory version of the above protocol for both two-party and multiparty cases.

Diffie–Hellman is the first public key distribution protocol that opened new directions in cryptography [5]. In this important protocol for key distribution, two entities *A, B* after having agreed on a prime number p and a generator g of the multiplicative group Zp, can generate a secret session key.

An essential property for the majority of cryptographic applications is the need for fast and precise arithmetic. Calculations over the set of real numbers are slow and inaccurate due to round-off error [6]. Finite arithmetic groups, such as

$$F_p, F_{2^m}.$$

which have a finite number of points, is used in practice. All practical public-key systems today exploit the properties of arithmetic using large finite groups. Additionally, elliptic curves can provide versions of public-key methods that, in some cases, are faster and use smaller keys, while providing an equivalent level of security. Consequently, the use of ECC can result in faster computations, lower power consumption, as well as memory and bandwidth savings. This is very useful for mobile devices, like the ones used in ad hoc networks, which face limitation in terms of CPU, power, and network connectivity.

An elliptic curve [7] consists of elements (x, y) satisfying the equation:

$$y^2 = x^3 + \alpha x + \beta (\mathrm{mod} p). \tag{1}$$

for two numbers α, β. If (x, y) satisfies the above equation then $P = (x, y)$ is a point on the elliptic curve.

The elliptic curve discrete logarithm problem (ECDLP) can be stated as follows:

Fix a prime p and an elliptic curve E. Let xP represent the point P added to itself x times. Suppose Q is a multiple of P, so that $Q = xP$ for some x, then the ECDLP is to determine x given P and Q.

The general conclusion of leading cryptographers is that the ECDLP requires fully exponential time to solve. The security of ECC is dependent on the difficulty of solving the ECDLP.

Research community has given considerable attention to the ECDLP. Like the other types of cryptographic problems, no efficient algorithm is known to solve the ECDLP. The ECDLP seems to be particularly harder to solve. Moderate security can be achieved with the ECC using an elliptic curve defined over Z_p where the prime p is several times shorter than 230 decimal digits.

An elliptic curve cryptosystem implemented over a 160-bit field currently offers roughly the same resistance to attack, as would a 1024-bit RSA [8]. However, there have been weak classes of elliptic curves identified such as super singular elliptic curves [9] and some anomalous elliptic curves [10]. Implementations, such as ECDSA [11], merely check for weaknesses and eliminate any possibility of using these "weak" curves [12].

3.2 Elliptic Curve Diffie–Hellman

The original Diffie–Hellman (D-H) algorithm is based on the multiplicative group modulo p. However the elliptic curve Diffie–Hellman (ECDH) protocol is based on the additive elliptic curve group as desribed below. We assume that two entities A, B have selected the underlying field, $GF(p)$ or $GF(2^k)$, the elliptic curve E with parameters a, b, and the base point P. The order of the base point P is equal to n. Also, we ensure that the selected elliptic curve has a prime order to comply with the appropriate security standards [11].

At the end of the protocol, the communicating parties end up with the same value K, which represents a unique point on the curve. A part of this value can be used as a secret key to a secret-key encryption algorithm. We give a brief description of the protocol.

Entity A selects an integer,

$$d_A : d_A \in [2, n-2] . \tag{2}$$

Entity B selects an integer

$$d_B : d_B \in [2, n-2]. \tag{3}$$

A computes

$$Q_A = d_A \times P . \tag{4}$$

The pair Q_A, d_A consists A's public and private key.

B computes

$$Q_B = d_B \times P.$$ (5)

The pair Q_B, d_B consists B's public and private key.

A sends Q_A to B,

$$A : Q_A \to B.$$ (6)

B sends Q_B to A,

$$B : Q_B \to A.$$ (7)

A computes

$$K = d_A \times Q_B = d_A \times d_B \times P.$$ (8)

B computes

$$K = d_B \times Q_A = d_B \times d_A \times P.$$ (9)

Quantity K is now the commonly shared key between A and B. Moreover, it can also be used as a session key. Quantity n is the order of the base point P.

3.3 D-Cube Protocols and Aggressive 3-D Cube Algorithm

For key establishment procedures in mobile ad hoc networks, where several entities are involved, multiparty authentication protocols should be applied. A lot of research has been done in this direction [13], [14]. Becker and Wille [15] presented a method very efficient in terms of number of authentication rounds. According to this method, also known as the d-cube protocol, all entities planning to participate in a network are initially arranged in a d-dimensional hypercube. Each potential network entity is represented as a vertex in the d-dimensional cube and it is uniquely assigned a d-bit address. The addresses are assigned in a way so that two vertices connected along the $i^t h$ dimension differ only in the $i^t h$ bit. There are 2^d vertices, each of which are connected to other d vertices.

In [17], a modified version of [16] called aggressive d-cube algorithm is presented, where faulty nodes are isolated from the ad hoc network during the early stages of the d-cube algorithm. According to the algorithm, the interaction of faulty-legitimate nodes and the chances a faulty node will enter the network by guessing the password are minimized. Moreover, their protocol protects legitimate nodes from unnecessary energy spending, which may be more important in case of thin clients.

To clearly demonstrate the differences between [17] and [16], we describe the algorithm of [17] through examples in 3-d case. In this case we assume that node G is the faulty partner. During the first round the DH key exchange procedure performed between (G:110) and (H:111) will fail, since node (G:110) is a faulty one. However, instead of remaining idle and wait for the next round (as in [16]), node (H:111) starts a DH key exchange with node (E:100). Meanwhile Node (E:100) has already performed a successful DH key exchange with (F:101), during the first half of the first round, so this key exchange will be the second successful one for this round. Node

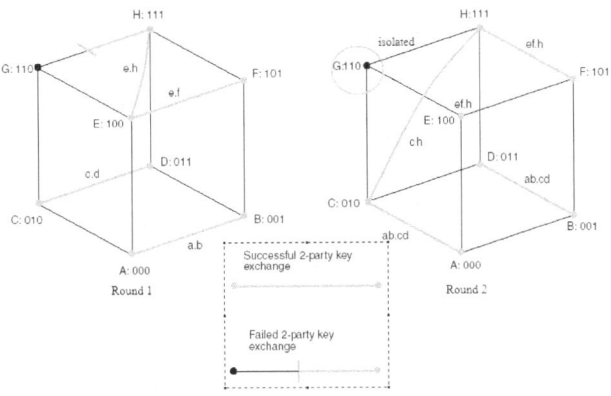

Fig. 1. Aggressive 3-d cube round 1 and 2

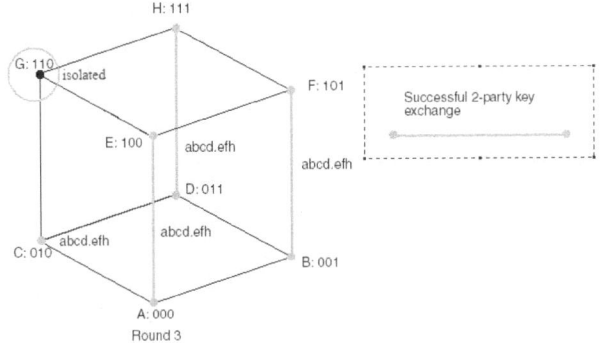

Fig. 2. Aggressive 3-d cube round 3

(E:100) having being notified by H that G is a faulty node will remain idle until the third round, instead of having attempted unnecessary DH exchanges with (G:110). In the next round (round 2) (H:111) performs a DH with node (F:101) and a DH with node (C:010). Given that (C:010) has performed two successful DH with (D:011) and (H:111) respectively, he will remain idle in the next round. However (C:010) has already performed a successful DH with (A:000), during round one.

In total node (C:010) has performed three successful DH, with three different nodes, which means that (C:010) has completed all the appropriate procedures. Thus it will remain idle for the next round, which is the last round in our case. Summarizing the description of this procedure, the upper bound of the total successful DH procedures for a node participating in an aggressive d-cube algorithm is equal to d. In this example $d = 3$. During the third and final round there will be three more successfully accomplished DH key exchanges. One between (H:111) and (D:011), one between (F:101) and (B:001), and one between (A:000) and (F:101).

Through this example it is clear that using the aggressive 3-d cube algorithm, the faulty partner is being isolated. He only participates in one DH key exchange, the one performed in round 1 with node (H:111), and since then he is excluded from all the subsequent DH key exchanges. Consequently, the faulty node loses the ability to have another change, during the generation process of the common session key.

4 The Proposed Architecture

The dynamic topology of mobile ad-hoc networks introduces challenging security issues. The continuous flow of incoming and departing nodes is a key issue for designing a key agreement mechanism. Furthermore, when a node publicly claims that it is leaving the network it does not mean that it looses its ability to "hear" the messages exchanged among the remaining nodes, unless action is taken.

We propose a cryptographic key agreement algorithm that initiates from an aggressive 2-d or 3-d algorithm. The proposed method is a modification of [18], however it provides a completely different solution.

In the case of more than 8 nodes, instead of moving to a higher degree of space (4-d or more) we exploit the face centered architecture by arranging the next 6 new nodes on the centers of the 6 faces of the 3-d cube. For simplicity, in the rest of the paper, each bond in the 3-d space corresponds to a two-party, password based, elliptic curve, Diffie-Hellman key exchange as described in chapter 3.

4.1 The Face Centered Cubic (FCC) Algorithm

The proposed algorithm is depicted in figure 3. The first 8 nodes (or less) are arranged in a 3-d cube as shown in the left side of figure 3. They perform an aggressive 3-d cube algorithm and obtain a common session key. The first 6 nodes that will arrive in a later phase will be arranged in the centers of the six faces of the cube as shown in the central picture of figure 3.

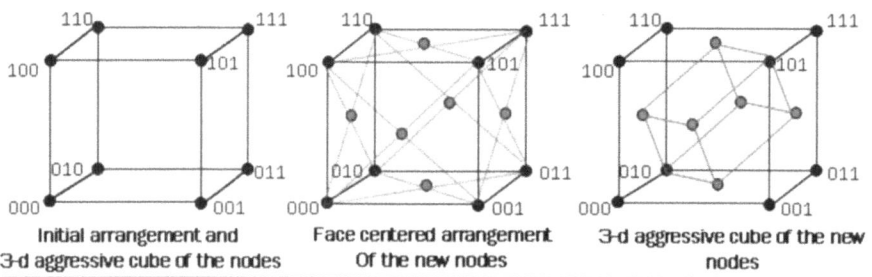

| Initial arrangement and 3-d aggressive cube of the nodes | Face centered arrangement Of the new nodes | 3-d aggressive cube of the new nodes |

Fig. 3. The Face Centered Cubic algorithm

The 6 new nodes together with nodes (010) and (101) that contributed to the initial cube, create a new cube and perform a new (second) aggressive 3-d cube algorithm. This way the inner cube creates a second common session key. After the set up of the second session key nodes (010) and (101) hold both session keys corresponding to both cubes. This privilege makes nodes (010) and (101) leading nodes for the established network since any communication between black and grey nodes should pass through them. If we wish to avoid this hierarchy in our network, during the set up of the common session key within the inner cube, nodes (010) and (101) propagate the common session key of the initial (black) cube to the new nodes. This way the first session key can be used by all nodes to communicate securely with each other, while the second can be used for the secure communication of the internal (grey) cube.

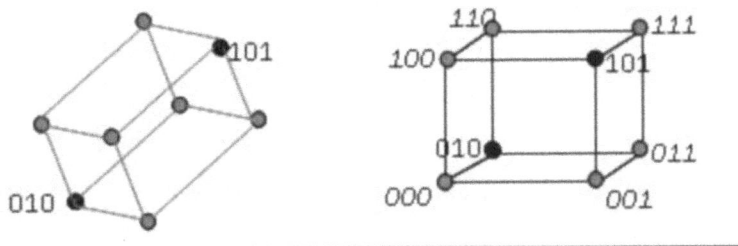

Fig. 4. Addressing of the new nodes

The addressing of the new nodes is shown in figure 4. We observe that the 2 old (black) nodes keep the same address with the one they had during the set up of the first session key. For the communication between nodes belonging to different cubes there is a separate metric (cube number) declaring the cube that the node is belonging to. In this example, black nodes are identified as cube 1 nodes, grey nodes are identified as cube 2 nodes and nodes (010) and (101) have both identifiers since they belong to both cubes.

4.2 Key Refreshment due to Departing Nodes

In the hierarchical model, where every cube has its independent session key, key refreshment due to departing nodes is easy. As soon as a node is leaving the network, the rest nodes of the common cube, perform a new aggressive 3-d cube algorithm and create a new session key. In case the leaving node is belonging to two consecutive cubes, a new aggressive 3-d cube algorithm is performed automatically to both cubes.

In the case where the previous session keys belonging to previous cubes are forwardly distributed to the next cubes, the key renewal should be performed to all previous cubes. This appears not to be a desired feature, since if there is a departure in the last cube all previous stages/cubes will be affected. However this can be also avoided if the set up is a combination of the two solutions. Periodically the key forwarding method is interrupted by the hierarchical solution. This way, we create isolated groups of concatenated cubes and any necessary key refreshment is bounded within these groups.

5 Conclusion

Our research was motivated from the requirement of certain groups to establish fast, reliable, efficient and secure MANET's without relying on pre-existing infrastructures. The actual operational environment and the very nature of the established networks impose further key issues (e.g. the ability to add or subtract nodes depending on operational and security considerations) that need to be taken into account.

We have reviewed existing proposals around two-party or multiparty authentication and introduced a new key establishment method. Our proposal overcomes some of the main issues (such as rapid deployment, accuracy, and dynamic and robust behaviour) of existing solutions and operational environments. The proposed solution introduces the use of elliptic curve cryptography in such a scenario. ECC computations require less storage, less power, less memory, and less bandwidth than other systems. This allows implementation of cryptography in constrained platforms such as wireless devices, handheld computers, smart cards, and thin-clients. For a given security level, elliptic curve cryptography raises computational speed and this is important in ad hoc networks, where the majority of the clients have limited resources.

We have also described known protocols for password authenticated multiparty DH key exchange and have chosen the aggressive cube algorithm due to its resilience against dictionary attacks. The proposed protocol meets all security requirements according the initial specification and it is stronger in terms of security. Finally, we have proposed a security architecture for dynamic MANETs, where the composition of the network changes in time with the arrivals and departures of nodes. The secure dynamic recomposition of the network could be proved very useful in battlefields where a soldier, under threat of capture, signs off the network on time.

The proposed FCC algorithm can be applicable in several other scenarios such as emergency situations, where rescue workers arrive at a disaster field, or for groups of people meeting in a room, i.e., in a classroom together with the teacher, etc. The password-based feature of our work could be used in cases where a group of people meets one another in person for the first time, and would like to go back home and set up a secure network among them.

The proposed algorithm leaves several open issues for future work. Formal analysis is necessary. The incorporation of several new password-based key agreement protocols, which do not require the use of asymmetric encryption, is a challenging consideration. The case where the number of network entities fluctuates unevenly, changing the network topology rapidly, is also very interesting.

References

1. Verikoukis, C., Alonso, L., Giamalis, T.: Cross-Layer Optimization for Wireless Systems: A European Research Key Challenge. IEEE Communications Magazine 43(7), 1–3 (2005)
2. Bonnefoi, P.-F., Sauveron, D., Park, J.H.: MANETS: an exclusive choice between use and security? Special Issue on Interactive Multimedia & Intelligent Services in Mobile and Ubiquitous Computing (MUC) of COMPUTING AND INFORMATICS 27(5) (2008)
3. Narayanan, A., Shmatikov, V.: Fast dictionary attacks on passwords using time-space trade-off Conference on Computer and Communications Security. In: Proceedings of the 12th ACM conference on Computer and communications security, Alexandria VA USA (2005)
4. Bellovin, S.M., Merrit, M.: Encrypted key exchange: Password based protocols secure against dictionary attacks. In: Proceedings of the IEEE Symposium on Research in Security and Privacy, Oakland, USA (May 1992)
5. Diffie, W., Hellman, M.E.: New directions in cryptography. IEEE Transactions on Information Theory 22, 644–654 (1976)

6. Cucker, F., Smale, S.: Complexity estimates depending on condition and round off error. Journal of the Association for Computing Machinery 46(1), 113–184 (2000)
7. Koblitz, N.: Elliptic curve cryptosystems. Mathematics of Computation 4(8), 203–209 (1987)
8. Rivest, R., Shamir, A., Adleman, L.M.: A Method for Obtaining Digital Signatures and Public-Key Cryptosystems. Communications of the ACM 21(2), 120–126 (1978)
9. Menezes, A., Okamoto, T., Vanstone, S.: Reducing elliptic curve logarithms to logarithms in a finite field. IEEE Transactions on Information Theory 39, 1639–1646 (1993)
10. Menezes, A., Teske, E., Weng, A.: Weak Fields for ECC. In: Okamoto, T. (ed.) CT-RSA 2004. LNCS, vol. 2964, pp. 366–386. Springer, Heidelberg (2004)
11. Johnson, D., Vanstone, S.: The elliptic curve digital signature algorithm (ECDSA). International Journal on Information Security 1, 36–63 (2001)
12. Kalele, A.A., Sule, V.R.: Weak keys of pairing based Diffie-Hellman schemes on elliptic curves. Cryptology ePrint Archive 2005/30 (2005)
13. Zheng, D., Chen, K., You, J.: Multiparty authentication services and key agreement protocols with semi-trusted third party. Journal of Computer Science and Technology archive 17(6), 749–756 (2002)
14. Ateniese, G., Steiner, M., Tsudik, G.: New Multiparty Authentication Services and Key Agreement Protocols. IEEE Journal of Selected Areas in Communications 18(4) (April 2000)
15. Becker, C., Wille, U.: Communication complexity of group key distribution. In: 5th ACM Conference on Computer and Communications Security, San Francisco, California (November 1998)
16. Asokan, N., Ginzboorg, P.: Key agreement in ad hoc networks. Computer Communications 23, 1627–1637 (2000)
17. Askoxylakis, I.G., Kastanis, D.D., Traganitis, A.P.: Elliptic curve and password based dynamic key agreement in wireless ad-hoc networks, Communications. In: Networks and Information Security CNIS 2006, Cambridge, USA (October 2006)
18. Askoxylakis, I.G., Sauveron, D., Markantonakis, K., Tryfonas, T., Traganitis, A.: A Body-Centered Cubic Method for Key Agreement in Dynamic Mobile Ad Hoc Networks. In: Second International Conference on Emerging Security Information, Systems and Technologies, Cap Esterel, France, August 25-29, pp. 193–202 (2008)

A Tree Based Self-routing Scheme for Mobility Support in Wireless Sensor Networks

Young-Duk Kim[1], Yeon-Mo Yang[2], Won-Seok Kang[1], Jin-Wook Kim[1], and Jinung An[1,*]

[1] Mobile Robot Lab., Daegu Gyeongbuk Institute of Science and Technology
{ydkim,wskang,jwkim,robot}@dgist.ac.kr
[2] Dept. of Electronics Engineering, Kumoh National Institute of Technology
yangym@kumoh.ac.kr

Abstract. Recently, WSNs (Wireless Sensor Networks) with mobile robot is a growing technology that offer efficient communication services for anytime and anywhere applications. However, the tiny sensor node has very limited network resources due to its low battery power, low data rate, node mobility, and channel interference constraint between neighbors. Thus, in this paper, we proposed a tree based self-routing protocol for autonomous mobile robots based on beacon mode and implemented in real test-bed environments. The proposed scheme offers beacon based real-time scheduling for reliable association process between parent and child nodes. In addition, it supports smooth handover procedure by reducing flooding overhead of control packets. Throughout the performance evaluation by using a real test-bed system and simulation, we illustrate that our proposed scheme demonstrates promising performance for wireless sensor networks with mobile robots.

Keywords: Wireless Sensor Networks, Handover, Self-routing, Mobile Robots.

1 Introduction

Recently, WSN (Wireless Sensor Network) [1] and mobile robot technology are the one of the most popular technologies for realization of ubiquitous networks. WSN can be widely used such as military, medical and industrial purpose. However, when we deploy WSN in multi-hop environments, a number of open problems can be observed because of limited bandwidth capacity and significant packet collisions by channel interference, and so on. In order to tackle these problems, [5] [6] [7] are proposed with BOP (Beacon Only Period) and LAA (Last Address Assignment) mechanisms. However, they do neither consider nodes mobility nor smooth route recovery mechanisms during the communication session. In addition, when the network traffic is significantly congested, existing schemes suffer severe packet collisions between beacons and other control packets. In this paper, we designed and

---

* Corresponding author.

F. Granelli et al. (Eds.): MOBILIGHT 2009, LNICST 13, pp. 114–124, 2009.
© ICST Institute for Computer Sciences, Social-Informatics and Telecommunication Engineering 2009

developed an efficient wireless sensor network system with autonomous mobile robot for smooth mobility support. In order to reduce the handover overhead and the latency of mobile robots which role mobile nodes in WSNs, we propose a tree based self-routing scheme. All sensor nodes and sink nodes are implemented on the TinyOS [2] system which is based on NesC [3]. In addition, we also developed a monitoring system which is able to collect and process data packets from every sensor node including the mobile robot.

The rest of this paper is organized as the follows. In Section 2, we review TinyOS architecture for our operating system platform and IEEE 802.15.4 MAC protocol as well as its improved versions with beacon scheduling. In Section 3, we illustrate the detail design architecture and implementation issues of our tree based routing scheme. Performance evaluation by real test-bed and simulation study is presented in Section 4. Finally, concluding remarks with future works are given in Section 5.

2   Related Works

2.1   TinyOS

TinyOS is developed in U.C. Berkeley and designed for exclusive operating system in wireless sensor networks. Most applications based TinyOS can be compiled into very tiny volume under 30Kbytes which is the optimal size for general specifications of wireless sensor nodes such as a small hardware device, small memory size, low CPU performance, and limited wireless channel resources. In addition, since TinyOS excludes unnecessary libraries and components, it can reduce extra overhead of the source code and produce minimum sized programs. For the more convenient application development, TinyOS is written in NesC language which is a component based architecture. The components of each application are connected to each other by using interfaces during the compile procedure. Although the grammar of NesC is similar to traditional C language, there are several differences between them such as types, development scheme, code size and etc. The other main features of NesC are as follows. It offers very convenient environment for application programming and the final code size is small enough to install on tiny sensor motes. However, NesC does not support a dynamic memory allocation mechanism, which may disturb intelligent computing processes.

In order to support simple scheduling service, the process of TinyOS defines a 2-level scheduling scheme which consists of tasks and events. The task is a process which is used for computing operations and procedure call operations. All tasks run in a FIFO (First In First Out) queue. When all tasks in the queue finish their processing, they minimize the CPU power consumption to reduce limited energy until other tasks are activated. Although a task is not able to be preempted by other tasks, it is able to be preempted by events. The event is a kind of process which has higher priority than task and is invoked usually when a hardware interrupt occurs or certain conditions are satisfied. When an event is produced by the interrupt, the related component is called and the wiring component in upper layer is also called in succession if it is connected to each other. At the same time, the related functions are transformed into tasks and stored in the FIFO queue.

2.2 IEEE 802.15.4 and Beacon Based Protocols

IEEE 802.15.4 [4] is the one of most representative protocols to support the communication between sensor nodes in wireless PANs (Personal Area Networks). In the basic mode, IEEE 802.15.4 usually operates in star network topology and requires a coordinator node to control the whole communication procedures between nodes by using beacon frames. However, it has severe limitations that it supports only 1 hop distance nodes from the coordinator, which is not suitable for multi-hop environments or multi-beacon enabled mesh networks. If we adopt the legacy IEEE 802.15.4 in the wireless mesh network with multiple paths, the network may suffer from significant performance degradation such as beacon collisions, failures of routing path and etc.

In order to tackle these limitations of IEEE 802.15.4, [5] [6] [7] proposed the BOP (Beacon Only Period) and the LAA (Last Address Assignment) algorithm for dynamic mesh networks. However, these schemes were not implemented with autonomous mobile robots and sensor nodes did not support a stable operating system such as TinyOS. Another limitation of [5] [6] [7] is that they show poor network performance because they do not solve packet collision problems between flooding packets for route discovery and beacon frames. Moreover, they do not suggest actual solution of node mobility support when the application requires seamless data services. Thus, throughout this paper, we propose an efficient network architecture for smooth mobility support with tree based the self-routing scheme.

3 Proposed Scheme

3.1 Association Process

It is necessary that each end node starts an association process to participate in PAN communication when it hears beacon frames from the coordinator. Our network scheme also uses the beacon policy like [5] [6] [7]. However, most WSNs have many-to-one communication paradigm, which means that all nodes transmit their sensing data upload direction. Thus, in order to make hierarchy architecture for efficient association, we define three node types, which are WC (Wireless PAN Coordinator), WR (Wireless Router), and WED (Wireless End Device). WC plays a role of a sink node and gateway by transmitting periodic beacon frames and collects data from WRs and WEDs. The collected data is forwarded to monitoring server for more specific processing such as management of alert message to user terminals. WR also periodically transmits beacon frames to neighbors and executes scheduling process with neighbors by exchanging beacon frames. WEDs are logically located in the end of the network and generate packets containing sensing data. Each packet of WEDs is forwarded to WC via WRs in every wakeup time of the superframe. In general IEEE 802.15.4 networks, there are only FFD (Full Function Device) and RFD (Reduced Function Device). However, in our work, we assume that FFD is able to be a not only WR but also WC. In addition, FFD can manage the PAN or make its own network without participating in other PANs.

In order to organize and synchronize the network, WC and WR transmit beacon frames to neighbors periodically. At first, one of WRs becomes WC if it does not hear any beacon frame from neighbors. Then WC starts to beacon with its network information such as beacon interval, identification, and information of its neighbors. When other WRs or WEDs try to scan the channel, they executes MLME_SCAN_request() process which is a MAC layer management entity in order to associate with parent node. In this situation, WR also can associate with another WR node and it calculates its own beacon schedule within the BOP length, which is executed by using received BTTSL (Beacon Tx Time Slot Length) information from its neighbors or parent node [7]. The channel scanning process in MAC layer is invoked by calling Network_Discovery_request() command, then each node records beacon information of accessible channels which is between 11 and 26. The scanning information is delivered to upper layer by SCAN_Confirm() function. Finally, by using MLME_SCAN_request() and MLME_SCAN_confirm(), a node transmits Assocation_request() primitive to the parent node with maximum signal strength which is derived from scanned beacon. This Assocation_request() is called by Network_Discovery_confirm() from network layer. Figure 1 illustrates the overall procedure of association.

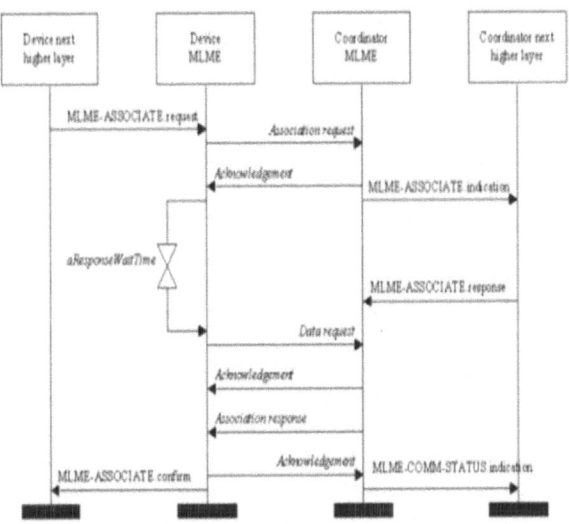

Fig. 1. Association process

3.2 Tree Based Self-routing Scheme

The traditional on demand routing protocols such as AODV (Adhoc On-demand Distance Vector routing protocol) [8] and DSR (Dynamic Source Routing protocol) [9] broadcast RREQ packets and receive RREP packets for route discovery. Even though the flooding scheme using these control packets is efficient for mobile ad hoc networks, it wastes network bandwidth and battery power in the wireless sensor network which consists of tiny sensors and motes. In addition, when relay nodes use the beacon frame for synchronization, it suffers significant packet collisions between beacons and other control packets. Moreover, when the duty cycle of each node increases, the flooding overhead also increases and it may result in network congestions. Thus, in order to reduce the control packet overhead for routing in network layer, we propose a self-routing approach by using association information between parents and child nodes in MAC layer.

When a node tries to participate in network communications, its parent node (WR) or coordinator (WC) may assign the address by using beacon frames with the LAA scheme. Therefore, after association procedure, the parent node can obtain the address of child node and the child node also can obtain the address of the parent node in tree based topology. The sharing address information between the parent and the child is stored in the simplified routing table which is described in table 1.

Table 1. An example of simplified routing table

Parent address		Child address	
Short address (16bit)	Long address (32bit)	Short address (16bit)	Long address (32bit)
7	0x0000000 000000007	8	0x0000000 000000008

Then, if a node receives incoming packets from the lower layer, it directly transmits to its parent node without using the RREQ flooding scheme. Consequently, the source node and relay node can guarantee rapid packet forwarding and reduce additional control overhead. In addition, since each node does not need to maintain and exchange the routing table information of whole network, it can not only resolve memory overhead but also accomplish self-routing.

3.3 Mobility Management and Route Recovery Process

When a node does not receive the expected beacon frame or a data packet in a certain interval, it believes that unexpected link failure or handover is taken placed in MAC layer. In this case, there are two desirable solutions to recover the routing path. The first one is that each node starts the process re-association to another parent node with our self-routing scheme mentioned in the previous section, which is simple and efficient approach from the fact that it does not require new route discovery process by using RREQ (Route Request) packet flooding. Consequently, this re-association scheme prevents unnecessary bandwidth wastes and prolongs the battery life time of each node. The other approach of path recovery is to use a route maintenance scheme by using network layer operation with RERR (Route Error) packet. Although this scheme is most common approach in mobile ad hoc networks, it significantly suffers from more route rediscovery delay and more bandwidth wastes.

Thus, we used the MAC layer re-association scheme and left the network layer approach as optional operation. After finishing the re-association procedure, MAC informs network layer with the updated route information and the node does not need to flood RREQ packets to the whole network. Since our network architecture intends to reduce the number of flooding of control packets, WR and WC execute the route discovery operation with association based tree routing. During the L2/L3 re-association procedure, WR acquires the relation information between child and parent. By using this information, the intermediate WR forwards the uplink data packet from the child node to the link of parent node. This relay process is continued until it arrives in WC. Consequently, the mobile node reduces handover latency and we can say that it is a self-routing scheme from the fact that the intermediate node does not depends on other routing information.

4 Performance Evaluation

4.1 Implementation of Test-Bed

As shown in figure 2, we developed sensor modules with CC2420 of TI Chipcon product as RF transceiver and ATMega128L as a main processor. The application was implemented for fire and atmosphere monitoring service such as temperature, gas, smoke, humidity and illumination. The gas information is classified into CO, CO_2, HCHO, SO_2, NO_2, and etc. This information is forwarded to WC in every seconds and KIP-AF (Knowledge Information Process Air/Fire data) shows the measured values in real time. If any emergent data arrive in KIP-AF, the server immediately transmits the alert message to the user terminal. For network entities, we used 1 coordinator, 8 relay WRs, and 50 WEDs with maximum 3hops. The transmission range was 30~40m and RF power control was set from 0 to -20dBm. The application used 40 bytes length packet and the duty cycle was maintained to 100%, which means all nodes transmit packets every interval in order to process real time data. All sensor nodes transmit their sensing data to coordinator and the gathered data is forwarded to KIP-AF which maintains the database for intelligent decision and further processing.

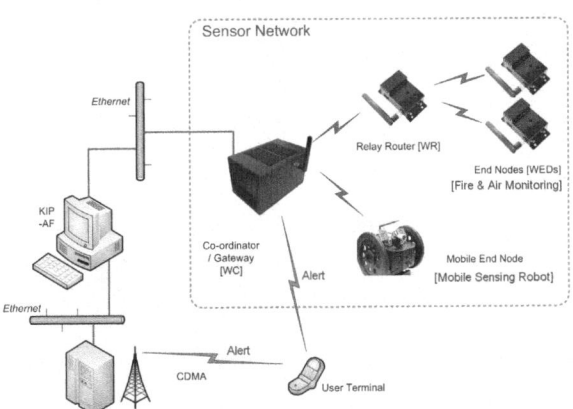

Fig. 2. Test-bed topology

For the mobility support, we used a mobile robot with sensing module which let the robot to be a mobile sensor node. The robot has 2 wheels and moved randomly with maximum 50cm/sec velocity. The robot platform is also designed and implemented on TinyOS and equipped with ATmega128L for the compatibility with sensing module. When the robot has mobility, there is an inevitable problem of link failure due to network handover or loss of LOS (Line of Sight). Then, the mobile robot tries to search another WR or WC with the best LQI value among the scanned candidates.

For the verification of reliable transmissions, we measured packet loss rate from end node to coordinator, which is logged and calculated in monitoring server. The measured results are shown in table 2 and the maximum loss rate is less than 4%. From the measurement our implemented network system is significantly reliable and the performance is well suitable for real time processing applications.

Table 2. Loss rate measurements

Performance Measurement			
Node ID	Loss Rate (%)	Node ID	Loss Rate (%)
A	3.38	L	2.94
B	1.88	M	3.71
C	2.93	N	0.91
D	2.62	O	1.91
E	2.63	P	1.97
F	1.87	Q	1.93
G	3.44	R	0.72
H	2.92	S	2.93
I	1.87	T	3.01
J	1.92	U	1.87
K	2.50	V	2.26

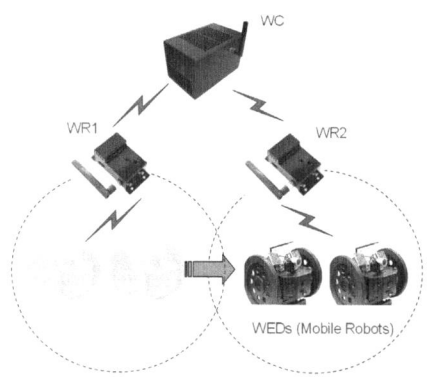

Fig. 3. Handover scenario for mobile robots

We also conducted the performance evaluation for mobility support by using "Sensor Network Analyzer (SNA)" of Daintree Networks [10], which is a commercial product for packet analysis. Figure 3 shows the route recovery scenario with mobile robots and the experiment is executed as follows. At first, we set up two WRs, named WR1 and WR2, which are associated to WC and two mobile WEDs associate with WR1. Then, WEDs move near to WR2, which means that they suffer the link failure.

After the re-association process, the mobile nodes have a new route to WC and they start to transmit data packets.

When the mobile node executes handover procedure in this experiment, the average handover latency, T_{HO}, is calculated as and figure 1 and expression (1).

$$T_{HO} = T_{beacon\_loss} + T_{asc\_req} + T_{ack} + T_{data\_req}$$
$$+ T_{ack} + T_{asc\_res} + T_{ack}$$

(1)

By using parameters as follow

T_{beacon_loss} : Interval of beacon loss due to handover
T_{asc_req} : Transmission time of association request frame
T_{asc_res} : Transmission time of association response frame
T_{data_req} : Transmission time of data request frame
T_{ack} : Transmission time of ACK frame

As shown in (1), since association duration is relatively short, the average handover latency is highly depends on beacon loss interval during the link failure. Hence, in order to minimize the beacon loss interval, we set the pending counter in MAC layer as 2. This means that after the mobile node does not hear beacon frame more than two times, it consider that the link is broken and executes the re-association procedure, immediately. In our implementation, we generated packets every second and set the beacon interval 1 sec. Thus, T_{beacon_loss} value is approximately 2 sec.

Table 3. Handover latency measurements

Trials	Handover starting time (min:sec:ms)	Handover finising time (min:sec:ms)	Handover latency (ms)
1	10:48:083	10:50:174	2,091
2	10:48:092	10:50:275	2,183
3	12:36:048	12:38:168	2,120
4	12:36:125	12:38:253	2,128
5	15:23:116	15:25:518	2,402
6	15:23:426	15:25:688	2,262

Table 3 shows the measurement results of handover latency of each trial. Around time 10:48:08 (min:sec:ms), both mobile robots which are associated to WR1, start to move to WR2. After the occurrence of route failure by handover, another route is established between WR2 and WEDs in 10:50:174 and 10:50:275, respectively. As shown in other results of trials, the average handover latency is under 2.5 sec. Thus we can say that it is possible to support mobility for communication between mobile nodes and the coordinator in wireless sensor networks.

4.2 Simulation Study

In addition to evaluation of our real test-bed, we also performed simulations to verify our proposed tree based self routing scheme comparing to original AODV protocol.

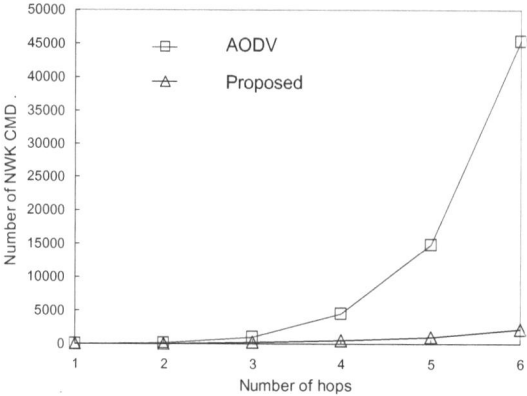

Fig. 4. The number of network commands

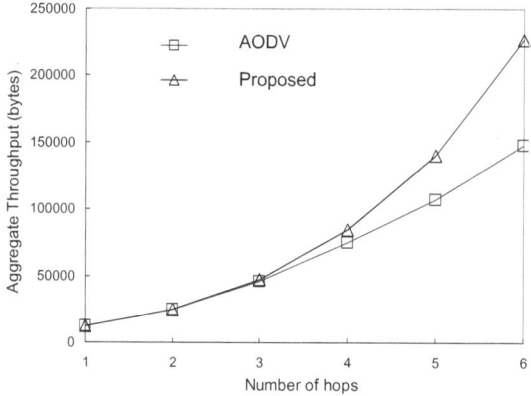

Fig. 5. Aggregated throughput

We used TOSSIM [11] with our beacon enabled MAC protocol and run the simulation for 1,000 seconds. All metrics are measured as a function of the number of hops and the network topology was the form of a perfect binary tree. For the traffic generation, we set the duty cycle at 50% with beacon order BO=8 and superframe order SO=7.

Figure 4 shows the number of network command packets as a function of the number of hops. When the topology is simple and the number of hops is smaller than 3, legacy AODV and our proposed scheme show similar performance. However, When the number of hops is higher than 4, our proposed scheme shows better performance because it does not need to flood the RREQ control packets and it only need to perform the association procedure. Hence, our proposed scheme has more opportunity to transmit data packets to neighbors with the limited wireless channel.

Figure 5 describes aggregated throughput during the simulation time. Since the intermediate node does not relay control packets such as RREQs and RREPs, it

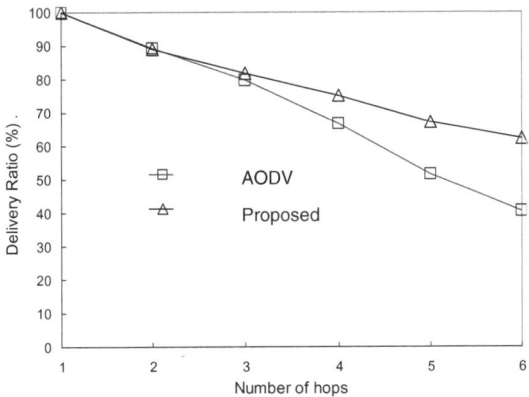

Fig. 6. Packet delivery ratio

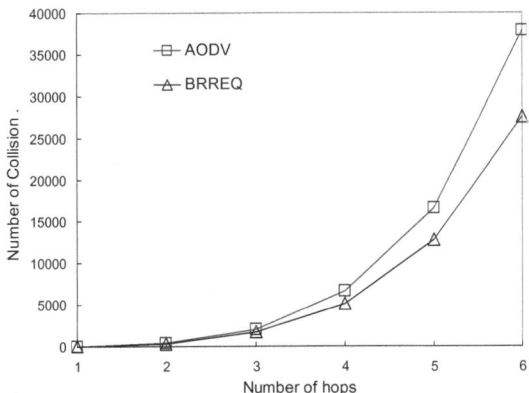

Fig. 7. The number of packet collisions

transmits more data packets during the channel access time. In case of AODV, it should wait another channel acquisition by the wireless contention after the route discovery procedure, which results in throughput performance degradation.

Figure 6 shows packet delivery ratios according to hop count and the result is correspondent to figure 5. This implies that when the number of hops increases, the performance gap between our scheme and AODV also increases. When AODV tries to establish the optimal route to the destination node, it should use RREQ flooding. These flooding packets may collide with data packets, which finally results in lower packet delivery ratio. Furthermore, the collision problem is more serious when the number of hops increases because the number of flooding increases exponentially in the tree based network topology. The number of packet collision is illustrated in figure 7. The collision performance shows a similar pattern with throughput results Therefore, in large wireless sensor networks, we can observe that the on-demand routing protocol like AODV is not suitable for communication with limited bandwidth.

5 Conclusion

In this paper, we have designed and implemented beacon mode based wireless sensor network system in TinyOS platform. For mobility support and smooth handover, we proposed a noble tree based self-routing scheme with association information between parent and child nodes. In other to verify the network performance, we implemented various sensing nodes as well as coordinator in real test-bed. Throughout performance evaluation with respect to loss rate, handover latency and several simulation studies, we showed that our network architecture accomplishes the reliable transmission for real-time processing service such as fire and emergency monitoring systems under heavy traffic environments.

As the future work, we plan to perform other extensive experiment to support QoS enabled packets such as voice and image. Then, we want to develop optimized and stable network architectures for WSN with mobility support.

Acknowledgement

This work is supported by DGIST and funded by MEST (Ministry of Education, Science and Technology) in Korea.

References

1. Akyildiz, I.F., Su, W., Sankarasubramaniam, Y., Cayirci, E.: Wireless sensor networks: a survey. Computer Networks 38 (March 2002)
2. Gay, D., Levis, P., Culler, D.: Software Design Patterns for TinyOS. LCTES (June 2005)
3. Gay, P.L., Culler, D.: The nesC Language: A Holistic Approach to networked Embedded Systems. In: ACM SIGPLAN 2003 (2003)
4. Draft IEEE Std. 802.15.4, Part 15.4, Wireless Medium Access Control (MAC) and Physical Layer (PHY) Specifications for Low-Rate Wireless Personal Area Networks (WPANs) (September 2006)
5. Jeon, H.-I., Kim, Y.S.: BOP and beacon scheduling for MEU devices. In: ICACT 2007 (Feburary 2007)
6. Jeon, H.-I., Kim, Y.S.: Efficient, Real-Time Short Address Allocations for USN Devices using LAA Algorithm. In: ICACT 2007 (Feburary 2007)
7. ISO/IEC JTC1/SC25, WiBEEM for Wireless Home Network Services - part 1, 2, 3, ISO standard (March 2008)
8. Perkins, C.E., Royer, E.: Ad-hoc on-demand Distance Vector Routing. In: Proc. 2nd IEEE Wksp. Mobile Comp. Sys. App. (February 1999)
9. Johnson, D.B., Maltz, D.A., Hu, Y.-C.: The Dynamic Source Routing Protocol for Mobile Ad Hoc Networks (DSR). In: Internet Draft, IETF Mobile Ad hoc Networks (MANET) Working Group
10. Sensor Network Analyzer (SNA),
    http://www.daintree.net/products/sna.php
11. Levis, P., Lee, N., Welsh, M., Culler, D.: TOSSIM: Accurate and Scalable Simulation of Entire TinyOS Applications. In: SenSys 2003 (November 2003)

Cooperative Spectrum Sensing for Cognitive Radios: Performance Analysis for Realistic System Setups and Channel Conditions

Marco Di Renzo[1], Laura Imbriglio[2], Fabio Graziosi[2], Fortunato Santucci[2], and Christos Verikoukis[1]

[1] Telecommunications Technological Center of Catalonia (CTTC)
08860 Castelldefels, Barcelona, Spain
{marco.di.renzo,cveri}@cttc.es
http://www.cttc.es
[2] University of L'Aquila, Dept. of Electrical and Information Engineering
and Center of Excellence DEWS – 67100 L'Aquila, Italy
{laura.imbriglio,fabio.graziosi,fortunato.santucci}@univaq.it
http://www.dews.ing.univaq.it

Abstract. In this paper, we propose an analytical framework for analysis and design of cooperative spectrum sensing methods over correlated Log–Normal shadow–fading environments, when each cooperative user makes use of a simple Amplify and Forward (AF) relaying mechanism to send the detected signal to a sink node. We will show that the framework requires efficient and accurate methods for modeling the power–sum of correlated Log–Normal Random Variables (RVs), which well describe shadowing phenomena, and propose novel approximation methods to efficiently solve this problem. Numerical results will be shown to substantiate the proposed framework.

Keywords: Cognitive Radio, Spectrum Sensing, Cooperative Communications, Correlated Log–Normal Shadowing, Performance Analysis.

1   Introduction

Cognitive Radio (CR) is commonly considered a key enabling technology to provide high bandwidth to mobile users via heterogeneous wireless architectures and Dynamic Spectrum Access (DSA) capabilities (see, e.g., [1], [2]). Broadly speaking, a CR can be defined as "an intelligent wireless communication device that exploits side information about its environment to improve spectrum utilization" [3], and is likely to consist of several components, but mainly of a sensing, decision, and execution unit. Various definitions of CRs exist in the literature. However, a promising solution, which is based on the idea of opportunistic communications, is the so–called *interweave* paradigm, according to which a CR is defined as "an intelligent wireless communication system that periodically monitors the radio spectrum, intelligently detects occupancy in the different parts

F. Granelli et al. (Eds.): MOBILIGHT 2009, LNICST 13, pp. 125–134, 2009.
© ICST Institute for Computer Sciences, Social-Informatics and Telecommunication Engineering 2009

of the spectrum and then opportunistically communicates over spectrum holes with minimal (i.e., no harmful) interference to the active users" [3].

A fundamental element for the successful exploitation of interweave CRs is the design of robust spectrum sensing methods to detect licensee users transmitting over a given frequency band. Accordingly, several spectrum sensing methods have been proposed to enable CR functionalities, and studied via analytical frameworks and experimental activities, see, e.g., [4] and references therein for a survey, and [5]–[9] for analysis and design of specific spectrum sensing methods. Among the various proposals, cooperative spectrum sensing methods using energy–based detectors are often considered a good candidate to enable CR functionalities, as they provide a good trade–off for keeping the complexity of every cooperative node at a moderate level, as well as counteracting the limitations of energy–based detection in the low Signal–to–Noise Ratio (SNR) regime via distributed diversity [10]. Accordingly, several studies have been conducted to analyze the performance of such a kind of spectrum sensing methods over a variety of fading channels (see, e.g., [5], [7]). In these papers, the authors have recognized the importance of including the characteristics of wireless propagation for system analysis and design. In particular, they have pointed out the importance of accurately modeling correlated Log–Normal shadow–fading phenomena to properly analyze the impact of distributed cooperation. It has been shown that shadowing correlation can significantly reduce the performance of cooperation, which results in an optimal number of cooperative users yielding the highest cooperative gain and lowest traffic overhead due to cooperation. Moreover, asymptotic analyzes based on a different kind of detector have explicit shown the performance limits set by correlated shadowing [6].

In the light of the above results, there is a common understanding about the importance of developing accurate and simple frameworks for the analysis and design of cooperative spectrum sensing methods over correlated Log–Normal shadow–fading environments. Moreover, the importance of accurately modeling correlation has also been reinforced by some recent experiments, which have proposed specific statistical models to well describe Log–Normal shadow–fading correlation for cooperative networks [11]. However, despite there is a significant number of studies for the analysis of energy–based cooperative spectrum sensing methods (see, e.g., [4], [5], [7]), as far as correlated Log–Normal shadow–fading environments are considered, performance metrics are typically obtained via extensive numerical simulations, which do not yield, in general, a solid basis for a systematic system analysis and optimization.

One of the main reasons for the absence of sound analytical frameworks to analyze the above mentioned scenario is due to the inherent analytical complexity of handling correlated Log–Normal Random Variables (RVs) if compared to other fading distributions. However, in [12] we have recently proposed a general and simple framework for the analysis of cooperative CR systems over correlated Log–Normal shadowing and analyzed its accuracy for several system setups, as well as compared our proposed method with other techniques so far available in the literature and assessed its superiority. However, the system setup described in [12] heavily relies on the not very realistic assumption that the signals sensed by several cooperative users can be sent via an error–free reporting channel to a fusion center, which can then combine them to improve system performance. So, the main aim of this contribution is to propose an advanced system setup that can remove this unrealistic assumption, as well as propose a simple yet accurate framework for its performance analysis and design

More specifically, the following contributions and results are claimed in the present paper: i) we propose a two–step method for Log–Normal power–sum approximation, which is based on the Improved Schwartz–Yeh (I–SY) and Pearson type IV approximation frameworks, and allows to handle correlation among all links of the cooperative network, ii) differently from typical unrealistic assumptions where data sensed by every cooperative node are sent to a common central unit via an error–free feedback channel [5], we consider a more realistic Amplify and Forward (AF) relaying mechanism for data gathering, and iii) we quantify the impact of shadowing correlation on the performance of distributed and decentralized energy–based spectrum sensing methods.

The remainder of the manuscript is organized as follows. In Section 2, system model and cooperative spectrum sensing protocol will be introduced. In Section 3, the cooperative spectrum sensing problem will be formulated. In Section 4, the novel method for Log–Normal power–sum approximation will be presented for a generic cooperative network with AF relying. In Section 5, numerical and simulation results will be compared to assess the accuracy of the proposed approximation to compute Detection Probability in CR scenarios, and the impact of correlated shadowing on system performance will be investigated. Finally, Section 6 will conclude the paper.

2   System Model

Let us consider a typical CR network that performs spectrum sensing operations in a distributed and cooperative fashion (see, e.g., [9, pp. 20, Fig. 3]). In general, cooperative spectrum sensing is composed by four main and subsequent steps: 1) every CR performs spectrum sensing locally and independently from each other, 2) every measurement is sent to a common band manager via an error–free reporting channel, 3) based on the collected measurements, the band manager makes a decision about the status of the sensed frequency band, and 4) the band manager broadcasts back the final decision to the cognitive users, thus enabling or not the transmission of one CR over that frequency band.

In the above standard procedure, step 2) relies on the unrealistic assumption that a noise–free channel is available by every cooperative user to send data to a common central unit. With the aim to overcome this idealistic assumption, we consider a more realistic setup where the data sensed during step 1) are forwarded to the band manager via an AF relying mechanism [13]. By this way, the reporting channel (i.e., relay channel) is not assumed to be error–free, but the relay mechanism accounts for noise accumulation due to dual–hop transmissions, as well as the effect of wireless propagation. For the sake of simplicity, but without loss of generality, we assume that the AF protocol is implemented in a time–scheduled fashion such that collisions are avoided. Furthermore, similar to [5], we assume that the band manager is equipped with a simple energy–based detector for spectrum sensing, and that a Square–Law–Combining (SLC) mechanism is used to combine the signals forwarded by every cooperative (secondary) user.

3   Problem Statement

According to the system model described in Section 2, the cooperative spectrum sensing problem with AF relaying can be modeled as the well–known dual–hop parallel relay channel [14, pp. 1002, Fig. 1]. In particular in [14, pp. 1002, Fig. 1], i) S (i.e., source) represents the primary user to be detected, ii) D (i.e., destination) denotes the common band manager that wants to get access to the wireless medium and performs spectrum sensing, and iii) $\{R_l\}_{l=1}^{L}$ are the L active secondary users (i.e., relays), which help D to detect the active transmission of S via AF relaying.

3.1   Notation

The following notation is used in what follows: i) G_l is the relay gain associated to relay R_l, ii) $\alpha_{l,SR}$ and $\alpha_{l,RD}$ are the fading amplitude of the source–to–relay and relay–to–destination hops in the l–th branch, respectively, iii) N_0 is the one–sided power spectral density of the Additive White Gaussian Noise (AWGN) at the input of $\{R_l\}_{l=1}^{L}$ and D, iv) $\gamma_{l,SR} = \alpha_{l,SR}^2 E_s / N_0$ and $\gamma_{l,RD} = \alpha_{l,RD}^2 E_s / N_0$ are the per–hop SNRs of the source–to–relay and relay–to–destination links in the l–th branch, respectively, and v) E_s is the average radiated energy in every transmission. In what follows, we will assume $\left\{ \alpha_{l,SR}^2, \alpha_{l,RD}^2 \right\}_{l=1}^{L}$, i.e., the channel power gains, to be Log–Normal distributed and generically correlated RVs, as a consequence of shadow–fading propagation.

3.2   Analytical Formulation

According to [5], the decision statistic of a SLC distributed detector can be written as follows:

$$y_{\mathrm{SLC}} = \sum_{l=1}^{L} y_l \tag{1}$$

where $\{y_l\}_{l=1}^{L}$ are the signals received by D from the L relays after square–and–integrate operation (i.e., energy detection).

Moreover, $\{y_l\}_{l=1}^{L}$ can be written as follows:

$$y_l = \frac{2}{\Psi_l} \int_0^T r_l^2 (t) \, dt \tag{2}$$

where $r_l(\cdot)$ is the signal received by D from R_l, Ψ_l is one–sided power spectral density of the noise component of $r_l(\cdot)$, and T is the observation window.

By relying on the AF relay mechanism, $r_l(\cdot)$ can be written as follows [13]:

$$r_l(t) = (\alpha_{l,SR}\alpha_{l,RD}G_l)\, s(t) + (\alpha_{l,RD}G_l)\, n_{R_l}(t) + n_D(t) \tag{3}$$

where $s(\cdot)$ is the signal transmitted by terminal S, and $n_{R_l}(\cdot)$, $n_D(\cdot)$ are the AWGNs at the input of terminals R_l and D, respectively.

According to (3), Ψ_l in (2) can be readily computed as follows:

$$\Psi_l = \left(\alpha_{l,RD}^2 G_l^2 + 1\right) N_0 \tag{4}$$

In particular, we consider the well–known Channel State Information (CSI–) assisted relay mechanism, thus assuming that every relay R_l has full CSI about the S–to–R_l link. In such a case, the relay gain is $G_l = G_l^{CSI} = 1/\alpha_{l,SR}$ [13].

Following similar analytical steps as described in [5], it is possible to show that the performance of the cooperative and distributed spectrum sensing network can be characterized by two performance measures, i.e., False Alarm Probability ($\mathrm{P_{fa}}$) and Detection Probability ($\mathrm{P_d}$), which can be computed as follows, when conditioning upon the fading channel statistics:

$$\begin{cases} \mathrm{P_{fa}} = \dfrac{\Gamma\left(\frac{LN}{2}, \frac{\lambda}{2\sigma^2}\right)}{\Gamma\left(\frac{LN}{2}\right)} \\ \mathrm{P_d} = \int\limits_0^{+\infty} Q_{\frac{LN}{2}}\left(\sqrt{\frac{a\xi}{\sigma^2}}, \sqrt{\frac{\lambda}{\sigma^2}}\right) f_{\gamma_t}(\xi)\, d\xi \end{cases} \tag{5}$$

where i) $\Gamma(\cdot)$ and $\Gamma(\cdot,\cdot)$ denote the Gamma [16, pp. 255, Eq. (6.1.1)] and incomplete Gamma [16, pp. 260, Eq. (6.5.3)] functions, respectively, ii) $Q_m(\cdot,\cdot)$ is the generalized Marcum Q–function [7, pp. 73], iii) N is the number of degrees of freedom of the system [5], iv) λ is the detection/decision threshold used by D in the binary hypothesis testing problem to discriminate between presence and absence of a licensee user, and v) $\sigma^2 = 1$, $a = 2$. Moreover, $f_{\gamma_t}(\cdot)$ is the PDF of the end–to–end SNR, γ_t, in D. According to the AF/CSI relay mechanism, γ_t can be explicitly written as $\gamma_t = \sum_{l=1}^L \gamma_{l,t}$, where [13]:

$$\gamma_{l,t} = \frac{\gamma_{l,SR}\gamma_{l,RD}}{\gamma_{l,SR} + \gamma_{l,RD}} = \left(\frac{1}{\gamma_{l,SR}} + \frac{1}{\gamma_{l,RD}}\right)^{-1} \tag{6}$$

As $\mathrm{P_{fa}}$ in (5) is independent from channel statistics, in the present contribution we are mainly interested in developing a simple but effective framework to compute $\mathrm{P_d}$ in (5), which requires a closed–form expression for the PDF of γ_t. In Section 4 we will show that the computation of $f_{\gamma_t}(\cdot)$ boils down to have accurate and simple methods for approximating the power–sum of generically correlated Log–Normal RVs.

4 A Novel Method for Log–Normal Power–Sum Approximation

By carefully looking at $\gamma_{l,t}$ defined in Section 3.2, we can easily figure out that the inverse of end–to–end SNR in every dual–hop cooperative link is given by the summation of correlated Log–Normal RVs. So, modeling the distribution of the SNRs in Section 3.2 is equivalent to find the distribution of the inverse of a linear combination (i.e., power–sum) of generically correlated Log–Normal RVs.

The SNR $\gamma_{l,t}$ can be re–written as $\gamma_{l,t} = \left[\sum_{n=1}^{2} X_{l,n}\right]^{-1} = \left[\sum_{n=1}^{2} 10^{0.1 Y_{l,n}}\right]^{-1}$, where $\{Y_{l,n}\}_{n=1}^{2}$ is a vector of Normal RVs with mean vector $(\boldsymbol{\mu}_{Y_l})$ and covariance matrix $(\boldsymbol{\Sigma}_{Y_l})$ given as follows[1]:

$$\begin{cases} \boldsymbol{\mu}_{Y_l}(1) = -\mu_{l,SR} - 10\log_{10}(E_s/N_0) \\ \boldsymbol{\mu}_{Y_l}(2) = -\mu_{l,RD} - 10\log_{10}(E_s/N_0) \\ \boldsymbol{\Sigma}_{Y_l}(1,1) = \sigma_{l,SR}^2 \\ \boldsymbol{\Sigma}_{Y_l}(2,2) = \sigma_{l,RD}^2 \\ \boldsymbol{\Sigma}_{Y_l}(1,2) = \boldsymbol{\Sigma}_{Y_l}(2,1) = \rho_{l,\{SR,RD\}}\sigma_{l,SR}\sigma_{l,RD} \end{cases} \tag{7}$$

where i) $\mu_{l,SR}$ and $\mu_{l,RD}$ are the mean values, ii) $\sigma_{l,SR}^2$ and $\sigma_{l,RD}^2$ are the variances, and iii) $\rho_{l,\{SR,RD\}}$ is the correlation coefficient of RVs $\chi_{l,SR} = 10\log_{10}\left(\alpha_{l,SR}^2\right)$ and $\chi_{l,RD} = 10\log_{10}\left(\alpha_{l,RD}^2\right)$, i.e., the Normal RVs associated to the Log–Normal power gains $\alpha_{l,SR}^2$ and $\alpha_{l,RD}^2$, respectively.

According to the above analysis, the problem of computing P_d boils down to have general and flexible methods for managing the power–sum of correlated Log–Normal RVs. In particular, two main problems need to be addressed: i) first of all, the PDF of $\{\gamma_{l,t}\}_{l=1}^{L}$ needs to be estimated, which results in the need to have efficient tools for approximating the power sum of generically correlated Log–Normal RVs, and ii) secondly, the PDF of $\gamma_t = \sum_{l=1}^{L} \gamma_{l,t}$ needs to be computed, which, in general, results in dealing with the power–sum of correlated either Log–Normal or non–Log–Normal RVs depending on the assumptions done to compute the PDF of $\{\gamma_{l,t}\}_{l=1}^{L}$ [15].

4.1   A Two–Step Approximation for Computing $f_{\gamma_t}(\cdot)$

We propose a simple yet accurate two–step procedure for computing $f_{\gamma_t}(\cdot)$, and then use it for the estimation of P_d in (5).

Step 1: Improved Schwartz–Yeh (I–SY) Approximation for $\{\gamma_{l,t}\}_{l=1}^{L}$.
The main idea of the Improved Schwartz–Yeh (I–SY) method [17] is to approximate the Log–Normal power–sum $\gamma_{l,t}$ with another Log–Normal RV, as follows:

$$f_{\gamma_{l,t}}(\xi) \cong \frac{10/\ln(10)}{\sqrt{2\pi\sigma_{l,I-SY}^2}\,\xi} \exp\left[-\frac{(10\log_{10}(\xi) - \mu_{l,I-SY})^2}{2\sigma_{l,I-SY}^2}\right] \tag{8}$$

where $\mu_{l,I-SY}$ and $\sigma_{l,I-SY}$ are the parameters of the approximating PDF, which are obtained via moment matching in the logarithmic domain between $\gamma_{l,t}$ and the approximating Log–Normal RV, i.e.:

---

[1] We denote with $\mathbf{v}(i)$ the i–th element of vector $\mathbf{v}$, with $\mathbf{M}(i,j)$ the element in the i-th row and j–th column of matrix $\mathbf{M}$.

$$\begin{cases} \mu_{l,I-SY} = m^{(1)}_{\gamma_l,t_{\mathrm{dB}}} \\ \sigma_{l,I-SY} = \sqrt{m^{(2)}_{\gamma_l,t_{\mathrm{dB}}} - \left(m^{(1)}_{\gamma_l,t_{\mathrm{dB}}}\right)^2} \end{cases} \tag{9}$$

where $m^{(n)}_{l,t_{\mathrm{dB}}} = (-1)^n E\left\{[10\log_{10}(1/\gamma_{l,t})]^n\right\}$, and $E\{\cdot\}$ denotes statistical expectation.

$$m^{(n)}_{l,t_{\mathrm{dB}}} = (-1)^n \left(\frac{10}{\ln(10)}\right)^n \sum_{p_1=1}^{N_p} \sum_{p_2=1}^{N_p} \cdots \sum_{p_Q=1}^{N_p} \Pi_{\gamma_l,t_{\mathrm{dB}}}(\mathbf{p}) \left\{\ln\left[\Omega_{\gamma_l,t_{\mathrm{dB}}}(\mathbf{p})\right]\right\}^n \tag{10}$$

$$\begin{cases} \Pi_{\gamma_l,t_{\mathrm{dB}}}(\mathbf{p}) = \prod_{i=1}^{2} \frac{H_{p_i}}{\sqrt{\pi}} \\ \Omega_{\gamma_l,t_{\mathrm{dB}}}(\mathbf{p}) = \sum_{i=1}^{2} \exp\left[\frac{\ln(10)}{10}\left(\sqrt{2}\sum_{j=1}^{2} \Sigma^{sq}_{Y_l}(i,j)x_{p_j} + \boldsymbol{\mu}_{Y_l}(i)\right)\right] \end{cases} \tag{11}$$

$$f_{\gamma_t}(\xi) \cong \frac{10}{\ln(10)} \frac{h}{\xi} \left[1 + \frac{(10\log_{10}(\xi)+u)^2}{d^2}\right]^{-m} \exp\left[-\nu\tan^{-1}\left(\frac{10\log_{10}(\xi)+u}{d}\right)\right] \tag{12}$$

From (8), (9), it turns out that the I–SY method requires the computation of the log–moments $m^{(n)}_{l,t_{\mathrm{dB}}}$ of the power–sum $1/\gamma_{l,t}$. These log–moments have been recently computed in [12], and can be obtained (with $Q = 2$) as shown in (10) and (11) on top of this page, where $\mathbf{p}$ is a vector with elements $\{p_j\}_{j=1}^2$, and $\{x_p\}_{p=1}^{N_p}$, $\{H_p\}_{p=1}^{N_p}$ are zeros and weights of the N_p–order Hermite polynomial [16, Table 25.10, pp. 924], respectively. Moreover, $\Sigma^{sq}_{Y_l} = \mathbf{UV}^{1/2}$, and $\mathbf{U}$ and $\mathbf{V}$ are the matrices containing the eigenvectors and eigenvalues of Σ_{Y_l}, respectively.

Step 2: Pearson Type IV Approximation for γ_t. As a result of the I–SY approximation for $\{\gamma_{l,t}\}_{l=1}^L$ in Step 1, the computation of $f_{\gamma_t}(\cdot)$ boils down to the estimation of the PDF of the power–sum of generically correlated Log–Normal RVs. To get very accurate results, we propose to use a non–Log–Normal approximation method to estimate $f_{\gamma_t}(\cdot)$. Moving from the excellent matching accuracy at the PDF level shown by the Pearson type IV method introduced in [12], [15], we rely on this method for Step 2.

Accordingly, the PDF of γ_t, $f_{\gamma_t}(\cdot)$, is approximated as shown in (12) on top of this page, where h is a normalization factor, and u, m, d, ν are the parameters that define the Pearson type IV distribution [15]. These latter parameters can be computed from the non–central moments of RV $\gamma_{t_{\mathrm{dB}}} = 10\log_{10}(\gamma_t)$. Due to space constraints, we dot report in the present contribution the formulas that allow to obtain u, m, d, ν from the non–central moments, but they can be found in [15]. On the other hand, the most complicated task in this approximation is the computation of the non–central moments $m_{\gamma_{t_{\mathrm{dB}}}}^{(n)} = E\{[10\log_{10}(\gamma_t)]^n\}$, which cannot be directly derived from [15], as a consequence of the particular form taken by the end–to–end SNR $\{\gamma_{l,t}\}_{l=1}^L$ in (6) for each cooperative dual–hop link. As the main objective of the present paper is to analyze the accuracy of the proposed approximation, we will compute these moments from Monte Carlo simulations, while the development of a framework for their computation is left to a future contribution.

5   Numerical and Simulation Results

The aim of this section is to analyze the accuracy of the proposed approximations to compute $P_m = 1 - P_d$ (i.e., the Miss Detection Probability), as a function of the number of cooperative dual–hop links (L), and shadowing correlation among them. In particular, P_m will be obtained, via straightforward numerical integration techniques, from (5) by approximating the PDF of the SNR γ_t by using the two–step method described in Section 4.1. Analysis will be compared with Monte Carlo simulations to assess its accuracy.

The following system setup is considered for performance analysis: i) the detection/decision threshold λ is computed according to a Constant False Alarm (CFA) criterion [7] by using the formula for P_{fa} in (5) with $P_{fa} = \{10^{-3}, 10^{-4}\}$; ii) without loss of generality, the Log–Normal RVs are assumed to be identically

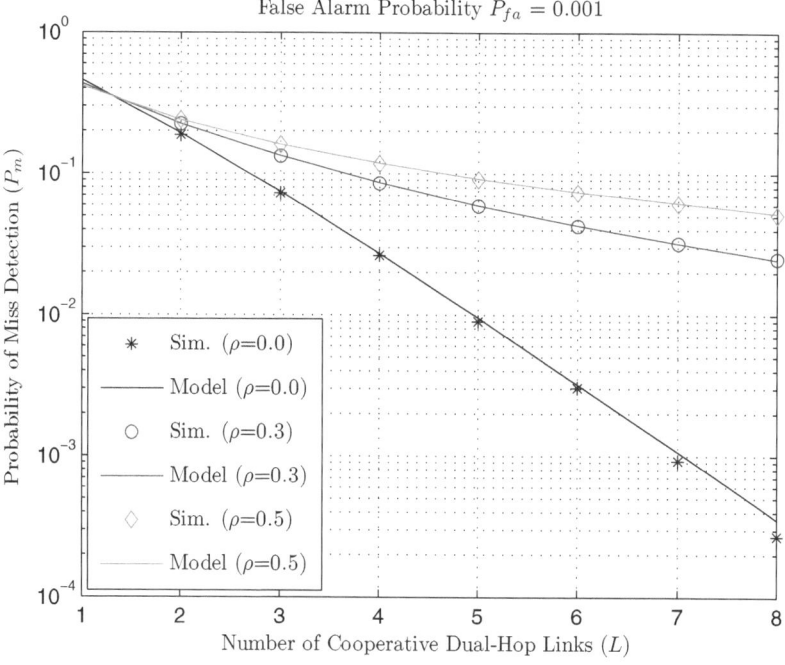

Fig. 1. P_m vs. number (L) of cooperative dual–hop links (CSI relays and $P_{fa} = 10^{-3}$)

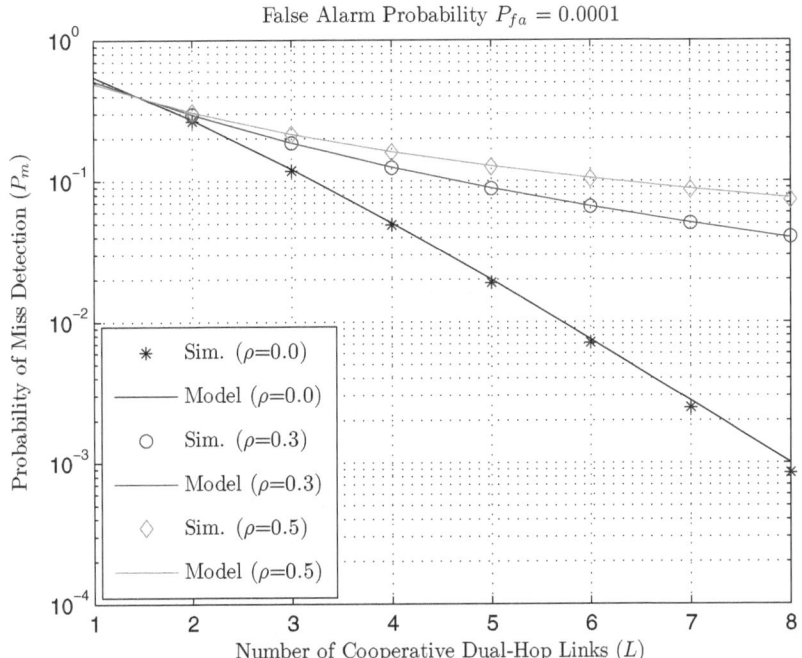

Fig. 2. P_m vs. number (L) of cooperative dual–hop links (CSI relays and $P_{fa} = 10^{-4}$)

distributed with parameters $\mu = 15$ dB, $\sigma = 6$ dB, and with equal correlation coefficient $\rho = \{0.0, 0.3, 0.5\}$; iii) $N = 10$, iv) $N_p = 7$, and v) $E_s/N_0 = 0$ dB.

The results shown in Figure 1 and Figure 2 clearly illustrate that the proposed two–step approximation is pretty accurate for several system setups, and a limited number of GQR points (i.e., $N_p = 7$) is also required to get these good accuracies. We can observe that, in general, increasing the number of cooperative dual–hop links improves spectrum sensing capabilities (i.e., P_m decreases). However, the net gain obtained with cooperation gets significantly down when the cooperative links are subject to correlated shadowing.

6   Conclusions

In this paper, we have provided an analytical framework for the analysis of co-operative spectrum sensing techniques over correlated Log–Normal shadowing environments. Novel approximation methods have been introduced to handle correlated scenarios, and their accuracy has been validated via Monte Carlo simulations. Our empirical investigations show that the proposed two–step approach based on jointly using I–SY and Pearson type IV approximations offers a general, simple yet adequately accurate framework for performance analysis and design of efficient collaborative spectrum sensing methods over realistic propagation environments.

Acknowledgment

This paper is supported, in part, by the Torres Quevedo 2008 Program's aid (PTQ–08–01–06437), and the research projects PERSEO (TEC2006–10459/ TCM), LOOP (FIT–330215–2007–8), and m:VIA (TSI–020301–2008–3).

References

1. Mitola III, J., Maguire Jr., G.Q.: Cognitive radio: making software radios more personal. IEEE Personal Commun., 13–18 (August 1999)
2. Akyildiz, I.F., Lee, W.–Y., Vuran, M.C., Mohanty, S.: A survey on spectrum management in cognitive radio networks. IEEE Commun. Mag. 46, 40–48 (2008)
3. Goldsmith, A., Jafar, S.A., Maric, I., Srinivasa, S.: Breaking spectrum gridlock with cognitive radios: An information theoretic perspective. In: Proc. of the IEEE (to appear, 2009)
4. Letaief, K.B., Zhang, W.: Cooperative spectrum sensing. In: Cognitive Wireless Communication Networks, pp. 115–138. Springer, Heidelberg (2007)
5. Digham, F.F., Alouini, M.-S., Simon, M.K.: On the energy detection of unknown signals over fading channels. IEEE Trans. Commun. 55, 21–24 (2007)
6. Ghasemi, A., Sousa, E.S.: Asymptotic performance of collaborative spectrum sensing under correlated Log–Normal shadowing. IEEE Commun. Lett. 11, 34–36 (2007)
7. Ghasemi, A., Sousa, E.S.: Opportunistic spectrum access in fading channels through collaborative sensing. IEEE J. of Commun. 2, 71–82 (2007)
8. Ganesan, G., Li, Y.: Cooperative spectrum sensing in cognitive radio, Part I: Two user networks; Part II: Multiuser networks. IEEE Trans. Wireless Commun. 6, 2204–2222 (2007)
9. Unnikrishnan, J., Veeravalli, V.V.: Cooperative sensing for primary detection in cognitive radio. IEEE J. Sel. Topics Signal Process 2, 18–27 (2008)
10. Tandra, R., Sahai, A.: SNR walls for signal detection. IEEE J. Sel. Topics Signal Process 2, 4–17 (2008)
11. Agrawal, P., Patwari, N.: Correlated link shadow fading in multi–hop wireless networks. Tech. Report arXiv:0804.2708v2 (April 2008)
12. Di Renzo, M., Graziosi, F., Santucci, F.: Cooperative spectrum sensing in cognitive radio networks over correlated Log–Normal shadowing. In: IEEE Vehic. Technol. Conf., Barcelona, Spain (April 2009)
13. Hasna, M.O., Alouini, M.-S.: End–to–end performance of transmission systems with relays over Rayleigh–fading channels. IEEE Trans. Wireless Commun. 2, 1126–1131 (2003)
14. Di Renzo, M., Graziosi, F., Santucci, F.: On the performance of CSI–assisted cooperative communications over generalized fading channels. In: IEEE Int. Conf. Commun., vol. 1, pp. 1001–1007 (2008)
15. Di Renzo, M., Graziosi, F., Santucci, F.: A general formula for log–MGF computation: Application to the approximation of Log–Normal power sum via Pearson type IV distribution. In: IEEE Vehic. Technol. Conf., May 2008, vol. 1, pp. 999–1003 (2008)
16. Abramowitz, M., Stegun, I.A.: Handbook of mathematical functions with formulas, graphs, and mathematical tables, 9th edn., NY (1972)
17. Di Renzo, M., Graziosi, F., Santucci, F.: Performance of cooperative multi–hop wireless systems over Log–Normal fading channels. In: IEEE Global Commun. Conf., New Orleans, LA, USA (November 2008)

Cross-Layer Optimization of Video Services over HSDPA Networks

Luca Superiori, Martin Wrulich, Philipp Svoboda, and Markus Rupp

Institute of Telecommunication and RF Engineering,
Vienna University of Technology,
Vienna, Austria
{lsuper,mwrulich,psvoboda,mrupp}@nt.tuwien.ac.at

Abstract. In the third generation networks a quality of service is guaranteed for key applications, such as video streaming. However, all the packets belonging to one application are handled in the same way, even though they have different impact on the perceived quality. In this article we present a standard compliant cross layer optimization for video services. The importance of a packet, signalized at application level, is used to filter packets into different logical channels with different qualities of service. Simulations performed using a HSDPA system level simulator show that our implementation increases the video quality of over 0.6 dB.

Keywords: HSDPA, Video Streaming, H.264/AVC, PDP Context.

1 Introduction

Beside voice call applications, the third generation wireless networks offers mobile broadband to the customers. The Universal Mobile Telecommunications System (UMTS) [1] has been standardized by the 3rd Generation Partnership Project (3GPP) in the year 2000 (*Release 99*). In the *Release 5*, 2002, two main features have been added to the standard: (i) the IP Multimedia System (IMS) describes a framework for delivering Internet Protocol (IP) multimedia services, (ii) the cell throughput has been increased by the introduction of the High Speed Download Packet Access (HSDPA) from the 384 kbit/s of Release 99 to 14.4 Mbit/s.

Increasing available downlink datarate and simplified infrastructures make IP traffic in the wireless network constantly increase. The traffic consists of a variety of different applications [2], with different degrees of interactivity and technical needs. In the 3GPP standards, four classes of IP traffic have been defined in [3]: (i) Conversational: usually reserved for Voice over IP (VoIP), it covers connections between two or more human users with stringent delay constraints, (ii) Streaming: reserved for multimedia streaming from a server to one or more humans, which requires variance in the delay to be avoided, (iii) Interactive: it refers to applications such as web browsing, (iv) Background: includes all the applications where the user is not expecting the content in a given time, such as File Transfer Protocol (FTP) download. This classification has been introduced in order to guarantee Quality of Service (QoS) to the customers. IP

F. Granelli et al. (Eds.): MOBILIGHT 2009, LNICST 13, pp. 135–146, 2009.
© ICST Institute for Computer Sciences, Social-Informatics and Telecommunication Engineering 2009

packets belonging to an application with stringent delay constraint are favoured over the ones without timing specifications.

The QoS in UMTS is established by guaranteeing a given bitrate, handling priority and transfer delay to a logical connection. However, the QoS does not necessarily reflect the perceived Quality of Experience (QoE) [4], i.e. the user satisfaction for the required service. In this work we consider the cross-layer optimization of video streaming over HSDPA networks, aiming at the maximization of the QoE as perceived by the end users. In video streaming, although a single QoS value is assigned to all the transmitted packets, different packets have a different impact on the reconstruction of the video sequence and, therefore, on the QoE. For this reason, we propose to signalize the importance of a packet from application layer to the IP layer in the DiffServ field of the IP header. The different packets will then be multiplexed depending on their DiffServ marking into different logical connection to the end user. An appropriate QoS class is assigned to each logical connection.

This paper is structured as follows. In Section 2 an overview over wireless video streaming is offered. The proposed method is discussed in Section 3, together with a description of the HSDPA features relevant for this work. The HSDPA system level simulator used for transmission is presented in Section 4. The results in Section 5 and the conclusions in Section 6 terminate the paper.

2   Video Streaming over Wireless Networks

Streaming applications in 3G networks have been specified by the 3GPP Technical Specifications (TS) belonging to the class of "Transparent end-to-end Packet-switched Streaming Service (PSS)". Since the seventh release of [5], defining the mandatory and suggested codecs, the H.264/AVC [6] is mentioned as *suggested* video codec. The H.264/AVC, jointly standardized by the International Standard Organization (ISO) Moving Picture Expert Group (MPEG) and International Telecommunication Union (ITU) Video Coding Expert Group (VCEG), is currently the state-of-the art video codec for commercial applications. For this reason, the following discussion will refer this codec.

The H.264/AVC belongs to the family of the hybrid block based video codecs. These kinds of video codecs exploit the correlation of small regions of the sequence pictures in space and in time. Each sample of the raw video sequence, a video frame, is subdivided into square blocks, called MacroBlocks (MB). Depending on the frame type, two encoding strategies are allowed. The macroblocks belonging to Intra (I) predicted frames are encoded using the neighboring blocks of the same picture as a source of prediction. For Inter (P) predicted frames, the prediction is searched in the previously encoded pictures. In both cases, the prediction is then refined by means of frequency transformed *residuals*, representing the difference between the original block and the best prediction.

The video encoder has been conceptually subdivided into two functional blocks. The Video Coding Layer (VCL) deals with the proper encoding functionalities whereas the Network Abstraction Layer (NAL) provides network friendliness to the produced data stream, managing the segmentation of the code into NAL Units (NALU) and reducing the dependency of the data stored in different packets. The maximum size of a NALU is specified depending on the network Maximum Transfer Unit (MTU). Therefore, a NALU contains a variable number of macroblocks (representing a picture *slice*) depending on the effectiveness of the considered prediction. For video streaming over 3G Networks, the NALUs are further encapsulated into RTP, UDP and, finally, IP.

Because of bad channel conditions and/or network congestion, some packet might not be correctly received. At the decoder side *error concealment* techniques reduce the impact of missing packets. Because of the temporal prediction, a missing packet does not only affect the reconstruction of the picture slice it contains but rather all the frames that use that slice as a source of prediction, as indicated in Fig. 1. We will refer to this effect as *temporal error propagation*. The

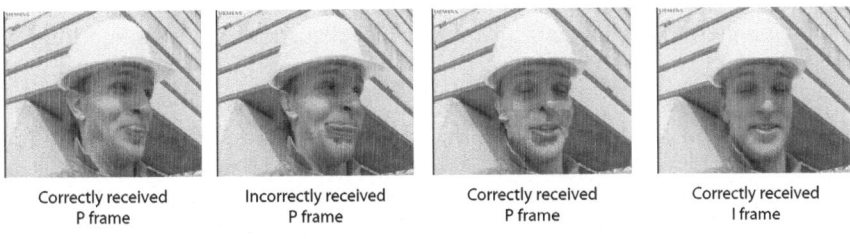

| Correctly received P frame | Incorrectly received P frame | Correctly received P frame | Correctly received I frame |

Fig. 1. Temporal error propagation

temporal error propagation ceases with the following I frame. Packets containing intra encoded information are self contained, i.e. the slice they contain can be reconstructed without the need of information stored in other packets.

The distance, in number of frames, between two consecutive I frames is defined as Group Of Picture GOP. The size of the GOP has a strong impact on the quality of the receiver side: small GOPs reduce the temporal propagation of the error. On the other hand, the spatial prediction used in the I frames is much less effective than the temporal one, particularly for high frequency patterns. The curve in Fig. 2 (left) shows the rate distortion behavior of the video quality in Luminance Peak to Signal Noise Ratio (Y-PSNR) depending on the GOP size. The transmission of the encoded 'Foreman' sequence in Common Intermediate Format (CIF) resolution (352×288 pixel) has been simulated considering different GOP sizes varying between 10 and 100 frames and fixed packet error rates (PERs) of one, three and five percent.

Since the error propagation is terminated by a correctly received I frame, preserving the payload of the packets containing encoded I slices is of major importance with respect to the one containing P slices. However, within the P frames of a GOP, one can differentiate between P packets of different importance. Different works in the literature, deal with the unequal error protection

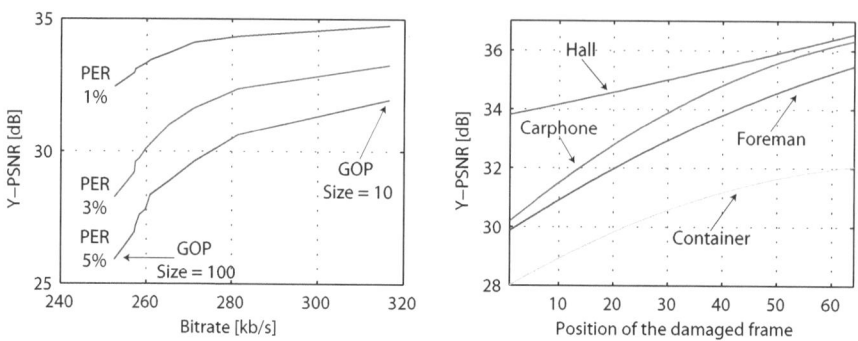

Fig. 2. Quality of corrupted sequences

of video packets, depending on the impact on the quality the slice they contain have. These methods often call for refined rate distortion analysis and limit the observation on the punctual frame the slice belongs to. In this work we consider a more generic approach, aiming at better protecting the packets belonging to the frames closer to the beginning of the GOP. Following the previous discussion, we do not consider only the "punctual" effect of the errors at frame level, but rather the overall impact they have on the reconstruction of the whole GOP. Errors affecting the first frames of the GOPs propagate more in time, degrading more the QoE if compared to errors occurring near the end of the GOP.

In order to analytically evaluate this effect, different standard test sequences have been encoded with a fixed GOP size of 65 frames. For each simulation performed, the packets containing a frame in a given GOP position have been removed. In this scenario, for obtaining generic results, the effect of the single packets has not been investigated, since it strongly depends on the specific characteristic of the considered sequence. The results are shown in Fig. 2 (right)

3   Video Packet Prioritisation in HSDPA

The importance of an encoded slice might be signalized exploiting the fields of the NALU header [7]. It consists of one byte signaling the correctness of the slice (one bit), the NAL reference indicator (NRI) (two bits) and the nalu type (five bits). The NRI value 00 is used for marking encoded slice not used as a reference by future inter frames. Values greater than 00 specify increasing packet priority, as indicated by the encoder.

At the video streaming server, this priority information is conveyed from the application layer to the IP layer. The IP header contains a byte originally thought for specifying the Type Of Service (TOS) and currently used for Differentiate Service (DiffServ) marking [8]. The DiffServ specifies how a packet has to be handled by each network element, i.e. a Per Hob Behaviour (PHB). Three main classes of DiffServ marking have been specified: (i) Default PHB: Best effort Traffic, (ii) Assured PHB: Ensures the forwarding of the packet as soon as a traffic threshold has not been exceeded, (iii) Expedited Forwarding PHB: Handles the packet with highest priority.

In the following we need a method to make use of this signaling information within the UMTS network. Other works in the literature, such as [9], already discuss this topic, but do not consider the possibility of assigning different DiffServ marking to packets belonging to the same data stream. This is a non trivial task as the IP packets carrying the user data are encoded into other transport protocols between GGSN and mobile terminal, e.g., GPRS Tunneling Protocol (GTP) and NodeB Application Protocol (NBAP). The data itself is only available at the borders of the UMTS network, i.e., GGSN and mobile terminal. However, there is a different and standard compliant way to use dynamic QoS settings for different packets. In a UMTS network a logical Packet Switched (PS) connection is setup by a PDP-context. Within this method, the user terminal at one side and the GGSN at the other side agree on several parameters, e.g., IP-address, PDP type and QoS profile for the following packets. In the UMTS network QoS only exists on this PDP-context entity, allowing for one setting at a time. To allow the system to serve independent QoS parameters for multiple applications running at the same mobile terminal, a feature called multiple PDP-contexts was introduced.

There exist two different categories of PDP-contexts, namely primary and secondary. Every mobile has to activate a primary context first. Then the mobile can either attach a secondary to the primary or initiate further primary ones. Multiple primary PDP-contexts have different IP addresses and are typically used for different applications, e.g., black berry in parallel to standard Internet access. Multiple primary PDP-contexts share the same IP address and must be attached to a previous initialized primary context. The following Figure 3 depicts the difference in the setup. As the IP address is the same for the primary and all the associated secondary PDP-contexts, a so called Traffic Flow Template (TFT) defines a split up of user data into the GTP tunnels.

A TFT can be seen as a packet filter applied onto each IP packet entering the GGSN. There exists one TFT per secondary PDP-context. The filtering rules are based on one or more of the following attributes: source address, IP protocol number, destination port (range), source port (range), IPsec security parameter index and the type of service. The last attribute in this list makes the TFT the perfect match for the proposed method as it is the same header field used for DiffServ. The following Figure 4 depicts the user of TFT for one user.

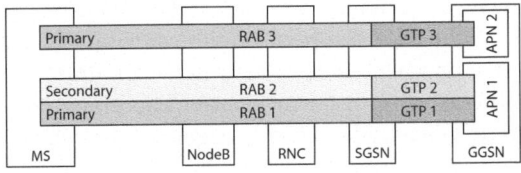

Fig. 3. Example for two primary PDP-Contexts and one secondary

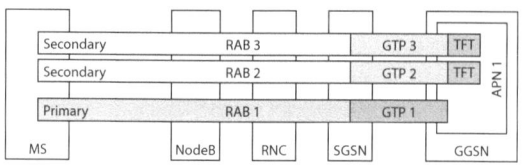

Fig. 4. Traffic Flow Tamplets

While in UMTS R99 the users are served with dedicated channels (DCH) only, HSDPA offers a shared channel. It does this by the implementation of new coding schemes and modulations techniques combined with scheduling techniques directly in the NodeBs. In other words the Spreading Factor (SF) is no longer variable and there is no more fast power control available. These two elements of Rel. 99 are replaced by Adaptive Modulation and Coding (AMC), Fast retransmission strategy (HARQ) and scheduling algorithms [10,11].

In the following we give a short discription of the new features introduced for the HSDPA MAC layer.

Scheduling Algorithms. The place of the scheduling systems in Rel. 99 is inside the RNC. In HSDPA the function has been moved into the NodeBs, which allows for faster scheduling as there is no more 'reaction' delay present. The scheduler in HSDPA also has additional task. Beside selecting the correct modulation and coding scheme and the HARQ process, it now schedules the transmission for all users. In Rel. 99 the scheduler was implemented on a per user base only. The available algorithms are Round-Robin, Proportional Fair and Maximum C / I.

Hybrid ARQ: A Fast Retransmission Strategy. The retransmission logic moved from the RNC entity into the NodeB. There exist two different error control and recovery methods to guarantee error free transmissions to and from the UE, namely Forward Error Correction (FEC) and Automatic Repeat reQuest (ARQ), see [12,13,14].

The disadvantage of these two methods is the delay which occurs in case of an packet error. This can be overcome by a combining the ARQ and the FEC method in a so called HARQ mode. The FEC is set to cover the most frequent error patterns and therefore will reduce the number of retransmissions necessary for the system. The ARQ part covers the less frequent error patterns, which allows to reduce the number of bit added by the FEC. There are different types of HARQ methods available and the performance in total depends on the channel conditions, receiver equipment and other related parameters. Considering the complexity of a UMTS radio implementation choosing the 'correct' or 'best' retransmission strategy is a wide field for ongoing research.

Adaptive Modulation and Coding (AMC). The original implementation of UMTS-Rel. 99 offered one fixed modulation scheme. The adaption to the actual radio channel is then achieved using a power control algorithm. The momentaneous data rate is set by choosing an appropriate spreading factor offering the needed gain for the given signal to interference situation.

In HSDPA the method was changed. Instead of relying on a fast power control the SF was fixed and the modulation now follows the channel conditions, both modulation and coding format adapt in accordance with variations in the channel conditions. This system is called Adaptive Modulation and Coding (AMC), or link adaptation. Compared to standard power control such methods deliver higher data rates. In HSDPA the AMR scheme assigns higher-order modulation with higher code rates, such as 16 QAM.

4   HSDPA System-Level Simulations

In order to assess the performance of enhanced scheduling algorithms, the implications in the context of network have to be evaluated. Therefore, standard physical-layer simulations are not sufficient, but rather system-level simulations are necessary [15,16]. One of the major difficulties of system-level analyses is the computational complexity involved in evaluating the performance of the radio links between all base-stations and mobile terminals. Performing such a large number of link-level simulations is clearly prohibitive. Thus, those evaluations have to rely on simplified link models that still must be accurate enough to capture the essential behavior [17,18,19].

In this paper, we conduct our simulations on a computationally efficient system-level simulator implemented in MATLAB [20]. The simulator is capable of simulating classical HSDPA networks as well as the enhanced version utilizing MIMO for increased data rates. In this work however, we restrict ourselves to the classical single antenna HSDPA without the possibility for spatial multiplexing in the downlink.

The physical-layer modeling utilized in the system-level simulator accounts for MMSE equalization at the receiver side and accurately reproduces the intercode interference in the multi-code operation of the shared downlink channel of HSDPA. The simulator is able to generate the cell deployment according to the desired configuration and deals with a large variety of user set-ups. A basic overview of the simulation methodology is depicted in Figure 5.

The HSDPA cell deployment considered in the simulator consists of 19 three-sector sites, corrsponding to the layout type 1 of [21]. The simulator allows for the power of the neighboring base stations to be controlled independently, such that the network to be simulated can be stripped down to seven three sector sites or even a single cell scenario. Different propagation models are available in the simulator, where in this work we stuck to the well known Walfish-Ikegami model [22] representing urban micro cell scenarios. Radio link control, as well as scheduling in the MAC-hs are simulated only for the target sector, thus keeping the computational effort manageable. This, however, requires the simulation to get along without handovers (because in the case of a handover, the associated

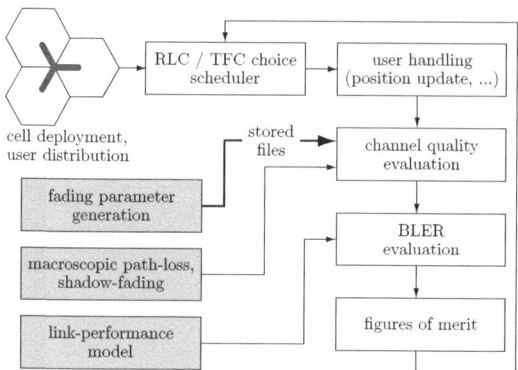

Fig. 5. Overview of the employed system-level methodology [20]

algorithms—residing in the RNC—would have to coordinate the radio link control of two sectors). In this work we set up the user mobility such that no handover will occur during the simulation of a video transmission.

The basic simulation procedure is as follows (see Figure 5): The first step of the simulation invokes the network generation, i.e. cell deployment, and the user generation according to the selected UE capability class together with their positioning. Also the fading parameters (describing the physical-layer) suitable for the scenario are loaded, the shadow fading traces are generated and the data necessary for the link-performance model (describing the decoding performance) is loaded. In the main simulation loop, according to the feedback of the UE in the target cell, the RLC and the MAC-hs scheduler decide upon the user to be served and the transmission settings of this transmission. After this decision an update of the user position takes place. With the position being known, the macro-scale pathloss and the effective antenna gain can be calculated. The SINR in the current transmission is then evaluated and consequently the correctness of the received packet is determined according to the link-performance model. The user feedback is then formed of the ACK/NACK report and the CQI for the current transmission evaluated pursuant to the mapping of the UE capability class. At the end of the simulation time, the resulting data is collected and statistically evaluated. The basic simulation settings for the investigations in this work are given in Table 1.

4.1   Content Aware Scheduling

To exploit the information available at the MAC-hs scheduler, i.e. the priority of the incoming IP packets, we implemented a scheduler that dynamically adapts the transmission settings. The basic idea behind this is to protect packets of high priority better against transmission errors, but also allow the scheduler to downgrade packets in case of less importance [23].

Table 1. System-level simulation parameters

Parameter	Value
network load	homogeneous
number of cells	19
Node-B distance	750 m
transmitter frequency	1.9 GHz
total power available at Node-B	20 W
spreading codes available for HS-DSCH	15
macro-scale pathloss model	urban micro [22]
channel type	PedA
active users in target sector	5
user mobility	3 km/h, random direction
UE capability class	10
UE receiver type	MMSE

HSDPA dynamically adapts the encoding and modulation—i.e. the transport block size γ—of a transmission according to the channel quality feedback information (CQI) of the UEs, c_{UE}, thus

$$\gamma = f(c_{\mathrm{UE}}). \tag{1}$$

Our scheduler now interferes with this mapping. Depending on the priority of the packets, we remap the transport block size γ,

$$\gamma_{\mathrm{new}} = \Phi_p\left[f(c_{\mathrm{UE}})\right], \tag{2}$$

where $\Phi_p[\cdot]$ denotes the remapping function depending on the packet priority p. Better protection can be achieved by remapping the transport block size to lower values, and vice versa. However, if we decrease the transport block size too often, the average throughput of the cell would also decrease notifiable, which is general undesired because spectral resources would be wasted.

In Section 2 we elaborated that packets belonging to I-frames contribute greatly to the video quality, whereas packets belonging to P-frames influence the quality not that prominent. In particular this holds for packets of P-frames that are at the end of a GOP. Since one I-frame is approximately equally large as four P-frames, we balance the loss in average throughput by increasing the transport block size for the packets belonging to these last four P-frames. Let us assign the following priority classes

- priority $p = 2$: packets belonging to I-frames,
- priority $p = 1$: packets belonging to the first $(l_{\mathrm{GOP}} - 4)$ P-frames of the GOP,
- priority $p = 0$: packets belonging to the last 4 P-frames of the GOP,

where l_{GOP} denotes the length of the GOP (in frames). Then the remapping $\Phi_p[\cdot]$ of the transport block size can be written as

$$\Phi_p[f(c_{\mathrm{UE}})] = f(c_{\mathrm{UE}} + 1 - p). \tag{3}$$

Table 2. Content-aware scheduler settings

Parameter	Value
user selection	round robin
max. nr. of HARQ retransmissions	1
transmission remapping	Equation (3)
remapping boundaries	CQI 0 and 30
HARQ recombining scheme	incremental redundancy

To ensure fairness against all active users in the cell, the scheduler selects the user according to a round robin strategy. In addition we limited the maximum delay that may occur in the transmission of the packet by setting the network to allow a maximum number of one HARQ retransmission. The basic scheduler parameters are summarized in Table 2.

5   Results

The simulations have been performed using the standard test sequence "Foreman" in QCIF resolution comparing the typical round robin (rr) approach with the content aware (ca) scheduling as described in Section 4.1.

In order to evaluate the performance of the proposed method in terms of preservation of the payload content, the error probability of the Transport Blocks (TBs) has been depicted in Fig. 6 (left). For the round robin scheduling, no distinction has been made between TBs containing I or P encoded frames. As expected, the error probability lies around 10 %. For the proposed content aware scheduling mechanism, the error probabilities associated to slice containing I and P frames have been presented separately. The proposed CQI mapping allow the error probability of the transport blocks containing the I frames to decrease of a factor 4, being it around 2.7 %. Aiming this work at the optimization of the quality of experience, the overall performance of the proposed method has been evaluated with respect to the Y-PSNR. Following the discussion in Section 2,

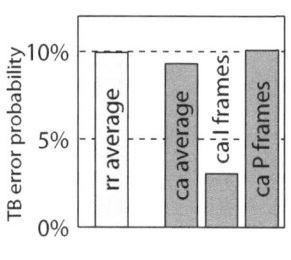
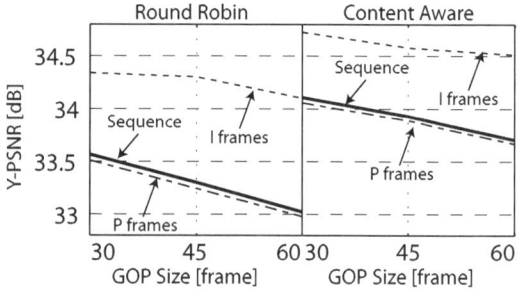

Fig. 6. Simulation results: TB error rate and video quality

three different GOP sizes have been investigated: 30, 45 and 60 frames. The results are shown in Fig. 6 (right).

As a consequence of the smaller TB error probability, as shown in Fig. 6 (left), the quality of the I frames has increased of over 0.5 dB when using content awareness. Such increase, however, is not only beneficial in terms of contribution to the average frame quality, but rather has to be considered advantageous for the quality of the following P frames. On the one hand, a valid source of prediction is offered to the following P frames, on the other hand, in case the previous GOP was damaged, the temporal error propagation of the error is terminated. The overall sequence quality measured using the proposed content aware scheduling mechanism is 0.6 dB higher than the one obtained with the typical round robin approach.

6 Conclusion

In this paper a standard compliant content-aware scheduling mechanism for HSDPA network has been presented. In a video streaming session, the application layer marks the importance of a packet into the NAL header. This information is used to select the appropriate DiffServ marking at IP level. The IP packets are then multiplexed at the GGSN in different logical channels with appropriate QoS settings. The performance of the method has been investigated using an HSDPA system level simulator. By means of the proposed solution, the video quality has been increased of over 0.6 dB.

Acknowledgments

The authors thank mobilkom austria AG for technical and financial support of this work. The views expressed in this paper are those of the authors and do not necessarily reflect the views within mobilkom austria AG.

References

1. Holma, H., Toskala, A.: WCDMA for UMTS – Radio Access For Third Generation Mobile Communications, 3rd edn. John Wiley & Sons, Ltd., Chichester (2005)
2. Svoboda, P., Ricciato, F., Pilz, R., Hasenleithner, E.: Composition of GPRS, UMTS traffic: snapshots from a live network. In: MOME 2006, Salzburg, Austria (Feburary 2006)
3. 3GPP, Quality of Service (QoS) concept and architecture (TS 23.107 V8.0.0) (2008), http://www.3gpp.org/ftp/Specs/html-info/23107.htm
4. Gomez, G., Sanchez, R.: End-to-End Quality of Service over Cellular Networks. John Wiley & Sons, Ltd., Chichester (2006)
5. 3GPP, Transparent end-to-end Packet-switched Streaming Service (PSS); Protocol and codecs (TS 26.234 V7.3.0) (2007), http://www.3gpp.org/ftp/Specs/html-info/26234.htm
6. ITU-T Rec. H.264 / ISO/IEC 11496-10, Advanced Video Coding. Final Committee Draft, Document JVTE022 (September 2002)

7. IETF, RTP Payload Format for H.264 Video (RFC 3984) (2005),
   www.ietf.org/rfc/rfc3984.txt
8. IETF, An Architecture for Differentiated Service (RFC 2475) (1998),
   www.ietf.org/rfc/rfc2475.txt
9. Agharebparast, F., Leung, V.C.M.: QoS support in the UMTS/GPRS backbone
   network using DiffServ. In: IEEE GLOBECOM 2002, November 2002, vol. 2, pp.
   1440–1444 (2002)
10. 3GPP, Multiplexing and channel coding (FDD) (TS 25.212 V4.0.0) (2006),
    http://www.3gpp.org/ftp/Specs/html-info/25213.htm
11. 3GPP, Physical Layer Aspects of UTRA High Speed Downlink Packet Access (TS
    25.848 V4.0.0) (2001), http://www.3gpp.org/ftp/Specs/html-info/25848.htm
12. 3GPP, Services provided by the physical layer (TS 25.302 V4.8.0) (2003),
    http://www.3gpp.org/ftp/Specs/html-info/25302.htm
13. 3GPP, Radio Resource Control (RRC); Protocol specification (TS 25.331 V6.2.0)
    (2008), http://www.3gpp.org/ftp/Specs/html-info/25331.htm
14. 3GPP, Physical layer procedures (FDD) (TS 25.214 V7.1.0) (2008),
    http://www.3gpp.org/ftp/Specs/html-info/25214.htm
15. Wrulich, M., Mehlführer, C., Rupp, M.: Interference aware MMSE equalization for
    MIMO TxAA. In: Proc. IEEE 3rd ISCCSP, pp. 1585–1589 (2008)
16. Wrulich, M., Weiler, W., Rupp, M.: HSDPA performance in a mixed traffic network.
    In: Proc. IEEE Vehicular Technology Conference Spring (VTC), May 2008, pp.
    2056–2060 (2008)
17. Wrulich, M., Eder, S., Viering, I., Rupp, M.: Efficient link-to-system level model for
    MIMO HSDPA. In: Proc. IEEE 4th Broadband Wireless Access Workshop (2008)
18. Staehle, D., Mader, A.: A model for time-efficient HSDPA simulations. In: Proc.
    IEEE 66th Vehicular Technology Conference Fall (VTC), pp. 819–823 (2007)
19. Wrulich, M., Rupp, M.: Efficient link measurement model for system level simu-
    lations of Alamouti encoded MIMO HSDPA transmissions. In: Proc. ITG Inter-
    national Workshop on Smart Antennas (WSA), Darmstadt, Germany (Feburary
    2008)
20. Wrulich, M., Rupp, M.: Computationally efficient MIMO HSDPA system-level eval-
    uation (submitted, 2009)
21. 3GPP, Technical specification group radio access network; spatial channel model
    for multiple input multiple output (MIMO) simulations (TS 25.996 V7.0.0) (2007),
    http://www.3gpp.org/ftp/Specs/html-info/25996.htm
22. Cichon, D.J., Kürner, T.: COST 231 – Digital Mobile Radio Towards Future Gen-
    eration Systems, ch. 4. COST (1998)
23. Superiori, L., Wrulich, M., Svoboda, P., Rupp, M., Fabini, J., Karner, W., Stein-
    bauer, M.: Content-aware scheduling for video streaming over HSDPA networks.
    submitted to IWCLD (2009)

Dynamic Resource Allocation for IEEE802.16e

Alberto Nascimento[1,2] and Jonathan Rodriguez[2,3]

[1] University of Madeira (UMa)
[2] Telecommunications Institute, Aveiro
[3] University of Aveiro (UA)
ajn@uma.pt, jonathan@av.it.pt

Abstract. Mobile communications has witnessed an exponential increase in the amount of users, services and applications. New high bandwidth consuming applications are targeted for B3G networks raising more stringent requirements for Dynamic Resource Allocation (DRA) architectures and packet schedulers that must be spectrum efficient and deliver QoS for heterogeneous applications and services. In this paper we propose a new cross layer-based architecture framework embedded in a newly designed DRA architecture for the Mobile WiMAX standard. System level simulation results show that the proposed architecture can be considered a viable candidate solution for supporting mixed services in a cost-effective manner in contrast to existing approaches.

Keywords: Mobile WiMAX, Dynamic Resource Allocation, Scheduler, Cross-Layer, Quality of Service, Fairness.

1 Introduction

The cost-effective delivery of Triple play applications (video, audio and data) in a ubiquitous and seamless manner are expected to be the main drivers of B3G networks, placing highly stringent constraints on both high data rate transmission and Quality of Service (QoS) demands. Therefore, it is important to investigate techniques that can maximize spectral utility by efficiently managing the radio resources for heterogeneous packet-based services. Fundamental to resource management is packet scheduling. An efficient scheduling policy must exploit cross-layer information from the PHY layer as well as from higher layers, to provide efficient mapping of data blocks onto the available transport channels within the Medium Access Control (MAC) layer, whilst providing the requested quality of service (QoS) in a heterogeneous service environment [1,2]. In order to cater for fairness, efficiency in resource usage and QoS provisioning, different variations of Opportunistic Scheduling have been proposed. The Proportional Fairness scheduler (PF) algorithm is widely used in 1xEV-DO, also named HDR networks [2]. The algorithm selects the user, among all backlogged users, which has the best feasible data rate normalized by the average throughput over a sliding window filter. Modified Largest Weighted Delay First Scheduler (M-LWDF) scheduler is a kind of modified PF, where the priority value to assign to the user is equal to that of the PF scaled by the weighted delay the user has endured. The algorithm attempts to satisfy the QoS statistically [4]. The Exponential

F. Granelli et al. (Eds.): MOBILIGHT 2009, LNICST 13, pp. 147–159, 2009.
© ICST Institute for Computer Sciences, Social-Informatics and Telecommunication Engineering 2009

Scheduler (EXP) attempts to equalize the weighted delays of all queues when their differences become large [5]. Utility schedulers [6] prioritize users according to their potential revenue if serviced. Different schedulers based on the notion of utility functions have been proposed in the literature [7-11]. But most of these invariably attempt to maximize the total system utility. In [7] the authors propose a scheduling algorithm principle similar to the concept of the utility function proposed in this work. However, they do not consider the channel state in the scheduling process and cannot benefit from a fully-compliant cross-layer scheduling design.

In this paper we propose an efficient packet scheduling algorithm fully aligned with this cross-layer paradigm, and specifically designed around an adaptive DRA architecture framework for the OFDMA multiple access scheme adopted by the IEEE 802.16e, also called Mobile WiMAX [12]. A key principle of our DRA is the exploitation of the inherent system diversities in various domains through the intelligent management of the allocation and access of users to the available radio resources.

This paper is organized as follows: section 2 summarizes our utility-based scheduling principle, section 3 describes the WiMAX Dynamic Resource Allocation architecture, section 4 presents the simulation scenarios and the performance results, and finally section 5 concludes the paper.

2 Utility-Based Scheduling

This family of schedulers prioritizes users according to their potential revenue if serviced. The notion of cost or revenue can be quantified by the means of "utility functions", a notion derived from utility theory in economics which can answer these economic scheduling criteria [6]. Fig. 1 provides an example of diverse utility functions for traffic applications with different delay constraints.

Typical utility schedulers attempt to maximize the total system utility. The typical approach is to select packets from queues in order to maximize the following metric [9, 11]:

$$k(n) = \arg \max_i \left(\left| U^{'}\left(\tau_i(n)\right) \right| . R_i(n) \right) \tag{1}$$

Where:

- $\left| U^{'}\left(\tau_i(n)\right) \right|$ is the absolute value of the derivative of the utility function for user i at the beginning of (TTI) n.
- $\tau_i(n)$ is the HOL packet delay of user i at the beginning of transmission time interval (TTI) n.
- $R_i(n)$ is the transmitted data rate offered by the channel of user i if it is scheduled for transmission.

Although conceptually simple, this solution may degenerate into a greedy algorithm such as the Max C/I because it only concerns a single user. Therefore, the transmission of data for a given user translates in a benefit to this particular user, but at the same time there is also a cost in deferring transmissions for other users. However, they do not consider the channel state in the scheduling process and cannot benefit from a fully-compliant cross-layer scheduling design.

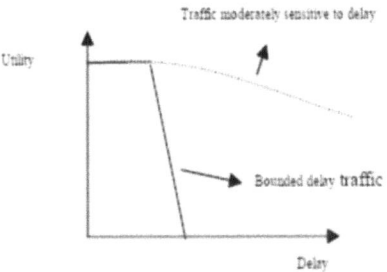

Traffic moderately sensitive to delay

Utility

Bounded delay traffic

Delay

Fig. 1. Different types of Utility Functions

2.1 Proposed Algorithm

The scheduler should maximize the total utility of the system expressed as the sum of utilities of each user. More precisely the basic idea behind the proposed utility based scheduler is described herein:

1. At the beginning of TTI n estimate the total amount of potential utility: $U_p(n)$

2. For a given user j, estimate the amount of utility that will be transferred by the network if the user is scheduled for transmission:

$$U_j\big(R_j(n),\mathbf{Q}_j(n)\big)= \sum_{k=1}^{L_j} U\big(\tau_k^j\big)(1-PER_j) \tag{2}$$

Where:

- $R_j(n)$ is capacity of the channel for user j during TTI n. It represents the number of packets (L_j) that can be transferred if this user is scheduled for transmission in this TTI.

- $\mathbf{Q}_j(n)$ is the vector representing the delay of each packet from the buffer of user j.

- PER_j is the packet error rate associated with the channel for user j during TTI n.

- 3. Assuming user j is scheduled for transmission, the packets not transmitted loose their utility (they have their delays increased by one TTI). This translates to a remaining potential utility: $U_p(n+1\,|\,j)$.

- 4. The selected user is the one that satisfies the following scheduling decision:

$$k(n) = \arg \max_{j}\big(U_j\big(R_j(n),\mathbf{Q}_j(n)\big)-L_j(n)\big) \tag{3}$$

Where:

$$L_j(n) = U_p(n)-U_p(n+1\,|\,j)-U_j(R_j(n),\mathbf{Q}_j(n)) \tag{4}$$

is the loss of utility incurred by choosing user j. This strategy means that one wants to maximize the utility transferred to a given user by subtracting from the transferred utility the losses incurred.

2.2 Utility Function Definition

The utility function of packet delay $U(\tau)$ must satisfy the following properties:

4. $U(\tau):[0,\tau_{max}[\rightarrow[0,1]$. If the network offers a delay higher than the maximum delay tolerated its utility is equal to zero in order to express the total user dissatisfaction.

- $\lim_{\tau\to0^+} U(\tau)=1$. This property reflects a maximum user satisfaction (100%) when the network offered delay tends to zero.
- $U(\tau):[0,\tau_{max}[\rightarrow[0,1]$. If the network offers a delay higher than the maximum delay tolerated its utility is equal to zero in order to express the total user dissatisfaction.
- $\lim_{\tau\to0^+} U(\tau)=1$. This property reflects a maximum user satisfaction (100%) when the network offered delay tends to zero.
- if $\tau_i > \tau_j \Rightarrow U(\tau_i)\le U(\tau_j))$. The utility function is monotonically decreasing with delay.
- $U'(\tau)\le0$ and $U''(\tau)>0$. The first order derivative is negative to reflect the decrease in the level of satisfaction of the user if the offered delay increases.
- The second order derivative must increase as the delay approaches the deadline.

One particular case of a function complying to these properties is the sigmoid function. Sigmoid functions have been used in the literature to approximate the user's satisfaction with respect to the quality of service provided by operators. We used the following modified sigmoid utility function:

$$U(\tau) = \begin{cases} 1 - \dfrac{2}{1+\exp(-\alpha(\tau-\beta))}, & \text{if } \tau \in [0,\tau_{max}] \\ 0 & , \text{ if } \tau \in [\tau_{max},+\infty[\end{cases} \tag{5}$$

The parameters α and β determine, respectively the steepness and the center of the curve. They can be tuned to customize the function for different types of service. In our case $\beta = \tau_{max}$ and $\alpha = 0,5$.

3 WiMAX DRA Architecture

A system level simulator that models the salient features of the WiMAX standard IEEE802.16e [12-14] was developed. The simulator implements the DL communication in a PMP configuration using the TDD mode of transmission. In the MAC frame, resources are available in both frequency (sub-channels) and time domains (OFDM symbols). The smallest granularity of resource allocation in the time and frequency domains is the slot. The size of the slot depends on the type of sub-channelization mode and on the direction of transmission. A burst is a rectangular allocation of a group of slots with the same modulation and coding scheme (MSC) belonging to the same or different users. In our simulations each packet is mapped into a group of contiguous slots forming a burst intended to a single user provided the same modulation and coding scheme is followed in the transmission of all packets allocated to it. Each burst contains only one MAC PDU (to which a single packet is mapped into). The MAC PDU is segment into forward error correction (FEC) blocks that are coded and interleaved within the burst.

In the implemented DRA, Partial Usage sub-channelization Scheme (PUSC) was used. In PUSC sub-channels are realized using a distributed sub-carrier permutation method that pseudo-randomly draws sub-carriers to form a sub-channel and used to achieve frequency diversity in cases where the mobile speed makes it difficult or inefficient to track frequency-selective channel variations. For further details regarding the SINR modeling, Link Level Interface and Link Adaptation schemes please refer to [14].

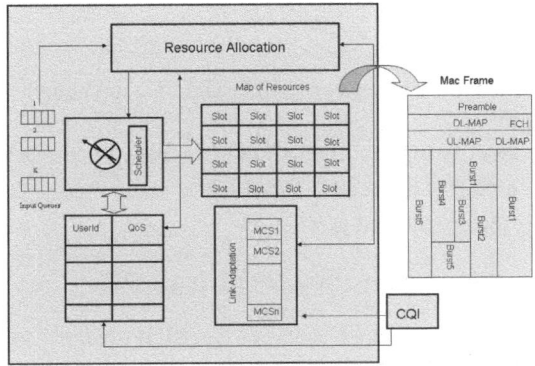

Fig. 2. Dynamic Resource Allocation Architecture

Figure 2 is a schematic representation of the proposed DRA scheme for the Wi-MAX standard. The packet scheduler creates a list sorted by the decreasing order of the priority metric. Packets achieving the maximum delay bound for the service are dropped in the BS. The Resource Allocator assigns slots for the packets remaining in the buffer of the selected user. The process continues until there are no packets for transmission or no slots available in the map of resources. Each burst is assigned a free HARQ channel that uses type II Chase Combining . The whole transmission process elapses along a cycle equivalent to two frames. Feedback channels (HARQ and CQICH) are assumed to be transmitted in the uplink sub-frame of the same frame in which the corresponding downlink transmission occurred.

In the simulations all users experiencing a bad channel condition, i.e., those users reporting a CQI value which do not allow the selection of the most robust MCS scheme, are not considered in the scheduling process. The idea is to limit the amount of packets received in error due to bad channel quality.

3.1 Channel Models

Fast fading is generated by a modified Jakes model where the carrier frequency and the mobile speed are used for fast generation of independent Rayleigh faders [15]. In the simulation a wideband SISO channel model is implemented by a six tapped delay model, according to the Pedestrian-B 3Km/h channel model for the serving sectors of the central BS. A flat fading channel is assumed for neighboring cells. Bi-dimensional log-normal shadowing is generated at the beginning of each run i [16]. The shadowing $SH(x, y, j)$ in DB between one $MS(x, y)$ and one BS j is the sum of two variables:

$$SH(x, y, j) = \sqrt{0.5(F_0(x, y) + F_j(x, y))} \tag{6}$$

Where $F_0(.)$ and $F_j(.)$ are spatial functions generated using the method described in [16]. They have a Gaussian distribution with zero mean and σ_{shad} standard deviation and a spatial correlation given by:

$$R(d) = e^{-\ln(2)d \, / \, DecorrLength} \tag{7}$$

Where d is the distance between two points and "DecorrLength" is the shadowing de-correlation length.

The path loss model is the modified COST231 Hata for the urban macrocell at a carrier frequency f [GHz] between 2 and 6 GHz, assuming the BS and MS heights of 32m and 1.5m respectively, and is given by:

$$Pl(d) = 35.2 + 35\log_{10}(d) + 26\log_{10}(f / 2) \tag{8}$$

Using the aforementioned channel models, the SINR (Signal-to-Interference-plus-Noise Ratio) of each OFDM sub-carrier is computed according to the approach:

$$\gamma_k = \frac{I_{or}}{I_{oc} + N_o} \cdot \frac{N_{used}}{N_d + PDR.N_p} \cdot H_k \tag{9}$$

Where N_{used} is the total number of sub-carriers, PDR is the Pilot-to-Data sub-carriers power ratio and N_d is the number of data sub-carriers per OFDM symbol, for the PUSC channel mode. N_o is the receiver thermal noise power and I_{oc} is the other-cell interference power density, assumed spatially and temporarily uncorrelated. It is assumed that neighboring cells transmit with maximum power, i.e. with full load.

The gain of the kth sub-carrier is computed according to the recommendations of as follows:

$$H_k = \left| \sum_{p=1}^{N_{paths}} M_p A_p e^{j\theta_p} e^{-2\pi k T_p} \right|^2 \tag{10}$$

Where p represents the multi-path index for the Ped. B channel model A_k is the amplitude value corresponding to the long-term average power for the pth path of this same channel model f_k is the relative frequency offset of the kth sub-carrier of the specific FUSC sub-channel and T_p is the relative time delay of the pth path.

In the receiver, post-processing of the signals received from the serving and interfering BSs is performed. For a Single-Input-Single-Output (SISO) architecture and a matched filter in the receiver, the received signal at the kth sub-carrier for the target user is computed according to:

The transmission of a coded block over different sets of sub-carriers results in a number of SINR measures that equals the number of sub-carriers sets, which can be quite high. Hence, data compression is mandatory. The coded symbol SINR in sub-carrier k is given by:

$$SINR^{(0)}[n] = \frac{P_{slot}^{(0)} P_{loss}^{(0)} \left| H^{(0)}[k] \right|^2}{\sum_{j=1}^{N} P_{slot}^{(j)} P_{loss}^{(j)} \left| H^{(j)}[k] \right|^2 + \sigma_n^2} \tag{11}$$

Where N is the number of interferers, $P_{slot}^{(j)}$ is the transmit power per slot for the jth cell, $P_{loss}^{(j)}$ is the distance dependent path loss, including shadowing and antenna gains/losses, $H^{(j)}[k]$ is the channel gain for the desired MS from the jth cell and for the or the kth sub-carrier, $X^{(j)}[k]$ is the transmitted symbol by the jth cell at the kth sub-carrier, $N^{(0)}[k]$ is the thermal noise at the received, modeled as AWGN with zero mean and variance σ_n^2.

The set of coded symbols SINRs are mapped onto a single value named the Effective SINR value. This value can be used to match AWGN Look-Up Tables (LUTs). The EESM expression determines how the effective SINR is obtained from the multiple SINR's on the different subcarriers:

$$SINR_{eff} = -\beta \ln \left(\frac{1}{P} \sum_{p=1}^{P} e^{-\frac{SINR_p}{\beta}} \right) \tag{12}$$

Where the parameter β is to be optimized for every link mode (MCS) based on link level simulation results.

3.2 User Quality Tracking

Periodically the SIR measurement performed by each MS, $y_n(MS_i)$, using the symbol of the preamble is reported and is available at the BS at nT_{CQI} (T_{CQI} is the reporting period). Every T_{CQI} the quality of the channel is updated combining the past information and the new value $w_n(MS_i)$ provided by these measurements, according to the time-smoothing filter:

$$CQI_n(MS_i) = 0.7 * w_n(MS_i) + 0.3 * CQI_{n-1}(MS_i) \tag{13}$$

3.3 Traffic Models

Simulations were carried on using the NRTV64 traffic model [17]. This traffic model periodically generates packets of variable sizes which compose a frame of video data arriving at a regular interval of T seconds. The NRTV traffic model is a bursty traffic model with an average source bits rate of 64Kbps. Table I lists the parameters used in the NRTV traffic model.

Table 1.

Characteristics	Distribution	Parameters
Inter-arrival time between frames (T)	Deterministic	100ms
Number of packets/frame	Deterministic	8
Packet size	Truncated Pareto (Mean: 50, Max: 125 (bytes))	K=20 bytes, $\alpha = 1.2$
Inter-arrival time between packets	Truncated Pareto (mean: 6, Max: 12.5 (ms))	K=2.5ms, $\alpha = 1.2$

3.4 Simulation Scenario

The system level performance evaluation methodology is based on the draft evaluation proposal for IEEE802.16. Simulations are carried out using a combined snapshot-dynamic mode, where in each simulation run (of a total of $N_{run} = 5$ independent dynamic runs) N_{user} users are randomly uniformly distributed over a hexagonal network of tri-sectored cells composed of three tiers of 19 BSs with 3 sectors each). Each simulation run lasts for $T_{run} = 75000$ frame periods. User positions were updated on every run and for each MS-BS pair, a random shadowing value is drawn whilst mobile positions, shadowing and path loss values are kept constant for the whole simulation duration. A full load scenario is assumed where all BSs around the central cluster are assumed as transmitting with maximum power all the time. They are considered for DL interference only on the central BS, which is the only one simulated.

In each frame interval the following events are performed: packets are generated according to the NRTV64 traffic model; the fast fading channel is updated; DRA executed and packet quality detection is performed. During the packet quality detection, the received block SIR of the packet is computed and mapped to the corresponding BLER using the link interface. A random variable u uniformly distributed between 0 and 1 is drawn. If $u < BLER$ the packet is assumed as erroneous and a NACK message is feedback to the serving BS on the CQICH of the uplink subframe, otherwise the packet is assumed as correct and an ACK message is sent. The ACK or NACK message is assumed to be received within the uplink portion of the same TDD frame (uplink sub-frame) in which data is initially transmitted or retransmitted

3.5 Performance Metrics

The following metrics were used to quantify system level performance:

Average Service Throughput per MS: $R_{Serv}^i = \dfrac{b_i}{T}$

Where b_i the total amount of bits is correctly received by the MS i over the whole simulated time and T is the simulation elapsed time.

Average Transfer Delay per MS: $T_D^i = \dfrac{1}{N} \sum\limits_{n=1}^{N} T_D^i(n)$

Where $T_D^i(n)$ is the transferred delay associated to the nth packet successfully received by user i. This is the average packet transfer delay for all packets received with success by the mobile.

95-percentile of the CDF of the transfer delay per user
This corresponds to the 95-percentile of the delay from the CDF of all packets transferred delays for each user in the system.

Average Packet Error Rate per user: $PER^i = \dfrac{N_{Correct}^i}{N}$

Where $N_{Correct}^i(n)$ is the total amount of packets received with success by the user.

3.6 Simulation Results

We performed simulations for 50, 60, 70, 80 and 90 users for the UTIL, CI, PF, M-LWDF and EXP schedulers.

Figure 3 show the evolution of the average packet delay vs. the number of users in the system. In the overloaded scenario the M-LWDF scheduler does not perform well because it prioritizes packets based on the HOL packet delay, without considering how close a packet's delay can be to its delay threshold. The EXP scheduler has a better performance because it attempts to equalize the HOL packet delays for all

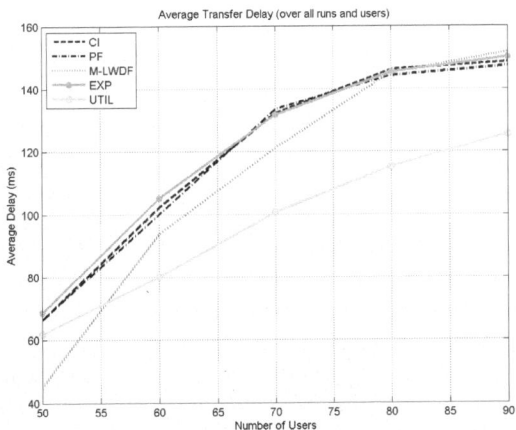

Fig. 3. Average Transfer Delay vs. total number of users

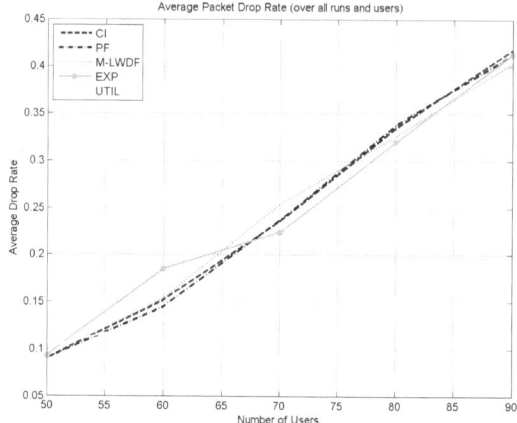

Fig. 4. Average Packet Drop Error Rate vs. total number of users

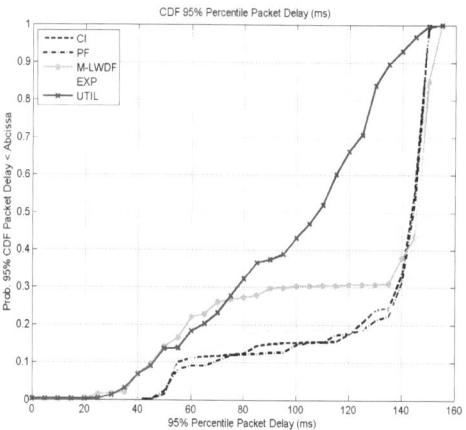

Fig. 5. CDF of the 95 Percentile of the Packet Delay

users. However, the highest gain can be observed with the UTIL scheduler. This is due to the sensitivity of the algorithm as the delay approaches its delay bound, which is determined by the according to the type of utility function used.

Figure 3 show the evolution of the average packet delay vs. the number of users in the system. In the overloaded scenario the M-LWDF scheduler does not perform well because it prioritizes packets based on the HOL packet delay, without considering how close a packet's delay can be to its delay threshold. The EXP scheduler has a better performance because it attempts to equalize the HOL packet delays for all users. However, the highest gain can be observed with the UTIL scheduler. This is due to the sensitivity of the algorithm as the delay approaches its delay bound, which is determined by the according to the type of utility function used.

Figure 4 is the plot of the average packet error rate vs. no. of users. It can be noticed that the gain achieved with the UTIL algorithm for a total amount of users is higher than 70. The better performance achieved with the MLWDF for a low a number of users smaller number of users is due to the higher percentage of packets dropped.

Figures 5 and 6 shows the CDF of the 95th percentile of the CDF of all packets delays and the average packet delay for all users respectively, for 70 users (for 70 users the average PER is below 20% and the average service throughput is around 50kbits. Note that the offered service data rate is 64 kbps). The performance gains of the UTIL algorithm are evident from both the graphs in Fig. 3 and 4.

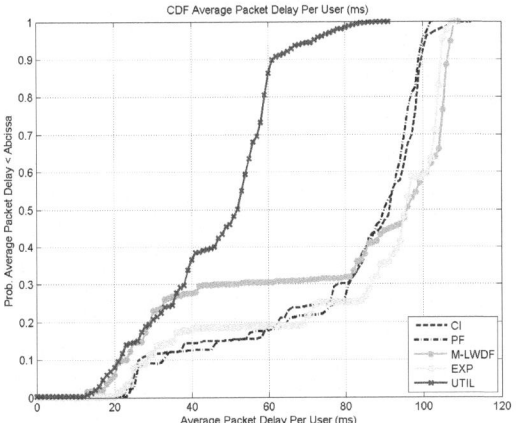

Fig. 6. CDF of the Average Packet Delay

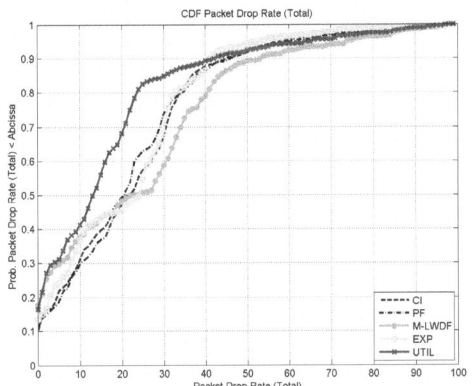

Fig. 7. CDF of the Average Packet Drop Rate

Figures 7 corroborate the gains achieved with the UTIL scheduler. The packet drop rate (including both packets dropped for quality reasons and those dropped due to delay bound violation) has a better performance for the UTIL scheduler.

Figure 8 shows the average user throughput versus the average sector throughput for the utility-based scheduler and for the max C/I scheduler as a benchmark for

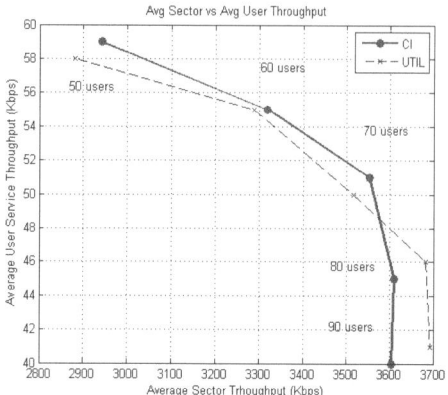

Fig. 8. Avg. User Throughput vs. avg Sector Throughput

different loads. It can be seen than although the Max C/I scheduler is more efficient in terms of resource utilization, the performance achieved with utility-based scheduling is quite close to that one achieved with the max C/I. Indeed, for higher amount of users, overloading results in the utility-based scheduler achieving even higher user throughputs for the same amount of sector throughput as for the Max C/I scheduler.

4   Conclusions

In this paper we propose an innovative cross-layer DRA architecture for the Mobile WIMAX standard using a packet based scheduler based on the notion of utility functions. Both the channel state as well as the packet delay are considered in the scheduling process. The simulation results show that the DRA scheme together with the utility-based scheduling algorithm has the capacity to increase the delivered QoS within WiMAX whilst utilizing spectrum efficiently.

Acknowledgement

This work presented in this paper was partially supported by the EUREKA CELTIC LOOP (CP4-016) project.

References

1. Gomez, J., Alonso, J., Verikoukis, Ch., Pérez-Neira, A., Alonso, L.: Cooperation on Demand Protocols for Wireless Networks. In: IEEE PIMRC 2007, Athens (September 2007)
2. Shakkottai, S., Rappaport, T.S., Karlson, P.: Cross-Layer Design for Wireless Networks. IEEE Comm. Magazine 4(10), 74–80 (2003)
3. Kopp, R., Humblet, P.: Information Capacity and Power Control in Single-Cell Multiuser Communications. In: Proc. Of ICC (June 1995)

4. Viswanath, P., Tse, D.N.C., Laroia, R.: Opportunistic beamforming using dumb antennas. IEEE Trans. Inform. Theory 48(6), 1277–1294 (2002)
5. Andrews, M., Kumaran, K., Ramanan, R., Stolyar, A., Vijayakumar, R., Whiting, P.: CDMA data QoS scheduling on the forward link with variable channel conditions, Bell Labs Technical Report preprint (April 2000)
6. Shakkottai, S., Stolyar, A.: Scheduling algorithms for a mixture of real-time and non-real time data in HDR. In: Proc. of the 17th ITC, Salvador, Brazil (2001)
7. Shenker, S.: Fundamental design issues for the future Internet. IEEE JSAC 13(7), 1176–1188 (1995)
8. Choi, Y.J.: Delay-Sensitive Packet Scheduling for a Wireless Access Link. IEEE Trans. on Mob. Computing 5(10), 1374–1382 (2008)
9. Choi, Y.J., Choi, J.G., Bahk, S.: Upper-level scheduling supporting multimedia traffic in cellular data networks. Computer Networks 51(3), 621–631 (2007)
10. Ryu, S., Ryu, B.-H., Seo, H., Dhin, M., Park, S.: Wireless Packet Scheduling Algorithm for OFDMA System Based on Time-Utility and Channel State. ETRI Journal 27(6), 777–787 (2005)
11. Sang, A., Wang, X., Madihian, M., Gitlin, R.: A Flexible Downlink Scheduling Scheme in Cellular Packet Data Systems. IEEE Trans. On Wireless Comm. 5(3) (March 2006)
12. Lei, H., Zhang, L., Yang, D.: A Packet Scheduling Algorithm Using Utility Function for Mixed Services in the Downlink of OFDMA Systems. In: IEEE VTC-2007 Fall, September 2007, pp. 1664–1668 (2007)
13. IEEE Std 802.16e, Part 16: Air Interface for Fixed and Mobile Broadband Wireless Access Systems (December 2005)
14. WiMAX forum, http://www.wimaxforum.org
15. Nascimento, A., Rodriguez, J., Gameiro, A.: Dynamic Resource Allocation Architecture for IEEE802.16e: Design and Performance Analysis. ICT Mobile Summit, Stockholm (June 2008)
16. Li, Y., Huang, X.: The generation of independent Rayleigh faders. In: Proc. ICC200 – International Conference on Communications, vol. 1, pp. 41–45
17. Cai, X., Giannakis, G.B.: A two-dimensional channel simulation model for shadowing processes. IEEE Transactions on Vehicular Technology 52(6)
18. UMTS, Selection procedures for the choice of radio transmission technologies of the UMTS, ETSI, UMTS TR 101.112 v.3.2.0. Tech. Rep. (April 1998)

Developing an Innovative Multi-hop Relay Station Software Architecture in the Scope of the REWIND European Research Programme

Konstantinos N. Voudouris[1], Ioannis P. Chochliouros[2], Panagiotis Tsiakas[1],
Avishay Mor[3], George Agapiou[2], Avner Aloush[3], Maria Belesioti[2],
and Evangelos Sfakianakis[2]

[1] Technological Educational Institution (T.E.I.) of Athens, Dept. of Electronics
Ag. Spyridonos & Milou 1 Street, 12210 Egaleo, Athens, Greece
kvoud@ee.teiath.gr
[2] Hellenic Telecommunications Organization (O.T.E.) S.A., Research Programs Section,
99, Kifissias Avenue, 15126 Maroussi, Athens, Greece
ichochliouros@oteresearch.gr
[3] DesignArt Networks Ltd., Ha'Haroshet Street, P.O. Box 2278, Ra'anana, Israel
avishaym@designartnetworks.com

Abstract. In the scope of the REWIND European Research Program we propose a Multi-Hop Relay network architecture, based on the recently developed IEEE 802.16j standard with the aim of enhancing throughput, network coverage and capacity density. In particular, we provide an essential description of the Relay Station Software architecture design, mainly focused on the PHY and MAC layers architecture, by analyzing the essentials of the downlink and uplink data flows, on the basis of the corresponding relay node to be included in the wireless network.

Keywords: Control and Data Flow, IEEE 802.16j standard, MAC layer, Multi-hop Relay, PHY layer, Relay Station (RS), WiMAX.

1 Introduction

Wireless communication has become a "hot topic" in IT (Information Technology) and CT (Communication Technology); personal broadband is currently considered as one of the *most emerging* and *most promising* services in the global area of wireless communications. Due to the introduction/deployment of innovative facilities (such as video-on-demand (VoD), on-line gaming and e-learning) and the continuously increasing abundance of numerous portable devices (such as PDAs (personal digital assistants) and smart-phones), the need for ubiquitous connectivity and coverage has now become a necessity. Towards fulfilling that purpose, WiMAX (Worldwide Interoperability for Microwave Access), as a robust and reliable innovative technology, enables users to enjoy the same experiences they have at home

F. Granelli et al. (Eds.): MOBILIGHT 2009, LNICST 13, pp. 160–172, 2009.

or in the office wherever they go, at affordable prices[1]. Activity in the relevant sector continues very fast: i.e. products are reaching in the market, networks are being rolled-out and service offerings are still developing, under various scenarios. WiMAX technology has the potential to be an important component of future converged (or ubiquitous) networks due to its reach and the relatively high-speed wireless connectivity and service availability. In fact, operators have to face (and are still struggling with) a number of major challenges, such as increased system capacity with better coverage; moreover, they have to deliver new and innovative, "value-added" services ever-improving performance features. Finally, all previous concerns are strongly bundled with lower network cost per subscriber.

The term "repeater" (or "relay") originated with telegraphy and referred to an electromechanical device, used to regenerate related signals. Use of the term has continued in telephony and data communications. So, a repeater is an electronic device that receives a signal and retransmits it at a higher level and/or higher power, or onto the other side of an obstruction, so that the signal can cover longer distances without degradation. In telecommunications, the term "repeater" has the following wider standardized meanings: (i) An analogue device that amplifies an input signal regardless of its nature (analogue or digital); (ii) A digital device that amplifies, reshapes, retimes, or performs a combination of any of these functions on a digital input signal for retransmission. Thus, in digital communication systems, a repeater receives a digital signal on an electromagnetic or optical transmission medium and regenerates it along the next leg of the medium. In electromagnetic media, repeaters overcome the attenuation caused by free-space electromagnetic-field divergence or cable loss. A series of repeaters make possible the extension of a signal over a longer distance. Repeaters can "remove" the unwanted noise in an incoming signal. Unlike an analogue signal, the original digital signal, *even if weak or distorted*, can be clearly perceived and restored. With analogue transmission, signals are re-strengthened with amplifiers, which unfortunately also amplify noise. Because digital signals depend on the presence (or absence) of voltage, they tend to dissipate faster than analogue ones and need more frequent repeating. Whereas analogue signal amplifiers are spaced at 18,000 meter intervals, digital signal repeaters are typically placed at 2,000-6,000 meter intervals.

As far as WiMAX is concerned, the repeater is a two-way radio transceiver system designed to provide coverage of "dark zones" not served by WiMAX base stations (i.e. areas where coverage gaps are detected), bypassing direct Line-of-Sight (LOS) obstacles between a base station (BS) and a number of remote terminals or extending coverage beyond the BS limits. A typical WiMAX repeater is composed of two outdoor transceivers (local and remote), specifically designed to minimize the cable running to each antenna and the implicit signal degradation, as well the induced signal delay. The only connection requirement from the outside is a power feed. Each transceiver executes a transparent retransmission of the signal on each way. A frequency offset mechanism avoids co-channel interference and "eases" the frequency planning process, while an ALC (Automatic Level Control) provides a stable signal regardless

---

[1] Related work is performed by the WiMAX Forum, an organization with more than 150 member-companies with the goals of promoting worldwide adoption of the IEEE 802.16 air interface standard and to develop a suite of conformance tests to ensure equipment interoperability. [www.wimaxforum.org]

of weather and other transient conditions. Thus, a repeater is used to enhance signals between mobile user equipment and a BS, to extend the coverage of a single BS and so to solve design problems and performance issues (for small areas).

The EU-funded "*REWIND*" Project (*ICT-FP7, Grant Agreement No.216751*) examines issues of relay station implementations for WiMAX, as such technology is currently the "most advanced" wireless one available for deployment, and many of its aspects are likely to be implemented in any 4G wireless technology. Relay stations standardization has already commenced by the IEEE under the "Multi-hop Relay Task Group" also known as *802.16j Task Group*. The intention of the REWIND Project is to support this standard process, through direct participation and through research and development of relay technology and products in order to assist the standard body with interoperability, laboratory and field information on possible implementations of the WiMAX relay. REWIND will proceed to the algorithmic research and technology development of the mobile multi-hop WiMAX relay networks in order to increase coverage and throughput issues. Part of its actual work is responsible for the design of the novelty software (SW) and hardware (HW) functional areas of the corresponding Relay Station (RS) product, which includes algorithmic research and simulations, system architecture and requirements specifications, and appropriate software code development and integration. In the scope of the present work, Section 2 deals with essential architectural features of relay-based networks. Section 3 focuses on fundamentals of the 802.16j protocol stack while Sections 4 emphasizes on control and data flow issues of the essential REWIND-based architecture.

2 Architecture of Relay-Based Networks

In order for the next generation of wireless technology to be able to deliver ubiquitous broadband content, the network is required to offer excellent coverage (both outdoor and indoor) and significantly higher bandwidth per subscriber. To achieve this at frequencies above 2 and 3 GHz (as these are "targeted" for future wireless technologies), network architecture must reduce significantly the cell size or the distance between the network and subscribers' antennas.

The principal element in all wireless networks is the "cell". A cell is a region of radio reception that is a part of a larger radio network and is served by a fixed transmitter called the Base Transceiver Station (BTS or BS) which is controlled by the Base Station Controller (BSC). As the cell size becomes smaller the number of subscribers served per unit area is increased or higher bit rates can be provided for services [1]. Additionally, the size of the BS equipment as well that of antennas is reduced, since the transmitted power is lower in order to keep the cell dimensions small, replacing bigger equipment and thus making installation easier. Another benefit compared to larger cells (e.g. macro-cells) is smaller site acquisition costs and easier to obtain planning permissions [2]. While micro-, pico- and femto-BTS technologies reduce the cost of BS equipment, they all still depend on a dedicated backhaul which results in significant capital expenditures (CAPEX) and operating expenditures (OPEX) to the network operator. The wireless multi-hop relay station solution is designed to overcome such limitations.

Although relay technologies are not an entirely new concept [3], large-scale deployment of standardised related technology is new. Thus, there is a need for new approaches to network planning [4]. Multi-hop Relay (MR) is a deployment that may be used to provide additional coverage or performance advantage in access networks. In MR networks, the Multi-hop Relay BS (MR-BS) is connected to several RSs, in a "multi-hop topology", to enhance the network coverage and capacity density [5].

Traffic and signalling between the Subscriber Station-SS[2] (i.e., the equipment set providing connectivity between subscriber equipment (potentially including customer premises equipment-CPE and a BS) and MR-BS are relayed by the RS thereby extending the coverage and performance of the system in areas where RSs are normally deployed. Each RS is under the supervision of an MR-BS. In a system with more than two hops, traffic and signalling between an access RS and MR-BS may also be relayed through intermediate RSs. The RS is fixed in location. The SS may also communicate directly with the MR-BS.

Fig.1 illustrates the MR-BS (connected to an access service network gateway (ASNGW) via a Giga-Ethernet connection) and two-hop RS deployment. For each one of the RSs there is an "access link" (i.e. the radio link between a MR-BS or a RS and a SS) that covers the current cell and a "backhaul link" to the next cell (i.e. the radio link between a MR-BS and RS or between a pair of RSs).

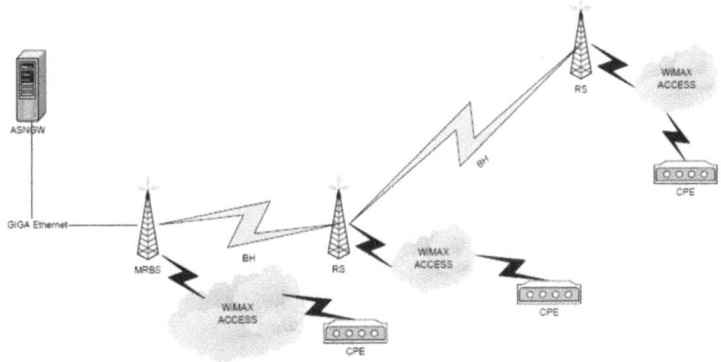

Fig. 1. MR-BS and RSs Deployment

2.1 Relay Scenarios for Network Coverage Extension

The current outlook for the overall broadband market worldwide shows that wireless broadband systems will account for 48% (568 million of 1.175 billion) of broadband subscribers worldwide by 2012, significantly higher than the present 17% (69 million of 407 million) [6]. Despite the prognosis that the number of wired broadband subscribers will continue to increase, it is also estimated that their penetration rate will slow down when emerging markets will be converted in developed ones. And this

---

[2] The term "subscriber station" can be also met as the "user equipment"-UE. It is also known as "mobile station" (MS).

will happen mainly due to the lack of availability of wired lines and due to their expensive deployment cost. On the other hand, this exactly constitutes an opportunity for the wireless broadband providers [7].

As it is well known in all mobile communications, at a given power and carrier frequency, the available data rate decreases when the distance from a BS increases. In order to achieve a satisfactory radio coverage, high capacity, and advanced service quality, the provider should deploy a network with high base stations density. But, in this case the overall cost will increase dramatically and, *consequently*, the end-user should be obliged to pay more expensive for services, especially in emerging markets. In order to meet the goal of low-cost radio network deployment for both, short-range and wide-area coverage, the deployment concept based on layer 2 (multi-hop) relay nodes appears to be one of the most promising solutions [8].

WiMAX relays can achieve more than simply extending the coverage similar to Wi-Fi Mesh or 3G's technologies. Due to OFDMA (Orthogonal Frequency-Division Multiple Access) waveform [9] -which offers network operators the advantage of being able to operate with the larger delay spread of the NLOS (non-LOS) environment [10] - combined with adaptive modulation, adaptive power management, adaptive sub-channelization methods [11] and finally an adaptive architecture of the network, relays can be a quite useful tool in the hand of the operators. When analyzing a business scenario both for fixed and wireless access, it is informative to breakout the end-to-end network as described below and look separately at its parts, i.e.: (i) Customer Premise Equipment (CPE); (ii) Base Station / Relay Station Infrastructure, and; (iii) Edge, Core, and Central Office (CO) Equipment.

There are several cases where repeaters could be used as an active "part" of the considered network. Forwarding traffic between wireless network nodes is an interesting method to fill "coverage holes" and to improve in-building coverage. For example, at an airport where the need for capacity is high at certain peak periods (e.g. departures, arrivals) a repeater could be used in order to "extend" the coverage area of a powerful BS. A case where coverage extension is essential would be during the set up of a "major" event, e.g. a big festival. Such events are usually set up in the borderlines of already established communities, where the required space would be available, but the network infrastructure might not be as extended to cover the extra space or may not be able to support the sudden temporary increase of users (as at the suburbs of a city). This is the case of "nomadic relay stations" which are brought into effect for temporary cover. These can also be used for emergencies and/or disasters.

The terrain morphology (such as height, density, separation, terrain type – hilly or flat) plays an important part for planning requirements, since in wireless radio planning the accurate prediction of electromagnetic field strengths over large areas, both land and sea and the regional clutter are required. For example, a village which is located at the foothill of a mountain might not be covered sufficiently by a BS at the top of the mountain. In this case a RS placed in between, could solve any coverage-related problem or inconsistency. Indeed, a RS can efficiently enhance coverage and so provides an economically satisfactory solution for the network provider, especially in remote areas where usually small communities are situated and a large investment for a corresponding BS could not be justified. In such cases, where buildings are sparse, the RS will mainly be used for coverage range extension.

In fact there are three sub-environments that should be taken under consideration: The first case concerns the vehicular environment, where high mobility and nomadic devices exist (i.e., the case of a traveller in a train who wants to access the Internet, to use multimedia services and to enjoy video streaming or other novel services). In this particular case a number of distributed repeaters may be located in the outside environment of the vehicle (e.g. on the roof of the train). The bandwidth/capacity offered by the network should be "adequate" to support throughput, compatible with data services required from such moving users. The second case relates to the indoor/outdoor environment, where there are static or medium/high mobility devices, and thus very few repeaters may be required to cover the group of establishments and houses and to form a small network, at which users should have a "proper" access for services. Last but not least, are the suburban areas, meaning little populated and scarcely networked areas (such as schools and private residential houses), where the installation of expensive equipment is not justified and so RS implicates the most appropriate solution. Furthermore, there are rural areas which are scarcely populated areas with insufficient coverage, usually of "outdoor" nature. These environments have often little (or no possibilities) to backhaul high bandwidth connections (e.g. via fixed line copper or fiber links or microwave radio links), and consequently, two alternatives are applicable: Either dedicated high bandwidth microwave connections per base station, or the formation of a mesh-like network by the BSs themselves to forward traffic between base stations with no dedicated backhaul connection. This option significantly reduces deployment costs (i.e. CAPEX and OPEX).

Some other notable attributes of relay-based networks is that they can deploy rapidly with flexibility in placing relays and, *at the same time*, they create multiple protection routes with optimized network load distribution, thus ensuring the ability for better quality of service (QoS). A WiMAX repeater provides cost-effective solution to extend the coverage of a WiMAX BS beyond its boundaries, as it does not hold as much infrastructure requirements (shelter, backhaul, energy consumption, etc.) as in the case of a pure WiMAX BS. In addition, the small size of the repeater reduces real estate requirements and therefore reduces deployment cost. It even can be co-located with already existing cell sites making the network management easier. Wireless low cost infrastructure based systems (such as the proposed WiMAX repeaters) may increase the penetration of communications technology in higher levels [12], resulted many end-users with different needs and demands.

3   Essentials of the IEEE 802.16j Protocol Stack

In WiMAX, the IEEE only defined the Physical (PHY) and Media Access Control (MAC) layers in 802.16 [13]. This approach has worked well for technologies such as Ethernet and WiFi, which rely on other bodies such as the IETF (Internet Engineering Task Force) to set the standards for higher layer protocols such as TCP/IP (Transport Control Protocol / Internet Protocol), SIP (Session Initiation Protocol), VoIP (Voice over Internet Protocol) and IPSec (Internet Protocol Security). The main objective of the IEEE 802.16 standards is to develop a proper set of specifications of the air interface, while the WiMAX Forum defines system's profile, i.e. a list of selected functionalities for a particular usage scenario and overall network architectures [14].

The sector continues to innovate in many ways; one is the development/promotion of new standards to solve "open" problems. Such an initiative (which is currently of extreme interest) is the development of the *Multi-hop Relay Standard, IEEE 802.16j*. This is being developed to provide low cost coverage in the initial stages of network deployment and increased capacity when there is high utilisation of the network. The standard has been identified with a better feasibility and efficiency due to the similarities in the MAC and PHY layers and the support of fast route change[3]. The standard is expected to have significant impact in new 802.16 rollouts.

When deploying an IEEE 802.16j based WiMAX network, this can be considered as a cellular network since they both have the same design principles [15]. The standard specifies a set of technical issues in order to enhance previous standards with the main objective of supporting relay concepts [16]. The 802.16j standard defines an air interface between a MR-BS and a RS [17]. Its most important technical issues are listed as follows: Centralized vs. distributed control; Scheduling; Radio resource management; Power control; Call admission and traffic shaping policies; QoS based on network wide load balancing and congestion control; Security, and; Management. Besides these issues, when it comes to network deployment the main objective remains the optimal placement of the RSs. Operators are mainly concerned about operating costs, revenues and the pay-off periods for their investments; but the quality of services offered, is also an important issue for them. The IEEE 802.16j protocol layering for a simple RS is shown in Fig. 2.

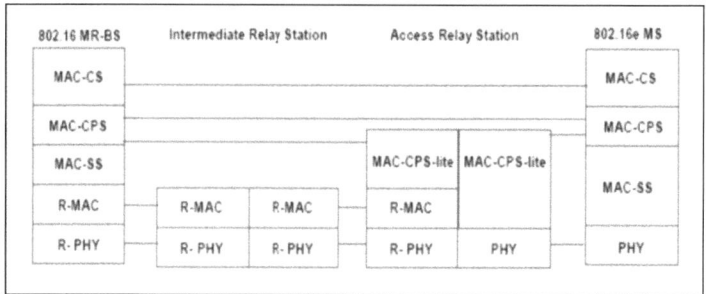

Fig. 2. Depiction of the IEEE 802.16j Protocol Layering for a Simple Relay Station

The principles of the above layers are briefly discussed as follows:

The *MAC Converge Sub-layer (CS)* is primarily used to "map" any data received from the upper layers (e.g. IP, Ethernet) to an appropriate MAC connection, to manage data flow from the upper layer and to ensure that QoS requirements are fulfilled. More specifically, the CS accepts data from the network layer through the CS Service Access Point (SAP) and performs data classification into the appropriate MAC Service Data Units (SDUs). A "classifier" is an entity which selects packets based on the content of packet headers and categorises them according to a set of

---

[3] Harmonized Contribution 802.16j (Mobile Multihop Relay) Usage Models.
http://grouper.ieee.org/groups/802/16/relay/docs/80216j-06-015.pdf

"matching criteria" (such as destination IP address). The external higher-layer Protocol Data Units (PDUs) entering the CS are checked against those criteria and accordingly delivered to a specific MAC connection.

The *MAC Common Part Sub-layer (CPS)* is a connection-oriented protocol. Contrarily to previous wireless network technologies (such as IEEE 802.11), it does provide QoS guarantees. It receives the MAC SDUs from MAC SAP; next, it delivers the MAC PDUs to a peer MAC SAP according to the requested QoS, to perform various transport functions (i.e. packing, fragmentation and concatenation). Each MAC PDU is identified by a unique connection identifier (CID). In the scope of past IEEE 802.16 standard versions, in 2004, an operation has been defined where multiple MAC PDUs could be concatenated and comprised into a single burst for transmission purposes. These PDUs had to be encoded and modulated, by using the same PHY (i.e. using the same burst profile); however they could be associated with subscriber stations. The position of each burst into the DL (downlink) frame has been specified by DL-MAP[4], which contained additional Information Elements (IEs). The IEs specify the CID of the receiver, the burst profile used, the start time of the burst and a bit indicating whether an optional preamble is present. The IEEE 802.16e standard has further extended the DL MAP IE of legacy IEEE 802.16, in order to carry the identifiers of multiple connections (CIDs) in a single IE. However, the last missing link for enabling efficient MAC PDU concatenation on relay link is the capability of supporting multiple connections using one uplink (UL) information element. Several approaches have been done to extend the UL MAP[5] IE for relay link.

The *MAC Security Sub-layer (MAC-SS)* handles security issues such as authentication, key exchange and privacy by encrypting the connections between BS and subscriber station. It is based on the Private Key Management (PKM) protocol, which has been enhanced to "fit" the IEEE 802.16 standard. At the time a subscriber connects to the BS, they perform mutual authentication with public-key cryptography using X.509 certificates [18]. The payloads themselves are encrypted by using a symmetric-key system, which may be either DES (data encryption standard) with cipher block chaining or triple DES with two keys.

The *Relay MAC (R-MAC) Sub-layer* has been introduced in the IEEE 802.16j standard. It provides efficient MAC PDU relaying/ forwarding and control functions, (such as scheduling, routing, and flow control). It is applicable to the links between MR-BS and RSs and between RSs.

The *Relay Physical (R-PHY) layer* provides definition of physical layer design, (i.e. sub-channelization, modulation, coding, etc.), for links between MR-BS and RS and between RSs. The IEEE 802.16j standard has extended the past IEEE 802.16e frame structure to support in-band BS-to-RS communication. A high level diagram of the 802.16j frame structure in TDD (Time Division Duplex) OFDMA PHY mode is shown in Fig.3. The frame structure supports a typical two-hop relay-enhanced communication, where some MSs are attached to a RS and communicating with a BS via the RS, and some MSs connected directly to the BS.

---

[4] Downlink map (DL-MAP) is a MAC message that defines burst start times for both Multihop concept in Time Division Multiplex and Time Division Multiple Access (TDMA) by a subscriber station (SS) on the downlink.

[5] Uplink map (UL-MAP): A set of information defining the entire access for a scheduling interval.

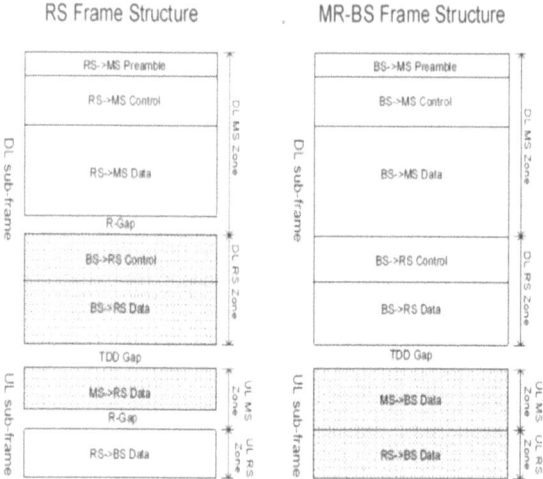

Fig. 3. OFDMA 802.16j Frame Structure

In Fig.3, the horizontal dimension denotes time and the vertical dimension denotes frequency. Frame sections in grey denote receive (Rx) operation, whereas sections in white denote transmit (Tx) operation. The BS and RS frames are subdivided into DL and UL subframes in order to support TDD operation. Both DL and UL subframes are further subdivided into MS and RS zones[6]. The MS zones, supported at both the BS and RS, are backwards compatible with the 802.16e standard. The RS transmits to MSs in its coverage in the DL MS zone and receives control and data from the BS in the adjacent DL RS zone[7]. Each MR-BS frame begins with a preamble followed by a FCH (Frame Control Header) and the DL-MAP and possibly UL-MAP. The DL subframe shall include at least one DL access zone and may include one or more DL relay zones. The UL subframe may include one or more UL access zones and it may include one or more UL relay zones. A relay zone may be utilized for either transmission or reception, but the MR-BS shall not be required to support both modes of operation within the same zone.

4   Control and Data Flow of the Proposed Architecture

The REWIND Project examines RS implementations for WiMAX networks, by proposing the design of the novelty software and hardware functional areas of a WiMAX

---

[6] For the DL case we can distinguish: (i) The DL access zone (i.e., a portion of the DL subframe in the MR-BS/RS frame used for MR-BS/RS to MS (or "transparent RS transmission"), and; (ii) the DL relay zone (i.e., a portion of the MR-BS/RS to RS transmission). In a similar way, for the UL case we can consider: (i) The UL access zone (i.e., a portion of the UL subframe in the MR-BS/RS frame used for MS to MR-BS/RS transmission), and; (ii) the UL relay zone (i.e., a portion of the UL subframe in the MR-BS/RS frame used for RS to MR-BS/RS transmission).

[7] Detailed structure of the RS zones is under consideration in the IEEE 802.16j Task Group.

Relay station (RS) product. Such a product is to be based on the design of a state-of-the-art SoC (System-on-Chip) silicon platform. The RS shall incorporate all the BS functionality, as well as the relay backhaul and routing functionality, in a single compact, low cost unit. Thus, the Relay node SW shall run all the PHY, MAC, scheduler and networking functionalities, required to operate a complete BS with relay functionality, and to integrate the node into the relay network. We so provide an overview of the RS SW architecture. The corresponding Relay node is based on a highly integrated SoC device, which incorporates all baseband processing (PHY, MAC and CS), networking and control processors required for the Relay station functionality. The RS shall incorporate all the BS functionality, as well as the Relay backhaul and routing functionality. The current section provides a brief overview of the required modules in the RS SW. The host SW is comprised of the following main packages:

- Relay Routing Package: It implements the relay routing connections.
- Frame Manager Package: It implements frame planning and map building.
- Management MAC Package: It implements the 802.16e MAC management procedures.
- DBS (Database) Package: It implements the BS, MSS and connections databases.
- CM (Connection Management) Package: It implements the Connections Management.
- Scheduler Package: It implements Data Delivery Services (QoS).
- Wireless Link Driver (WLD): It implements the drivers to the MAC/PHY machines.
- Backhaul Package: It implements the backhaul links between the MR-BS and RS.
- Configuration Management: It implements configuration parameters.
- RF (radio-frequency) Driver Package: It implements the RF driver.
- ARQ (Automatic-Repeat-Request) Package: Implements the ARQ mechanism.
- HARQ (Hybrid Automatic-Repeat-Request) Package: It implements the HARQ mechanism.
- Network Control package: Implements the R6 control plane.
- Management Package: It implements the unit management (SW upgrade, Telnet[8] and so on).
- Buffer Management Package: It implements the buffers allocation/free management.
- Inter-process Communication: It implements the communication between tasks that reside on different ARM (Advanced RISC Machine)[9] cores.
- Inter-task Communication: It implements the communication between tasks that reside on the same core.
- Boot Loading Package: It implements the boot loading process.
- OSA: It implements the Operating System Abstraction (OSA) Layer.
- ThreadX: High-performance real-time kernel.

Fig.4 describes the Data Flow Diagram (DFD) and the Control Flow Diagram (CFD), as both performed in the scope of the proposed relay station SW architecture

---

[8] A "Telnet" is a terminal emulation program for TCP/IP networks, such as the Internet.

[9] The ARM architecture (previously, the Advanced RISC Machine and prior to that Acorn RISC Machine) is a 32-bit RISC processor architecture that is widely used in embedded designs. Because of their power saving features, ARM CPUs are dominant in the mobile electronics market, where low power computation is a critical design goal.

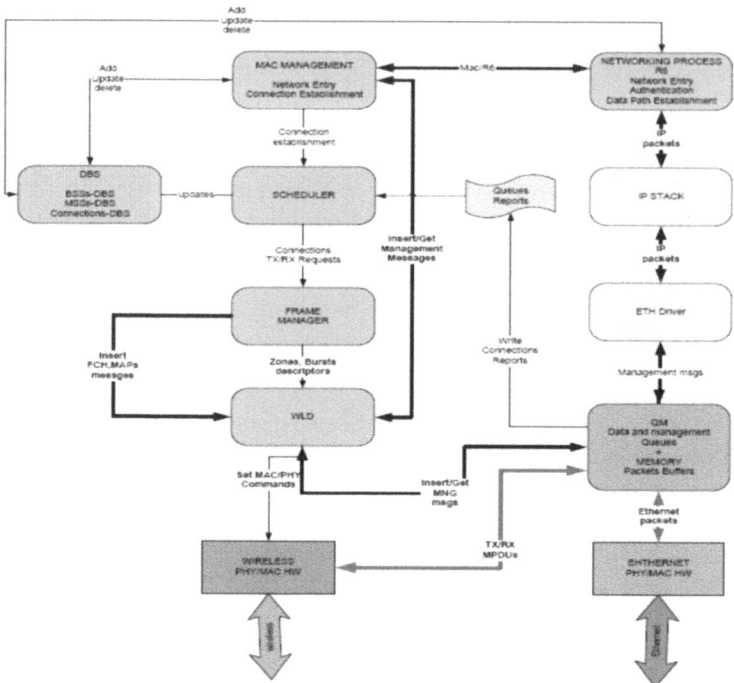

Fig. 4. Data and Control Flow Diagram

for the design of relevant product. In the continuity of the present work, we discuss the SW architecture used. The data critical path is run by the HW and controlled by the Scheduler, the Frame Manager and the Wireless Driver. All other missions are performed by other tasks, which have lower priority. DFD is demonstrated by green arrows; CFD is demonstrated by the black thin arrows; management messages are demonstrated by black wide arrows.

4.1 Downlink Data Flow

In the *downlink data flow*, packets are received via the Ethernet (ETH) port and classified by the ETH Rx Parser, which delivers the received packets together with their CID to the Queue Manager (QM). The latter inserts them to the proper connections queues, according to the specified CIDs and updates connections' reports according to the report groups defined by the Scheduler. The Scheduler polls the connection reports (every frame), and defines how many bytes should be transmitted from each DL connection. The Scheduler builds a list of requests for the Frame Manager. The Frame Manager, for a certain time frame (i.e. each 5ms), tries to "satisfy" Scheduler's requests, and plans (and builds) segment zones and bursts that contain MPDUs (MAC Protocol Data Units) for each connection, according to the specified rate (modulation/FEC (Forward Error Correction) and SISO (single-input, single-output) / MIMO (multiple-input, multiple-output) definition). After filling the frame, the Frame Manager builds both the DL MAP (which describes the transmitted bursts and the zones in the frame) and the UL MAP (which contains the Tx allocations for the MSs). The Frame Manager creates Tx commands for each transmitted MPDU; it builds Tx burst descriptors and zone descriptors and it delivers them to the WLD. The WLD sets the zone descriptors and the PHY descriptors to the PHY, and sets the Tx commands of each MPDU to the QM.

4.2 Uplink Data Flow

In the *uplink data flow*, for every frame, the Scheduler polls the bandwidth requests (as sent by the MSs) and other QoS UL connection attributes, and defines how many bytes are granted for each MS, for each UL connection. The Scheduler builds a list of UL requests to the Frame Manager. The letter, for the prescribed time of frame, tries again to "satisfy" Scheduler's requests, by using a method identical to the previous DL-related case. After filling the UL subframe, the Frame Manager builds the UL MAP; this describes the UL allocations and the zones in the UL subframe. The Frame Manager builds Rx burst descriptors and zone descriptors and delivers them to the WLD. According to the UL MAP, the MSs transmit their MPDUs, as per their UL Tx allocations. The WLD sets the UL zone descriptors and the Rx burst descriptors to the PHY. The PHY/MAC layer receives the UL mobile subscriber station transmission bursts in the allocated buffers memory. The hardware MAC extracts MPDUs from the received bursts into MPDU buffers, and inserts the MPDU buffers into the connection queues. The QM updates the report lists according to the received messages. The Scheduler reads the reports and updates the WLD about the received messages. The WLD then gets the received management messages and informs the MAC management about the management messages. The MAC handles the received management MPDUs and the QM inserts the user data traffic into the data user connection queues and creates report lists. Finally, the Scheduler reads the reports and orders the WLD to transmit the MPDUs to the Ethernet port, according to predefined directions for each queue.

5 Conclusion

Relay technology has focused significant attention due to several important and quite identifiable reasons i.e., simplicity, flexibility, deployment efficiency and cost effectiveness, as relays can permit a faster network rollout. Conformant to the scope of the European REWIND Research Project aiming to develop an effective relay station implementation for WiMAX technology, we have discussed several architecture principles and benefits for relay-based networks and then we have presented an essential description of the related RS software architecture design (PHY and MAC layers architecture) for the realization of a proper relay node required for the relay station functionality. Our approach has been based on the core concept of the Multi-Hop relay network architecture, conformant to the IEEE 802.16j standard.

Acknowledgments. The present work has been performed in the scope of the *REWIND ("RElay based WIireless Network and StandarD")* European Research Project and has been supported by the Commission of the European Communities - *Information Society and Media Directorate General* (FP7, Collaborative Project, *ICT-The Network of the Future*, Grant Agreement No.216751).

References

1. Tse, D., Viswanath, P.: Fundamentals of Wireless Communication. Cambridge Press (2005)
2. Rappaport, T.S.: Wireless Communications: Principles and Practice, 2nd edn. Prentice Hall, Upper Saddle River (2002)
3. Pabst, R., Walke, B., Schultz, D.C., et al.: Relay-Based Deployment Concepts for Wireless and Mobile Broadband Radio. IEEE Communications Magazine 42(9), 80–89 (2004)
4. Yu, Y., Murphy, S., Murphy, L.: Planning Base Station and Relay Station Locations, in IEEE 802.16j Multi-hop Relay Networks. In: IEEE International Conference on Communications (ICC), pp. 2586–2591. IEEE Press, New York (2008)
5. Hoymann, C., Klagges, K., Schinnenburg, M.: Multihop Communication in Relay Enhanced IEEE 802.16 Networks. In: 17th Annual IEEE International Symposium on Personal, Indoor and Mobile Radio Communications, Helsinki, Finland (2006)
6. Soldani, D., Dixit, S.: Wireless Relays for Broadband Access. IEEE Communications Magazine, 58–66 (2008)
7. Organization for Economic Coordination and Development (OECD): The Implications of WiMAX for Competition and Regulation (JT03204793). OECD, Paris (2006)
8. Walke, B., Wijaya, H., Schultz, D.: Layer 2 Relays in Cellular Mobile Radio Networks. In: IEEE VTC-2006 Spring, Melbourne, Australia (2006)
9. Van Nee, R., Prasad, R.: OFDM for Wireless Multimedia Communications. Artech House (2000)
10. Tucker, E.: Can Voice be the Killer App for WiMAX? The Basestation e-Newsletter, Aperto Networks (November 2006),
    http://www.openbasestation.org/Newsletters/
    November2006/Aperto.htm
11. Syputa, R.: The Impact of Mobile Multi-Hop Relay on WiMAX: Above and Beyond. Maravedis Newsletter 2(2) (November 7, 2007),
    http://www.maravedis-bwa.com/article-32.html
12. Andrews, J.G., Ghosh, A., Muhamed, R.: Fundamental of WiMAX - Understanding Broadband Wireless Networking. Prentice Hall, Englewood Cliffs (2007)
13. IEEE 802.16-2004: IEEE Standard for Local and Metropolitan Area Networks. Part 16: Air Interface for Fixed and Mobile Broadband Wireless Access Systems (2004),
    http://standards.ieee.org/getieee802/download/
    802.16e-2005.pdf
14. Nakamura, M., et al.: Standardization Activities for Mobile WiMAX. Fujitsu Sci. Tech. J. 44(3), 285–291 (2008)
15. Lin, B., Ho, P.-H., Xie, L.-L., Shen, X.: Optimal relay station placement in IEEE 802.16j networks. In: The 2007 International Conference on Wireless Communications and Mobile Computing (IWCMC 2007), Honolulu, Hawaii, pp. 25–30 (2007)
16. Zeng, H., Zhu, C.: System-Level Modeling and Performance Evaluation of 802.16j Multi-Hop Relay Systems. In: IWCMC, Next Generation Mobile Networks Symposium (2008)
17. Okuda, M., Zhu, C., Viorel, D.: Multihop Relay Extension for WiMAX Networks - Overview and Benefits of IEEE 802.16j Standard. Fujitsu Sci. Tech. J. 44(3), 292–302 (2008)
18. Chokhani, S., Ford, W.: Internet X.509 Public Key Infrastructure: Certificate Policy and Certification Practices Framework (RFC 2527). Internet Engineering Task Force (IETF), Sterling, VA (1999)

Full Scale Software Support on Mobile Lightweight Devices by Utilization of All Types of Wireless Technologies

Ondrej Krejcar

VSB Technical University of Ostrava, Center for Applied Cybernetics, Department of measurement and control, 17. Listopadu 15, 70833 Ostrava Poruba, Czech Republic
Ondrej.Krejcar@remoteworld.net

Abstract. New kind of mobile lightweight devices can run full scale applications with same comfort as on desktop devices only with several limitations. One of them is insufficient transfer speed on wireless connectivity. Main area of interest is in a model of a radio-frequency based system enhancement for locating and tracking users of a mobile information system. The experimental framework prototype uses a wireless network infrastructure to let a mobile lightweight device determine its indoor or outdoor position. User location is used for data prebuffering and pushing information from server to user's PDA. All server data is saved as artifacts along with its position information in building or larger area environment. The accessing of prebuffered data on mobile lightweight device can highly improve response time needed to view large multimedia data. This fact can help with design of new full scale applications for mobile lightweight devices.

Keywords: Mobile Lightweight Device; Localization; Prebuffering; Response Time; Area Definition.

1 Introduction

The usage of various mobile wireless technologies and mobile embedded devices has been increasing dramatically every year and would be growing in the following years. This will lead to the rise of new application domains in network-connected PDAs (Personal Digital Assistants) that provide more or less the same functionality as their desktop application equivalents. The idea of full scale applications pursuable on mobile lightweight devices is based on current hi-tech devices with large scale display, large memory capabilities, and wide spectrum of network standards plus embedded GPS module. Example of such devices is HTC Touch HD.

Users of these portable devices use them all time in context of their life (e.g. moving, searching, alerting, scheduling, writing, etc.). Context is relevant to the mobile user, because in a mobile environment the context is often very dynamic and the user interacts differently with the applications on his mobile device when the context is different [1].

F. Granelli et al. (Eds.): MOBILIGHT 2009, LNICST 13, pp. 173–184, 2009.
© ICST Institute for Computer Sciences, Social-Informatics and Telecommunication Engineering 2009

My recent research of context-aware computing has been restricted to location-aware computing for mobile applications using a WiFi network (LBS Location Based Services). The information about basic concept and technologies of user localization such as LBS, Searching for WiFi AP) can be found in my article [2]. On localization basis, I created a special framework called PDPT (Predictive Data Push Technology) which can improve a usage of large data artifacts of mobile devices [3]. I used continual user position information to determine a predictive user position. The data artifacts linked to user predicted position are prebuffered to user mobile device. When user arrives to position which was correctly determined by PDPT Core, the data artifacts are in local memory of PDA. The time to display the artifacts from local memory is much shorter than in case of remotely requested artifact.

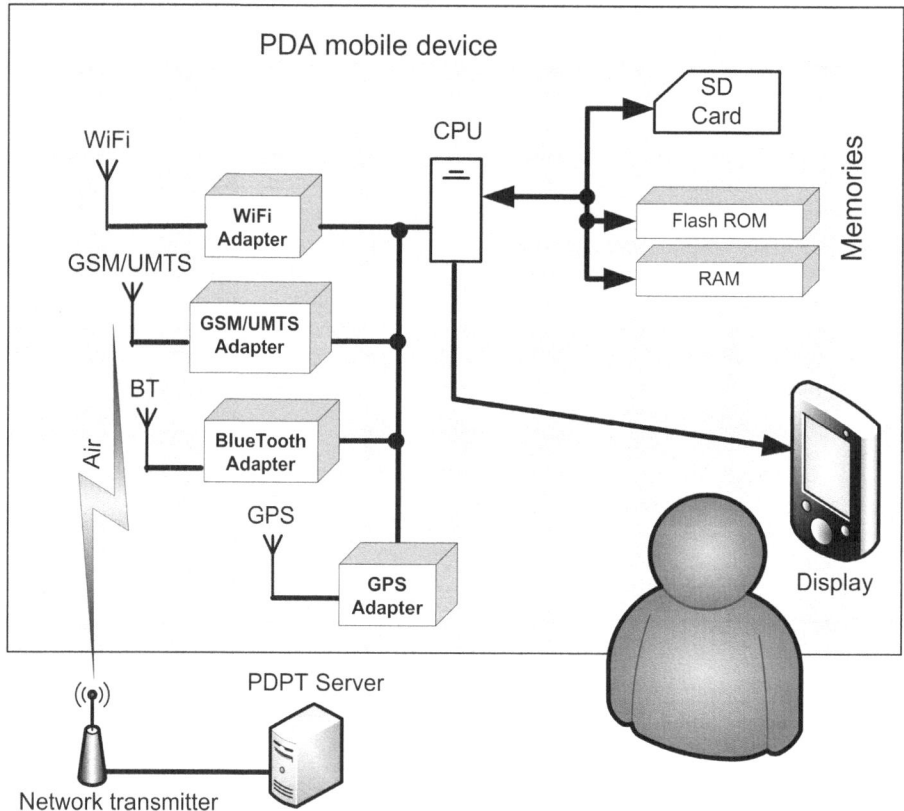

Fig. 1. Wireless networks and GPS sensor localization possibilities on mobile devices

The idea of prebuffering may not be only one application method for user position knowledge. As well as WiFi is not only one wireless network to use for localization of user device. WiFi has advantage in speed in indoor positioning therefore the GSM/UMTS can be used in outdoor [Fig. 1]. The GPS sensor is also embedded in several types of current mobile devices, or it can be plugged by SDIO or BT interface.

I would like to describe a position obtaining from wireless networks background in the beginning of next chapter to give a reader more information about these themes.

2 The PDPT Framework and PDPT Core

The general principle of my simple localization states that if a WiFi-enabled mobile device is close to such a stationary device – Access Point (AP) it may "ask" the provider's location position by setting up a WiFi connection. If position of the AP is known, the position of mobile device is within a range of this location provider. This range depends on type of WiFi AP. The Cisco APs are used in my test environment at Campus of Technical University of Ostrava. I performed measurements on these APs to get signal strength (SS) characteristics and a combination of them called "super ideal characteristic". More details can be found in chapter 2.3 [5]. The computed equation for Super-Ideal characteristic is taken as basic equation for PDPT Core to compute the real distance from WiFi SS.

From this super ideal characteristic it is also evident the signal strength is present only to 30 meters of distance from base station. This small range is caused by using of Cisco APs. These APs has only 2 dB WiFi omnidirectional antenna. Granularity of location can be improved by triangulation of two or more visible WiFi APs. The PDA client will support the application in automatically retrieving location information from nearby location providers, and in interacting with the server. Naturally, this principle can be applied to other wireless technologies like Bluetooth, GSM or WiMAX.

To let a mobile device determine its own position is needed to have a WiFi adapter still powered on. This fact provides a small limitation of use of mobile devices. The complex test with several types of battery is described in my article [4] in chapter (3). The test results with a possibly to use a PDA with turned on WiFi adapter for a period of about 5 hours.

2.1 The Need of Predictive Data Push Technology

PDPT framework is based on a model of location-aware enhancement, which I have used in created system. This technique is useful in framework to increase the real dataflow from wireless access point (server side) to PDA (client side). Primary dataflow is enlarged by data prebuffering. PDPT pushes the data from SQL database to clients PDA to be helpful when user comes at final location which was expected by PDPT Core. The benefit of PDPT consists in time delay reducing needed to display desired artifacts requested by a user from PDA. This delay may vary from a few seconds to number of minutes. Theoretical background and tests were needed to determine an average artifact size for which the PDPT technique is useful. First of all the maximum response time of an application (PDPT Client) for user was needed to be specified.

Nielsen [6] specified the maximum response time for an application to 10 seconds [7]. During this time the user was focused on the application and was willing to wait for an answer. The book is over 20 years old (published in 1994). I suppose the modern user of mobile devices is more impatient so the stated value of 10 second will be shorter. This is for me even better, because my framework is more usable. I used this time period (10 second) to calculate the maximum possible data size of a file transferred from server to client (during this period). If transfers speed wary from 80 to 160 kB/s the result file size wary from 800 to 1600 kB. More details about the facts of slow transfer speed on mobile devices can be found in chapter 2.5 [5].

The next step was an average artifact size definition. I use a network architecture building plan as sample database, which contained 100 files of average size of 470 kB. The client application can download during the 10 second period from 2 to 3 artifacts. The problem is the long time delay in displaying of artifacts in some original file types. It is needed to use only basic data formats, which can be displayed by PDA natively (bmp, jpg, wav, mpg, etc.) without any additional striking time consumption.

The final result of our real tests and consequential calculations is definition of artifact size to average value of 500 kB. The buffer size may differ from 50 to 100 MB in case of 100 to 200 artifacts.

2.2 From Data Collection to Localization

A first key step of the PDPT is a data collection phase. I record information about the radio signals as a function of a user's location. The signal information is used to construct and validate models for signal propagation. Among other information, the WaveLAN NIC makes the signal strength (SS) available. SS is reported to units of dBm. Each time the broadcast packet is received the WaveLAN driver extracts the SS information from the WaveLAN firmware. Then it makes the information available to user-level applications via system calls.

If the mobile device knows the position of the stationary device (transmitter), it also knows that its own position is within a range of this location provider. The typical range wary from 30 to 100 m in WiFi case, respectively 50 m in BT case or 30 km

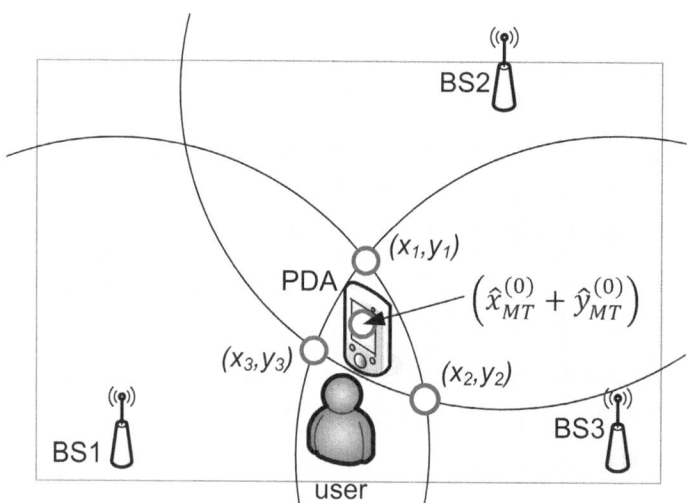

Fig. 2. Localization principle – triangulation

for GSM. Granularity of location can be improved by triangulation of two or more-visible APs (Access Points). The PDA client currently supports the application in automatically retrieving location information from nearby WiFi location providers, and in interacting with the PDPT server. Naturally, this principle can be applied to other wireless technologies like BT, GSM, UMTS or WiMAX. The application (locator) is implemented in C# language using the MS Visual Studio .NET with .NET Compact Framework and a special OpenNETCF library enhancement. Schema on figure [Fig. 2] describes a localization process. The mobile client gets the SS info of three BSs (Base Stations) with some inaccuracy. Circles around the BSs are crossed in red points on figure. The intersection red point (centre of three) is the best computed location of mobile user. The user track is also computed from these measured SS intensity levels and stored in database for later use by PDPT Core. This idea is applicable in case of WiFi as well as BT and GSM networks.

In previous research, I focused only to use of WiFi networks while the other wireless possibilities remained without a proper notice. Now I made an enhancement of Locator component of PDPT framework [Fig. 5] to allow operate with BT and GSM networks.

In BT network case, the position of BT APs must be known to allow the position determination. To collect BT APs position info in outdoor environment, the GPS can be used. For indoor area, the GIS (Geographic Information System) software with buildings map must be used to measure exact position of BT AP against to local environment. To manage with BT hardware of mobile device another library InTheHand 32Feet.NET is used. The source code has a simple implementation:

Example of a Locator Source Code – Scanning the nearby for BT APs.

```
using InTheHand.Net.Bluetooth;

BluetoothClient bc = new BluetoothClient();
BluetoothDeviceInfo[] bdi = bc.DiscoverDevices();

foreach (BluetoothDeviceInfo BTdi in bdi)
{
    drDataRow = dtVisibleAP.NewRow();
    drDataRow["AP_name"] = BTdi.DeviceName.ToString();
    drDataRow["MAC_AP"] = BTdi.DeviceAddress.ToString();
    drDataRow["Signal_Strength"] = BTDi.Rssi;
    drDataRow["Date_Time"] = DateTime.Now;
    drDataRow["AP_type"] = AP_type.Bluetooth;
    dtVisibleAP.Rows.Add(drDataRow);
}
```

GSM network is not local network but a cellular network. The problem is in position information of GSM BTSs (Base Transceiver Stations). The operator doesn't provide any such information. One of possible solutions is based on unofficial BTSs lists which can be found on internet. The lists are typically available in HTML, TXT or CSV formats. The medium rate for BTs with GPS position information is about

90 % of all BTs in European countries. In case of PDPT Framework, the list must be converted to PDPT server database – GSM_BTS table [Fig. 3].

In all three described cases of nearby BSs scanning, the data are saved to Locator Table in PDPT server DB [Fig. 3]. Data are processed from Locator Table throw the PDPT Core to Position Table. The processing techniques depend on selected wireless network. WiFi and BT network provide all visible APs nearby the user. From list of these APs is computed actual position (by triangulation [Fig. 2]).

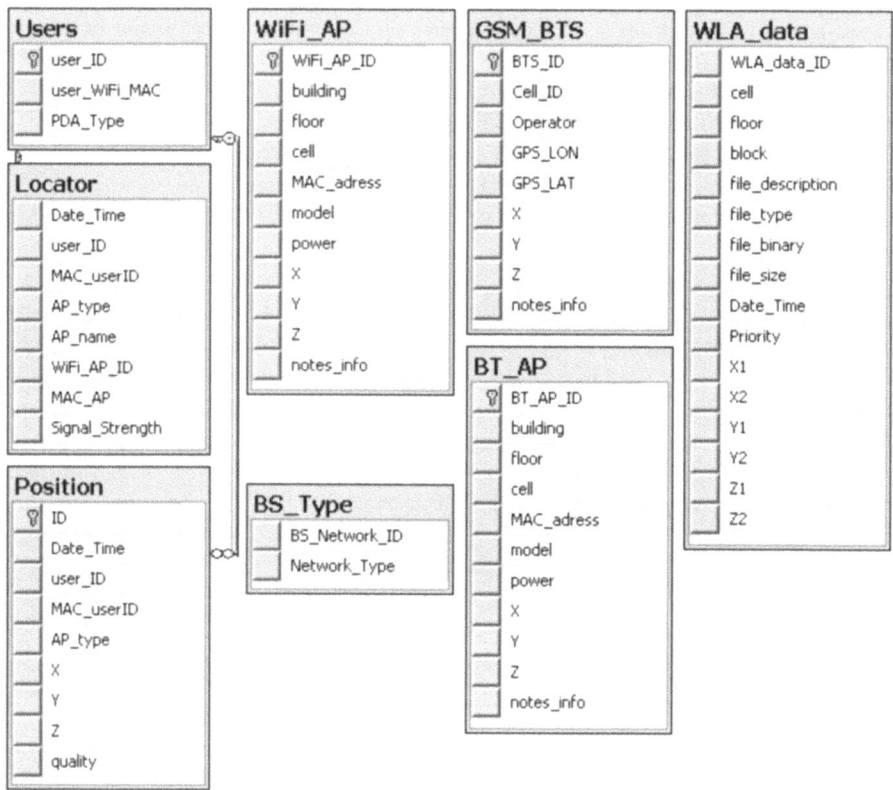

Fig. 3. PDPT server DB (Data Base) - New database architecture

Mobile devices with windows mobile operation system do not provide any GSM info to .NET Compact Framework. Even any special framework as in previous two cases is not known for me until now. Only possibility is in use of RIL (Radio Inter-face Layer) library. This library is divided into two separate components, a RIL Driver and a RIL Proxy. The RIL Driver processes radio commands and events. The RIL Proxy performs arbitration between multiple clients for access to the single RIL driver. When a module calls the RIL to get the signal strength, the function call im-mediately returns a response identifier. The RIL uses the function response callback to convey signal strength information to the module [Fig. 4].

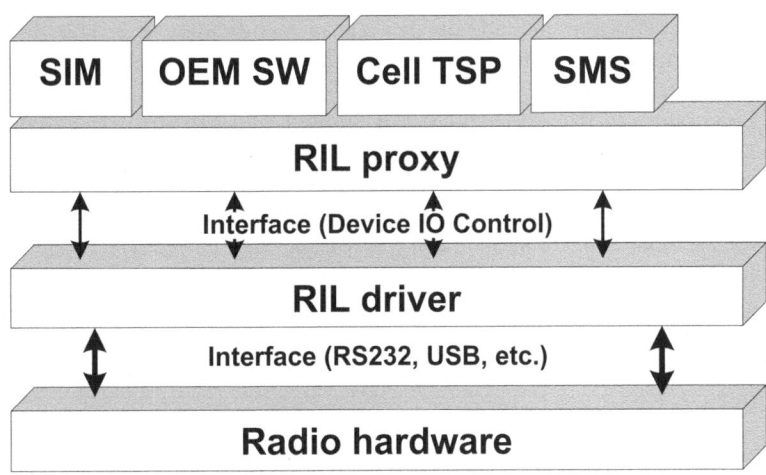

Fig. 4. Radio Interface Layer Architecture

Example of a Locator Source Code – retrieving the GSM BTSs info with LIR.

```
[DllImport("ril.dll")]
public static extern IntPtr RIL_GetCellTowerInfo(IntPtr
                                                  lphRil);
[DllImport("ril.dll")]
public static extern IntPtr RIL_Deinitialize(IntPtr
                                              lphRil);
[DllImport("ril.dll")]
public static extern IntPtr RIL_GetSignalQuality(IntPtr
                                                  lphRil);
  res = RIL_GetCellTowerInfo(hRil);
  res = RIL_GetSignalQuality(hRil);

RILCELLTOWERINFO rci = new RILCELLTOWERINFO();

result += String.Format("MCC: {0}, MNC: {1}, LAC: {2},
CID: {3}, ", rci.dwMobileCountryCode,
rci.dwMobileNetworkCode, rci.dwLocationAreaCode,
rci.dwCellID);

RILSIGNALQUALITY rsq = new RILSIGNALQUALITY();

result += String.Format("Signal Quality: {0}, MinSig {1},
MaxSig {2}, LowSig {3}, HighSig {4}",
rsq.nSignalStrength, rsq.nMinSignalStrength,
rsq.nMaxSignalStrength, rsq.nLowSignalStrength,
rsq.nHighSignalStrength);
```

The GSM network provide only one BS info in each search cycle. This BS has the highest signal strength. The more BTSs info is collected by a several iteration cycles. During 10 cycles (per 10 seconds) the 4 BTS info is obtained on average.

The important info from BTSs is Signal Strength and Time Advance (TA). SS is refreshed every several seconds (in every scan) whereas TA is provided only during some type of communication with selected BTS (e.g. request to talk, move to another area - Location Area Code (LAC)). The list of these BTSs with info is further processed as in previous case for WiFi and BT networks. Only change is in usage of TA if it is accessible.

Another possibility to get the user position in outdoor space is in GPS [8]. GPS provide a position by LONgitude and LATitude (X and Y). Only simple conversion is needed to transform a GPS coordinates to S-JTSK, which is used in PDPT Framework.

2.3 The PDPT Framework Design

The PDPT framework design is based on the server-client architecture. The PDPT framework server is created as a web service to act as a bridge between MS SQL Server (other database server eventually) and PDPT PDA Clients [Fig. 5].

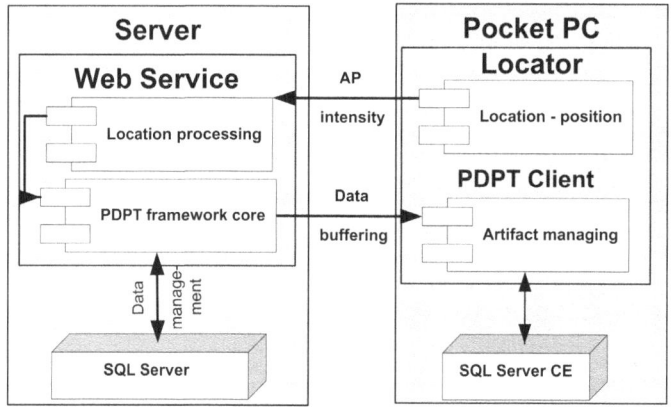

Fig. 5. PDPT architecture – UML design

Client PDA has location sensor component, which continuously sends the information about nearby AP's intensity to the PDPT Framework Core. This component processes the user's location information and it makes a decision to which part of MS SQL Server database needs to be replicated to client's SQL Server CE database [9][10]. The PDPT Core decisions constitute the most important part of PDPT framework, because the kernel must continually compute the position of the user and track, and predict his future movement. After doing this prediction the appropriate data are prebuffered to client's database for the future possible requirements. This data represent artifacts list of PDA buffer imaginary image [Fig. 6].

2.4   PDPT Core - Area Definition

The PDPT buffering and predictive PDPT buffering principle is shown in [Fig. 6]. Firstly the client must activate the PDPT on PDPT Client. This client creates a list of artifacts (PDA buffer image), which are contained in his mobile SQL Server CE database. Server create own list of artifacts (imaginary image of PDA buffer) based on area definition for actual user position and compare it with real PDA buffer image.

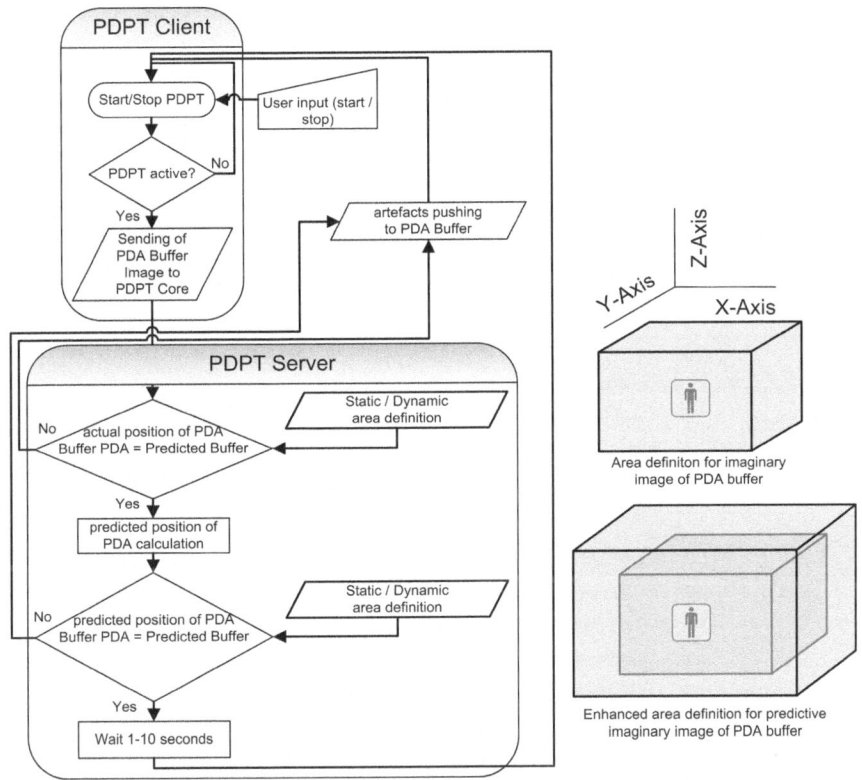

Fig. 6. Object diagram of PDPT prebuffering and predictive PDPT prebuffering. Right part shows the area definition for imaginary image of PDA buffer.

The area can be defined as an object where the user position is in the object centre. I am using the cuboid as the object in present time for initial PDPT buffering. This cuboid has static area definition with a size of 10 x 10 x 3 (high) meters. The PDPT Core will continue with comparing of both images. In case of some difference, the rest artifacts ale prebuffered to PDA buffer. When all artifacts for current user position are in PDA buffer, there is no difference between images. In such case the PDPT Core is going to make a predicted user position. On base of this new user position it makes a new predictive enlarged imaginary image of PDA buffer. The size of this new cuboid is statically defined area of size 20 x 20 x 6 meters. The new cuboid has a center in direction of predicted user moving and includes a cuboid area for current user position. The PDPT Core compares the both new images (imaginary and real PDA buffer) and it will continue with buffering of artifacts until they are same. In real case of usage is better to create an algorithm to dynamic area definition to adapt a system to user needs more flexible in real time. For additional info please refer to [11].

2.5 PDPT Framework Data Artifact Management

The PDPT Server SQL database manages the information (artifacts) in the context of their location in building environment. This context information is same as location information about user track. The PDPT Core selecting the data to be copied from PDPT server to PDA client by context information (position info). Each database artifacts must be saved in database along the position information, to which it belongs.

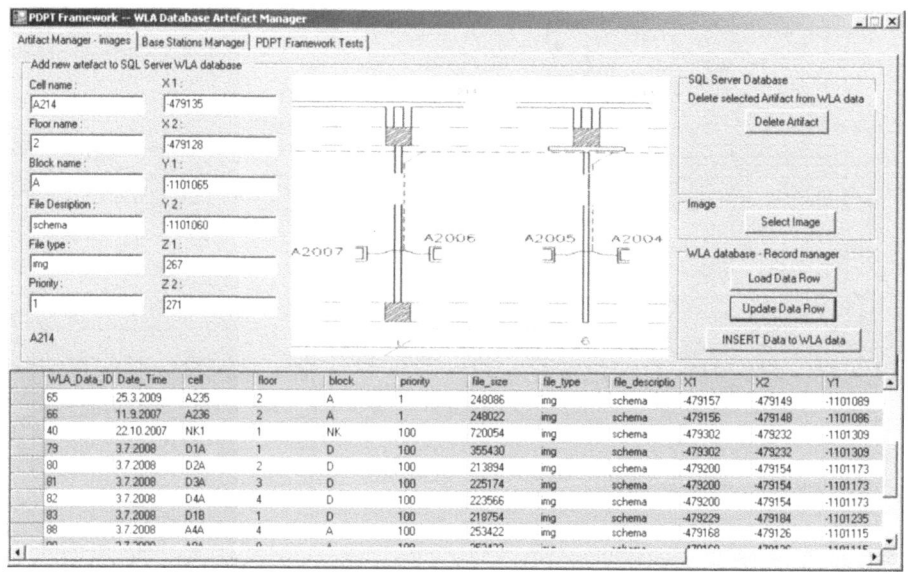

Fig. 7. PDPT Framework Data Artifact Manager

During the creating process of PDPT Framework the new software application called "Data Artifacts Manager" was created. This application manages the artifacts in WLA database (localization oriented database). User can set the priority, location, and other metadata of the artifact [Fig 7]. The Manager allows creating a new artifact from multimedia file (image, video, sound, etc.), and work with existing artifacts. More info can be found in chapter 3.1 [5].

The needs of interface to operate with APs info arose out of the developing process of PDPT Framework. The enhancement of Artifact manager was created on that ground. Now the Artifact Manager contains a new tab "Base Stations Manager" to operate with APs or BSs of selected networks [Fig. 7]. This manager is connected directly to PDPT Server database, to tables WiFi_AP, BT_AP, GSM_BTS [Fig. 3].

2.6 The PDPT Client Application

The PDPT Client application realizes thick client to the server side and an extension by PDPT and Locator modules. Figure [Fig. 8] shows three screenshots from the mobile client. The first one [Fig. 8a] shows the Locator module with selected GSM scanning. The info text box "Locator AP ret." Provide info about last founded GSM BTSs and number of recognized BTSs (BTSs with GPS position). In current case the 6 BTSs was founded and 5 of them was recognized by PDPT Framework. Figure [Fig. 8b] shows the classical view of the data artifact presentation from MS SQL CE database to user (in this case the image of Ethernet plan of the current area). The PDPT tab [Fig. 8c] presents a way to tune the settings of PDPT Framework. The middle section shows the logging info about the prebuffering process. The right side means measure the time of one artifact loading ("part time") and full time of prebuffering in millisecond resolution. More screens and details of PDPT Client can be found in chapters 2.7 and 2.8 [5].

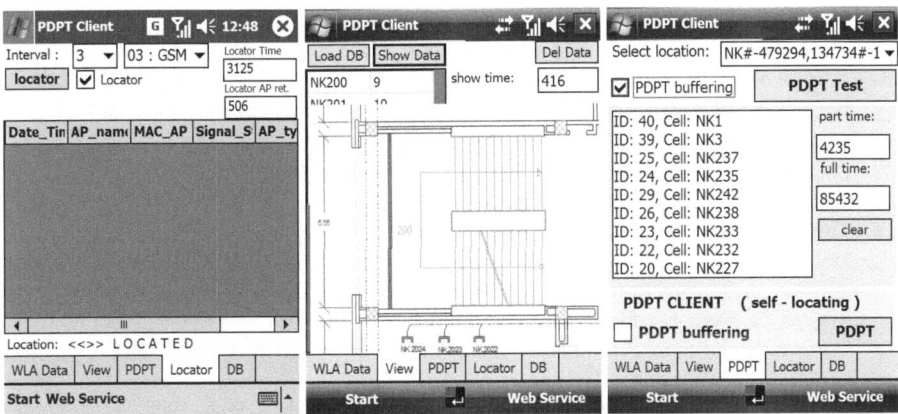

Fig. 8. PDPT Client – Left one figure 8a – Locator component with GSM scanning. Middle one figure 8b – View of classical client data representation. Right one figure 8c - The PDPT options screen allow to start and control the PDPT buffering.

3 Conclusions

I am focused on the real usage of the developed PDPT Framework on a wide range of wireless mobile devices and its main issue at increased data transfer. For testing purpose, five mobile devices were selected with different hardware and software capabilities. The high success rate found in the test data surpassed our expectations. This rate varies from 84 to 96 %. Please see the chapter 4 [5] for more info.

The PDPT prebuffering techniques can improve the using of medium or large artifacts on wireless mobile lightweight devices connected to information systems. The localization part of PDPT framework is currently used in another project of biotelemetrical system for home care named Guardian to make a patient's life safer [12]. Another utilization of PDPT consists in use of others wireless networks like BT,

GSM/UMTS, WiMAX, or in GPS. This idea can be used inside the information systems like botanical or zoological gardens where the GPS navigation can be used in outdoor. The BT and GSM data collecting and processing is described in this article along with sample code. Some improvements of Locator module or Artifact Manager are described as well as improved architecture of PDPT server database. The larger area of PDPT utilization can improve importance of PDPT Framework in wireless mobile lightweight systems.

Acknowledgment. This work was supported by the Ministry of Education of the Czech Republic under Project 1M0567.

References

1. Abowd, G., Dey, A., Brown, P., et al.: Towards a better understanding of context and context-awareness. In: Gellersen, H.-W. (ed.) HUC 1999. LNCS, vol. 1707, p. 304. Springer, Heidelberg (1999)
2. Krejcar, O.: User Localization for Intelligent Crisis Management. In: AIAI 2006, 3rd IFIP Conference on Artificial Intelligence Applications and Innovation, Boston, USA, pp. 221–227 (2006)
3. Krejcar, O., Cernohorsky, J.: Database Prebuffering as a Way to Create a Mobile Control and Information System with Better Response Time. In: Bubak, M., van Albada, G.D., Dongarra, J., Sloot, P.M.A. (eds.) ICCS 2008, Part I. LNCS, vol. 5101, pp. 489–498. Springer, Heidelberg (2008)
4. Krejcar, O.: PDPT Framework - Building Information System with Wireless Connected Mobile Devices. In: ICINCO 2006, 3rd International Conference on Informatics in Control, Automation and Robotics, Setubal, Portugal, August 1-5, pp. 162–167 (2006)
5. Krejcar, O., Cernohorsky, J.: New Possibilities of Intelligent Crisis Management by Large Multimedia Artifacts Prebuffering. In: I.T. Revolutions 2008, Venice, Italy, December 17-19. LNICST. Springer, Heidelberg (2008)
6. Nielsen, J.: Usability Engineering. Morgan Kaufmann, San Francisco (1994)
7. Haklay, M., Zafiri, A.: Usability engineering for GIS: learning from a screenshot. The Cartographic Journal 45(2), 87–97 (2008)
8. Evennou, F., Marx, F.: Advanced integration of WiFi and inertial navigation systems for indoor mobile positioning. In: Eurasip journal on applied signal processing, Hindawi publishing corp., New York, USA (2006)
9. Arikan, E., Jenq, J.: Microsoft SQL Server interface for mobile devices. In: Proceedings of the 4th International Conference on Cybernetics and Information Technologies, Systems and Applications/5th Int Conf on Computing, Communications and Control Technologies, Orlando, FL, USA, July 12-15 (2007)
10. Jewett, M., Lasker, S., Swigart, S.: SQL server everywhere: Just another database? Developer focused from start to finish. DR DOBBS Journal 31(12) (2006)
11. Krejcar, O.: Utilization Possibilities of Area Definition in User Space for User-Centric Pervasive-Adaptive Systems. In: First International Workshop on User-Centric Pervasive Adaptation, UCPA 2009, Berlin, Germany, April 27. LNICST. Springer, Heidelberg (2009)
12. Janckulik, D., Krejcar, O., Martinovic, J.: Personal Telemetric System – Guardian. In: Biodevices 2008, Insticc Setubal, Funchal, Portugal, pp. 170–173 (2008)

QoS-constrained Energy Minimization in Multiuser Multicarrier Systems

Qing Bai, Michel T. Ivrlač, and Josef A. Nossek

Institute for Circuit Theory and Signal Processing,
Technische Universität München, Munich, Germany
bai.qing@nws.ei.tum.de
http://www.nws.ei.tum.de/

Abstract. In this paper the QoS-constrained resource allocation problem in multicarrier systems is considered. Within the established cross-layer framework, parameters for subchannel assignment, adaptive modulation and coding, and ARQ/HARQ protocols are jointly optimized. Instead of the conventional transmit power minimization, the total energy consumption for the successful transmissions of all information bits is set as the optimization goal. The nonconvex primal problem is solved by using Lagrange dual decomposition and the ellipsoid method. Numerical results indicate that the recovered primal solution is well acceptable in performance, and efficient in terms of computational effort.

Keywords: resource allocation, multicarrier systems, cross-layer optimization, energy minimization.

1 Introduction

Resource allocation in wireless communication networks is both important and challenging not only because of the scarcity of radio resources and time-variant channel conditions, but also due to the increasing demand to support heterogeneous *quality of service* (QoS) requirements of various applications. From a mathematical point of view, one specific resource allocation corresponds to a mapping from the available radio resources to a set of QoS values. When parameters from different protocol layers are jointly taken into account in the mapping, the optimizations we do, either optimizing QoS with limited resources or minimizing the amount of resources required to achieve certain QoS, are referred to as *cross-layer optimizations*, and the resource allocation itself is termed as *cross-layer assisted resource allocation*. In this paper, a QoS provisioning resource minimization problem at the downlink of a multicarrier system is investigated, where the cross-layer framework adopted integrates PHY and MAC layer functionalities such as subchannel assignment, adaptive modulation and coding, and retransmission protocols.

In most studies on resource allocation for wireless communication systems, the objective for the QoS-constrained resource minimization is to minimize the sum transmit power, *e.g.*, [1][2]. Since retransmission protocols are taken into account

F. Granelli et al. (Eds.): MOBILIGHT 2009, LNICST 13, pp. 185–195, 2009.

in this work, it is of interest and necessity to consider the transmit power spent
over time, *i.e.*, *energy*, instead of to merely consider the power consumption
for the first transmission, because on the long run, what is consumed at the
transmitter is energy. Based on this analysis, we formulate the minimization
goal as the sum energy consumption required to transmit a certain number of
information bits within respective latency times for a group of end users.

Though having different physical interpretations, structurally similar optimizations can be found in the literature such as in [3], [4] and [1]. However,
due to the discontinuity and nonconvexity of our objective, the methods therein
to solve the optimizations and the optimality conditions derived can not be directly applied. Exploiting the discontinuity, we set up a look-up table to lessen
the computational burden for the dual methods employed, and a primal recovery scheme is developed to give primarily feasible resource allocations from the
obtained dual optimal solutions.

2   System Model

We consider the downlink scenario of an isolated single-cell multicarrier system
with K users, each having one data stream to be served. The resource allocation is done on a per *slot* basis, where a *slot* is a short time period of length T
during which the wireless channel is assumed to stay constant. As information
bits loaded onto consecutive slots are independently modulated and coded, a
slot can formally be referred to as a *Transmission Time Interval* (TTI), and
the bit-loading procedure inherently includes *packetization* of the information
bits. For every TTI, each data stream has a number of information bits to be
transmitted, depending on its *throughput* requirement. The other relevant QoS
parameter characterizing the data streams, the *latency*, is defined as:

Definition: The latency τ_k of a packet from user k is the delay it experiences
until received correctly with an outage probability of no more than the predefined value $\pi^{(\text{out})}$. Let $f_k[m]$ be the probability that it takes exactly m TTI's
to transmit a packet error-free, then $\tau_k = (M_k - 1)(\text{RTD} + T) + T$ where RTD
represents *round trip delay*, and

$$M_k = \min_M M \quad \text{s.t.} \quad \sum_{m=1}^{M} f_k[m] \geq 1 - \pi^{(\text{out})}.$$

In the following subsections, the mathematical descriptions of the regarded system components are derived which lay the basis for cross-layer optimization.

2.1   Channel Model

The downlink broadcast channel is modeled as frequency-selective fading over the
total system bandwidth and frequency-flat fading over each *subchannel*, which
is consist of N_c adjacent subcarriers. FDMA is employed meaning the assignment

of every subchannel is exclusive to one user, and *intercarrier interference* (ICI) is not taken into account. Moreover, we restrict ourselves here to the single-antenna case both at the base station (BS) and at the mobile stations (MS).

Let $H_{k,n}$ and $\sigma^2_{k,n}$ be the channel coefficient and Gaussian noise variance of user k on the nth subchannel, and p_n be the amount of power allocated on subchannel n. When assigned to user k, the *signal-to-noise-ratio* (SNR) on subchannel n can be computed as

$$\gamma_{k,n} = \frac{|H_{k,n}|^2}{\sigma^2_{k,n}} \cdot p_n. \tag{1}$$

Note that throughout this work the index k refers to users and index n refers to subchannels. And as in the remaining part of this chapter, the focus is on any one of the subchannels which is assigned to one user, we drop the subscripts k and n for simplicity.

We choose the TTI to be of length $T = 2$ ms. The WiMAX standard suggests a symbol duration of 102.9 μs in a system with 10 MHz bandwidth and an FFT size of 1024. Based on this number we assume that one TTI contains $N_s = 16$ symbols for data transmission.

2.2   FEC Coding and Modulation

We assume that modulation and coding across the subchannels are done independently, and with reference to the WiMAX standard 8 modulation and coding schemes (MCS) are chosen as candidates, which are listed in Table 1.

Table 1. Modulation and Coding Schemes (MCS)

Index	Modulation Type	Alphabet Size A	Code Rate R	$R \log_2 A$
1	BPSK	2	1/2	0.5
2	QPSK	4	1/2	1
3	QPSK	4	3/4	1.5
4	16-QAM	16	1/2	2
5	16-QAM	16	3/4	3
6	64-QAM	64	2/3	4
7	64-QAM	64	3/4	4.5
8	64-QAM	64	5/6	5

Since with the help of cyclic prefix or an equalizer, intersymbol interference is not present in the system, each subchannel can be modeled as a *discrete memoryless channel* (DMC) over which the *noisy channel coding theorem* [5] can be applied. Let the modulation alphabet and the coding rate on the subchannel under consideration be $\mathcal{A} = \{a_1, \ldots, a_A\}$ and R respectively. The *cutoff rate* of the subchannel with SNR γ can be expressed as

$$R_0(\gamma, A) = \log_2 A - \log_2\left[1 + \frac{2}{A}\sum_{m=1}^{A-1}\sum_{l=m+1}^{A} e^{-\frac{1}{4}|a_l - a_m|^2\gamma}\right]. \qquad (2)$$

The noisy channel coding theorem states that there always exists a block code with block length l and binary code rate $R\log_2 A \le R_0(\gamma, A)$ in bits per subchannel use, such that with maximum likelihood decoding the error probability $\tilde{\pi}$ of a code word satisfies

$$\tilde{\pi} \le 2^{-l(R_0(\gamma, A) - R\log_2 A)}. \qquad (3)$$

In order to apply this upper bound on code word error probability to the extensively used turbo decoded convolutional code, quantitative investigations have been done in [2] and an expression for the *equivalent block length* is derived based on link level simulations. The result from [2] shows that the performance of a turbo decoded convolutional code applied to a coded packet of length L in a very good approximation equals the performance of a block code with block length

$$n_{\text{eq}} = \beta \ln L, \qquad (4)$$

where parameter β is used to adapt this model to the specifics of the employed turbo code, and $L = N_c N_s \log_2 A$. Consequently, the transmission of L bits is equivalent to the sequential transmission of L/n_{eq} blocks of length n_{eq} and has an error probability of

$$\pi = 1 - (1 - \tilde{\pi})^{\frac{L}{n_{\text{eq}}}} \le 1 - \left(1 - 2^{-n_{\text{eq}}(R_0(\gamma, A) - R\log_2 A)}\right)^{\frac{L}{n_{\text{eq}}}}. \qquad (5)$$

2.3 Protocol

At the MAC layer both *automatic repeat request* (ARQ) and *incremental redundancy hybrid ARQ* (IR HARQ) protocols are studied. The data sequence transmitted in one TTI on one subchannel, *i.e.*, a *packet*, is used as the retransmission unit.

ARQ: The corrupted packets at the receiver are discarded. Hence we assume that the *packet error probability* (PEP) of a retransmitted packet is the same as that of its original transmission, *i.e.*,

$$f[m] = \pi^{m-1}(1 - \pi), \quad m \in \mathbb{Z}^+.$$

When the number of transmissions M is given, the maximum allowable PEP can be obtained as

$$\sum_{m=1}^{M} f[m] = 1 - \pi^M \ge 1 - \pi^{(\text{out})} \quad \Rightarrow \quad \pi \le \sqrt[M]{\pi^{(\text{out})}}.$$

HARQ: The corrupted packets at the receiver are combined and jointly decoded using rate-compatible punctured convolutional codes. For the particular

IR scheme where the retransmissions contain pure parity bits of the same length as the first transmission, the code rate for the mth transmission can be expressed as

$$R[m] = \frac{B}{m \cdot L} = \frac{1}{m}R[1] = \frac{1}{m}R. \tag{6}$$

Let $\tilde{m}$ denote the maximum number of transmissions determined by the mother code. The equivalent block length n_{eq} is then given by

$$n_{\text{eq}} = \beta \ln(\tilde{m}L). \tag{7}$$

Plugging (6)(7) into (5) gives the PEP expression for the mth transmission as

$$\pi[m] = 1 - \left(1 - 2^{-\beta \ln(\tilde{m}L)(R_0(\gamma) - \frac{1}{m}R \log_2 A)}\right)^{\frac{mL}{\beta \ln(\tilde{m}L)}}. \tag{8}$$

When $R_0(\gamma) \leq \frac{1}{m}R \log_2 A$, (8) suggests that $\pi[m] = 1$. And when $R_0(\gamma)$ increases from $\frac{1}{m}R \log_2 A$, $\pi[m]$ approaches 0 very fast. As a result, given the number of transmissions M, $\pi[m]$ can be approximated by

$$\pi[M] = 1 - \pi^{(\text{out})}, \quad \pi[m] = 0, m = 1, \dots, M - 1, \tag{9}$$

where $R_0(\gamma)$ satisfies $\frac{1}{M}R \log_2 A < R_0(\gamma) \leq \frac{1}{M-1}R \log_2 A$.

The quantities mentioned in this section, their notations, as well as their simulation values are summarized in Table 2.

Table 2. System Parameters

Total bandwidth		10 MHz
Center frequency	f_c	2.5 GHz
FFT size		1024
Number of data subcarriers		720
Number of subchannels	N	30
Number of subcarriers per subchannel	N_c	720/30 = 24
Transmission Time Interval (TTI)	T	2 ms
Number of data symbols per TTI	N_s	16
Round Trip Delay (RTD)	RTD	10 ms
Maximum number of transmissions allowed	$\tilde{m}$	5
Turbo code dependent parameter	β	32
Outage probability $\pi^{(\text{out})}$		0.01

3   Problem Formulation

Suppose for the current TTI, the number of information bits intended for user k is b_k, and the maximum latency time for the transmission is $\tau_k^{(\text{rq})}$. The energy minimization problem can be formulated as

$$\min_{\boldsymbol{B}} \quad \sum_{k=1}^{K}\sum_{n=1}^{N} \eta_{k,n}\left(B_{k,n}, \tau_k^{(\mathrm{rq})}\right)$$

$$\text{s.t.} \quad \sum_{n=1}^{N} B_{k,n} = b_k, \ k = 1, \ldots, K, \tag{10}$$

$$\boldsymbol{B} \in \mathcal{B},$$

where $\boldsymbol{B} \in \mathbb{Z}_{+,0}^{K \times N}$ represents the bit-loading matrix with its entry $B_{k,n}$ as the number of information bits for the kth user loaded onto the nth subchannel, and $\eta_{k,n}(B_{k,n}, \tau_k^{(\mathrm{rq})})$ is the minimum energy consumption needed for the successful transmission of $B_{k,n}$ bits within the latency time $\tau_k^{(\mathrm{rq})}$. The first constraint in (10) is the completeness of bit-loading for the K users, and the second constraint comes from FDMA in which $\mathcal{B} \subset \mathbb{Z}_{+,0}^{K \times N}$ represents the set of matrices that have only one nonzero entry in each of their columns.

3.1 The η Function

We define a tuple (A, R, M) which is a modulation type, FEC code rate, and number of transmissions combination as one *mode of operation*. With 5 as the maximum number of transmissions for each packet and 8 available MCS, we have in all 40 different modes of operation, denoted by set $\mathcal{M}$. For a fixed B, each mode of operation (A, R, M) leads to a (latency, expected energy consumption) pair (τ, E) with

$$\tau = (M - 1)(\mathrm{RTD} + T) + T,$$

$$E = \left\lceil \frac{B}{R \log_2 A} \right\rceil \cdot T_\mathrm{s} \cdot \gamma(A, R, M) \cdot \sum_{m=1}^{M} f[m]\left(\frac{\sigma^2}{|H|^2} + \frac{(m-1)\sigma^2}{|H^{(\mathrm{avg})}|^2} \right)$$

$$\overset{!}{=} \phi \cdot \sum_{m=1}^{M} f[m]\left(\frac{\sigma^2}{|H|^2} + \frac{(m-1)\sigma^2}{|H^{(\mathrm{avg})}|^2} \right),$$

where $|H|^2$ and $|H^{(\mathrm{avg})}|^2$ are the instantaneous and average channel gains, and σ^2 is the noise power on one subcarrier. $\gamma(A, R, M)$ is the SNR required to convey the packet within M transmissions when MCS (A, R) is employed, which can be obtained from a binary search on the cutoff rate curve. Note that ϕ as defined is independent of the channel realizations. $\eta(B, \tau^{(\mathrm{rq})})$ is then given by

$$\eta(B, \tau^{(\mathrm{rq})}) = \min_{(A,R,M)\in\mathcal{M}} E(A, R, M) \quad \text{s.t.} \quad \tau(A, R, M) \le \tau^{(\mathrm{rq})}. \tag{11}$$

Limited by the highest MCS, the number of information bits that can be loaded onto one subchannel in one TTI is upper bounded by $B^{(\mathrm{u})} = 5 \cdot N_\mathrm{s} N_\mathrm{c}$. Let (τ_1, E_1) and (τ_2, E_2) be two (latency, energy) pairs. Analytical derivations show that if $\tau_1 < \tau_2$ and $\phi_1 < \phi_2$, then $E_1 < E_2$. That means, only those modes of operation

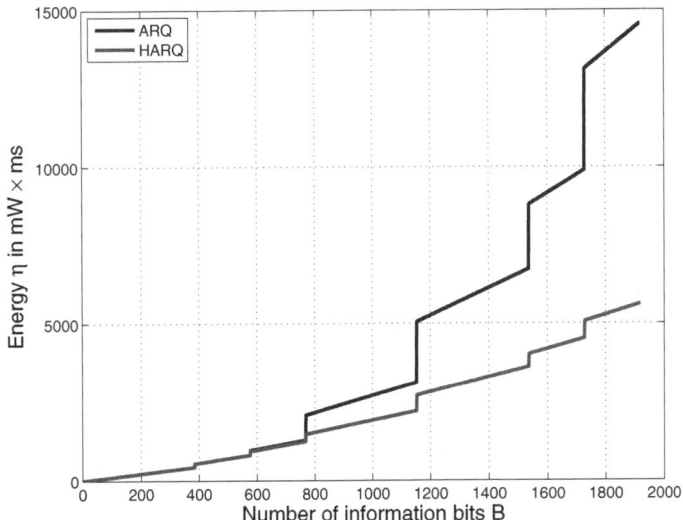

Fig. 1. An exemplary η function for ARQ and HARQ protocols

that are Pareto efficient in $(\tau, \phi)^1$ can lead to the solution of (11). Therefore, for each $B \in [1, B^{(u)}]$, finding and storing the (τ, ϕ) Pareto efficient points via an enumeration of all modes of operation are sufficient to solve (11) given the instantaneous channel realizations, which is to say, an offline computable look-up table can be established beforehand. At run time, only some simple calculations are needed to compute $\eta(B, \tau^{(\mathrm{rq})})$. An exemplary η function is shown in Fig. 1, where $\tau^{(\mathrm{rq})}$ is set to infinity.

From the visualization some of the expectations of the η function are verified: it is monotonically increasing with the number of information bits B, the energy increments for the same increment in B become larger with increasing B, and HARQ consumes less energy than ARQ for fairly large B. However, the η function is not convex due to the discrete inputs and changes of the optimum mode of operation at some B. As a result, the optimization (10) is not convex in both objective and constraints. Therefore when dual methods are applied, the solution is bound to suffer from the duality gap.

4   The Resource Allocation Algorithm

4.1   Dual Methods

The Lagrange dual decomposition method and the ellipsoid method are employed to solve the optimization problem (10), following a similar procedure as

---

[1] A mode of operation $(\tilde{A}, \tilde{R}, \tilde{M})$ is called Pareto efficient in (τ, ϕ) if the pair $(\tilde{\tau}, \tilde{\phi})$ it leads to is Pareto efficient, *i.e.*, there $\nexists (\tau, \phi)$ resulting from other modes of operations in $\mathcal{M}$ such that $\tau \leq \tilde{\tau}$ and $\phi \leq \tilde{\phi}$.

proposed in [4]. Introducing Lagrange multipliers $\boldsymbol{\lambda} \in \mathbb{R}^{K \times 1}$ to the K bit-loading constraints in (10) gives the Lagrangian

$$L(\boldsymbol{B}, \boldsymbol{\lambda}) = \sum_{k=1}^{K} \sum_{n=1}^{N} \eta_{k,n}(B_{k,n}, \tau_k^{(\mathrm{rq})}) + \sum_{k=1}^{K} \lambda_k \left(\sum_{n=1}^{N} B_{k,n} - b_k \right), \qquad (12)$$

and the dual function $g(\boldsymbol{\lambda})$ follows as

$$
\begin{aligned}
g(\boldsymbol{\lambda}) &= \inf_{\boldsymbol{B} \in \mathcal{B}} L(\boldsymbol{B}, \boldsymbol{\lambda}) \\
&= \inf_{\boldsymbol{B} \in \mathcal{B}} \sum_{n=1}^{N} \left(\sum_{k=1}^{K} \eta_{k,n}(B_{k,n}, \tau_k^{(\mathrm{rq})}) + \sum_{k=1}^{K} \lambda_k B_{k,n} \right) - \sum_{k=1}^{K} \lambda_k b_k \\
&= \sum_{n=1}^{N} \inf_{\boldsymbol{B} \in \mathcal{B}} \sum_{k=1}^{K} \left(\eta_{k,n}(B_{k,n}, \tau_k^{(\mathrm{rq})}) + \lambda_k B_{k,n} \right) - \sum_{k=1}^{K} \lambda_k b_k \\
&\overset{!}{=} \sum_{n=1}^{N} g_n(\boldsymbol{\lambda}) - \sum_{k=1}^{K} \lambda_k b_k,
\end{aligned}
$$

where $g_n(\boldsymbol{\lambda}), n = 1, \ldots, N$ are N independent optimization problems resulting from the decomposition of minimizing $L(\boldsymbol{B}, \boldsymbol{\lambda})$. In solving the dual problem, *i.e.*, $\max g(\boldsymbol{\lambda})$, the update of the dual variable $\boldsymbol{\lambda}$ is done efficiently using the ellipsoid method. We denote the optimal value and solution to the dual problem as d^* and $\boldsymbol{\lambda}^*$ respectively, and the bit-loading matrix obtained with $\boldsymbol{\lambda}^*$ as $\tilde{\boldsymbol{B}}$. By weak duality, d^* gives a lower bound on the primal optimal value. However, $\tilde{\boldsymbol{B}}$ is not necessarily primal-feasible, which makes primal recovery necessary.

4.2   Primal Recovery Scheme

Due to the nonconvexity of the objective function of (10), the conclusion drawn in [3] that the duality gap vanishes when the number of subchannels approaches infinity is not valid anymore. Consequently, the subchannel assignment (SA) implicitly given by $\tilde{\boldsymbol{B}}$ ($\tilde{B}_{k,n} > 0$ indicates that the nth subchannel is assigned to the kth user) can not be assumed optimum. In fact, as $B_{k,n}$ is limited by $B^{(\mathrm{u})}$ from above, the dual optimum SA can be infeasible, especially when the total number of information bits to be loaded is large. Therefore, in order to perform primal recovery based on the dual optimum SA, we have to assure its feasibility first.

The minimum number of subchannels needed by user k can be computed as $N_k^{(1)} = \left\lceil \frac{b_k}{B^{(\mathrm{u})}} \right\rceil$. Let the set of subchannels assigned to user k by the dual optimum SA be $\mathcal{S}_k$, *i.e.*, $\mathcal{S}_k = \{n : \tilde{B}_{k,n} > 0\}$. If $\exists k$ with $|\mathcal{S}_k| < N_k^{(1)}$, then the dual optimum SA is infeasible. Denote the set of users with $|\mathcal{S}_k| > N_k^{(1)}$ as $\mathcal{K}_\mathrm{o}$[2]. One adjustment scheme can be to solve

---

[2] An empty set $\mathcal{K}_\mathrm{o}$ indicates the infeasibility of (10), the case of which should be tested and excluded at the beginning of the whole program. In order to provide the resource allocation entity with appropriate traffic loads, a scheduling component on its top is necessary.

$$(k^*, n^*) = \underset{k' \in \mathcal{K}_o, n \in \mathcal{S}_{k'}}{\operatorname{argmin}} \left(\eta_{k,n}(B^{(\mathrm{u})}, \tau_k^{(\mathrm{rq})}) - \eta_{k',n}(B^{(\mathrm{u})}, \tau_{k'}^{(\mathrm{rq})}) \right),$$

and reassign subchannel n^* to user k instead of its former possessor k^* by updating $\mathcal{S}_k$ and $\mathcal{S}_{k^*}$ accordingly.

Fixing the feasible SA, we have K decoupled minimization problems, one for each user, as

$$
\begin{aligned}
\underset{\{B_{k,n}:n\in\mathcal{S}_k\}}{\min} \quad & \sum_{n\in\mathcal{S}_k} \eta_{k,n}(B_{k,n}, \tau_k^{(\mathrm{rq})}) \\
\text{s.t.} \quad & \sum_{n\in\mathcal{S}_k} B_{k,n} = b_k,
\end{aligned}
\tag{13}
$$

which can again be solved in the dual domain. Let the dual optimal bit-loading be $\{B_{k,n}^* : n \in \mathcal{S}_k\}$. If $\sum_{n\in\mathcal{S}_k} B_{k,n}^* \neq b_k$, we can load or unload the extra bits one by one on the subchannel that leads to the minimum energy increment or the maximum energy decrement. Mathematically, we iteratively find

$$
n^* = \begin{cases}
\underset{n\in\mathcal{S}_k}{\operatorname{argmin}} \left(\eta_{k,n}(B_{k,n}^* + 1, \tau_k^{(\mathrm{rq})}) - \eta_{k,n}(B_{k,n}^*, \tau_k^{(\mathrm{rq})}) \right), & \sum_{n\in\mathcal{S}_k} B_{k,n}^* < b_k, \\
\underset{n\in\mathcal{S}_k}{\operatorname{argmax}} \left(\eta_{k,n}(B_{k,n}^*, \tau_k^{(\mathrm{rq})}) - \eta_{k,n}(B_{k,n}^* - 1, \tau_k^{(\mathrm{rq})}) \right), & \sum_{n\in\mathcal{S}_k} B_{k,n}^* > b_k,
\end{cases}
\tag{14}
$$

and update B_{k,n^*}^*, until $\sum_{n\in\mathcal{S}_k} B_{k,n}^* = b_k$ is satisfied. Such a recovery scheme is simple, but greedy and performance-degrading.

5 Simulation Results

For simulations, $K = 10$ users uniformly located in a cell of radius 2 km are assumed. The wireless channel is modeled as a frequency-selective fading channel consisting of six independent Rayleigh multipaths with an exponentially decaying power profile. The delay spreads are uniformly distributed within 1 μs, resulting in a rms delay spread of about 0.3 μs which is consistent with the assumed channel coherence bandwidth. The path loss in dB is computed as $PL(d) = 140.6 + 35.0\log_{10} d$ following the COST-Hata model, where d is the distance between MS and BS in km, and the receiver noise level is assumed to be -174 dBm/Hz.

Each user's information bits to be served and latency requirements are listed in Table 3, where the unit for b_k is bit and the unit for τ_k is ms, and α is a scalar that takes values from $\{0.5, 1, 1.5, 2, 2.5, 3, 3.5, 4\}$. Besides the test results of the algorithm discussed in the previous sections, a static resource allocation scheme is simulated for comparison purpose. The static scheme first assigns each user with a fixed set of subchannels and then performs the greedy bit-loading, in the same way as used for primal recovery. Each test scenario has been simulated under 1000 independent channel realizations.

In Fig. 2 the statistics of energy consumptions are shown, where Fig. 2(a) shows the cumulative distributions of the energy spent under different retransmission protocols and resource allocation schemes, and Fig. 2(b) illustrates the

Table 3. QoS requirements of 10 users for simulation

User	b_k	τ_k	User	b_k	τ_k	User	b_k	τ_k
1-4	$512{\cdot}\alpha$	20	5-7	$800{\cdot}\alpha$	40	8-10	$1600{\cdot}\alpha$	80

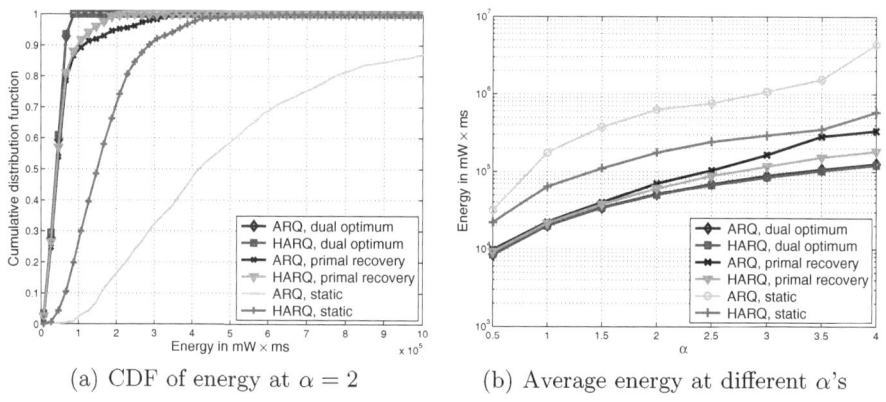

(a) CDF of energy at $\alpha = 2$ (b) Average energy at different α's

Fig. 2. Energy Comparisons

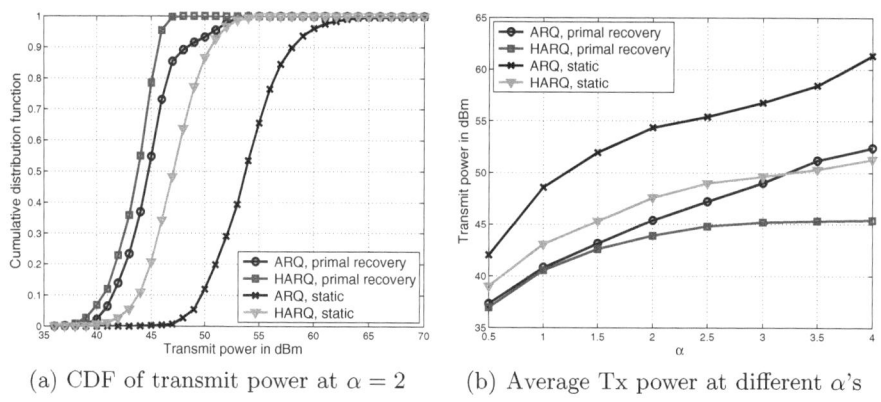

(a) CDF of transmit power at $\alpha = 2$ (b) Average Tx power at different α's

Fig. 3. Transmit Power Comparisons

average energy consumption over 1000 simulations at each α value. Note that the actual optimal energy curves lie between the dual optimum and the primal recovery curves. The corresponding statistics for the transmit power spent for the first transmission are drawn in Fig. 3.

It is clear from the figures that the algorithm developed in this paper greatly outperforms the static resource allocation scheme, and HARQ protocol outperforms ARQ, in reducing energy consumption as well as transmit power consumption. The higher density the traffic has, the more obvious the advantages. However, with increasing traffic density, the deviation from the primal recovered objective to the dual optimum also gets larger, *e.g.*, the ratio of the deviation to the dual optimum increases from 8% at $\alpha = 0.5$ to 50% at $\alpha = 4$ on average for the HARQ case. On the one hand, this could be caused by possibly larger optimal duality gaps at higher traffic densities, while on the other hand, the more frequent situation at higher traffic densities that infeasible SA is obtained from solving the dual problem, which has to be heuristically adjusted, may deteriorate the performance of primal recovery and in turn, deteriorate the performance of the whole algorithm.

6 Conclusions

A novel energy minimization problem for QoS provisions in multicarrier systems has been formulated and solved, within a cross-layer framework that involves adaptive modulation and coding and retransmission protocols. By using the cutoff rate theorem, the channel and user independent parameters are connected to the time-varying resource allocation parameters with the SNR, which provides the means to reducing computations by setting up offline-computable look-up tables. Though the algorithm has been proved efficient at low to medium traffic densities, there are more issues to be studied: first, the optimal duality gap of the optimization should be estimated; second, more delicate primal recovery schemes are necessary to further improve system performance; last but not the least, the transmit power constraint at the BS should be integrated into the optimization. The three-fold work will be left to our future research.

References

1. Wong, C.Y., Cheng, R.S., Letaief, K.B., Murch, R.D.: Multiuser OFDM with adaptive subcarrier, bit and power allocation. IEEE Journal on Selected Areas of Communications 17, 1747–1758 (1999)
2. Zerlin, B., Ivrlač, M.T., Utschick, W., Nossek, J.A., Viering, I., Klein, A.: Joint optimization of radio parameters in HSDPA. In: IEEE 61st Vehicular Technology Conference VTC 2005-Spring, vol. 1, pp. 295–299 (2005)
3. Yu, W., Lui, R.: Dual Methods for Nonconvex Spectrum Optimization of Multicarrier Systems. IEEE Transactions on Communications 54(7), 1310–1322 (2006)
4. Seong, K., Mohseni, M., Cioffi, J.M.: Optimal Resource Allocation for OFDMA Downlink Systems. In: IEEE Int. Symposium on Information Theory, July 2006, pp. 1394–1398 (2006)
5. Gallager, R.G.: Information Theory and Reliable Communication. John Wiley and Son, Chichester (1968)

Performance Analysis of Power Saving Class of Type 1 with Both Downlink and Uplink Traffics in IEEE 802.16e

Sangkyu Baek and Bong Dae Choi

Department of Mathematics and Telecommunication Mathematics Research Center
Korea University, Seoul 136-701, Korea
{sangkyubaek,queue}@korea.ac.kr

Abstract. We investigate power consumption of a mobile station with the power saving class of type 1 in the IEEE 802.16e. We deal with stochastic behavior of mobile station during not only sleep mode period but also awake mode period with both downlink and uplink traffics. Our methods for investigating the power saving class of type 1 are to construct the embedded Markov chain and the semi-Markov chain generated by the embedded Markov chain. To see the effect of the sleep mode, we obtain the average power consumption of a mobile station and the mean queueing delay of a message. Numerical results show that the larger size of the sleep window makes the power consumption of a mobile station smaller and the queueing delay of a downlink message longer.

Keywords: power saving class, IEEE 802.16e, sleep mode, embedded Markov chain, semi-Markov chain.

1 Introduction

The IEEE 802.16 standard [1] has been designed to provide communication paths between subscriber stations and the base station as an emerging broadband wireless access system. An amendment to the standard, the IEEE 802.16e, has been concluded in 2006 [2] with mobility so that subscriber stations can move during services. This amendment[2] supports handover procedure and power saving of the mobile station(MS).

Due to the mobility of stations, power saving is one of the significant issues for the battery-powered MSs. The IEEE 802.16e defines sleep mode operations called power saving classes of type 1, 2 and 3. Power saving class of type 1 is recommended for connections of Best Effort (BE) and non-real time variable rate (NRT-VR) types, power saving class of type 2 for connections of unsolicited grant service (UGS) and real time variable rate (RT-VR) types, and power saving class of type 3 for multicast connection and managements operations.

As for the sleep mode operation of the power saving class of type 1 in the IEEE 802.16e, a few studies have been done to evaluate its performance. Xiao[3] proposed a simple model for power saving class of type 1 in the IEEE 802.16e

F. Granelli et al. (Eds.): MOBILIGHT 2009, LNICST 13, pp. 196–209, 2009.

for downlink traffic by focusing on sleep mode period. Xiao[4] and Zhang and Fujise[5] extended Xiao's model[3] to cover uplink and downlink traffics. Nejatian et al.[6][7] and Zhang[8] developed analytical models for non-Poisson traffic arrivals. Nejatian et al.[6][7] assumed that the interarrival times of packets follow Erlang distribution, whereas Zhang[8] assumed that the interarrival time of downlink packet follows hyper-Erlang distribution. They[3][4][5][6][7][8] obtained the average power consumption and the average delay during the sleep mode period. They[3][4][5][6][7][8] focused only on the sleep mode period, so they didn't model any stochastic behaviors of packets during the awake mode period.

Seo et al.[9], Park et al.[10], Han et al.[11] and Kong et al[12] proposed analytical models considering both awake mode period and sleep mode period. In Seo et al.[9] and Park et al.[10], the mobile station was modeled as an M/G/1/K queue. They[9][10] obtained the power consumption, mean downlink packet delay and packet blocking probability. Han et al.[11] proposed an analytical model using semi-Markov chain. Kong[12] mathematically analyzed power saving classes of both type 1 and type 2. Their models[9][10][11][12] focused only on the downlink traffic, i.e., they didn't model any uplink traffic.

We investigate analytical performance on sleep mode operation of power saving class-type 1 in the IEEE 802.16e with both downlink and uplink traffics. The main contribution of this paper is that we deal with stochastic behavior of mobile stations during not only sleep mode period but also awake mode period with both downlink and uplink traffics. In the literature mentioned above, they analyzed either only downlink traffic [9][10][11][12], or only sleep mode without awake mode[3][4][5][6][7][8]. Additionally, we find the mean queueing delay to see how sleep mode affects on queueing delay. Our model also takes account of practical setup times such as switching time from awake mode to sleep mode, switching time from sleep mode to awake mode and close-down time.

We organize the rest of this paper as follows. In Section 2, we describe sleep mode operation of power saving class of type 1 in the IEEE 802.16e. In Section 3, we present a mathematical model for the power saving class of type 1. Embedded Markov chains and semi-Markov chains generated by the embedded Markov chain of MS are constructed. We obtain the average power consumption. In section 4, we give numerical results and evaluate the performance of the power saving class of type 1.

2   Sleep Mode Operation of Power Saving Class of Type 1 in IEEE 802.16e

A mobile station(MS) has two modes: awake mode and sleep mode. While the MS is in the awake mode, it can send or receive data according to the BS's bandwidth scheduling. The MS sends a sleep mode request message (MOB_SLP-REQ) to the BS when there is no arrival message destined to the MS during the following close-down time (called *idle frame threshold*[11]) after both the downlink buffer at the BS and the uplink buffer at the MS become empty. The MS receives a sleep mode response message (MOB_SLP-RSP) from the BS, which contains the sizes of the initial sleep window, the final sleep window and the listening window, and the MS goes into the sleep mode. A sleep mode consists of sleep windows and listening windows, which are switched alternatively until the MS is notified of the buffered data at the BS, or an uplink data message arrives at the MS. During the sleep windows the MS may power down physical operation components. If no messages arrive during a sleep window and a following listening window, the size of the next sleep window is doubled, but not greater than the final sleep window. A traffic indication message (MOB_TRF-IND) shall be broadcasted by the BS during the listening window to alert the MS of the appearance of downlink traffic demand. When the MS receives a positive MOB_TRF-IND, the MS terminates the sleep mode and receives the pending data after the switching time from sleep mode to awake mode. During the switching time from sleep mode to awake mode, the BS requests the MS to send a bandwidth request header and confirms that the MS is in awake mode. The sleep mode is also terminated immediately if an uplink message arrives at the MS. Fig. 1 illustrates power saving class of type 1.

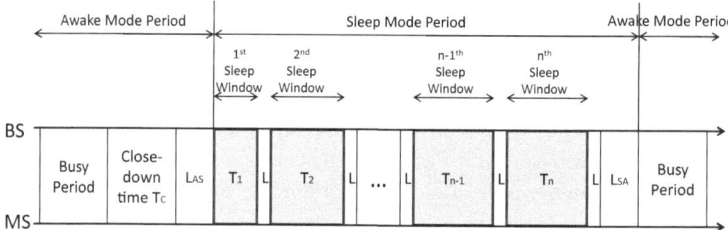

Fig. 1. Power saving class of type 1

3   Analytical Model

3.1   Assumtions

We assume that the downlink message arrivals at the BS toward the tagged MS and the uplink message arrivals at the MS toward the BS follow independent Poisson processes of rate λ_d and λ_u, respectively. The number of packets in a downlink message has a geometric distribution with parameter ρ_d and the number of packets in a uplink message has a geometric distribution with parameter ρ_u. Let T_j denote the length of the j-th sleep window. Let L denote the fixed length of a listening window. The MS stays in the awake mode period for an additional close-down time T_C after the buffers become empty Let L_{AS} denote the switching time from awake mode to sleep mode. It is known that L_{AS} needs at least 4 frames. During the switching time, the MS sends MOB_SLP-REQ and receive MOB_SLP-RSP. Let L_{SA} denote the switching time from sleep mode to awake mode. It is known that L_{SA} needs at least 3 frames.

Even though uplink arrivals and downlink arrivals are independent and are served by uplink subframe and downlink subframe, separately, they share a common sleep mode period. Thus, two traffics interact each other. We assume that capacity of each queue (in the BS for downlink message and in the MS for uplink message, respectively) is infinite.

3.2   Analysis of Sleep Mode Period

We calculate the necessary probabilities for finding the steady-state probabilities of the embedded Markov chain. We find the probability that the MS enters sleep mode period from awake mode period after both queues become empty. If there are neither downlink arrivals nor uplink arrival during a close-down time T_C, the MS sends a MOB_SLP-REQ to the BS. After getting approval(MOB_SLP-RSP) at the last frame of switching time L_{AS} from awake mode to sleep mode, the MS goes into sleep mode period and powers down its physical operation components. If any message arrives during the close-down time or the switching time L_{AS} from awake mode to sleep mode, the MS keeps on the awake mode period and sends/receives the message. However, the BS cannot cancel the MS's initiation of the sleep mode period due to downlink arrivals during the last frame of L_{AS}. Let q_1 be the probability that there are neither downlink nor uplink arrivals during $T_C + L_{AS} - 1$ frames. Let q_2 be the probability that there are no uplink arrivals during the last frame of L_{AS}.

$$
\begin{aligned}
q_1 &= e^{-(\lambda_d + \lambda_u)(T_C + L_{AS} - 1)} \\
q_2 &= e^{-\lambda_u}.
\end{aligned}
\tag{1}
$$

After both queues become empty, the MS enters the sleep mode period with probability $q_1 q_2$.

The instants in terminating sleep mode period are different for uplink traffic and downlink traffic. In case that downlink traffic occurs during sleep mode period, sleep mode period is ended in the next listening window. In contrast, if there are uplink arrivals during sleep mode period, the sleep mode period is ended immediately.

We consider two cases separately, and find the probability that there are i downlink messages in the BS and j uplink messages in the MS at the beginning of awake mode period, i.e. the end of sleep mode period.

– Case 1 - termination by uplink messages: If an uplink message arrives at the MS during a sleep mode period, the MS terminates the sleep mode period at the arrival frame and sends the message after the switching time L_{SA} from sleep mode to awake mode. Note that the arrivals of downlink messages during the same sleep window and the listening window do not affect the termination of the sleep mode period. Let α_n^u denote the probability that a sleep mode period is terminated by uplink arrivals at the MS during the n-th sleep window or the n-th listening window.

$$\alpha_1^u = 1 - e^{-\lambda_u(T_1+L)}$$
$$\alpha_n^u = e^{-\lambda_d(1-L+\sum_{k=1}^{n-1}(T_k+L))}e^{-\lambda_u\sum_{k=1}^{n-1}(T_k+L)}(1-e^{-\lambda_u(T_n+L)}) \quad n = 2,3,\cdots.$$
$$(2)$$

Let $h_{i,j}^u$ be the probability that a sleep mode period is terminated by uplink arrivals and there are i downlink messages in the BS and j uplink messages in the MS at the beginning of the awake mode period. The probability $h_{i,j}^u$ is obtained by

$$
h_{i,j}^u = \alpha_1^u \sum_{k=1}^{T_1+L} \frac{e^{-\lambda_u(k-1)}(1-e^{-\lambda_u})}{1-e^{-\lambda_u(T_1+L)}} \frac{(\lambda_d(1+k+L_{SA})^i e^{-\lambda_d(1+k+L_{SA})})}{i!}
$$
$$
\frac{e^{-\lambda_u(1+L_{SA})}((\lambda_d(1+L_{SA}))^j - (\lambda_d L_{SA})^j)}{j!}
$$
$$
+ \sum_{n=2}^{\infty} \alpha_n^u \sum_{k=1}^{T_n+L} \frac{e^{-\lambda_u(k-1)}(1-e^{-\lambda_u})}{1-e^{-\lambda_u(T_1+L)}} \frac{(\lambda_d(L+k+L_{SA})^i e^{-\lambda_d(L+k+L_{SA})})}{i!}
$$
$$
\frac{e^{-\lambda_u(1+L_{SA})}((\lambda_d(1+L_{SA}))^j - (\lambda_d L_{SA})^j)}{j!} \quad for\ i \geq 0\ and\ j \geq 1
$$
$$(3)$$

– Case 2 - termination by downlink messages: If a downlink message arrives in the BS during a sleep mode period, a positive traffic indication message (MOB_TRF-IND) is sent at the beginning of the following listening window and the sleep mode period is terminated. Note that the case that any uplink messages arrive in the MS until the listening window in which the positive traffic indication message is sent belongs to *Case* 1. Let α_n^d denote the probability that a sleep mode period is terminated by downlink arrivals in the BS during the $(n-1)$th listening window or the n-th sleep window.

$$\alpha_1^d = (1 - e^{-\lambda_d(1+T_1)})e^{-\lambda_u(T_1+L)}$$
$$\alpha_n^d = e^{-(\lambda_d+\lambda_u)\sum_{k=1}^{n-1}(T_k+L)-\lambda_d(1-L)-\lambda_u(T_n+L)}(1-e^{-\lambda_d(L+T_n)}), \quad n = 2,3,\cdots.$$
$$(4)$$

Let $a_{i,0}$ denote the probability that a sleep mode period is terminated by downlink messages and there are i downlink messages in the BS at the end of the last sleep window. By the assumption that the sleep mode period is terminated by downlink messages, any uplink messages cannot arrive in the MS until the end of the last sleep window. $a_{i,0}$ is obtained by

$$
a_{i,0} = \alpha_1^d \frac{1}{1-e^{-\lambda_d(T_1+1)}} \frac{(\lambda_d(T_1+1))^i e^{-\lambda_d(T_1+1)}}{i!}
$$
$$
+ \sum_{n=2}^{\infty} \alpha_n^d \frac{1}{1-e^{-\lambda_d(T_n+L)}} \frac{(\lambda_d(T_n+L))^i e^{-\lambda_d(T_n+L)}}{i!}
$$
$$(5)$$

Let $\tilde{a}_{i,j}$ denote the probability that i downlink message arrive during the last listening window(the listening window during which a positive traffic indication message is sent) and the switching time L_{SA} from sleep mode to awake mode and j uplink messages arrive during the switching time L_{SA}.

$$\tilde{a}_{i,j} = \frac{(\lambda_d(L+L_{SA}))^i e^{\lambda_d(L+L_{SA})}}{i!} \frac{(\lambda_u L_{SA})^j e^{\lambda_u L_{SA}}}{j!} \tag{6}$$

Let $h_{i,j}^d$ be the probability that a sleep mode period is terminated by downlink arrivals and there are i downlink messages in the BS and j uplink messages in the MS at the beginning of the awake mode period. Since we consider no uplink arrivals during the last listening window, $h_{i,j}^u$ is obtained as follows

$$h_{i,j}^d = \sum_{k=1}^{i} a_{k,0}\tilde{a}_{i-k,j} \quad for \; i \geq 1. \tag{7}$$

Thus the probability $h_{i,j}$ that there are i downlink messages in the BS and j uplink messages in the MS at the beginning of awake mode period is given by

$$h_{i,j} = h_{i,j}^u + h_{i,j}^d. \tag{8}$$

3.3   Embedded Markov Chain

We consider an embedded Markov chain representing the numbers of downlink and uplink messages, respectively, immediately after the following embedded points:

- time epoch where an uplink or downlink message completes its transmission,
- the beginning of the transmission of a uplink message at the MS after the uplink queue becomes empty,
- the beginning of the transmission of a downlink message at the BS after the downlink queue becomes empty.

Let X_n be the number of downlink messages and Y_n the number of uplink messages immediately after the n-th embedded point. Then, $\{(X_n, Y_n)\}$ is a discrete time embedded Markov chain with state space $\{(0,0), (i,0), (0,j), (i,j)|i \geq 1, j \geq 1\}$. Let $\{\pi_{i,j}\}$ be the steady-state probability of the embedded Markov chain. Before we find one-step transition probability, we calculate the distribution of the residence time of state (i,j) for the semi-Markov chain generated by the embedded Markov chain.

Distribution of Residence Times. Let $b_{i,j}(n)$ be the probability that the residence time at state (i,j) is n frames.

- $b_{0,0}(n)$: We obtain the distribution of the residence time at state $(0,0)$ as follows.

$$b_{0,0}(n) = \begin{cases} e^{-(\lambda_d+\lambda_u)(n-1)}(1-e^{-(\lambda_d+\lambda_u)}) \\ \quad for \ 1 \le n \le T_C + L_{AS} - 1 \\ e^{-\lambda_d(T_C+L_{AS}-1)}e^{-\lambda_u(T_C+L_{AS}-1-L_{SA})}(1-e^{-\lambda_u}) \\ \quad for \ n = T_C + L_{AS} \\ 0 \qquad for \ T_C + L_{AS} + 1 \le n \le T_C + L_{AS} + L_{SA} + 1 \\ e^{-\lambda_d(T_C+L_{AS}-1)}e^{-\lambda_u(n-1-L_{SA})}(1-e^{-\lambda_u}) \\ \quad for \ T_C + L_{AS} + L_{SA} \le n \le T_C + L_{AS} + T_1 + L - 1 + L_{SA} \\ e^{-\lambda_d(T_C+L_{AS}-1)-\lambda_u(T_C+L_{AS}+T_1+L-1)}(1-e^{-\lambda_d(T_1+1)-\lambda_u}) \\ \quad for \ n = T_C + L_{AS} + T_1 + L + L_{SA} \\ e^{-\lambda_d(T_C+L_{AS}+\sum_{k=1}^{N(n)} T_k+(N(n)-1)L)-\lambda_u(n-1-L_{SA})}(1-e^{-\lambda_u}) \\ \quad for \ T_C + L_{AS} + \sum_{k=1}^{N(n)} T_k + (N(n)-1)L - 1 + L_{SA} \le n \\ \qquad \le T_C + L_{AS} + \sum_{k=1}^{N(n)+1} T_k + N(n)L - 1 + L_{SA} \\ e^{-\lambda_d(T_C+L_{AS}+\sum_{k=1}^{N(n)-1} T_k+(N(n)-2)L)-\lambda_u(n-1-L_{SA})}(1-e^{-\lambda_d(T_{N(n)}+L)-\lambda_u}) \\ \quad for \ n = T_C + L_{AS} + \sum_{k=1}^{T_{N(n)}} T_k + N(n)L + L_{SA} \end{cases}$$
$$(9)$$

where $N(n) = max\{m | T_C + L_{AS} + \sum_{k=1}^{m}(T_k + L) + L_{SA} \le n\}$, the number of the sleep window during a sleep mode period.

- $b_{0,j}(n)$, $j \ge 1$: Let $b_{0,j}^u(n)$ be the probability that the residence time at state $(0,j)$ is n frames and the next embedded point is the completion of a transmission of an uplink message. Let $b_{0,j}^d(n)$ be the probability that the residence time at state $(0,j)$ is n frames and the next embedded point is the arrival epoch of a downlink message during an uplink transmission. $b_{0,j}^u(n)$ and $b_{0,j}^d(n)$ are obtained by

$$b_{0,j}^u(n) = e^{-\lambda_d(n-1)}(1-\rho_u)^{n-1}\rho_u \qquad (10)$$

$$b_{0,j}^d(n) = e^{-\lambda_d(n-1)}(1-e^{-\lambda_d})(1-\rho_u)^n \qquad (11)$$

Additionally, the distribution of the residence time from state $(0,j)$ is obtained by

$$b_{0,j}(n) = b_{0,j}^u(n) + b_{0,j}^d(n) \quad for \ j \ge 1. \qquad (12)$$

- $b_{i,0}(n)$, $i \ge 1$: Let $b_{i,0}^d(n)$ be the probability that the residence time at state $(i,0)$ is n frames and the next embedded point is the completion of a transmission of a downlink message. Let $b_{i,0}^u(n)$ be the probability that the residence time at state $(i,0)$ is n frames and the next embedded point is the arrival epoch of an uplink message during a downlink transmission. $b_{i,0}^d(n)$ and $b_{i,0}^u(n)$ are obtained by

$$b_{i,0}^d(n) = e^{-\lambda_u(n-1)}(1-\rho_d)^{n-1}\rho_d \qquad (13)$$

$$b_{i,0}^u(n) = e^{-\lambda_u(n-1)}(1-e^{-\lambda_u})(1-\rho_d)^n. \qquad (14)$$

Additionally, the distribution of the residence time at state $(i, 0)$, $i \geq 1$ is obtained by

$$b_{i,0}(n) = b_{i,0}^d(n) + b_{i,0}^u(n) \qquad for \ i \geq 1 \tag{15}$$

- $b_{i,j}(n)$, $i, j \geq 1$: The residence time at state (i, j) is the minimum of transmission time of a uplink message and a downlink message from the beginning of the embedded point. Since we assume that the number of packets in downlink and uplink messages have geometric distributions with parameter ρ_d and ρ_u, respectively, the minimum of two independent geometric distributions is also a geometric distribution with parameter $\rho = \rho_d + \rho_u - \rho_d \rho_u$. The distribution of the residence time at state (i, j) is obtained by

$$b_{i,j}(n) = (1 - \rho)^{n-1} \rho \qquad for \ i \geq 1, j \geq 1 \tag{16}$$

One-step Transition Probabilities. To find the one-step transition probabilities $p_{(i,j),(k,l)}$ from state (i, j) to state (k, l), we consider all possible transitions at the following each state:

- $p_{(0,0),(k,l)}$: If a message arrives in the BS or the MS during a close-down time or a switching time except the last frame ($T_C + L_{AS} - 1$ frames), the MS stays in awake mode period and receives or sends the message. In this case, the conditional probability $g_{k,l}$, given that messages arrive during a close-down time or the switching time except the last frame, that there are k downlink messages and l uplink messages at the beginning of the frame where the transmission starts again after the queues become empty is given by

$$g_{k,l} = \frac{1}{1 - e^{-(\lambda_d + \lambda_u)}} \frac{\lambda_d^k e^{-\lambda_d}}{k!} \frac{\lambda_u^l e^{-\lambda_u}}{l!}. \tag{17}$$

If an uplink message arrives in the MS during the last frame of switching time, the MS doesn't enter the sleep mode period, and it sends the message to the BS. In this case, the conditional probability $g'_{k,l}$, given that uplink messages arrive during the last frame of switching time, that there are k downlink messages and l uplink messages at the beginning of the frame where the transmission starts again after the queues become empty is given by

$$g'_{k,l} = \frac{1}{1 - e^{-\lambda_u}} \frac{\lambda_d^k e^{-\lambda_d}}{k!} \frac{\lambda_u^l e^{-\lambda_u}}{l!}. \tag{18}$$

The probability $s_{k,l}$ that there are k downlink messages and l uplink messages in each queues at the beginning of the frame where the transmission start again after the queues become empty is obtained by

$$s_{k,l} = (1 - q_1)g_{k,l} + q_1(1 - q_2)g'_{k,l} + q_1 q_2 h_{k,l}. \tag{19}$$

In other word, $p_{(0,0),(k,l)}$, the one-step transition probability from $(0, 0)$ to (k, l) is $s_{k,l}$.

- $p_{(0,j),(k,l)}$: The probability that the next embedded point is the completion of a transmission of an uplink message and k downlink messages and l uplink messages arrive in a state $(0, j)$ is given by

$$v_{k,l}^u = \sum_{n=1}^{\infty} b_{0,j}^u(n) \frac{\lambda_d^k e^{-\lambda_d}}{k!} \frac{(n\lambda_u)^l e^{-n\lambda_u}}{l!}. \tag{20}$$

The probability that the next next embedded point is the arrival epoch of a downlink message during an uplink transmission and k downlink messages and l uplink messages arrive in a state $(0, j)$ is given by

$$v_{k,l}^d = \sum_{n=1}^{\infty} b_{0,j}^d(n) \frac{1}{1 - e^{-\lambda_d}} \frac{\lambda_d^k e^{-\lambda_d}}{k!} \frac{(n\lambda_u)^l e^{-n\lambda_u}}{l!} \quad for \ k \neq 0 \tag{21}$$

$$v_{0,l}^d = 0.$$

The one step transition probability $p_{(0,j),(k,l)}$, $j \geq 1$ from state $(0, j)$ to state (k, l) is obtained by

$$p_{(0,j),(k,l)} = \begin{cases} v_{k,l-j+1}^u + v_{k,l-j}^d, & l \geq j \\ v_{k,0}^u, & l = j - 1 \end{cases} \tag{22}$$

- $p_{(i,0),(k,l)}$: The probability that the next embedded point is the arrival epoch of an uplink message during a downlink tranmission and k downlink messages and l uplink messages arrive in a state $(i, 0)$ is given by

$$w_{k,l}^u = \sum_{n=1}^{\infty} b_{i,0}^u(n) \frac{1}{1 - e^{-\lambda_u}} \frac{(n\lambda_d)^k e^{-n\lambda_d}}{k!} \frac{\lambda_u^l e^{-\lambda_u}}{l!} \quad for \ l \neq 0 \tag{}$$

$$w_{k,0}^u = 0.$$

The probability that the next embedded point is the completion of a transmission of a downlink message and k downlink messages and l uplink messages arrive in a state $(i, 0)$ is given by

$$w_{k,l}^d = \sum_{n=1}^{\infty} b_{i,0}^d(n) \frac{\lambda_d^k e^{-\lambda_d}}{k!} \frac{(n\lambda_u)^l e^{-n\lambda_u}}{l!} \tag{23}$$

The one step transition probability $p_{(i,0),(k,l)}$, $i \geq 1$ from state $(i, 0)$ to state (k, l) is obtained by

$$p_{(i,0),(k,l)} = \begin{cases} w_{k-i,l}^u + w_{k-i+1,l}^d, & k \geq i \\ w_{0,l}^d, & k = i - 1 \end{cases} \tag{24}$$

- $p_{(i,j),(k,l)}$, $i,j \geq 1$: The probability that k downlink messages and l uplink messages arrive in a state (i, j) is given by

$$r_{k,l} = \sum_{n=1}^{\infty} (1 - \rho)^{n-1} \rho \frac{(n\lambda_d)^k e^{-n\lambda_d}}{k!} \frac{(n\lambda_u)^l e^{-n\lambda_u}}{l!}. \tag{25}$$

A transition can occur by the transmission of a downlink or uplink message. The probability that a transition occurs by the transmission of a downlink message is given by

$$p_d = \frac{\rho_d - \rho_d\rho_u}{\rho_d + \rho_u - \rho_d\rho_u} \qquad (26)$$

The probability that a transition occurs by the transmission of an uplink message is given by

$$p_u = \frac{\rho_u - \rho_d\rho_u}{\rho_d + \rho_u - \rho_d\rho_u} \qquad (27)$$

The probability that a transition occurs by the transmission of both downlink and uplink messages is given by

$$p_{du} = \frac{\rho_d\rho_u}{\rho_d + \rho_u - \rho_d\rho_u} \qquad (28)$$

The one step transition probability $p_{(i,j),(k,l)}$, $i, j \geq 1$ from state (i,j) to state (k,l) is obtained by

$$p_{(i,j),(k,l)} = \begin{cases} r_{k-i+1,l-j}p_d + r_{k-i,l-j+1}p_u + r_{k-i+1,l-j+1}, & k \geq i \ and \ l \geq j \\ r_{0,l-j}p_d + r_{0,l-j+1}p_{du}, & k = i-1 \ and \ l \geq j \\ r_{k-i,0}p_u + r_{k-i+1,0}p_{du}, & k \geq i \ and \ j = l-1 \\ r_{0,0}p_{du}, & k = i-1 \ and \ j = l-1 \end{cases} \qquad (29)$$

Balance Equations By (19)-(29), the steady-state probabilities $\{\pi_{i,j}\}_{i,j=0}^{\infty}$ of the embedded Markov chain satisfy the following balance equations

$$\pi_{k,l} = \sum_{i=0}^{\infty}\sum_{j=0}^{\infty} \pi_{i,j}p_{(i,j),(k,l)} \qquad (30)$$

and the normalization condition $\sum_{i=0}^{\infty}\sum_{j=0}^{\infty}\pi_{i,j} = 1$. In calculation, we restrict the state space by large i and j.

3.4   Power Consumption of Mobile Station

Now we consider a semi-Markov chain generated by the embedded Markov chain to find the mean residence time of the semi-Markov chain at state (i,j) and to derive sleep mode ratio which is defined as the proportion of the sleep mode period. The mean residence time $\eta_{i,j}$ of the semi-Markov chain at state (i,j) is given by

$$\eta_{i,j} = \sum_{n=1}^{\infty} nb_{i,j}(n) \qquad (31)$$

We obtain the expected length of a sleep mode period as follows.

$$R_{sleep} = \sum_{n=1}^{\infty}\alpha_n^u[\sum_{j=1}^{n-1}(T_j+L) + \sum_{k=1}^{T_n+L} k\frac{e^{-\lambda_u(k-1)}(1-e^{-\lambda_u})}{1-e^{-\lambda_u(T_n+L)}}] + \sum_{n=1}^{\infty}\alpha_n^d\sum_{j=1}^{n}(T_j+L) + L_{SA} \qquad (32)$$

Sleep mode ratio is given by

$$Sleep\ Mode\ Ratio = \frac{\pi_{0,0}q_1q_2R_{sleep}}{\sum_{a=0}^{\infty}\sum_{b=0}^{\infty}\pi_{a,b}\eta_{a,b}} \tag{33}$$

The MS is assumed to consume its energy by E_{awake} (mJ) per one frame in the awake mode period and $E_{powersaving}$ per one frame in the sleep window. The energy of the MS is consumed additionally by E_{change} when the MS switches from a sleep window to a listening window. The average energy consumption during a sleep mode period is obtained by

$$\begin{aligned}
E_{sleep} = \sum_{n=1}^{\infty} \alpha_n^u \Big[\sum_{j=1}^{n-1}(T_j E_{powersaving} + LE_{awake}) + \sum_{k=1}^{T_n} kE_{powersaving}\frac{e^{-\lambda_u(k-1)}(1-e^{-\lambda_u})}{1-e^{-\lambda_u(T_n+L)}} \\
+ \sum_{k=1}^{L}(T_n E_{powersaving} + kE_{awake})\frac{e^{-\lambda_u(T_n+k-1)}(1-e^{-\lambda_u})}{1-e^{-\lambda_u(T_n+L)}} + nE_{change}\Big] \\
+ \sum_{n=1}^{\infty} \alpha_n^d \sum_{j=1}^{n}(T_j E_{powersaving} + LE_{awake} + E_{change}) + L_{SA}E_{awake}.
\end{aligned} \tag{34}$$

The average energy consumption per one frame during a sleep mode period is given by

$$E_S = \frac{E_{sleep}}{R_{sleep}} \tag{35}$$

Thus the average power consumption with both uplink and downlink traffics can be derived as follows:

$$Power\ Consumption = (Sleep\ mode\ Ratio)E_S + (1 - (Sleep\ mode\ Ratio))E_{awake} \tag{36}$$

3.5   Queueing Delay

The steady-state probability $\pi_{i,j}^*$ for the semi-Markov chain generated by the embedded Markov chain in subsection 3.3 is given by

$$\pi_{i,j}^* = \frac{\pi_{i,j}\eta_{i,j}}{\sum_{a=0}^{\infty}\sum_{b=0}^{\infty}\pi_{a,b}\eta_{a,b}}. \tag{37}$$

Let $X(t)$ be the number of downlink messages and $Y(t)$ the number of uplink messages at time t. To find the queueing delay for both uplink and downlink traffics, the steady-state probability $p_{k,l}$ for a stochastic process $(X(t), Y(t))$ is given as follows[13]

$$\begin{aligned}
p_{k,l} &= \sum_{i=0}^{k}\sum_{j=0}^{l}\frac{\pi_{i,j}^*}{\eta_{i,j}}\sum_{n=1}^{\infty}P(k-i\ downlink\ arrivals, l-j\ uplink\ arrivals\ in\ [0,n])b_{i,j}(n) \\
&= \sum_{i=0}^{k}\sum_{j=0}^{l}\frac{\pi_{i,j}^*}{\eta_{i,j}}\sum_{n=1}^{\infty}\frac{(\lambda_d n)^{k-i}e^{-\lambda_d n}}{(k-i)!}\frac{(\lambda_u n)^{l-j}e^{-\lambda_u n}}{(l-j)!}b_{i,j}(n)
\end{aligned} \tag{38}$$

Thus, the downlink mean queueing delay and uplink mean queueing delay are respectively obtained by

$$
mean\ queueing\ delay\ for\ downlink\ message = \frac{1}{\lambda_d}\sum_{i=0}^{\infty}\sum_{j=0}^{\infty} i p_{i,j} - \frac{1}{\rho_d}
$$
$$
mean\ queueing\ delay\ for\ uplink\ message = \frac{1}{\lambda_u}\sum_{i=0}^{\infty}\sum_{j=0}^{\infty} j p_{i,j} - \frac{1}{\rho_u}.
$$
(39)

4   Numerical Results

For numerical results, we use the following parameters:

- Listening Window L : 2 frames
- Close-down time T_C : 5 frames
- L_{AS} : 4 frames
- L_{SA} : 3 frames
- E_{awake} : 10 mJ
- $E_{powersaving}$: 1 mJ

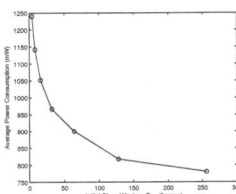 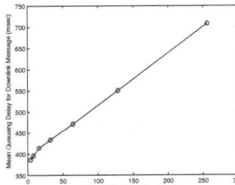 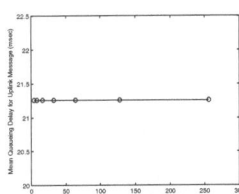

(a) Power consumption   (b) Mean queueing delay (c) Mean queueing delay
for a downlink message   for an uplink message

Fig. 2. Effect of the Initial Sleep Window

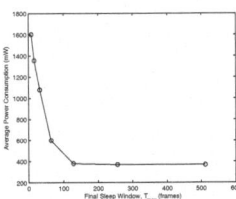 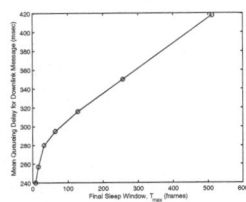 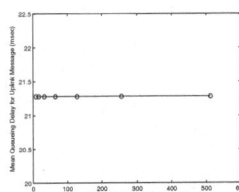

(a) Power consumption   (b) Mean queueing delay (c) Mean queueing delay
for a downlink message   for an uplink message

Fig. 3. Effect of the Final Sleep Window

- E_{change} : 10 mJ
- λ_d : 0.01 /frame
- λ_u : 0.01 /frame

The length of a message is assumed to be geometrically distributed with mean 10 frames.

Fig. 2 depicts (a) the average power consumption, (b) the mean queueing delay for a downlink message and (c) the mean queueing delay for an uplink message as the initial sleep window increases from 4 frames to 256 frames. In Fig. 2, we assume $T_{max} = 256$ frames. We see that the average power consumption decreases and the mean queueing delay for a downlink message increases, as the initial sleep window increases. However, the mean queueing delay for an uplink message does not depend on the initial sleep window size.

Fig. 3 depicts (a) the average power consumption, (b) the mean queueing delay for a downlink message and (c) the mean queueing delay for an uplink message as the final sleep window increases from 8 frames to 512 frames. In Fig. 3, we assume $T_{min} = 8$ frames. We also see that the average power consumption decreases and the mean queueing delay for a downlink message increases, as the final sleep window increases. The final sleep window does not influence the mean queueing delay for an uplink message.

Acknowledgment

This research is supported by the MIC (Ministry of Information and Communication), Korea, under the ITRC (Information Technology Research Center) support program supervised by the IITA (Institute of Information Technology Assessment).

References

1. IEEE 802.16-2004, Part 16: Air Interface for Fixed Broadband Wireless Access Systems, Standard for Local and Metropolitan Area Networks (2004)
2. IEEE 802.16e-2005, Part 16: Air Interface for Fixed and Mobile Broadband Wireless Access Systems - Amendment for Physical and Medium Access Control Layers for Combined Fixed and Mobile Operation in Licensed Bands (2006)
3. Xiao, Y.: Energy saving mechanism in the IEEE 802.16e wireless MAN. IEEE Communications Letters 9(7), 595–597 (2005)
4. Xiao, Y.: Performance Analysis of an Energy Saving Mechanism in the IEEE 802. 16e Wireless MAN. In: Consumer Communications and Networking Conference (CCNC) 2006, vol. 1, pp. 8–10 (2006)
5. Zhang, Y., Fujise, M.: Energy management in the IEEE 802.16e MAC. IEEE Communications Letters 10(4), 311–313 (2006)
6. Nejatian, N.M.P., Nayebi, M.M., Ashtiani, F.: Effect of different traffic patterns on power consumption of sleep mode in the IEEE 802.16e MAC. In: IEEE ICT-MICC 2007, pp. 649–653 (2007)

7. Nejatian, N.M.P., Nayebi, M.M.: Evaluating the Effect of non-Poisson Traffic Patterns on Power Consumption of Sleep Mode in the IEEE 802.16e MAC. In: IFIP WOCN 2007, pp. 1–5 (2007)
8. Zhang, Y.: Performance Modeling of Energy Management Mechanism in IEEE 802.16e Mobile WiMAX. In: IEEE WCNC 2007, pp. 3205–3209 (2007)
9. Seo, J., Lee, S., Park, N., Lee, H., Cho, C.: Performance analysis of sleep mode operation in IEEE 802.16e. In: IEEE VTC 2004-Fall, vol. 2, 26-29, pp. 1169–1173 (2004)
10. Park, Y., Hwang, G.U.: Performance Modelling and Analysis of the Sleep-Mode in IEEE802.16e WMAN. In: IEEE VTC 2007-Spring, pp. 2801–2806 (2007)
11. Han, K., Choi, S.: Performance Analysis of sleep mode Operation in IEEE802.16e Mobile Broadband Wireless Access Systems. In: IEEE VTC 2006-Spring, vol. 3 (2006)
12. Kong, L., Tsang, D.H.K.: Performance Study of Power Saving Classes of Type I and II in IEEE 802.16e. In: IEEE Conference on Local Computer Networks, pp. 20–27 (2006)
13. Takagi, H.: Queueing Analysis. Vacation and Priority Systems, vol. 1. North-Holland, Amsterdam (1991)
14. Gross, D., Harris, C.: Queueing Theory. Wiley, Chichester (1998)

Testing Cooperative Communication Schemes in a Virtual Distributed Testbed of Wireless Networks

George Kormentzas[1], Luis Alonso[2], and Christos Verikoukis[3]

[1] University of the Aegean (AEG)
[2] Technical University of Catalonia (UPC)
[3] Telecommunications Technological Centre of Catalonia (CTTC)
gkorm@aegean.gr, luisg@tsc.upc.edu, cveri@cttc.es

Abstract. It is expected that Next Generation Networks (NGNs) will offer seamless interoperability among heterogeneous access technologies in order to provide ubiquitous access. In such settings, short range technologies may be used in order to extend the coverage area of cellular systems while cooperative diversity can improve the efficiency of the wireless systems. An advanced, backward compatible, with the 802.11 standard, MAC protocol for cooperative ARQ scenarios in NGNs sets the research framework for this work. The functionalities of the RCSMA protocol 1 will be enhanced and the derived analytical models will be validated at the UNITE Virtual Distributed Testbed (VDT) (2).

Keywords: Ad hoc networks, medium access control protocols, cooperative communications, next generation networks, Cooperative ARQ.

1 Introduction

The network of the future seems to be a network of diverse wireless access technologies where the end-users will be able to attain any service, at any time, at the access that is optimised for the specific service resulting in a "flexibility and choice" that enhances the quality of life of the individuals. To this vision, the ubiquitous access in a cost-effective way in an era where spectral resources are important, constitutes an emerging challenge.

Focusing on the network of the future, the Long Term Evolution (LTE) that evolves HSDPA is going to offer a communication framework with higher-data-rate, low-latency and packet-optimized radio-access technology for both uplink and downlink. The 3GPP System Architecture Evolution (SAE) encompasses an IP core network infrastructure with IP interfaces targeting multiple RATs (Radio Access Technologies), which will not only be restricted to 3GPP legacy systems such as UMTS Terrestrial Radio Access Networks (UTRAN) or GSM/EDGE Radio Access Network (GERAN), but will also have inherent open interfaces to accommodate wireless networks (WLAN, WMAN) through the IEEE 802.21 Media Independent Handover Function. In this context, the research of new transmission paradigms and techniques as the so-called cooperative communication schemes, have been getting special interest in the very last time.

F. Granelli et al. (Eds.): MOBILIGHT 2009, LNICST 13, pp. 210–219, 2009.

The proposal herein presented addresses cooperative ad-hoc networking and considers heterogeneity with the final aim to ensure undiminished service perception whilst the subscriber terminal migrates freely between networks that might belong to different operators. We will consider enhancement techniques, which include networking to provide "virtually" collocated cellular and wireless coverage reflecting a converged network infrastructure. In this case the ad-hoc network, comprising ad hoc relay-capable nodes can be used to "fill" (guarantee the coverage of) the regions that are not covered by a cellular (WiMAX or UMTS-LTE) or an infrastructure-based WLAN. In such a kind of scenarios we identify the characteristics of a MAC protocol for cooperative Automatic Repeat reQuest (ARQ) and we define framework for a realistic performance evaluation.

The paper is organised as follows: Initially, the reference scenario of the EU funded project PASSENGER (Provision of optimum radio AcceSS at the Emerging Next GEneration NetwoRks) is presented as an indicative scenario of the network of the future. Then, a literature review of cooperative communications is presented. Section 4 analyses the UNITE Virtual Distributed Testbed (VTD) that will be employed for the validation of the proposed cooperative communication schemes. The way how the variants of the advanced MAC protocol for cooperative ARQ and optimization mechanisms will be uploaded at the UNITE framework is discussed at Section 5. Finally, Section 6 concludes the paper.

2 PASSENGER Scenario

In the context of PASSENGER project an innovative scenario is proposed considering the joint application of session roaming and ad-hoc networking with relay-based connectivity. In this case the ad hoc network comprising relay-capable nodes can be used to "fill" the regions that are not covered by the WiMAX/UMTS-LTE or the infrastructure-based WLAN. This scenario requires the existence of an additional WR (Wireless Router) on the WLAN side which will keep connections with both the infrastructure-based and the ad hoc networks. For a node belonging to the ad hoc network the WiMAX/UMTS-LTE and the infrastructure-based WLAN networks are virtually collocated in the sense that the test node can use any of them to gain access to the "outside" world. Central entity in this scenario is a multi-radio wireless router (WR), equipped with both WLAN and WiMAX/UMTS-LTE network air-interfaces. This entity is not just a conventional node in a general ad hoc network but possesses enhanced functionality by being able to communicate with the WiMAX/UMTS-LTE network thus providing internet connectivity to the rest of the ad hoc nodes. Technical challenges in this scenario are, among others, the functionality of the WR, network discovery and association mechanism, the multi-hop routing protocol used in the ad-hoc network, and cooperative ARQ protocols.

In this work, we focus on the research framework of MAC protocols for cooperative ARQ schemes. Preliminary results of the Relay Carrier Sensing Multiple Access (RCSMA) 1 protocol in ARQ scenarios in wireless systems call for more extensive research in heterogeneous wireless networks. RCSMA makes an advance on the state of the art by introducing the petitioner concept in the operation of the protocols, where cooperation can be claimed by the receiver. In this work, the operation of the

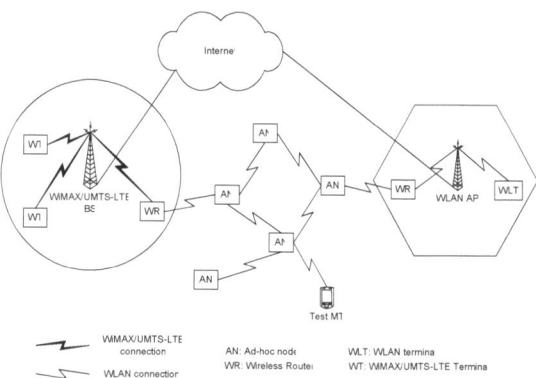

Fig. 1. Ad-hoc networking to provide "virtually" collocated WiMAX/UMTS-LTE and WLAN

protocol is enhanced by introducing the persistency concept. In addition, the use of advanced Radio Resource Management techniques based on Cross-Layer algorithms in the selection of the adequate set of relays in order to guarantee certain QoS and energy saving mechanisms will further optimize the performance of the protocol. Likewise, the overhead introduced by the cooperation phase will be also considered as an advanced in the state of the art since all the proposed in the literature MAC protocols only focus on the cooperation process.

3   State-of-the-art

The main reasons for introducing ad hoc networks in existing wireless systems are to increase and improve the services for packet users, as well as to reduce the cell planning difficulties for operators. The potential of wireless ad hoc networks was investigated by several EU research projects. For example, the FleetNet project developed a wireless multi-hop ad hoc network for inter-vehicle communication to improve the driver's and passenger's safety and comfort. The IST-MobileMAN project investigated the potential of the Mobile ad hoc network paradigm. Specifically, the project aimed at the definition and development of a metropolitan area, self-organizing, and totally wireless network called Mobile Metropolitan Ad hoc Network (MobileMAN). In the project IST-ROMANTIK, multi-hop architectures for capacity and coverage enhancement for 3G and beyond-3G wireless systems were investigated. Finally, exploitation of ad-hoc-networking aspects for coverage enhancement is currently being studied in the scope of the IST-WINNER project.

The broadcast nature of wireless communication systems makes possible to improve their performance by allowing users cooperating with each other. In multi-user environments, the use of distributed diversity seems to offer an interesting alternative to overcome the practical implementation drawbacks found out when experimenting with Multiple Input Multiple Output (MIMO) techniques using relatively small devices. The basis of cooperative communications is to exploit the fact that in the wireless channel, any transmission can be overheard by all the nodes within a certain transmission range; once a message is transmitted from a source to a given destination node, all nodes in the transmission range of the source station become potential helpers, referred to as relays, that could help out in the communication link. Since nodes in the network may be spatially distributed, the different copies received at each node can be used to create a spontaneous time and/or spatial diversity scheme. The improvement induced by exploiting cooperation in wireless networks can be achieved in terms of higher transmission rate, lower transmission delay, more efficient power consumption, or even increased coverage range.

The fundamental theory behind the concept of cooperation has been deeply studied among researchers during the last years 1-4 while significant effort and huge amount of innovative results for cooperation in the physical layer have been presented in the literature. However, significant less effort has been dedicated to the development of higher layer protocols.

Automatic repeat request (ARQ), which requests the data link layer of the transmitter to repeat the packet when a packet is erroneously received, had been widely used in wireless communications systems. Therefore, if multiple nodes had also received a copy of the transmitted, they could collaborate on the retransmission of such packet when needed and could increase the coverage area of a cellular network. This is known as Cooperative ARQ (C-ARQ) and introduces the cooperative transmission concept into the Date Link Control. C-ARQ topics have been addressed in several works in the literature. In 7, the SNR gain and average number of retransmissions of a single source cooperative ARQ protocol is studied. In 5 the saturation throughput of three double-source cooperative ARQ protocols is presented. Cerutti and al. present in 6 a delay model for single-source and single-relay cooperative ARQ protocols. They propose a simple set of retransmissions rules and their aim is to reduce the signalling and control overhead in the network, the hardware and algorithm complexity. In 8 three ARQ protocols are presented. In the first protocol the relay node always retransmits the packet. In the second, only the one with the better channel conditions between the relay and the transmitter is requested to repeat the packet. Likewise, space-time codes are used in order to repeat simultaneously the packet transmission in the third protocol. Finally, in 9 Morillo et all propose a collaborative ARQ protocol that exploits diversity through collaboration in wireless networks. They demonstrate that when M neighbouring nodes collaborate using the proposed algorithm can get the same efficiency as an array of M antennas. In most of the previous works on cooperative transmission focus is put on analyzing the gains of cooperation from a fundamental point of view and simple TDMA schemes are considered. Therefore, the design of MAC protocols which coordinate the cooperative retransmissions in is an interesting research challenge.

Finally, some projects funded by the 6th and 7th framework programs include cooperative communications in their objectives. The aim of IST-COOPMAC Project is to provide the advances that will permit the deep understanding of both the theoretical and practical aspects of cooperative and opportunistic communications. The IST-FIREWORKS project aims to improve the current IEEE 802.16 by an enhanced multi-hop relaying deployment concept using cooperative technology. The ICT-CODIV will try to solve open problems on cooperative transmission both in the Physical and MAC layers while the ICT-REWIND will address cooperative relaying schemes in the framework of WIMAX-based systems.

Regarding distributed test-beds, several EU funded projects such as the IST-UNITE, the ICT-PII and the ICT-One Lab2 propose different solutions for the inter-connection of federated testbeds for Future Internet Research and Experimentation.

4   Overview of the UNITE Framework

The UNITE VDT system aims to address two main functional requirements. The first relates to the derivation of a simulation facility for composite wireless networks by interconnecting existing simulators of diverse radio technologies running on different platforms. The second has to do with the incorporation of facilities for joint Radio Resource Management functions (called URRM functions).

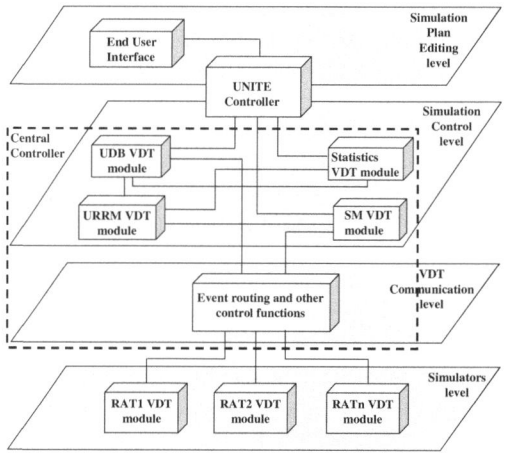

Fig. 2. The UNITE VDT logical levels

These requirements can be realized by the functional architecture depicted in Fig.2. This architecture relies on the identification of four conceptual levels for the distributed virtual testbed and the definition of discrete building blocks residing at these levels. At the topmost level, the end-user interface provides access to the UNITE controller, which is responsible for connecting the VDT platform with the external world. Using this interface and the facilities provided by the UNITE controller (e.g., identification of available simulators, retrieval of simulator parameters etc.), end-users compile simulation plans and submit them for execution (plans are created through a Scenario Editor, called VDT Simulation Plan Editor). Additional function-alities of this component include management of user authentications and permis-sions, retrieval of past simulation plans from the repository, validation of simulation plans, scheduling of the simulation instance execution and storage of simulation plans to the repository for future use (the repository is implemented as part of the UNITE Data Base – UDB).

When a simulation is eligible for execution, the UNITE controller hands over con-trol to the Central Controller. The latter is comprised of four modules:

- the UDB module responsible for storing simulation results and simulation plans,
- the statistics module that performs statistical processing on the data of the UDB,
- the module that implements the joint URRM algorithm.

The Scenario Manager (SM) that undertakes functions like terminal management, service and traffic stream management and simulator clusters time management. At the lowest level of the VDT architecture, there are a number of modules responsible for simulating radio technologies. Legacy simulators are attached to the distributed virtual platform by forming VDT modules. These modules are comprised of three entities (Fig.3):

- the actual simulator (in general, the federated service),
- the federated gateway which manages the simulation cluster and translates simulator-specific messages to VDT messages and vice versa,
- the VDT module API, which provides the interface (functions and parameter definitions), through which VDT modules communicate with the components of the Central Controller.

The Federated Gateway (FG) communicates with the other entities of the distributed testbed and transforms messages received from the testbed to simulator-related actions. The simulator calls specific functions from the Federated Gateway to send VDT Events. On the other hand, a specific function from the simulator is called each time a VDT Event is received from the VDT Central Controller, so that the event can be processed.

The communication framework is supported by an event-based middleware implemented using SOAP over HTTP. The control of the events and their routing towards the distributed components registered for them is undertaken by the Central Controller. The whole translation of events (SOAP messages) to simulator-specific functions is implemented through an API, providing functions and parameter definitions, in order simulator developers to easily attach their simulation engine to the VDT.

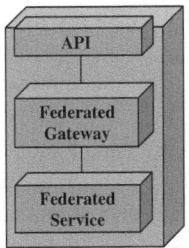

Fig. 3. The structure of a VDT module

The VDT module is the basic building block of the VDT architecture and it can be used for any function that needs to be implemented in a separate hardware/software environment. In general, the VDT module can be considered as an abstract implementation of a desired function serving a specific purpose. In this regard, the notion of the VDT module can perfectly apply as well to the simulation control functions of the

Central Controller (in fact, Fig.2 has been sketched with this assumption in mind). Additionally, the federated service of a VDT module can also correspond to an emulated or physical networking testbed.

5 Research Methodology

In PASSENGER the proposed research methodology includes analysis, using Markov chains, and validation, by means of simulation using the UNITE VDT, of the proposed protocols in different scenarios and conditions.

5.1 Development of PASSENGER Algorithms

Several infrastructure technologies such as WiMAX (802.16 e//j), UMTS-LTE, WLAN (802.11 b/g/n) will be taken into account in the proposed scenario while WLAN (802.11 b/g/n) will be considered for the ad hoc network.

The proposed MAC protocol will be based on the RCSMA protocol for cooperative ARQ. In RCSMA a new control packet was defined at the MAC layer in order to Claim For Cooperation (CFC). This CFC could be implemented following the structure of the RTS packet considered in 802.11-based systems, but indicating that both the source and destination of the data packet are the address of the node asking for cooperation. When a destination node needs to ask for cooperation, it must listen to the channel for a Short Inter Frame Space (SIFS) time. If the channel remains idle for that time, it will be able to broadcast the CFC. It is worth mentioning that the operation of the MAC protocol defined in the IEEE 802.11, sets a longer IFS (named DIFS) before getting access to the channel for the transmission of a data packet. Therefore, cooperation processes in RCSMA get higher priority than regular transmissions. At the reception of the CFC, the source node knows that has to wait for a cooperation procedure instead of executing the back-off procedure triggered by the lack of an ACK reception. On the other hand, the relay set initiates the cooperation process with the destination node. During the cooperation phase, the set of relays use the access rules defined in the 802.11 standard. Once the destination node is able to have a correct copy of the original packet, it is responsible for indicating to the source node and to its neighbouring nodes that the required cooperation procedure has finished. To do so, an ACK packet is transmitted.

In the context of PASSENGER, persistency is used, where the relays nodes continuously transmit the requested packet in order to overcome the problem of non sufficient number of relays 1. It is worth mentioning, that the proposed protocol is backward compatible with the existing MAC protocols in 802.11 standards in order to be easily exploited in the near future. Analysis is based on Makov Chains and existing models such as the ones proposed by Bianchi and Wu have been adapted to evaluate the efficiency of the proposed protocol. Both saturated and non-saturated network conditions for the ad hoc nodes are considered. In the former case, all ad hoc nodes in the system have already a packet for transmission in their buffer and when cooperation is claimed they use the existing back-off window. In the later case, we consider that all nodes reset their back-off window and initiate a random back off period in order to avoid collision in the first transmission whenever cooperation is claimed.

The protocol will be further optimized by minimizing the overheads due to the co-operation phase. Moreover, we will address the relays selection problem by using advanced Radio Resource Management algorithms based on cross-layer techniques. Therefore, channel conditions, loss-rates, energy consumption, interference, traffic load and status of node's battery may be considered, among other variables, in the selection of the adequate set of relays in order to optimize the average contention time and/or the total energy consumption. Finally, the aforementioned protocol will be extended in order to cover various schemes of cooperative relaying and various options for relay nodes (decode and forward, amplify and forward etc...). The previous scenarios will be studied both with fixed and mobile relays. Depending on the scenario and the application the previous schemes will be studied under QoS and energy constraints.

The proposed protocol will be evaluated in terms of the required delay to recuperate the original packet, the total throughput of the system and the total energy consumption, for different values of the back-off window and number of cooperative relay nodes.

5.2 Validation of PASSENGER Algorithms

PASSENGER algorithms are intended to run on separate URRM modules that will be VDT modules attached to the VDT system. At this case, the Federated Service of these VDT modules will not be simulators but the algorithms themselves. The following issues have to be addressed:

Implementation of the URRM

- Implementation of the URRM VDT modules
- Attaching of the URRM VDT modules to the central controller

Storing of simulation results

- Definition of results to be stored into the UDB
- Definition of the structure of these results when sent through the UpdateValues event
- Implementation of storing of these results to the UDB
- Testing

Retrieving of simulation results

URRM must retrieve simulation results from the UDB in order to take decisions. This is done through the getSystemInformation or getMobileInformation events. The following have to be addressed:

- Definition of the parameters that each event will have
- Definition of the structure of the information that each event will contain. For example, getMobileInformation will contain all the simulation results concerning a specific mobile. How will these results be extracted from the event in the Federated Gateway of the URRM VDT?

- Implementation of retrieving of the results from the UDB, packing them in a predefined structure and embodying them in the getSystemInformation or getMobileInformation events, when requested.
- Testing

Service IDs

Three services are supported so far with static parameters. Service parameters must be dynamic:

- Definition of more parameters in the SendStreamToMobile and StartTxToMobile events, so as to support dynamic configuration of their parameters, which will be dependent to each service
- Change in the definition of the scenario story so as to support dynamic configuration of service parameters
- Implementation of the new scenario story to the VDT editor
- Implementation of reading of the new scenario story from the UNITE controller and embodying the service ID parameters in the SendStreamToMobile and StartTxToMobile events.
- Testing

QoS parameters

Each service ID should be accompanied with QoS parameters, the values of which should be able to be configured dynamically by the user:

- The steps for the case of service IDs stand in this case as well.

After all the above issues are addressed, the steps needed to integrate the algorithm are the following:

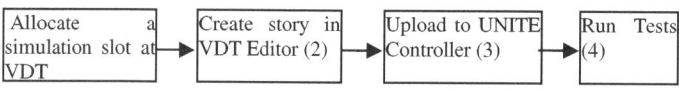

Fig. 4. The structure of a VDT module

6   Conclusions

The UNITE Virtual Distributed Testbed constitutes an ideal environment where co-operative communications schemes could be validated. Building at this potential, the paper discusses how the PASSENGER schemes are going to be uploaded and tested at VDT. In this context a synergy between two EC funded projects is achieved.

Acknowledgments. This work is funded by the European Commission under the umbrella of the PEOPLE programme in the context of the project PASSENGER (219561), by the Spanish Government in the in the context of Celtic project LOOP (FIT-330215-2007-8)

References

1. Gomez, J., Alonso, J., Verikoukis, Ch., Pérez-Neira, A., Alonso, L.: Cooperation on Demand Protocols for Wireless Networks. In: IEEE PIMRC 2007, Athens (September 2007)
2. Vassis, D., Sarakis, L., Kormentzas, G., Verikoukis, Ch.: A Distributed Testbed for Performance Evaluation of Inter-system Optimization Schemes in Heterogeneous Wireless Networks. In: The Middleware Technologies for Enabling Next-Generation Network Services and Applications multicon lectures notes published by Multicon Verlag, ISBN 978-3-930736-12-6
3. Cover, T.M., Gamal, A.E.: Capacity Theorems for the relay channel. IEEE Transactions on Information Theory 25(5), 572 (1979)
4. Sendonaris, A., Erkip, E., Aazhang, B.: User cooperation diversity-part I: System Description. IEEE Transactions on Communications 51(11), 1927–1938 (2003)
5. Sendonaris, A., Erkip, E., Aazhang, B.: User cooperation diversity-part II: Implementation aspects and performance analysis. IEEE Transactions on Communications 51(11), 1939–1948 (2003)
6. Gupta, P., Cerruti, I., Fumagalli, A.: Three transmission scheduling policies for a cooperative ARQ protocol in radio networks. In: Proc. WNGG conference (October 2004)
7. Cerruti, I., Gupta, P., Fumagalli, A.: Delay Model of a One-Way Cooperative ARQ Protocol in Slotted Radio Networks with Poisson Frames Arrivals. IEEE Transactions on Networking (January 2007)
8. Zimmermann, E., Herhold, P., Fettweis, G.: The impact of cooperation on diversity-exploiting protocols. In: Proc. Of 59th IEEE Vehicular Technology Conference (2004)
9. Yu, C., Zhang, Z., Qiu, P.: Cooperative ARQ in Wireless Networks: Protocolos Description and Performance Analysis. In: IEEE ICC 2005 (2005)
10. Morillo-Pozo, J., García-Vidal, J., Pérez-Neira, A.I.: Collaborative ARQ in Wireless Energy-Constrained Networks. In: Proceedings of the workshop on Foundations of Mobile Computing 2005 (DIAL-POM 2005) (2005)

RFID-Based Identification:
A Measurement Study

Marina Buzzi, Marco Conti, and Enrico Gregori

IIT-CNR,
via Moruzzi 1, I-56124 Pisa, Italy
{marina.buzzi,marco.conti,enrico.gregori}@iit.cnr.it

Abstract. In this paper we investigate the feasibility of using UHF RFID as a tool for identifying individuals. We first developed a demo application for greeting conference speakers by incorporating a passive UHF tag into a user badge. Next, based on participant feedback we carried out additional tests to gauge the reliability of the user identification. Results of our experiments are very encouraging. If certain technical precautions are kept in mind, UHF RFID has a high percentage of success and is more widely accepted in the user community than is active sensor technology.

Keywords: UHF RFID, identification, measurement, safety.

1 Introduction

In the late 1990s, the Internet revolutionized ICT and our entire society. The Web has changed the world profoundly, introducing millions of users to the network, with great economic, social and cultural impact. Today ubiquitous computing promises to revolutionize our society even further, providing "last mile" technology to bring applications and services to everyone.

In 1991 Mark Weiser, the father of ubiquitous computing wrote: "The most profound technologies are those that disappear. They weave themselves into the fabric of everyday life until they are indistinguishable from it." [16]. In his prophetic vision, Weiser imagines that computers vanish into the background, technology is embedded in the environment and user interaction becomes simple and natural. The future is thus "disappearing computing" and "calm technology", no effort is required of users to use the technology [16]. According this philosophy, RFID technology (Radio-frequency identification) could be a valuable tool for building context and position-aware services if it allows people to behave naturally without imposing constraints (i.e. requiring a short read distance or the use of active tags). In this paper we describe an experimental study that investigates the feasibility of UHF RFID as a tool for identifying individuals.

As ubiquitous computing becomes increasingly common, new applications begin to permeate every aspect of daily life. Today RFID and sensor technologies embedded in everyday objects are used to sense environmental data, convey information and enable services. Specifically Radio-frequency identification is a technology for

F. Granelli et al. (Eds.): MOBILIGHT 2009, LNICST 13, pp. 220–229, 2009.
© ICST Institute for Computer Sciences, Social-Informatics and Telecommunication Engineering 2009

automatic identification of objects, persons and animals, using radio waves. It consists of two components: readers and tags. Tags store information (usually a unique identifier) that can be retrieved by readers. This kind of distributed information transmitted via wireless networks and elaborated from resource-rich devices, makes services available everywhere and under any conditions. Thus RFID has become a key technology for enabling "object-based" services.

Tags can be passive, active (battery-powered) or semi-active. Passive tags transmit their unique id by using the energy induced by the reader. Passive tags are especially convenient since they are small, cheap and do not pose maintenance problems, i.e. have a potentially infinite life. Tags and readers operate in four frequency ranges, which are usually applied in different fields ([17]):

- Low-frequency (LF), i.e. 125/134 kHz and 140/148.5 kHz, used for tracking animals, human implants (to store personal medical info), and electronic keys (to access hotel rooms, controlled areas);
- High-frequency (HF) 13.56 MHz which follows two standard specifications with different aims:
  - ISO 15693, which optimizes performance of tag reading, is used for inventory and tracking of food, goods, etc. and also for automating libraries;
  - ISO 14443 which supports high levels of security, for e-passports, non-contact payments, credit cards, e-tickets, etc.
- Ultra-high frequency (UHF) 915 MHz (US) and 868 MHz (Europe), applied in logistics and inventory systems;
- 2.4 GHz and higher (Microwave tags) are commonly used for mobility and interports.

Radio-frequency identification may be a valuable tool for building new ubiquitous applications, if it allows people to behave naturally without imposing any awkward constraints. This paper investigates the feasibility of using UHF RFID technology for identification of people by inserting a passive tag in a badge, creating a kind of "intelligent badge". As previously discussed we aim to remove user constraints, and thus chose to focus on ultra-high frequency, which allows greater read distance of tags compared to high frequency (HF). Hereafter, we only refer to UHF RFID systems and passive tags.

RFID passive tags include an integrated circuit (for storing information, modulating/demodulating a signal) and one or more antennas, for receiving and transmitting the signal (from/to the RFID reader) [17]. Each tag comes with a unique id stored in its memory (usually a 24-character code). New generation tags are programmable, which means it is possible via SW to write product codes and other information in their memories. Therefore, RFID tags form a kind of distributed memory surrounding persons and environments, which may be advantageously used in a great number of new applications [9]. Considering that tag memory is constantly increasing, new and powerful applications will be designed in the near future.

The reader utilizes one or more antennas (linearly or circularly polarized). Depending on the frequency used and antenna features (i.e. size, polarization) the read distance of tags varies from one-half to seven meters. The availability in the market of different reader families (small, medium, large) offers a vast range of choices.

Readers integrated in handheld devices are interesting due to their effectiveness and simple use. Obviously an antenna integrated in a handheld device is small so the read distance is reduced as well. For example, at the moment the maximum declared read range of the Psion palm device, which integrates an RFID reader, is 80 cm or 1.50 m depending on the frequency (868 or 915 MHz) [12].

2   Issues in UHF RFID Technology

When applying RFID technology to a real case, reliability of the reading is crucial. RFID data are large-volume streams characterized by inaccuracy; duplicate, missed and ghost reads due to interference or temporary malfunction of some components make RFID data noisy and unreliable [14]. Thus reliable SW for data filtering and aggregation is crucial for the successful introduction of RFID technology.

Different frequencies can present different challenges. In the following we only consider problems of UHF RFID technology.

Tag reading may be challenging depending on features of tagged items: i.e. size, composition of materials, packaging, and tag placement. Three issues affect the reading: RF reflection, shadowing and absorption [15]:

- Metal reflects UHF; however an appropriate insulation separating the tag from the surface can improve reading;
- Shadowing occurs when several tags are placed very close and their antennas mask each other, by decreasing the read rate;
- Liquids (such as water) absorb RF and hinder tag reading. Since the human body is mostly composed of water and thus absorbs RF, this means that a person placed between the reader antenna and the tagged item prevents the reading (i.e. breaks RF propagation).

Furthermore, tag placement (i.e. its physical position in space) is crucial: in the worst case, if antenna and tag are perpendicular, the reading may fail. Lastly, motion (handheld readers or tagged items) facilitates tag reading [15].

3   Examples of Presence-Aware Applications

In the following we briefly outline three possible uses of RFID-based identification, conceived for our work environment. An extensive analysis of RFID state-of-art which also includes several examples of applications and future developments to enable pervasive computing is addressed by [13].

Public Safety

One possible application of the intelligent badge is in the field of public safety, such as in the case of fire. In our Research Area there are two very large 3-floor buildings containing offices with one, two or three desks. There are only a few labs with four or more people. Every building has numerous exits every 100 m. Thus, in case of evacuation people should be evenly distributed throughout the area. The only crowded place might be the 300-seat conference room located on the ground floor, with three emergency exits opening directly onto the lawn.

At entry/exit points of the building all persons would be identified by their intelligent badge and the relative data inserted in a database. In the event of evacuation and reunion at a meeting point (1 for each institute), a portable reader (connected to a wireless network) belonging to the security guard receives the list of all present in the institute. As individuals arrive at the meeting point they are crossed off the list. Anyone missing is thus quickly identified without resorting to a roll-call.

Speaker Introduction

Another use of the intelligent badge and vocal synthesis (as shown in the following in our SW prototype) could be automatic speaker presentations in conference sessions, to speed up the process. A few sentences could be loaded by the speaker (describing his/her activities) via web. When speakers wear the intelligent badge, the session chair is aware of their presence without needing to ask if all speakers are present, who they are, etc. Furthermore, if one speaker arrives late during the session, the chair knows exactly who is entering the room, identifying him visually and leading to greater flexibility in last-minute scheduling.

When the speaker is approaching the speech area, the reader gets the tag id and, if the person is the next speaker (the time schedule of talks is checked via SW) his/her description starts automatically after a few seconds (or alternatively can be activated by one click).

Traffic Gate Control

The intelligent badge could be used to control access to the CNR area, automatically opening/closing a traffic gate.

Each worker wears an intelligent badge. The UHF reader reads the unique id and opens the traffic gate. This would have an advantage over inserting the badge into the magnetic band reader (as currently used), by activating the bar from a configurable distance, for instance inside a car (in case of rain) or in a pocket, wallet or bag, if the user is on a bike or walking, thus shortening access time.

Compared to using an HF RFID reader which requires the badge to be a short distance away (5-10 cm), UHF technology allows greater reading distance, with all its related advantages.

4 Study Start-Up

In order to understand how introducing RFID is perceived, we set up a demo application to show the potential of this technology, then stimulated a discussion about designing new ubiquitous applications. Previous studies showed that systems are more successful when associated with a wide range of organizational issues dealt with throughout the development process and ensuring that members of the user community are actively involved [2].

We created a simple application for greeting conference speakers and committee members. Inside additional participant badges we inserted the Alien 9540 passive tags, which have a bi-directional circle-polarized antenna. The application interacts with the Alien ALR-8800 Enterprise RFID Reader (http:// www. alientechnology. com/), which is EPC Gen 2 and ETSI-Compliant (EN 302-08 and EN 300-220 compliant) and operates at 915 MHz. The Alien evaluation kit includes two circularly polarized antennas (called in the following antenna 0 and antenna 1) which are connected to the reader via cables. In this configuration the read distance covers up to 2 meters.

The application is developed by using the Alien Java SDK (which interfaces the ALR-8800 RFID Reader) and relies on open source SW:

- Java SPEECH APIs which relies on the Java voice synthesizer FreeTTS (http://freetts.sourceforge.net/);
- JDBC APIs which interacts with the MySQL DB running on a Linux server is used for storage.

In order to simplify deployment, the application was split into two parts:

- a java code for checking the RFID reader and for writing tag information in a mySQL database; for the demo the database was populated with 30 entries;
- a graphical UI which is actually a web page including details about observed tags, which reads entries from the database. The interface is built in PHP and refreshes every 3 seconds.

The reader runs in autonomous mode and, when it sees tags, it sends the Java application a table containing all tag identifiers and other associated information. The reader was set up in inventory command mode for multiple contemporary readings. The reading time was set up at 1 second.

The two antennas were located 4 m apart on the same side of a wall. Access to the demo room consisted of a single door, thus 2 antennas were sufficient to perceive entering and exiting individuals. When there is a transition from antenna 0 to 1 (i.e. the reader first sees the tag_i with antenna 0 and then with antenna 1) the voice synthesizer announces "Hello $Name_i$ $Surname_i$;" of the person wearing the badge containing the tag_i. This info was previously stored in the db with other data (role, organization, email). Vice versa, when a person leaves the room the SW announces the appropriate phrase, i.e. "Bye bye $Name_i$". Simultaneously to vocal synthesis, the Name, Surname and Role of individuals were shown in a web interface (Fig. 1).

By performing the demo we learned a great deal. First, our SW suffered from a lack of synchronization between voice (i.e. greetings) and the text shown in the web user interface since speech consisting of several words is intrinsically sequential while their visualization is parallel. If a group of attendees came close to the reader antenna, there was a delay in the announcement of their names since the speaking queue was longer. This could be annoying.

We also encountered one reading problem (no tags were perceived) when a person carried a cell phone next to the RFID tag (i.e. in a shirt pocket). Holding the tagged badge in one's hand facilitated the reading but a feeling of unreliability was perceived. As remarked in [7] trustiness of service depends on the reliability and robustness of the application and its ability to manage any event that might occur in a production environment. In addition, one individual was not recognized in a group crossing the room, possibly due to RF emission broken by the person's body, since the only two available antennas were located on the same side of the wall in order to track the subject's entry/exit. Thus position as well as number of antennas is crucial for successful reading.

Fig. 1. A screenshot of the web UI shown in the Demo

5 Measurements

The demo was designed to stimulate user curiosity, encourage discussion, gather new ideas, and verify acceptance of automating some basic organizational processes and services. After the demo we decided to perform some tests to verify the amplitude of the abovementioned problems. At the beginning we performed a pre-test to verify RF adsorption by the human body and potential interference with handheld devices and in general with devices which produce electro-magnetic fields.

We verify that RF is absorbed independently of body size. A hand covering the entire badge (and thus the tag) was enough to make the reading impossible. Instead, the reading was successful if one of the tag antennas remained uncovered.

Then we performed a test with a G17 Sharp cell phone. Interference when the cell phone was ringing or a person was talking was not observed, but the reading was impossible if the phone covered both antennas of the tag. Partial covering of one antenna did not impact reading reliability. Furthermore, placing the phone behind the tag (from the keyboard side) with the antenna in front hid the tag from the reader.

In order to verify reliability of the tag reading, we then carried out two tests with 10 colleagues wearing the "intelligent badge". We also split the group into two subsets of five persons each to determine whether any variation occurs according to the sample size.

Pilot Test and Set-Up

We arranged the 2 antennas of the Alien reader in different positions on the walls of our Institute building, to verify whether and how tag-reading percentage varies in different configurations. Antennas 0 and antenna 1 have the same features (i.e. size, polarization, etc.) and configurations (RF attenuation, etc.).

We registered tag read details (i.e. tag id, antenna, data, time) in a file and then processed collected data, calculating the average of readings for each tag. Specifically, single percentages were calculated for each tag as $N_i/20$ values where N_i is the number of readings for each tag$_i$ during 20 passages between antennas.

Since the reading was set at 1 second, a slow subject may be identified more than once during each passage. The total read percentage, shown in Table 1 is the mean of read percentages of each tag, for each experiment.

A pilot test was carried out before starting the experiments. Analyzing the data we found that not all tags performed in the same way. Specifically, one tag was read only once in this test. Thus during setup we checked all tags involved in the tests before starting data collection and defective tags were discarded. In the following our data refers to reliable tags; however on a large scale defective tags might be found, which would affect the reading.

Tag size also influences the success of a reading. We used squiggle tags, which are small and thin. Other tags may present different reading percentages. Last, as already mentioned, movements and positioning of participants may also cause results to vary.

Lateral Antennas

In this experiment, the two antennas were located on two parallel walls 2 m apart, at 1.25 m from the ground. Then we instructed the groups of subjects to walk between the reader antennas 10 times in one direction (until exiting the antennas' range of action) and 10 times in the opposite direction, for a total of 20 passages. In order to simulate behavior in case of building evacuation, we asked people to walk rapidly, talking and trying to remain as close to each other as possible. Table 1 summarizes results.

By using two antennas, results showed 100% identification of a 5-member as well as the 10-member groups. Identification with only one antenna decreased to 85.25% for the 10-person group and to 86.5% for the 5-person group, depending on antenna position (right or left) and the location and movement of participants.

Orientation of tags may lead to a missed reading when the badge is perpendicular to the antennas (parallel to RF flow). In addition, unless people are in single file, one body may block another and absorb the radio emission flow. The number of readings varies depending on the physical distribution of group components: people in the external lines are read more frequently than subjects on the inside.

As previously mentioned, a slow individual might be identified more than once during each passage. The maximum number of readings for a single tag (by both antennas) was 56 while the minimum was 27. With only one antenna, the maximum number of readings for a single tag was 37 (or 28) while the minimum was 8 (or 7) for antenna 0 (for antenna 1).

Frontal, Inclined Antennas

In this experiment the two antennas were placed on both sides of a wall over the door at a height of 2.10 m, and inclined at a $30°$ angle from the wall ($60°$ angle from the floor) in order to "see" people entering the door from the front. In this case the tag was parallel to the antennas (i.e. perpendicular to RF flow). In this set-up the reliability of identification was 100% for both groups (Table1).

With only one antenna identification decreased to 89.25% for the 10-member group and to 95% for the 5-person group depending on antenna position and the location and movement of participants (Table 1). Probably to permit easy walking, each individual was physically somewhat distant from the next in line, so antenna coverage increased compared to the previous test.

The maximum number of readings for a single tag was 64 and the minimum was 23 for two antennas. With only one antenna, the maximum number of readings for a single tag was 37 (or 28) while the minimum was 8 (or 7) for antenna 0 (or antenna 1).

Table 1. Percentage of identification of two groups (of 5 or 10 persons) with different locations and number of antennas

Persons	1 antenna h. 1.25 m. (lateral)	2 parallel antennas at 2 m.; h. 1.25 (lateral)	1 antenna h. 2.10 m. (frontal, inclined)	2 antennas h. 2.10 m. (frontal, inclined)
5	86.5%	100%	95%	100%
10	85.25%	100%	89.25%	100%

6 Discussion

Concerning tag reading, we believe that in order to obtain a reliable degree of identification, four antennas should be placed at every gate. Two or more antennas may offer a good degree of identification, however a 100% reading is not always guaranteed due to several factors: 1) cooperation of people is essential for allowing tag reading 2) position of tag in relation to the antenna is unpredictable 4) absorption or reflection is possible 5) rarely, a tag may be defective. With only one antenna the percentage of identification greatly decreases. Results of our experiments showed the technical feasibility of UHF-based identification if there is user collaboration. Thus user acceptance is a key factor in the success of presence-awareness services.

By simplifying processes and increasing efficiency, ubiquitous computing enables new ways of working but user acceptance plays a critical role in widespread adoption of ubiquitous applications. The perception of losing both privacy and control of the RFID environment is a great concern [4]. Although security and privacy issues are widely addressed in numerous studies and surveys [6], [10], [5], [3], [11], classic security data techniques such as user authentication, symmetric and asymmetric ciphering algorithms, public key infrastructure combined with ad hoc techniques such as killing, sleeping or blocking tags, tag password, tag pseudonyms, or the proxy approach are not enough to disperse the black cloud that surrounds the use of RFID systems. Thus, security and privacy issues should be carefully considered and included in the design of new pervasive services.

In our application, RFID-based identification may pose a serious privacy problem since including tags in worker badges could allow people within the Institute to be tracked without user knowledge (if readers were scattered throughout the Institute area). Several colleagues have expressed doubts about this.

Another common concern was related to the safety level of exposure to RF emitted from antennas, for human beings. It is remarkable that people are not equally afraid of using their cell phones or computers! As stated in the reader manual, there is a safe distance (at least 20 cm) which should be maintained by workers in daily contact with antennas. In our empirical experience we noticed that women were more sensitive to this problem. Two pregnant women at our Institute were extremely worried about having contact with RF. To our knowledge the very few studies in this field have not proved any health damage in the short term but we are unable to exclude it over time since reliable results should include very long-range data and observations (at least one or two decades). In the context of our application, exposure to RF is sporadic, limited to crossing the traffic gate of the Institute, and thus is not a problem.

However, in real applications, improved efficiency may overcome user fear. Kourouthanassis et al. by implementing and testing a prototypal application with users in a retail context, verified enhanced user experience in supermarket shopping [8]. Furthermore, Konomi et al. [7] observed that after an initial skepticism the introduction of smart shelves in a shoe shop gained wide acceptance from sale staff, due to improved efficiency and work simplification. In our experience, after people's initial fear of being tracked and having their privacy invaded, we observed a general interest in this technology as a way to simplify tasks and automate work. For instance, its potential for tracking objects, such as inventory and document workflow, appears to be quite useful and interesting.

7 Conclusion

This paper investigates the feasibility of using UHF RFID readers and passive tags for identification purposes. We performed this experimental study in order to better understand the limits, costs and performance of UHF RFID as a tool for identifying individuals.

First, we developed a simple application for greeting speakers at a conference. We then carried out several tests to verify the feasibility of using UHF RFID technology in "intelligent" badges in terms of percentage of successful readings. Results showed the technical feasibility of this approach, as long as the user collaborates. Despite its simplicity, and keeping in mind certain technical precautions, UHF RFID has a high percentage of success and is more widely accepted in the user community than active sensor technology.

In conclusion, UHF RFID is technically suitable for creating a range of useful new applications beyond the usual fields of logistics and inventory; however privacy and security issues should be carefully considered even in the design phase, to encourage general acceptance and ensure user collaboration.

References

1. Alien Technology: Tag Feng Shui. A Practical Guide to Selecting and Applying RFID Tags, http://www.alientechnology.com/docs/WP_Alien_Tag.pdf
2. Doherty, N.F., King, M., Al-Mushayt, O.: The impact of inadequacies in the treatment of organizational issues on information systems development projects. In: International Journal of Information & Management, vol. 41, pp. 49–62. Elsevier, Amsterdam (2003)

3. Dritsas, S., Gritzalis, D., Lambrinoudakis, C.: Protecting privacy and anonymity in pervasive computing: trends and perspectives. Telematics and Informatics Journal 23(3), 196–210 (2006)
4. Günther, O., Spiekermann, S.: RFID: tagging the world: RFID and the perception of control: the consumer's view. Communications of the ACM 48(9), 73–76 (2005)
5. Juels, A.: RFID Security and privacy: a research survey. IEEE Journal on Selected Areas in Communications 24(2), 381–394 (2006)
6. Hyangjin, L., Jeeyeon, K.: Privacy threats and issues in mobile RFID. In: First International Conference on Availability, Reliability and Security, pp. 510–514 (2006)
7. Konomi, S., Roussos, G.: Ubiquitous computing in the real world: lessons learnt from large scale RFID deployments. Personal and Ubiquitous Computing 11(7), 507–521 (2007)
8. Kourouthanassis, P.E., Giaglis, G.M., Vrechopoulos, A.P.: Enhancing user experience through pervasive information systems: The case of pervasive retailing. International Journal of Information Management 27, 319–335 (2007)
9. Mamei, M., Quaglieri, R., Zambonelli, F.: Making tuple spaces physical with RFID tags. In: Proceedings of the 2006 ACM symposium on Applied computing, pp. 434–439. ACM Press, New York (2006)
10. Paar, C., Weimerskirch, A.: Embedded security in a pervasive world - Information Security Technical Report, vol. 12, pp. 155–161. Elsevier, Amsterdam (2007)
11. Peris-Lopez, P., Hernandez-Castro, J.C., Estevez, J.M.: RFID Systems: A Survey on Security Threats and Proposed Solutions. In: Cuenca, P., Orozco-Barbosa, L. (eds.) PWC 2006. LNCS, vol. 4217, pp. 159–170. Springer, Heidelberg (2006)
12. Psion Teklogix. WorkAbout Pro Second Generation Hand-Held Computer, http://www.psionteklogix.com/
13. Roussos, G., Kostakos, V.: RFID in pervasive computing: State-of-the-art and outlook. Pervasive and Mobile Computing 5, 110–131 (2009)
14. Sheng, Q.Z., Li, X., Zeadally, S.: Enabling Next-Generation RFID Applications: Solutions and Challenges. Computer 41(9), 21–28 (2008)
15. Vega, V.: Common RFID Implementation Issues: 10 Considerations for Deployment, http://www.alientechnology.com/blog/
16. Weiser, M.D.: The Computer for the 21st Century. Scientific American Special Issue on Communications, Computers, and Networks, 66–75 (1991)
17. Wikipedia. Radio-frequency identification, http://en.wikipedia.org/wiki/RFID

Self-management in Future Internet Wireless Networks: Dynamic Resource Allocation and Traffic Routing for Multi-service Provisioning

Ioannis P. Chochliouros[1], Nancy Alonistioti[2], Anastasia S. Spiliopoulou[3], George Agapiou[1], Andrej Mihailovic[4], and Maria Belesioti[1]

[1] Hellenic Telecommunications Organization (O.T.E.) S.A., Research Programs Section, 99, Kifissias Avenue, 15126 Maroussi, Athens, Greece
ichochliouros@oteresearch.gr
[2] Dept. of Informatics and Communications, University of Athens, 15784, Panepistimioupolis, Ilissia, Athens, Greece
nancy@di.uoa.gr
[3] Hellenic Telecommunications Organization (O.T.E.) S.A., General Directorate for Regulatory Affairs, 99, Kifissias Avenue, 15126 Maroussi, Athens, Greece
aspiliopoul@ote.gr
[4] King's College London (KCL) Centre for Telecommunications Research, London, UK
andrej.mihailovic@kcl.ac.uk

Abstract. Evolution towards the Future (Internet) networks necessitates inclusion of self-management capabilities in modern network infrastructures, for a satisfactory provision of related services and for preserving network performance. We have considered a specific targeted methodology, in the form of the *generic cognitive cycle model,* which includes three distinct processes (i.e. *Monitoring, Decision Making and Execution*), known as the "MDE" model, able to support dynamic resource allocation and traffic routing schemes. For further understanding of the issue we have examined two essential use-cases of practical interest, both in the context of modern wireless infrastructures: The former was about dynamic spectrum re-allocation for efficient use of traffic, while the latter has examined intelligent dynamic traffic management for handling network overloads, to avoid congestion.

Keywords: Autonomic communications, cognitive networks, Future Internet, generic cognitive cycle model, self-configuration, self-management, self-organization, spectrum re-allocation, traffic routing, WiMAX.

1 Introduction

The scope of the global Internet implicates much more than a "simple communication" system. The current Internet practically "unlocks" the global wealth of information and knowledge while the offered feature of "universality" allows formerly unconnected

F. Granelli et al. (Eds.): MOBILIGHT 2009, LNICST 13, pp. 230–241, 2009.
© ICST Institute for Computer Sciences, Social-Informatics and Telecommunication Engineering 2009

people and organisations with similar and diverse interests to find each other, thus resulting in new and wide-ranging communities of interest, supply chains, and markets (and networks) for several intents [1]. Today's Internet has become the indispensable means for networked innovation and a proper "highway" to globalisation and circulation of services/facilities and knowledge. Its size, complexity and role in modern society has far exceeded the original expectations of its creators. With the further deployment of wireless and mobile technologies, the number of users is expected to jump to some 4 billion in a few years.

In this fast evolutionary context, networks and services of the future continue to generate new economic opportunities with new classes of networked applications, whilst reducing corresponding operational expenditures [2]. The actual challenge faced by all relevant sectors (i.e. the state, public sector, private businesses, manufacturers, operators and service/application providers, users, etc.) is to "properly deliver" the next generation of ubiquitous and converged networks and services for communication, computing and media [3], as several societal and commercial usages are strongly "pushing" the current Internet architecture to its "limits". As the Internet expands its effectiveness, a multiplicity of novel and innovative services is introduced, demanding an environment able to support innovation, creativity and economic growth. Thus, the issue of the *"Future Internet (FI)"* is attracting more and more attention, as it impacts all underlying network technologies.

The Future Internet will provide the means to share and distribute new multimedia content and services, together with appropriate network resources and related facilities, with superior quality and striking flexibility, in a trusted and personalized way, improving quality of life and safety. Such an option also entails overcoming the scalability, flexibility, dependability and security bottlenecks, as today's network and service architectures are primarily "static" and able to support a limited number of devices, service features and limited confidence. Novel infrastructures will permit the emergence (or "re-shaping") of a large variety of business and economic models capable of dynamic and seamless end-to-end composition of resources across a multiplicity of devices, networks, providers and service domains.

2 Self-management Activities in the Future Internet

An increasing number of processes (many of which have explicit business impacts) need to be automated in the networks of the future, i.e. they have to be performed upon relevant autonomous decisions, without any human intervention. Today's ICT systems are not inherently able to "learn" from past (or current) experience and cannot contextualize and adapt to evolutionary processes, based on their own observation and learning processes. However, many ICT applications cannot be developed further, if there are no new breakthroughs in systems' intelligence and engineering [4]. Overcoming such technology "roadblocks" can permit the emergence of a wide range of opportunities in novel application fields, also including management (and re-configuration) of communications networks. In the context of the Future Internet, networks possessing cognitive features are often quoted as being one the "key" next-generation communications technologies [5] and are expected to lead to a much improved communication service, while providing efficient solutions to various problems currently experienced by market "actors". A prime challenge of the Future Internet is to provide means that will enable cognitive network management through dynamic, ad-hoc and optimized resource allocation and control, fault tolerance and robustness associated with real-time trouble shooting capabilities.

Thus, the attribute of autonomic network management (or "self-management") becomes an essential issue as it affects variable domains, most of which are essential for the effective network's behaviour, under several (predicted or even unpredicted) circumstances. The unified network infrastructure of the future (composing the "*Future Internet*") should support a number of parallel network architectures with distinctive traffic and quality characteristics. The anticipated traffic rates and new usage patterns challenge the current static and peak/best effort-based network configurations and/or dimensioning, calling for more efficient and flexible infrastructures that can support dynamic resource allocation and new traffic routing (i.e. "intelligent routing") schemes. Such options are further discussed in the continuity of the work, in the scope of two fundamental use-cases, dealing with network management.

2.1   Defining a Generic Cognitive Cycle for Self-management Purposes

In the context of the present work we propose a specific targeted methodology, intending to conceptualize and to develop novel concerns, for an effective realisation of the evolution towards to Future Internet networks/structures. The main aspect of the advancements of the current approach lies upon the introduction of cognition and autonomy in networks (also at the level of individual network elements-NEs) as well as in the collective ability of corresponding management functionalities. This presents the major driving concept based on the cognition in various system aspects and the realization of the Generic Cognitive Cycle Model (*Monitoring-Decision Making-Execution* or "*MDE*") as shown in Fig.1.

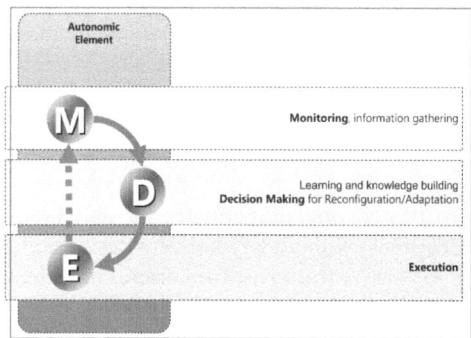

Fig. 1. Description of the Generic Cycle Cognitive Model

The above Generic Cognitive Cycle Model consists of three distinct phases, i.e. Monitoring (M), Decision-Making (D) and Execution (E). "*Monitoring*" involves gathering of information about the environment, and the internal state of any entity considered as an *"autonomic element"*[1]. *"Decision-Making"* includes learning, knowledge building and decision-making for reconfiguration and/or adaptation, by utilizing the developed knowledge model and situation awareness. Finally, the *"Execution"* process involves (self-) reconfiguration, software-component replacement or reorganization and optimisation actions. The model is present in several network elements (as well as in the collective management facilities) and constitutes advancements of current systems. It can ease the "transition" to the Future Internet by addressing multitudes of challenges (e.g. explicit protocol design for a wireless world, integrated functional design, alternative stacks, data-aware network equipment and handling service and network complexity) [6].

3 Autonomic Management of Future Internet

Defining and validating appropriate self-management methods (especially for a variety of network-related management issues) can facilitate, *significantly*, any process for making real Future Internet's vision. However, challenges emerge in terms of scalability, mobility, flexibility, security, trust and robustness of networks. In particular, *at the network level*, a major instigation lies on the proper "incorporation" of flexible and ad-hoc management capabilities.

Traditional network management involves complex labor-intensive processes performed by experts (humans and dedicated tools/systems). For example, some important configuration tasks such as installing or reconfiguring a system, provisioning network services and allocating resources typically engage a large number of activities involving multiple NEs. The latter may be associated with proprietary configuration management instrumentation and may also be spread across heterogeneous network domains thereby increasing the complexity of the relevant management activities. Human-guided (or "manual") processes currently involved with network management are rapidly reaching their "limits", as networks become more complex and implement a plethora of modern and innovative services. On the other hand, Future Internet networks will require minimal human involvement in the network planning and optimisation tasks; consequently, self-management can provide the proper countermeasures to that purpose [7]. Thus, related technologies are expected to pervade the next generation of network management systems and so to affect considerations for their deployment and exploitation, while the essential aim is to ensure that network continues to function, even if one or more nodes (or other elements) fail.

The rapid growth of Internet, both in terms of data traffic and in terms of diversity of services, has led to a high complexity of network architectures, which are even harder to manage. Challenges resulting from the current complex Internet architectures are manifold, and the goals for the envisioned architectures are sometimes conflicting. Future networks are complex systems with a large number of control mechanisms and

---

[1] An *"autonomic element"* may be a network element (e.g., router, base station (BS), and mobile device), a network manager, or any software element that lies at the service layer. Such an element, equipped with embedded cognition, has a process for monitoring and perceiving internal and environmental conditions, and then for planning, deciding and adapting (self-reconfiguring) on relevant conditions.

parameters, acting at diverse time scales (*i.e. from milliseconds to hours, or even more*); noteworthy interdependencies can occur among them, usually accompanied by constraints on measurements, signalling and processing. Understanding, interpreting, analyzing, categorizing and handling such potential intricacies, can affect the design and the usage of effective network management functionalities. For example, recent actions demonstrate that newly added base stations are self-configured in a "*plug-and-play*" fashion, while existing base stations continuously self-optimize their operational algorithms and parameters in response to changes in network, traffic and environmental conditions. Adaptations and interventions are so desirable, in order to provide the targeted service availability and quality, as efficiently as possible. In the same approach, it should be expected that several flexible and cognitive network management and operation frameworks should be developed, to enable dynamic, ad-hoc and optimised resource allocation and control, administration with accounting that ensures expansion of usage, differentiated performance that can be accurately monitored, fault-tolerance and robustness, associated with real-time trouble shooting capabilities. It is essential for the novel management architectures to target self-organised operations and support cooperative network composition as well as service support across multiple operator and business domains. There is a variety of external and influencing available definition on self-management related work. In addition, many of the aspects of self-management are not defined as guiding definitions with practical realisation but as "light" visions that provide conceptualization of the features in a system.

In the following sections, we examine two distinct examples of how a self-management conformant to the *M-D-E* scheme (i.e., including *Monitoring* actions, *Decision-Making* complexities and *Execution* options) can be implemented in specific wireless network environments. Both cases depict examples of "practical" importance, as they are usually concerned by network operators. The first case relates to the performance management of a wireless network, where spectrum reallocation should take place, either when traffic requirements at a certain link exceed the maximum achievable limit or in order to avoid interference effects. The second use-case refers to the intelligent dynamic traffic management of a wireless network environment (also considering multi-service provisioning); the aim there is to suitably "handle" any potential network overloads and to avoid (or overcome) congestion phenomena that may affect network performance. The proposed analysis is based on high autonomy of NEs in order to allow distributed management, fast decisions, and continuous local optimization. The collection of issues presented indicates the complexity of the framework associated with cognitive processes in systems. The process is useful to formulate relevant criteria for defining "degrees of success", when handling situations of interest. The suggested use-cases can help to "add" novel operational capabilities on the underlying system, mainly by introducing distinctive self-management features. Benefits can be considered either in reducing (time-based or actions), removing or instructing human interventions for tackling situations in the relevant system. In fact, self-management methods deal with operational aspects of the system for which the reduction (or even the "removal") of human intervention is not the dominant criteria of success but for which there is a "collection" of evaluation criteria such as various delays, diverse consideration for throughputs, reduction or minimization of packet losses, thus affecting performance.

3.1 Dynamic Spectrum Reallocation for Traffic Handling

New wireless access technologies will continue to emerge, including new versions of 802.11 (WiFi), 802.16 (WiMAX), 3G cellular, ad-hoc mesh networks, and more. All these offer enhanced opportunities to the end-users: they provide more efficient coverage, make available higher data rates at greater distances, and improve the quality and capacity of communications, while offering customers greater choice and flexibility. However, current networking technology cannot always respond, *efficiently*, to complex problems which arise from increasingly bandwidth-demanding applications/services competing for scarce resources. When this correlates to complex network structures with variable parameters and network performance objectives, the task of selecting the ideal network *"operating state"* may seem difficult. In order to upgrade network's performance, self-management may be considered. This usually incorporates a number of distinct operations with pre-defined roles in the overall system. Currently, the immense majority of the wireless systems are using fixed spectrum blocks between different wireless links. While the vast majority of the frequency spectrum is licensed to different organizations, recent observations provide evidence that usage of the licensed spectrum is by far not complete neither in the time domain nor the spatial domain [8].The most important issue which concerns operators worldwide is the fact that these spectrum blocks are "pre-arranged" or "pre-assigned" and do not have the option of dynamic change, when traffic requirements at a certain link exceed the maximum achievable limit [9]. The same happens when there is interference at a link between a user and a Base Station (BS) or between two BSs – either man-made interference or interference caused due to natural phenomena (e.g. storms, rains, snow, etc.). In order to assure the proper functioning of the system as for both the above distinct cases, spectrum should be dynamically reallocated among the existing links either by "removing" a part of it from one link and then by reassigning it to another link, or by reallocating the whole spectrum that is normally used, conformant to the actual technical/regulatory requirements. Furthermore, modern applications (such as VoD (video-on-demand), HDTV (high definition TV), IPTV, etc.) require additional spectrum and, *therefore*, a dynamic spectrum management is appropriate. The need of introducing/deploying new wireless applications, necessitate the use of dynamic spectrum access to turn existing networks into dynamic spectrum access ones.

The following use-case relates to the occurrence of significant spectrum interference between cognitive radio links supporting wireless communications. In such cases, either a network (or a wireless node) "modifies" its transmission/reception parameters in order to efficiently communicate by avoiding interference, when the latter occurs. This parameters' alteration is based on the active monitoring of several factors in the external and internal radio environment such as RF spectrum, user behavior and actual network state. Such radios, able to have an adequate knowledge of their surrounding electromagnetic spectrum environment, can facilitate the entire system to make any proper adjustments/modifications to the transmission characteristics when necessary, to diminish interference [10] to the minimum possible extent and so to preserve its normal functioning [11]. The purpose is to achieve a dynamically coordinated spectrum sharing & to facilitate interoperability between systems ([12,13]) thus optimizing network capacity, coverage and QoS and achieving reductions on operational or capital expenses of the system [14]. Thus, detecting any unused spectrum and sharing it without harmful interference with users when there is a need, becomes an important network requirement [15,16]. Future wireless communications can have major benefits from such functionalities operated by reconfigurable networks and related terminals [17]. In the present case, the interconnected NEs can form dynamically collaborative structures in order to identify/solve specific configuration or optimization problems (e.g., optimal spectrum usage) to improve their behaviour locally, taking into account the global behaviour of the network compartment where they participate [18]. Each network element (or user equipment) can cooperate and exchange local information in a peer-to-peer mode in order to achieve global properties/behaviour. This means that if, for example, frequency 2 for the link between two BSs is running large applications (and so requires extension of its spectrum usage), then a block of spectrum from frequency 3 will be removed and reassigned to frequency 2, to fulfill extra requirements (Fig.2).

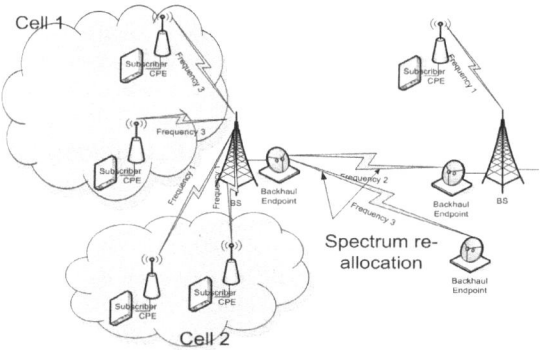

Fig. 2. Dynamic Spectrum Reallocation for efficient Traffic Use

The above example essentially relates to the network elements level (i.e. access points and BSs), where information has to be identified, collected, stored and treated, and then subsequent decisions have to be made. The case affects traffic transmission management, the decision for spectrum selection and for allocation capabilities of individual network elements (as well as "clusters" of them). We consider an approach conformant to *M-D-E* model.

Monitoring (M) data may include a variety of QoS and/or general wireless service performance indicators (i.e. data traffic, required link/service bandwidth, interference, signal coverage, (dynamic) network topology, available spectrum, utilization of spectrum, etc.). Data is to be collected and measurements are to be performed on a "conformant" basis, via suitable network counters. Measurements of channel characteristics and/or traffic aspects may then be processed to provide information for the related broader self-management activities: Here, the case essentially addresses the requirements of "*self-optimization*" and "*self-configuration*". The required accuracy (and periodicity) of collected information depends on specific mechanisms used. Each network elements (or mobile device), through the monitoring mechanisms collaborate with neighboring NEs, to exchange local monitoring data and detect inefficiencies or optimization opportunities. In more dynamic environments, such mechanisms are necessary to discover the neighboring network elements.

The *Decision Making (D)* process should refer to the identification of the "critical" physical parameters of operated wireless links (e.g. interference and network coverage) concurrently with the required bandwidth for supported services and the autonomic dynamic spectrum re-allocation between links to optimize network performance. The aim is to adapt, *appropriately*, the corresponding resource management parameters to sudden or to gradual variations in system traffic and/or propagation conditions. Performance measurements that derive either from individual NEs or are collectively calculated may prompt links to initiate autonomously a re-organization query in their vicinity, and decide a re-configuration action (i.e., spectrum re-allocation). The identification of the re-organization opportunities is the first step of the "negotiation phase"; it involves identification of available NEs and of spectrum resources, correspondingly. The negotiation among NEs includes policies and other local information exchange, to define the collaboration agreement among the involved entities and their interaction principles. After conclusion of that phase where available spectrum resources have been identified, then the involved elements "decide", in a distributed way, the new formation (i.e. spectrum re-allocation actions).

The *Execution (E)* phase mainly includes the processes of dynamically/ autonomously re-allocating spectrum between network links, to optimize traffic transmission and/or diminish harmful interference. Furthermore, policy rules and knowledge models are updated in the new formation, by considering the occurred reconfiguration action. Further evaluation can focus on determining whether wireless service provisioning can be improved (or preserved at a prescribed quality level considered as "adequate") by using cognitive dynamic spectrum re-allocation procedures for a flexible spectrum management and/or by examining spectrum utilization, QoS and network capacity criteria. In the same frame it is appropriate to evaluate the time required, success rate and gain of "self-management" actions, especially by comparing the effectiveness of the new formation to the previous one(s), and so "re-initiate" the *M-D-E* cycle.

3.2 Traffic Management for Multi-service Provisioning

The main issues traffic management [19] has to deal with are controlling and allocating network bandwidth, reducing delay, and minimizing congestion on networks [20]. The efficient management of network resources is very important for user requirements, especially in terms of bandwidth and service levels, for both operators and end-users. Almost all networks are subject to failure, in the sense that occasionally, for certain reasons, nodes can fail and so become unavailable to carry existing flow, thus leading to enormous service disruptions. A primary aim of network administration is the proper monitoring of routers/switches for "anomalous" traffic behaviour like outages, configuration changes, flash crowds and abuse. Identifying anomalous behavior is often based on ad-hoc methods developed from years of experience in managing networks [21]. Congestion can typically occur where multiple links feed into a single one. During periods of heavy congestion, directives can be dispatched to appropriate network modules to step-down traffic load to the network. As network congestion subsides, such directives can be dispatched to step-up traffic load until a normal level of traffic is restored [22]. Rate controls derived from traffic classification/prioritization can facilitate a meaningful congestion management [23], including prevention, avoidance, and finally recovery [24, 25].

Up-to-now, service providers desired to maintain their network(s) performance at satisfactory levels and so to keep an advanced QoS. The fast rising and gradual penetration of streaming services in the marketplace necessitates more enhanced reliability. By definition, a cognitive network is required to provide, over an extended period of time, better end-to-end performance than a non-cognitive one [26]. Cognition methods like *"self- configuration"* and *"self-organization"* can be used for improvement of the network performance via the proper handling of network traffic.

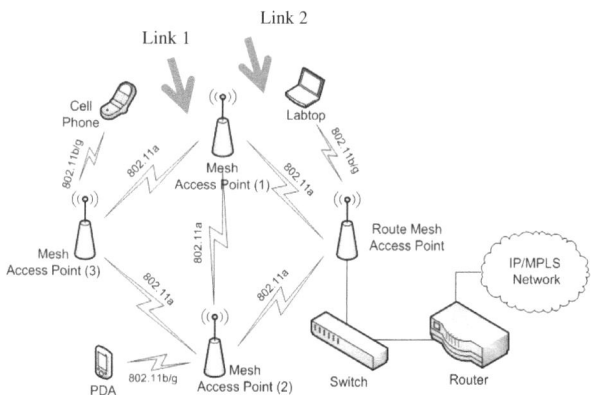

Fig. 3. Traffic Handling for Multi-Service Provisioning

The present use-case considers the improvement of congestion management in a large wireless network, supporting multiple services through dynamic traffic routing and source throttling. Congestion management is a very frequent and serious issue that network operators have to encounter; it directly impacts the overall network performance and, *correspondingly*, the customers' experience and satisfaction. The case concerns the situation when there is a bandwidth bottleneck at a certain wireless route directly affecting network performance, or when a certain link between two network nodes (e.g. routers) becomes suddenly congested. When this happens, then the load traffic should be routed to another appropriate link that can handle the extra traffic; otherwise, traffic may be distributed between a number of different links; or a notification can be sent to the relevant causal sources so that the latter to *"throttle"* their actual flows. The approach can use intelligent cognitive routers suitably programmed to acquire the traffic data of each link and, *when a congested link is detected*, to autonomously "re-route" traffic appropriately or to "signal"/notify the causal sources (as mentioned above). The case will take into account relevant traffic types and necessary resources. For example, Fig.3 illustrates the instance where the capacity of Link 1 is overloaded, for congestion reasons, and so extra traffic is routed to Link 2, to maintain overall network performance.

Monitoring (M) concerns measurement of current (and of average) loads to produce a variety of traffic (and occasionally of congestion) indicators, also in respect of QoS and service performance. Suitable parameters can be taken into account, which may include, *inter-alia*, identifying the controlled variable, the control structure, convergence and stability, and parameter configuration. Data should be collected appropriately, so that periodic measurements may be able to provide a dynamic "image" of the network loads, at proper time intervals. Once again, the required accuracy (and periodicity) of the collected information may depend on the specific mechanisms used. This use-case addresses the Future Internet requirements of "*self-optimization*" and "*self-configuration*". In several instances it could also deal with "*self-healing*", when network's performance is seriously threatened and immediate remedy is required, to avoid any potential corruption (or undesired "alteration") of network operation. The corresponding triggering event concerns the local recognized overload of a link for intermediate action and is initiated by performance data indicating high congestion between certain links in the network and/or associated performance degradation (in respect to resource allocation). Network overloads can seriously degrade the quality or availability of telecommunications services if they are not effectively controlled. Autonomic mechanisms will be triggered by procedures monitoring both network traffic and the performance of wireless service provisioning.

The Decision Making (D) process will react, autonomously, in an intermediate and longer term sense. Intermediate reactions will only have a few possibilities where in longer terms a general route re-configuration can take place. Requirements for appropriate overload controls and network traffic management can be considered, in terms of the conditions under which network links must operate and the behavior they should exhibit.

Execution (E) may implicate some among the following actions, that is: reconfiguration of routing tables, signaling of overload through causal sources, fair packet dropping and bandwidth reallocation. It should be expected that the appearance of network congestion phenomena should be reduced, while the overall network performance should be improved (or at least "preserved" at normal conditions).

In traditional networks, an overload situation over the shortest path results in significant packet loss. Self-management capabilities results in re-routing of packets along a path that has suitable resources available. This may occur either at network element level (where a proactive distribution of alternative routes allows local decision based on current load) or at network level (where overload indications can be the triggering event). Alternatively, suitable dynamic communication protocol mechanisms may use overload indications to reduce the amount of incoming traffic. In the previously examined use-case the system becomes able to "assess" its functional situation and to provide various operations targets that indicate a "desired operational status" (in this case for "solving" congestions).

4 Conclusion

Future Internet's vision affects an immense multiplicity of issues, and implicates major challenges for various domains of activities as it rapidly expands its effectiveness and penetrates several related environments (mainly network and service technologies). The perspective of autonomic network management (or "self-management") becomes a matter of "extreme priority" and of "prime importance" for the wider electronic communications sector, as it strongly influences network's behaviour. We have proposed a specific targeted methodology based on the introduction and the deployment of both cognition and autonomy in networks (also applicable at individual NEs). This methodology appears in the form of *the generic cognitive cycle model (M-D-E)* able to facilitate a satisfactory transition to the Future Internet, when coupled with available technological advancements at the network level. In order to further analyze that concept, we have proposed and presented two essential use-cases for study, which both deal with practical network management situations: The first one relates to dynamic spectrum re-allocation for the purposes of the efficient use of traffic in a modern wireless network. The second refers to the intelligent dynamic traffic management of a wireless network environment (also considering multi-service provisioning) with the aim of suitably handling any potential network overloads and in order to avoid (or to overcome) congestion phenomena that may affect network performance. The network can optimize, repair, configure and protect itself on its own, without external intervention.

Acknowledgments. The present work has been performed in the scope of the *Self-NET* (*"Self-Management of Cognitive Future Internet Elements"*) European Research Project and has been supported by the Commission of the European Communities - *Information Society and Media Directorate General* (FP7, Grant Agreement No.224344).

References

1. The Netherlands Ministry of Economic Affairs: The Internet: A Shared Future (Publication Number 08ET13), The Hague (2008)
2. Commission of the European Communities: Communication on i2010 - A European Information Society for Growth and Development [COM(2005) 229 final, 01.06.2005], Brussels (2005)
3. DG Information Society and Media of the European Commission: A Compendium of European Projects on ICT Research Supported by the EU 7th Framework Programme for RTD, Brussels (2008)
4. Pfeifer, R., Scheier, C.: Understanding Intelligence. MIT Press, Cambridge (1999)
5. Elliott, C., Heile, B.: Self-Organizing, Self-Healing Wireless Networks. In: IEEE International Conference on Personal Wireless Communications, pp. 355–362 (2000)
6. Biskupski, B., Dowling, L., Sacha, J.: Properties and Mechanisms of Self-organizing MANET and P2P Systems. ACM Transactions on Autonomous and Adaptive Systems (TAAS) 2(1), 1–34 (2007)
7. Blumenthal, M., Clark, D.D.: Rethinking the Design of the Internet: The End-to-End Arguments vs. the Brave New World. ACM Transactions on Internet Technology 1(1), 70–109 (2001)
8. Weiss, T., Jondral, F.: Spectrum Pooling: An Innovative Strategy for the Enhancement of Spectrum Efficiency. IEEE Communications Magazine 42, S8–S14 (2004)
9. Pioro, M., Mehdi, D.: Routing, Flow and Capacity Design in Communication and Computer Networks. Morgan Kaufmann, San Francisco (2004)
10. Mitola, J.: Cognitive Radio. An Integrated Agent Architecture for Software Defined Radio. Ph.D. Thesis. Sweden (2000)
11. Thomas, R.W., DaSilva, L.A., MacKenzie, A.B.: Cognitive Networks. In: IEEE DySPAN 2005, November 2005, pp. 352–360 (2005)
12. Haykin, S.: Cognitive Radio: Brain-Empowered Wireless Communications. IEEE Journal on Selected Areas in Communications 23(2), 201–220 (2005)

13. Buddhikot, M.M., Ryan, K.: Spectrum Management in Coordinated Dynamic Spectrum Access Based Cellular Networks. In: Proceedings of IEEE Dishpans 2005, November 2005, pp. 299–307 (2005)
14. Akyildiz, I.F., Lee, W., Vuran, M., Mohanty, S.: Next Generation/Dynamic Spectrum Access/Cognitive Radio Wireless Networks: A Survey. Computer Networks: The International Journal of Computer and Telecommunications Networking 50(13), 2127–2159 (2006)
15. Akyildiz, I.F., Altunbasak, Y., Fekri, F., Sivakumar, R.: AdaptNet: Adaptive Protocol Suite for Next Generation Wireless Internet. IEEE Communications Magazine 42(3), 128–138 (2004)
16. Zander, J., Kim, S.-L., Almgren, M.: Radio Resource Management for Wireless Networks. Artech House (2001)
17. Buddhikot, M.M., Kolodzy, P., Miller, S., Ryan, K., Evans, J.: DIMSUMNet: New directions in wireless networking using coordinated dynamic spectrum access. In: Sixth IEEE International Symposium on World of Wireless Mobile and Multimedia Networks, June 2005, pp. 78–85 (2005)
18. Zhao, O., Sadler, B.: A Survey of Dynamic Spectrum Access. IEEE Signal Processing Magazine 79(3), 79–89 (2007)
19. Ahuja, R.K., Magnanti, R.L., Orlin, J.B.: Network Flows. Prentice-Hall, Englewood Cliffs (1993)
20. Ahlswede, R., Cai, N., Li, S.-Y.R., Yeung, R.W.: Network information flow. IEEE Trans. on Information Theory 46(4), 1204–1216 (2000)
21. Barford, P., Plonka, D.: Characteristics of Network Traffic Flow Anomalies. In: ACM SIGCOMM Internet Measurement Workshop, San Francisco, pp. 69–73 (2001)
22. Estfan, C., Varghese, G.: New Directions in Traffic Measurement and Accounting. In: ACM SIGCOMM Internet Measurement Workshop, San Francisco, pp. 75–80 (2001)
23. Fowler, H.J., Leland, W.E.: Local Area Network Traffic Characteristics, with Implications for Broadband Network Congestion Management. IEEE Journal of Selected Areas in Communications 9(7), 1139–1149 (1991)
24. Sarachik, P., Panwar, S., Liang, P., Papavassiliou, S., Tsaih, D., Tassiulas, L.: A Modeling Approach for the Performance Management of High Speed Networks. In: Network Management and Control, vol. 2, pp. 149–163. Plenum Press, New York (1994)
25. Whitehead, M.J., Williams, P.M.: Adaptive Network Overload Controls. BT Technology Journal 20(3), 31–54 (2002)
26. Thomas, R.W., Friend, D.H., DaSilva, L.A., MacKenzie, A.B.: Cognitive Networks: Adaptation and Learning to Achieve End-to-End Performance Objectives. IEEE Communications Magazine 44(12), 51–57 (2006)

Self-organizing Mobile Ad Hoc Networks: Spontaneous Clustering at the MAC Layer

J. Alonso-Zárate[1], E. Kartsakli[2], P. Chatzimisios[3],
L. Alonso[2], and Ch. Verikoukis[1]

[1] Centre Tecnològic de Telecomunicacions de Catalunya (CTTC)
Av. del Canal Olímpic s/n, CTTC, 08860, Castelldefels, Barcelona, Spain
{jesus.alonso,cveri}@cttc.es

[2] Dept. of Signal Theory and Communications, Universitat Politècnica de Catalunya
(EPSC-UPC)
Av. del Canal Olímpic s/n, EPSC, 08860, Castelldefels, Barcelona, Spain
{ellik,luisg}@tsc.upc.edu

[3] Dept. of Technology Management, University of Macedonia
GR-59200, Nousa, Greece
pchatzim@uom.gr

Abstract. We present in this paper a master-slave, self-organized, spontaneous, passive, and dynamic clustering algorithm embedded into the Medium Access Control (MAC) layer for Mobile Ad hoc Networks. Any mobile station gets access to the channel by executing a contention-based mechanism similar to the IEEE 802.11 Standard. However, once it seizes the channel, it establishes a temporary cluster to which closer neighbors can get synchronized. Within each cluster, any infrastructure-based MAC protocol can be executed. Link-level computer simulations have been carried out to show that this approach can remarkably improve the performance of ad hoc networks at the MAC layer.

Keywords: MAC, DQMAN, DQCA, Clustering, Ad hoc, Self-organizing.

1 Introduction

Mobile ad hoc networks (MANETs) offer a set of advantages that look very promising to suit some of the hard and challenging requirements of a great number of new applications, which range from low-cost commercial systems or in-home applications, to rescue operations [1]. The need for continued connectivity and spontaneous networking has turned mobile ad hoc networking into a hot research topic over the last years.

MANETs consist of a set of mobile stations that can freely move and want to communicate with each other. Since there is no central point of coordination, all the decisions should be made in a distributed manner and communications must be done by establishing peer-to-peer links between pairs of source and destination stations. In some cases, the source of information and its intended destination might not be within the same transmission range and, in this case,

F. Granelli et al. (Eds.): MOBILIGHT 2009, LNICST 13, pp. 242–253, 2009.

cooperation is a must. Stations might relay packets in a multi-hop fashion in order to establish end to end links through a variable number of hops. The management of this kind of spontaneous scenarios is very challenging and has attracted many researchers. The goal is to design efficient protocols that can bring to life all the potential advantages of ad hoc networking.

Regarding the unpredictable nature of ad hoc networks, the design of those protocols should take into account the dynamics of these systems. First, the capability of the terminals to freely move in the network makes difficult to consider a frozen topology. Second, the radio channel is time-varying and thus the connectivity of the network is hardly predictable. Third, the variety of stations and services with different data traffics, requirements of quality of services, and heterogeneous constraints in terms of delay and throughput, is growing day by day. In order to deal with such complex network architecture, next-generation protocols should be self-configurable and dynamically adaptable to the conditions of the environment or system.

On the other hand, and regarding the strict performance requirements posed by new applications, those protocols should also attain high-performance in highly dynamic environments. Unfortunately, the design of such protocols has been mainly focused on infrastructure-based scenarios.

Having these two concerns in mind, a novel self-configuring technique is presented in this paper to extend the use of high-performance medium access control (MAC) protocols to distributed mobile networks. The main idea is to integrate a dynamic clustering algorithm into the MAC layer to create a spontaneous, temporary, and dynamic virtual backbone wherein any infrastructure-based MAC protocol could be used.

The remainder of the paper is organized as follows. Section 2 is devoted to overview the concept of self-organization in mobile wireless networks. Clustering techniques as a means of self-organization are presented in Section 3. The concept of integrating a spontaneous clustering technique into the MAC layer is introduced in Section 4. A case study with the Distributed Queueing Collision Avoidance (DQCA) protocol [2] is also analyzed in this section. Finally, Section 5 concludes the paper and gives some final remarks.

2   Self-organizing a Mobile Wireless Network

The concept behind self-configuration is the capability of a system to detect changes in the local environment and react to them in order to either minimize or obtain the most of their potential consequences. Regarding communication networks, a more specific definition could be the one expressed in [3]; a system is self-organized if it is organized without any external or central dedicated control entity. Mobile stations may exchange local information in a peer-to-peer fashion to achieve an overall organization. Among other possibilities, stations might decide, in a distributed manner, to establish a virtual hierarchical structure in which some of them act as directors or coordinators for their local neighborhood.

Although the design of efficient techniques to self-organize a hardly predictable and mobile wireless network is not a trivial task, the potential benefits of self-organization make the effort worth it. Among these benefits we can empha-size the increase of the capacity and coverage range of the network by allow-ing relay communication, the reduction of planning and deployment costs, the flexibility and scalability of the network to add or remove terminals, and the robustness towards station failures.

The concept of self-organization is very heterogeneous and multidisciplinary. In [4], the concept of self-organization in cellular wireless networks is presented as a mechanism to minimize the costs of both network planning and deployment. As indicated by the ETSI (European Telecommunications Standards Institute), it is convenient to work on systems that avoid the need for frequency planning and make easier the addition or removal of any base station. This was, for example, one of the motivations for selecting Code Division Multiple Access (CDMA) as the air-interface for UMTS (Universal Mobile Telecommunications System), instead of the previously used Time Division Multiple Access (TDMA) in GSM. In [5], self-configuration mechanisms are used as a way to improve the scalability of hybrid networks by means of combining cellular and relay ad hoc wireless networks. Another approach of self-configuration is the IP address configuration. In the past, an administrator should provide each of the stations of a network with an IP address. With the DHCP (Dynamic Host Configuration Protocol), and, furthermore, with the development of a stateless auto-configurable IPv6 [6], this burden is avoided. The congestion control of the Transport Control Protocol (TCP) is another bite of self-configuration mechanisms that demonstrates the effectiveness and the scalability of decentralized systems.

In the field of MANETs, since the operation of each layer is tightly coupled with the other layers due to the high complexity and dynamism of such kind of networks, the self-configuring paradigm applies to the whole protocol stack at once. This means that the concept of self-configuration should be applied at the MAC layer, at the routing mechanisms, and at the network protocols in conjunction. This is the approach presented in this paper, where a sponta-neous clustering mechanism is integrated into the MAC layer. Before getting into the details of the proposed self-organizing mechanism later in Section IV, the fundamental concepts of clustering are briefly overviewed in the next section.

3   Clustering: A Method for Self-organizing MANETs

Recently, experimental research has demonstrated that flat structures applied to MANETs perform well to a certain extent. When the size of a network grows in number of stations or in amount of traffic, the achievable performance in terms of throughput and delay is considerably reduced. However, by defining a hier-archical structure in which stations are classified into groups following certain rules, the management of the network can be made easier and the overall per-formance can be improved. This is the main motivation of clustering techniques by which the stations of a network get self-organized into groups or clusters.

Within a cluster, each station can play different roles depending on either its position or function. A station might be set to cluster head if it is the principal

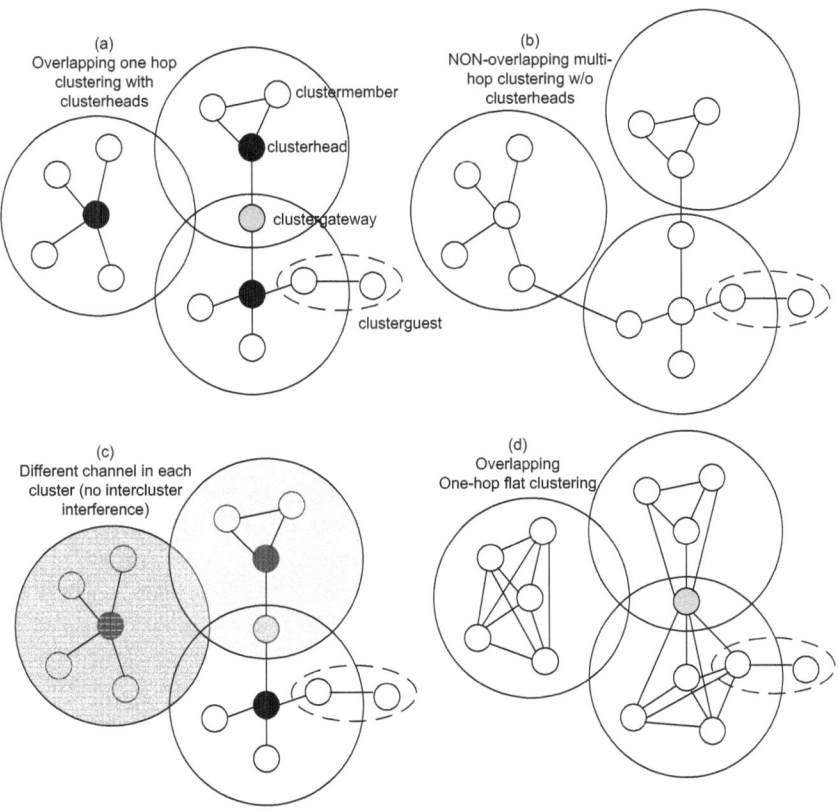

Fig. 1. Clustering Taxonomy

station in the cluster and it manages its one-hop neighborhood or to cluster member if it is connected to a cluster head. Furthermore, a station may act as a cluster gateway, if it is connected to more than one cluster head at the same time and, therefore, it is a potential relay for the inter-cluster communication, or as a cluster guest, if it is able to connect to a cluster member but not to a cluster head. This discussion is depicted in Figure 1.

One of the benefits of clustering is channel spatial reuse, which is strongly determined by the size of the clusters. Although small clusters can lead to a better spatial reuse, the end to end delay can be dramatically increased due to the potential occurrence of long paths, in terms of number of hops, between some pairs of stations. On the other hand, the use of big cluster sizes allows reducing these delays, but the spatial reuse is diminished and the interference range is increased. Therefore, a trade-off has to be managed when designing the size of the clusters, and thus the spatial reuse factor of the clustering scheme. In addition, typically the stations forming a MANET have a finite battery and thus the cluster-size must be properly selected. Big clusters imply higher transmission power, and therefore, higher power consumption. Moreover, as this power consumption does not grow linearly with the increase of the distance, the total energy cost of the transmission is also affected by the increase of the cluster size, which may be conveniently limited.

The set up and maintenance of a cluster topology may have a cost in terms of loss of the available resources, which may be caused by either *i)* the need for the explicit exchange of clustering information, *ii)* the time required to execute the clustering mechanisms, or *iii)* the costly process of re-clustering as the mobility of the network yields frequent changes in the topology and connectivity among the stations.

On the other hand, there is another trade-off between the benefits and drawbacks of applying some kind of cluster head centralization in a network; while the creation of a virtual backbone may be very useful for the overall network operation, cluster heads may become a bottleneck of the network, thus compromising its scalability.

Regarding the classification of clustering techniques, lots of clustering algorithms have been proposed in the literature and most of the strategies differ in the dynamic mechanisms to set up and maintain the cluster architecture. A survey on different clustering proposals can be found in [7] where authors classify them into six categories: dominating set-based clustering, low-maintenance clustering, mobility-aware clustering, energy-efficient clustering, load-balancing clustering, and combined-metrics-based clustering, depending on their criteria to define the cluster set. The main conclusion of that survey is that although different proposals are well suited for certain scenarios, it cannot be guaranteed that any of them is the best for all situations. Some examples of clustering schemes can be found in [8]-[13].

Taking into account this background, we present in this paper an approach to apply a spontaneous clustering mechanism at the MAC layer for networks without infrastructure.

4   Dynamic Clustering at the MAC Layer

4.1   Motivation

Traditionally, clustering algorithms have been based on the idea that the more stable the cluster set, the better the network will perform. The process of re-clustering a part of the network may entail a high cost in terms of resources due to the fact that one cluster head reassignment could trigger the re-configuration of the entire network. This could happen, for example, when the topology changes due to the mobility of the stations. This is known as the ripple effect of re-clustering and it has been traditionally avoided, especially in the case of large MANETs. However, when mobility is present, cluster stability is difficult to attain. In addition, if some stations are to be selected as cluster heads, it is difficult to design efficient criteria for selecting cluster heads in an extremely dynamic and changing environment as in the case of mobile wireless networks. As demonstrated in [14] the optimal cluster head set problem is NP-complete, and therefore, suboptimal clustering must be carried out in a highly dynamic environment.

Taking into account these considerations, and in order to extend the use of high-performance infrastructure-based MAC protocols to totally distributed networks, a dynamic clustering algorithm integrated into the MAC layer is presented in this paper. Up to our knowledge, the approach of integrating a spontaneous clustering mechanism into the MAC layer has never been tackled before in the literature.

Clusters are spontaneously created without explicit control information exchange whenever a station has data to transmit. The cluster structure is maintained for as long as there are data pending to be transmitted among all the stations associated to a cluster. Therefore, the cluster structure is spontaneously established and broken up according to the aggregate traffic load and the mobility of the network. Furthermore, cluster membership is soft-binding in the sense that there are no explicit association and disassociation processes, but a station belongs to a cluster as long as it simply receives the control packets broadcast by the master. Computer simulations discussed later in this section demonstrate that the performance of this spontaneous and dynamic mechanism outperforms legacy standard MAC protocols in mobile ad hoc networks by balancing a low-cost re-clustering mechanism with the benefits of employing a high-performance infrastructure-based MAC protocol in a distributed network. The clustering protocol is described in the next subsection.

4.2   Clustering Description

The proposed one-hop hierarchical clustering algorithm is based on a master-slave architecture wherein any station can operate in one of the following three modes; **master, slave,** or **idle.** Any station should be able to switch from one mode of operation to another one according to the dynamics of the network.

Despite the hierarchical cluster structure, all the communications can be done in a peer-to-peer fashion between any pair of source and destination stations. Note that the term destination in this context refers to the next-hop destination of a packet (which will be specified by the routing protocol), and not necessarily to its final destination station. For the description of the protocol, a slotted time scale is considered.

The mechanism works as follows. Any idle station with data to transmit listens to the channel for a deterministic period of time, as in [15], to determine whether the channel is idle or busy. If the channel is sensed idle, then the station attempts to establish a Master Service Set (MSS) by setting itself to master mode and starts broadcasting a clustering beacon (CB) every T_{frame} seconds. This transmission defines a MAC frame structure and allows neighboring stations to get synchronized with the master and become slaves. Slaves are responsible for transmitting an in-band busy tone (BT) upon the reception of each CB transmitted by the master. As long as any idle station willing to become master listens to the channel for at least T_{frame} seconds, these BTs ensure a minimum distance of three hops between masters and allow combating the hidden terminal

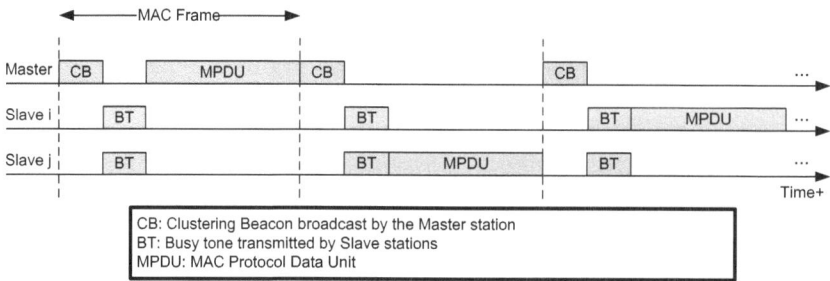

Fig. 2. Clustering Frame

problem. In addition, they constitute a collision detection mechanism for those stations which attempt to become master. The time elapsed between the BTs and the next CB is devoted to the exchange of data and control MAC Protocol Data Units (MPDU), wherein, indeed, any MAC protocol could be executed. This MAC frame structure is illustrated in Figure 2.

On the other hand, if an idle station senses the channel busy when attempting to establish a cluster, it initiates a Master Selection Phase (MSP) and sets its Master Selection Silent Interval (MSSI) counter to a random value within a MSSI window measured in time slots (as in the IEEE 802.11 Standard [15]). Likewise, any station performing a MSP listens to the channel and decrements the MSSI counter by one after each time slot as long as the channel is sensed idle. When the counter gets to zero, the station attempts to establish its cluster.

The value of the MSSI counter is determined as the sum of a deterministic period of time β and a randomized value selected within a constant-size contention window of length α. The purpose of the fixed period of time is to reduce the probability that a station becomes master twice consecutively in time, while the randomized time interval is required to reduce the probability of collision. It is worth mentioning that the size of the contention window should be properly selected as a function of the number of active stations in order to avoid either unnecessary wasted time in idle periods or a high probability of collision. A collision occurs when more than one station attempts to establish a MSS simultaneously. Any station which attempts to establish its MSS interprets that a collision with at least another station has occurred if no busy tones from slaves are received after the transmission of the CB. The stations involved in the collision reinitiate a new MSP by selecting a new value for their respective MSSI counters as soon as the collision is detected. Whenever a cluster is successfully established, its life time depends on the traffic load of the network. On the one hand, any station operating in master mode breaks its MSS and reverts to idle whenever there are no more pending data transmissions among all the stations associated to its MSS (including its data traffic). Whenever a slave mishears a number of CBs from the master, it reverts to idle mode. On the other hand,

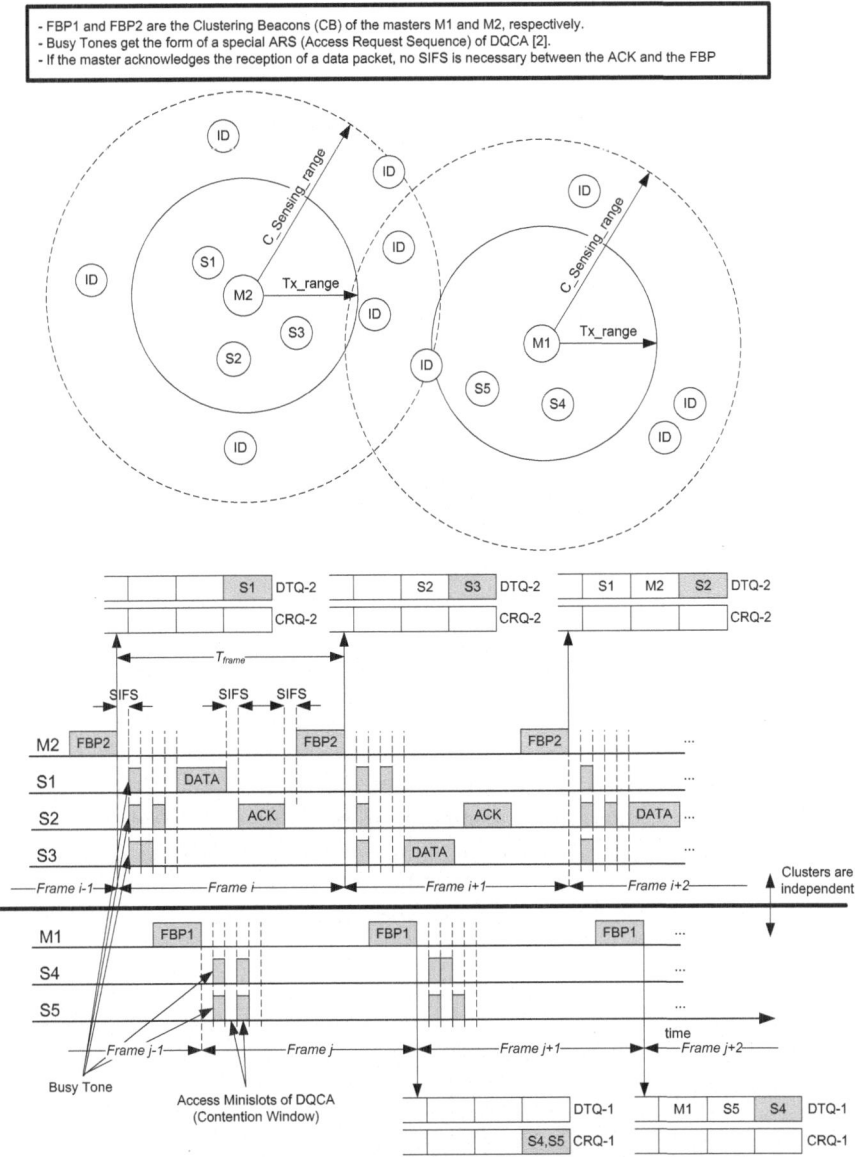

Fig. 3. Example: DQMAN Operation

and in order to avoid static cluster sets under heavy traffic conditions, any station operating in master mode also reverts to idle after a certain time, referred to as the Master Time Out (MTO), regardless of the pending data to transmit within its cluster.

4.3 Performance Evaluation

A performance evaluation of this spontaneous mechanism embedded into the MAC layer is presented in this section. In this case, it is considered that the high-performance DQCA MAC protocol [2] is executed within each cluster. The integration of the aforementioned dynamic clustering into the DQCA protocol is named Distributed Queueing MAC Protocol for Ad Hoc Networks (DQMAN). Masters act as access points and use the CB slot to transmit a Feedback Packet (FBP). An example of such a configuration is illustrated in Figure 3. Two stations, M1 and M2, have established their independent MSS (clusters). These stations periodically send the beacons FBP1 and FBP2, respectively, defining a time frame structure which allows stations S1 to S5 to get synchronized with their closest master. The time between beacons is used for the transmission of data where the MAC frame structure of DQCA is executed. This frame structure is divided into three parts:

1. A contention window further divided into access minislots wherein stations with data to send transmit an Access Request Sequence (ARS).
2. An almost collision-free data transmission part.
3. A control information part reserved for the transmission of ACK packets from any destination which receives a data packet and the FBP broadcast by the master. the FBP attaches feedback information on the state of each of the access minislots of the current frame. This is the only control information required by all the stations to execute the protocol rules described in [2].

Short Inter Frame Spaces (SIFS) are left after the transmission of either data or control information to tolerate non-negligible propagation delays and turn around times to switch from transmitting to receiving mode. Those stations operating in idle mode, labeled with ID in the figure, should wait until the cluster structure is modified due to the mobility of the network, the occurrence of an idle period

Table 1. System Parameters

Parameter	Value	Parameter	Value
Data packet length	1500 bytes	Message length	15000 bytes
Data Tx. Rate	54 Mbps	Control Tx. Rate	6 Mbps
MAC header	34 bytes	PHY preamble	96 μs
ACK length	14 bytes	SlotTime	10 μs
DQMAN			
FPB length	14 bytes	(α, β)	(32,10)
Access minislots	3	ARS	10 μs
MTO	100 (frames)	Busy Tones	10 μs
IEEE 802.11			
DIFS	50 μs	SIFS	10 μs
CW_{min}	32	CW_{max}	256
RTS length	20 bytes	CTS length	14 bytes

(no data packets pending to be transmitted within a MSS), or the execution of a MTO mechanism. In any of these cases, a reclustering process is triggered.

Link-level computer simulations in a custom-made C++ simulator have been performed to evaluate the performance of both DQMAN and the IEEE 802.11 Standard MAC protocol with both the basic and the collision avoidance access methods. A single-hop scenario has been considered and the emphasis has been put on calculating the saturation throughput of the two protocols as a function of the number of stations. The parameters used for the simulations are summarized in Table 1.

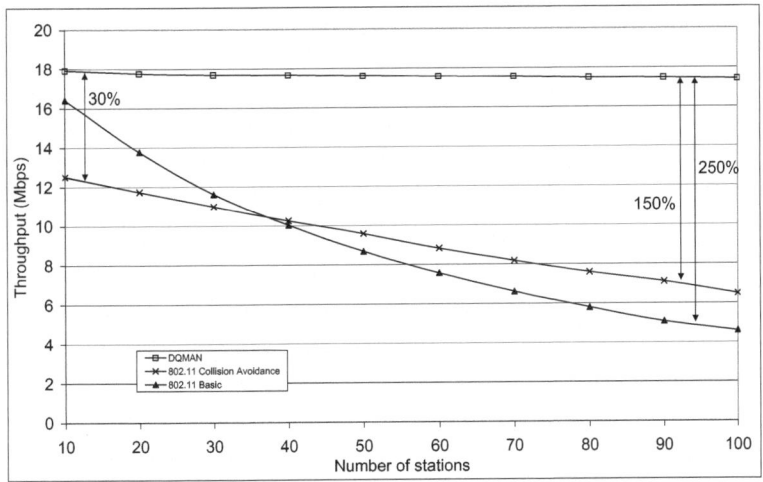

Fig. 4. Throughput Comparison (DQMAN vs. IEEE 802.11)

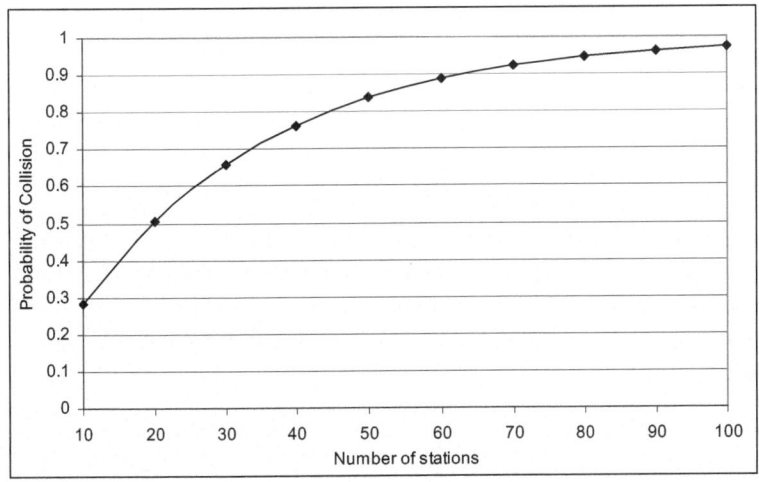

Fig. 5. Probability of Collision

4.4   Results

As illustrated in Figure 4, DQMAN outperforms IEEE 802.11 in all the considered scenarios. No only the saturation throughput is superior, but it is also more independent of the number of active stations. The saturation throughput of the IEEE 802.11 Standard drops dramatically as the number of active stations grows. For example, when 100 stations are considered, DQMAN attains 150% and 250% superior performance than the collision avoidance and basic access modes of the Standard, respectively. This is mainly due to the fact that the higher the number of stations, the higher the probability of collision. The probability of collision as a function of the number os stations is illustrated in Figure 5. Results show that the probability of collision can get close to 1 if the contention window size is not tuned dynamically. This is the reason why the 802.11 MAC protocol yields extremely poor performance for high number of stations. On the other hand, since the time devoted to contention in DQMAN is confined to clustering phases, a high probability of collision only causes a slight reduction of the saturation throughput. Therefore, the performance of DQMAN is less dependent of the configuration of the contention window.

In addition, it is worth noting that when the number of active stations is low, e.g. 10 stations, a minimum performance improvement of at least 30% and 16% in terms of throughput is still present due to the lower overhead of DQMAN compared to the IEEE 802.11 Standard in the collision avoidance and basic access modes, respectively.

5   Conclusions

The dynamic and spontaneous nature of MANETs poses hard challenges in their management and efficient operation. Since there is neither previous infrastructure nor any central point of coordination, all the decisions must be made in a distributed manner among the stations forming the network. Self-configuring techniques constitute a powerful tool to improve the performance and scalability of such kind of networks. The most common implementation of self-configuration in the field of MANETs is the use of clustering algorithms through which mobile stations get organized into groups or clusters.

We have presented in this paper the integration of a spontaneous and dynamic clustering algorithm into the MAC layer. Stations get access to the channel following a distributed contention-based access method. However, once they gain the control of the channel, they establish a temporary cluster structure where any infrastructure-based MAC protocol can be executed. Computer-based simulations show that the approach can enhance the performance of legacy 802.11 Standard networks up to 250%.

Acknowledgments. This work has been partially funded by the research projects NEWCOM++ (IST-216715), LOOP (FIT-330215-2007-8), PERSEO (TEC2006-10459/TCM), COOLNESS (218163-FP7-PEOPLE-2007-3-1-IAPP), R2D2 (CP6-013), and CENTENO (TEC2008-06817-C02-02).

References

1. Perkins, C.E.: Ad Hoc Networking. Addison-Wesley, Reading (2001)
2. Alonso-Zarate, J., Kartsakli, E., Cateura, A., Verikoukis, C., Alonso, L.: A Near-Optimum Cross-Layered Distributed Queuing Protocol for WLAN. IEEE Wireless Communications Magazine Special Issue on Medium Access Control Protocols for Wireless LANs 15(1), 48–55 (2008)
3. Prehofer, C., Bettstetter, C.: Self-Organization in Communication Networks: Principles and Design Paradigms. IEEE Communications Magazine 43(7), 78–85 (2005)
4. Spilling, A.G., Nix, A.R., Beach, M.A., Harrold, T.J.: Self-organization in future mobile communications. Electronics & Communications Engineering Journal, 133–147 (June 2000)
5. Dixit, S., Yanmaz, E., Tonguz, O.K.: On the design of Self-Organized Cellular Wireless Networks. IEEE Communications Magazine 43(7), 86–93 (2005)
6. Tobella, J.J.S., Stiemerling, M., Brunner, M.: Towards Self-Configuration of IPv6 Networks. In: IEEE Network Operations and Management Symposium 2004, vol. 1, pp. 895–896 (2004)
7. Yu, J.Y., Chong, P.H.J.: A Survey of Clustering Schemes for Mobile Ad Hoc Networks. IEEE Communications Surveys & Tutorials 7(1), 32–48 (First Quarter 2005)
8. Krishna, P., Vaidya, N.H., Chatterjee, M., Pradhan, H.K.: A Cluster-based Approach for Routing in Dynamic Networks. In: Proc. of the ACM SIGCOMM Computer Communication Review, pp. 49–65 (1997)
9. McDonald, A.B., Znati, T.F.: A Mobility-Based Framework for Adaptive Clustering in Wireless Ad Hoc Networks. IEEE Journal on Selected Areas in Communications 17(8), 1466–1487 (1999)
10. Haas, Z.J., Pearlman, M.: The zone routing protocol (ZRP) for Ad Hoc Networks, Internet Draft, Tech. Rep. (November 1997)
11. Ibriq, J., Mahgoub, I.: Cluster-Based Routing in Wireless Sensor Networks: Issues and Challenges. In: Proc. of the ACM Telecommunications, pp. 759–766 (2004)
12. Habetha, J., Wiegert, J.: Analytical and simulative performance evaluation of cluster-based multihop ad hoc networks. Journal of Parallel and Distributed Computing 63(2), 166–181 (2003)
13. Lin, C.R., Gerla, M.: Adaptive clustering for mobile wireless networks. IEEE Journal on Selected Areas in Communications 15(7), 1265–1275 (1997)
14. Li, C.: Clustering in Packet Radio Networks. In: Proc. of the IEEE International Conference on Communications (ICC 1985), pp. 283–287 (1985)
15. IEEE, Part 11: Wireless LAN Medium Access Control (MAC) and Physical Layer (PHY) Specifications. IEEE Std. 802.11-99 (August 1999)

Wireless Location Positioning Based on WiMAX Features - A Preliminary Study

Azmi Awang Md Isa, Garik Markarian, and Kamarul Ariffin Noordin

Department of Communication Systems,
Lancaster University, United Kingdom
{a.awangmdisa,g.markarian,k.noordin}@lancaster.ac.uk

Abstract. In this paper, we exploit the potential of positioning technologies in wireless broadband communications, which are based on worldwide interoperability for microwave access (WiMAX), in particular the IEEE802.16* standards. By utilizing the additional features in WiMAX including multiple input multiple output (MIMO), adaptive modulation and coding (AMC), beamforming, relay station and power control, we believe that the features can be used for enhancing the location estimation accuracy in location services.

Keywords: WiMAX features, positioning.

1 Introduction

As the wireless technologies develop further, many useful and promising applications come into our lives. In the recent years, location and positioning based services are among the most promising applications which have brought us with great convenience. For instance, by knowing a user's location many new applications often called as Location Based Services (LBS) can be enabled. However, one of the current key issues for LBS is the positioning technologies in wireless broadband communications.

Most current positioning systems are not functioning properly where people spend much of their time, meaning that coverage in these systems is either constrained to outdoor environments or limited to a particular building or campus with installed location infrastructure. For an example, the most popular positioning system, Global Positioning System (GPS) operates worldwide, but it requires a clear view of the skies and signals from at least four satellites. It does not work indoors and operates poorly in many cities where the so called "urban canyons" formed by buildings prevent GPS receiver units from seeing enough satellites to get a position lock [1]. Meanwhile, purpose-built systems such as Active Badge, Cricket, The Bat etc., can be used in indoor environments [2]. However, for cost reasons, people prefer to use existing infrastructure such as mobile phone networks, and wireless LAN (WLAN).

Predictions regarding wireless broadband communications and wireless internet services are cultivating visions of unlimited services and applications that will be available to the user "anywhere and at anytime" [3]. Users expect to surf the Web, access e-mail, download files, have several multimedia applications, such as real-time audio and video streaming, multimedia conferencing, interactive gaming and perform

F. Granelli et al. (Eds.): MOBILIGHT 2009, LNICST 13, pp. 254–262, 2009.
© ICST Institute for Computer Sciences, Social-Informatics and Telecommunication Engineering 2009

a variety of other tasks through a wireless communication link. The user further expects a uniform user interface that will provide access to the wireless link whether shopping at the mall, waiting at the airport, walking around town, or even driving in the car. Current wireless infrastructures, however, as well as next-generation proposals cannot furnish the necessary bandwidth and capacity to provide these services to mobile stations (MSs) [4]. Unfortunately, mobile users will likely be the most demanding of bandwidth and wireless services. Clearly, a broadband wireless solution is needed to provide MSs the high-bandwidth mobile service they demand at a low cost.

Of these, WLAN (also known as 'Wi-Fi') that supports wireless broadband communications can be implemented with the least effort, as its associated consumer hardware is the most readily available. However, Wi-Fi works within a limited range of an access point and suitable only for fixed wireless broadband. Besides that, to ensure the success of delivering high bandwidth and less interference from Base Station (BS) to users especially for MSs, the system depends on accurate positioning systems.

With the arrival of WiMAX recently, the technology is able to fullfill the criteria of the next generation wireless systems. WiMAX is a wireless standard to enable mobile broadband services at a vehicular speed of up to 120 km/h [5]. WiMAX complements and competes with Wi-Fi and the third generation (3G) wireless standards in terms on coverage and data rate. More specifically, WiMAX supports a much larger coverage area than WLAN. On the other hand, it operates at both outdoor and indoor environments as well, does not require line of sight (LOS) for a connection, and is significantly less costly and provides higher data rate as compared to the current 3G cellular standards. Although the WiMAX standard supports both fixed and mobile broadband data services, the latter have a much larger market. In addition, WiMAX supports additional features that can be used for enhancing location and positioning technologies, such as MIMO, AMC, power control, relay station and Beamforming.

The remainder of this paper is organized as follows. Section 2 explains briefly literature review about existing location and positioning techniques. Section 3 discusses proposed techniques to enhance location and positioning in WiMAX. Finally section 4 concludes the paper.

2 Existing Location and Positioning Techniques

The location of the mobile users can be determined in several different ways. Generally, there are three main groups of location and positioning technologies available in the market, namely Satellite based Positioning, Network-based Positioning and Indoor Positioning [6].

The most popular satellite-based positioning is Global Positioning System (GPS). The positioning technique used in GPS is known as *Trilateration* (distance measurement) by which the user's position is calculated using the intersection of the spheres determined by the distances between the each satellite and the GPS receiver. Having signals from at least three or four satellites, it is sufficient to compute the physical location of a device equipped with a GPS receiver with the location error varies from a couple of meters to several tens of meters depending on the propagation environment.

An example for Network-based Positioning is Global System for Mobile Communications (GSM). The most common positioning methods for GSM are Cell ID, Angle of Arrival (AOA), Received Signal Strength (RSS), Time of Arrival (TOA), and Time Difference of Arrival (TDOA).The Cell ID is used to determine the serving Base Station (BS) and then use the position of the BS as an estimate of the mobile's position. Its accuracy is directly proportional to the cell size in the network. For many location-based services, the accuracy of the Cell ID technique is not sufficient. The AOA involves measuring the angle of arrival of a signal from a BS at mobile phone or vice versa. In either case, a measurement produces a straight-line locus from a BS to a mobile phone. Another AOA measurement will yield a second straight line; the intersection gives the estimation of the mobile's position. The RSS method determines the user's position by signal strength received based on the known mathematical model for the relation of the signal strength and the distance between the mobile phone and the BS.

The TOA technique on the other hand determines the distance between mobile phone and BS by measuring the propagation time (absolute time) of radio wave between them. Using the same concept of trilateration, at least three BS are required to determine mobile user's position. The TDOA is a hyperbolic position determining technique. At two base stations, the time difference of arrival of the signal from a mobile unit is measured. Then the possible solutions where the time difference is constant lie on a hyperbola with each base station located in one of its foci. Forming those hyperbolas between different pairs of base stations, the position of the mobile unit is determined by intersection of all hyperbolas.

Finally, the common positioning technology used in indoor environment is Wireless Fidelity (Wi-Fi). Wi-Fi is based on the IEEE802.11 family of standards and is primarily a local area networking (LAN) technology designed to provide in-building broadband coverage [9]. In general, current Wi-Fi systems based on IEEE802.11a/g support a peak physical layer data rate of 54 Mbps and typically provide indoor coverage over a distance of 100 feet. Obviously, Wi-Fi is not designed and deployed for the purpose of positioning. However, it has found that the most current solutions to determine the location of any mobile users are based on the utilisation of signal strength. The signal strength values from the reference stations (access points (APs)) are measured by the positioning device. And based on the use of signal information (signal quality, signal strength, SNR and so on), there have been two possible implementations – the fingerprinting approach and the propagation approach.

3   Proposed Location and Positioning Based on WiMAX

There are many approaches in determining a user's location – some of which are explained in [1, 2, 6-8] – however, location and positioning (L&P) technologies based on WiMAX is not widely investigated yet although there exist some proposal such as in [10, 11]. WiMAX has some useful features that can be employed to enhance the positioning accurracy including MIMO, AMC, Beamforming, Relay Station and power control.

Based on the concept of 'Trilateration' that are used in many existing L&P technologies, we proposed an idea to determine WiMAX user's location as illustrated in figure

1. According to figure 1, each MIMO BS and MS are equipped with MIMO antennas that consist of multiple transmitters and receivers (multiple antennas) on both sides which is N_t transmit antenna on the BS and N_r receive antenna on the MS. In this case, N_t different signals are transmitted simultaneously over minimum of N_t x N_r transmission paths and each of those N_r received signals is a combination of all the N_t transmitted signals and distorting noise. Hence, the multiple of transmission path may be achieved by adopting MIMO systems as compared to conventional 1 x 1 systems that use single antenna at both ends of the link with the same requirement of power and bandwidth. On the other hands, multiple number of simultaneous signals can be transmitted from a MIMO BS, and by applying trilateration method, more signals will be detected by MS. Therefore, we believe that MIMO will not only improve the capacity and the throughput of a wireless link significantly but can also be used to enhance the accuracy of a user's location. Furthermore, each MIMO BS will transmit the signals to the Network Management Systems (NMS) which in turn monitors the network accurately and provides L&P services for updating the user's location.

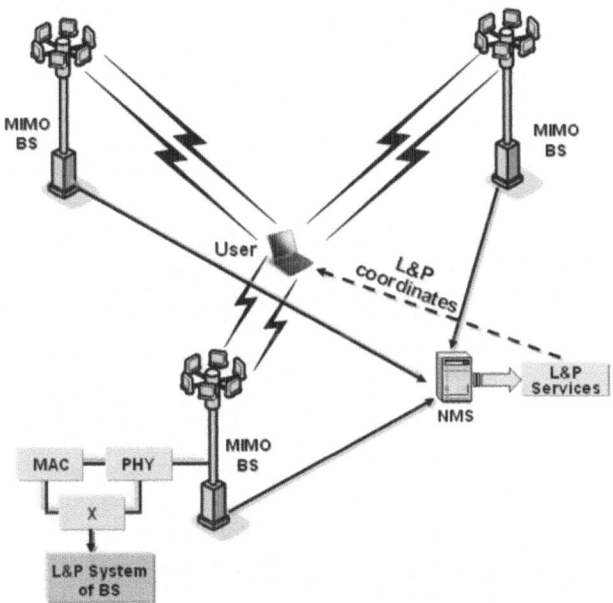

Fig. 1. Proposed System Architecture of Location and Positioning in WiMAX Services

Meanwhile, AMC allows WiMAX system to adjust the signal modulation scheme (64QAM, 16QAM, QPSK or BPSK) depending on the signal-to-noise (SNR) condition of the radio link. The idea behind the AMC is to dynamically adapt the modulation and coding scheme to the channel condition so as to achieve highest spectral efficiency at all times. However, this causes the signal level to be almost the same throughout a BS coverage area, so that the signal level measurement cannot be used to estimate the location of mobile users. In addition, WiMAX uses power control to adjust the signal quality based on SNR. As a result, the same scenario as above can be seen in the signal level. Nonetheless, by taking the information from both the physical layer (the type of modulation scheme used and power control reading) and MAC layer at WiMAX BS, the data can be used to determine the MS's location as shown in figure 2.

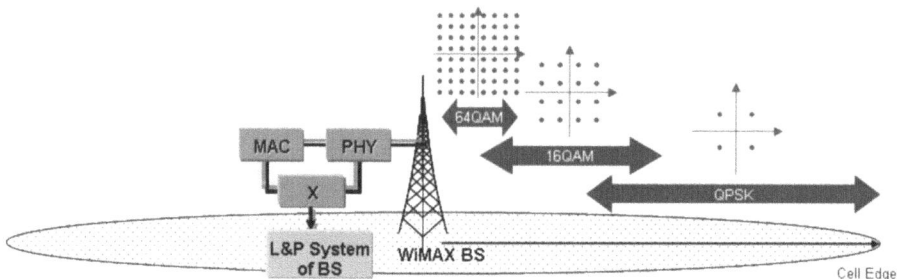

Fig. 2. Scheme for the Utilization of AMC

In order to determine the area covered by each modulation scheme, the maximal distance, R_i between BS and MS must be calculated first using a corresponding modulation. This distance is determined using the maximal SNR a user should received without data loss. Different values of received SNR for different modulation/coding scheme have been calculated in [13] and are shown in Table 1. Then, R_i can be calculated using information in Table 1. Without loss of generality, we study AMC in the presence of path loss for free space only (other sources of interference are negligible) and the model is given by [9]:

$$PL_i[dB] = -10log\left[\frac{\lambda^2 G_t G_r}{(4\pi R_i)^2}\right] \tag{1}$$

where λ is the wavelength, G_t and G_r is tranmitter and receiver antenna gain, respectively and R_i is the distance between the transmitter and the receiver. Equation (1) is also equal to:

$$PL_i[dB] = P_t[dBm] - SNR[dBm] - N[dBm] \tag{2}$$

where P_t is the transmitter power and N is the thermal noise which is given by:

$$N[dBm] = 10log(\tau TW) \tag{3}$$

where $\tau = 1.38 \times 10^{-23} J K^{-1}$ is the Boltzman constant, T is the temperature in Kelvin ($T = 290$) and W is the transmission bandwidth in Hz.

Using the above equations, we can calculate the relationship between the distance and the SNR as follows:

$$R_i = \frac{\lambda \times 10^{\frac{P_t[dBm]+10\log(G_t)+10log(G_r)-SNR[dB]-N[dm]}{20}}}{4\pi} \tag{4}$$

Table 1. Receiver SNR Assumptions

Modulation	Coding Rate	Receiver SNR (dB)
BPSK	½	3.0
QPSK	½	6.0
	¾	8.5
16-QAM	½	11.5
	¾	15.0
64-QAM	2/3	19.0
	¾	21.0

And the area of each region for each modulation is given by:

$$S_i = \pi \times \left(R_i^2 - R_{i-1}^2 \right) \tag{5}$$

It is worth noting that wireless relay has been proposed as a solution to extend the coverage of a single base station. In 2006, the IEEE approved a project called P802.16j (802.16j), for a mobile multihop relay (MMR) specification to extend BS reach and coverage without the backhaul requirement [13]. The MMR-BS provides the primary area of coverage. It also has a backhaul connection, such as leased copper or fiber optics. The relay station (RS) extends the BS coverage. A MS can connect to BS, an MMR-BS or a RS. Therefore, multiple relay stations, in addition to a BS are not only to be used for enhancing the throughput and improving the range of the BS, but they can be used for positioning purpose. In this feature, the concept to determine MS position is the same with other positioning methods by applying trilateration method based on at least three number of base station. However, in the case of Wi-MAX positioning, we propose to use only one BS with assisted relay stations which means that the MS can be estimated within a cell by using the serving BS with assisted of relay stations as shown in figure 3.

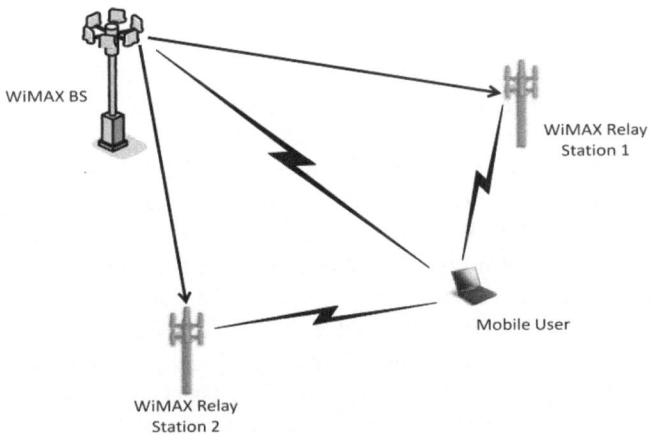

Fig. 3. WiMAX BS with Assisted Relay Stations

4   A Simulation Example

This section presents preliminary results from the simulation of a simple MIMO communication system and the effect of AMC on coverage area. In the case of MIMO simulation, we compare the location estimation accuracy at different types of base antenna mode between SISO and MIMO antennas. We consider the antenna as a diversity antenna, so that only the time of arrival (TOA) measurements are taken into account. For simplicity, we make assumption that transmitter and receiver are perfect-ly synchronized.

In the TOA measurements, the range data are created by calculating the true dis-tance from an MS position to known BS positions and measurement noise are added to the true calculated range to get the measured range data. The measurement noise is assumed to be Gaussian distribution with standard deviation of $\sigma = 100m$. Then the estimate of an MS position can be determined by using trilateration method as shown in Figure 1. The scenario consists of three synchronized BSs, organized in a cell with radius of about 500m and the true position of an MS is located near the center of the all BSs coverage. Table 2 compares the estimated error of the location between SISO and MIMO antenna. From the results, we can see that the MS location can be esti-mated with high accuracy as the number of antenna increases.

Table 2. Comparison of Average RMSE with Different Antenna Mode Configurations

Antenna Mode Configurations	Mean Distance Error [meter]	Standard Deviation [meter]
SISO	105.73	64.14
2x1 MIMO	87.32	49.89
2x2 MIMO	75.67	40.13
4x2 MIMO	71.73	35.45
4x4 MIMO	66.21	31.23

To illustrate the effect of coverage area from the BS upon the usage of AMC, let us consider the following example based on the licensed band for WiMAX outdoor envi-ronment which has carrier frequency and system bandwidth equal to 3.4GHz and 20 MHz, respectively. At this transmission bandwidth, the thermal noise can be equal to -100.97 dBm. According to the maximum allowed Effective Isotropic Radiated Power (EIRP) of 1W, the transmitters are assumed to have transmission power P_t of 1W which equals 30 dBm. We consider the case of antennas in BS and MS without gain. Based on Table 1, we run 500 samples for different SNR values and the result can be seen in figure 4. From the figure, we can see that a particular modulation and coding scheme (MCS) will be employed if the MS is within a certain distance from the BS. Therefore by using this logic, we can say that if an MS is using a particular MCS, we can approximate its distance from the BS. This is achieved by assuming that the selection of the MCS is based solely on the distance from the BS and other factors such as channel condition do not influence the selection of that particular MCS.

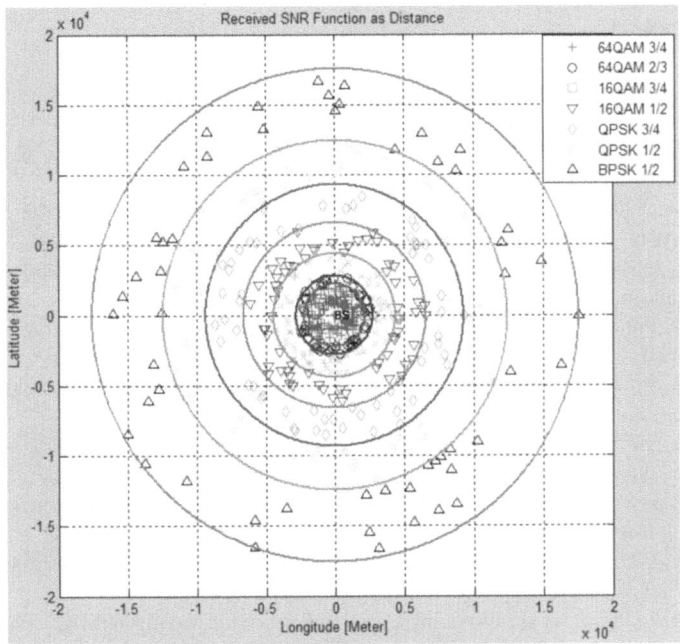

Fig. 4. Received SNR Function as Distance

5   Conclusion

In this paper, we presented the potential of wireless broadband communications usage in particular the WiMAX or the IEEE 802.16* standards for positioning technologies. Depending on the required accuracy and network topology, different positioning methods such as trilateration/triangulation, round trip delay, Cell ID, fingerprinting etc can be employed for WiMAX [14]. We have proposed a location and positioning service for WiMAX based on trilateration concept by taking into consideration the offered features of WiMAX so it can enhance the accuracy of user's location. Preliminary simulations for MIMO and AMC have been carried out. The MIMO results validate that the positioning accuracy of the MIMO antenna modes is better than that of SISO antenna modes, while the AMC could be used to approximate the distance of MS from BS. Our next step shall be on employing more accurate channel model to replace the gaussian channel model used in this work. Furthermore, we will include the parameters from the features of WiMAX such as relay, beamforming and power control in the calculation of the location of the MS.

Acknowledgments. The authors wish to thank the Government of Malaysia for the sponsorship.

References

1. Brännström, F.: Positioning Techniques Alternative to GPS, Master thesis, Lulea University of Technology (2002)
2. Hightower, J., Borriello, G.: Location systems for ubiquitous computing. IEEE Computer 34(8), 57–66 (2001)
3. Gavrilovich Jr., C.D., Ware, G.C., Freidenrich, L.L.P.: Broadband Communication on the Highways of Tomorrow. IEEE Communications Magazine, 146–154 (2001)
4. Correia, L.M., Prasad, R.: An Overview of Wireless Broadband Communications. IEEE Communications Magazine, 28–33 (1997)
5. WiMAX Forum: Mobile WiMAX Part 1: A technical Overview and Performance Evaluation (2006)
6. Kemppi, P.: Database Correlation Method for Multi-System Location, Master's thesis, Helsinki University of Technology (2005)
7. Lee, K., Lee, Y.: A LBS for Cellular Phones. In: 2007 International Conference on Convergence Information Technology, pp. 222–228. IEEE, Los Alamitos (2007)
8. Schwieger, V.: Positioning within the GSM Network. In: 6th FIG Regional Conference, San José, Costa Rica (2007)
9. Andrews, J.G., Ghosh, A., Muhamed, R.: Fundamental of WiMAX: Understanding Broadband Wireless Networking. Prentice Hall, Englewood Cliffs (2007)
10. Wenhua, J., Pin, J., et al.: Providing Location Service for Mobile WiMAX. In: IEEE International Conference on Communications (ICC 2008) (2008)
11. Mayorga, C.L.F., et al.: Cooperative Positioning Techniques for Mobile Localization in 4G Cellular Networks. In: IEEE International Conference in Pervasive Services (2007)
12. IEEE 802.16 Broadband Wireless Access working Group, Channel Model for Fixed Wireless Applications (2003)
13. IEEE 802.16 Broadband Wireless Access working Group, Part 16: Air Interface for Fixed and Mobile Broadband Wireless Access Systems – Multihop Relay Specification (2008)
14. White Paper, Alcatel Location-Based Services Solution – A Key Fixed/Mobile Convergence Enabler (2005)

IEEE 802.11s Wireless Mesh Networks: Challenges and Perspectives

Aggeliki Sgora[1], Dimitris D. Vergados[1,2], and Periklis Chatzimisios[3]

[1] University of the Aegean
Department of Information and Communication Systems Engineering
GR-832 00, Karlovassi, Samos, Greece
{asgora,vergados}@aegean.gr
[2] University of Piraeus
Department of Informatics
GR-185 34, Piraeus, Greece
vergados@unipi.gr
[3] University of Macedonia
Department of Technology Management
GR-59200, Naousa, Greece
pchatzimisios@ieee.org

Abstract. A promising solution for wireless environments is the wireless mesh technology that envisages supplementing wired infrastructure with a wireless backbone for providing Internet connectivity to mobile nodes (MNs) or users in residential areas and offices. The IEEE 802.11 TGs has started to work in developing a mesh standard for local area wireless networks. Although a lot of progress has been made and a few new drafts have been released recently, there exist many issues that demand enhanced or even new solutions to 802.11s mesh networking. This paper aims to overview the latest version of the IEEE 802.11s protocol (Draft 2.02), especially the MAC and routing layers, and to point out the challenges that these networks have to overcome in these layers.

Keywords: Wireless mesh network, IEEE 802.11s, routing, MAC, power management.

1 Introduction

The IEEE 802.11 protocol family has become the dominant solution for Wireless Local Area Networks (WLANs), due to its high performance, the low cost and its easiness in deployment. However, the increasing demand for wireless broadband access and high speed rates create new challenges for local area wireless networking, due to the fact that in order to increase the data rate the Access Points (APs) of the WLANs ought to decrease their transmission range in order to support a range of innovative services and access to the mobile users. Although, the interconnection of WLANs with a fixed infrastructure could be a solution, the cost is an inhibitory factor for this implementation.

F. Granelli et al. (Eds.): MOBILIGHT 2009, LNICST 13, pp. 263–271, 2009.

A promising solution for wireless environments is the wireless mesh technology that envisages supplementing wired infrastructure with a wireless backbone for providing Internet connectivity to mobile nodes (MNs) or users in residential areas and offices, and could be called the Web-in-the-sky [1]. A Wireless Mesh Network (WMN) consists of mesh routers and mesh clients [1]. Mesh routers have minimal mobility and form the mesh backbone for mesh clients. Furthermore, in order to further improve the flexibility of mesh networking, a mesh router is usually equipped with multiple wireless interfaces built on either the same or different wireless access technologies. A comprehensive survey regarding the mesh technology can be found in [2].

In order to provide wireless Internet connectivity at lower cost than the classic Wireless Fidelity (WiFi) networks, several companies are developing their proprietary WMN solutions, such as Motorola [9] with "MOTOMESH" and Proxim Wireless Corporation [10] with "ORiNOCO". Also, in order to develop a mesh standard for local area wireless networks the *Extended Service Set (ESS)* Mesh Networking Study Group (SG), authorized by IEEE 802.11 Working Group (WG), has been established in 2003. Its main task was the definition of the Project Authorization Request (PAR) and Five Criteria (5C), which are needed to request formation of a new Task Group (TG). From July 2004, "Mesh Networking" became TG "s".

A call for proposals was issued in May 2005, which resulted in the submission of 15 proposals submitted to a vote in July 2005. After a series of eliminations and mergers, the proposals dwindled to two (the "SEE-Mesh" and "Wi-Mesh" proposals), which became a joint proposal in January 2006. This merged proposal was accepted as draft version D0.01 after a unanimous confirmation vote in March 2006. Although a lot of progress has been made and a few new drafts have been released recently, there exist many issues that demand enhanced or even new solutions to 802.11s mesh networking [8]. As of September 2008 the draft is at version D2.02 [3].

Since the standardization of the 802.11s in an ongoing recent work, many changes have made from the one draft version to the other. The scope of this paper is to give an overview of the IEEE 802.11s draft (version 2.02), and especially of the MAC and routing layers, as well as, to point out the challenges that these networks have to be overcome in these layers. It should noticed that in this paper we follow the latest terminology for the 802.11s mesh networking, the one defined in draft (version 2.01) [4].

The rest of the paper is organized as follows: The new key terms of the IEEE 802.11s mesh networks are defined in Section 2. In Section 3 and 4 the proposed IEEE 802.11s MAC and routing layer enhancements are presented, respectively. Section 5 discusses challenging research issues that still exist in the current 802.11 standard at MAC and routing layer. Finally, Section 6 concludes the paper.

2 Terms and Definitions

In the IEEE 802.11s standard, all the devices that have mesh functionalities are called Mesh Stations (Mesh STAs or MSTAs). The term "mesh capabilities" means that the device can participate in the mesh routing protocol and forward data on behalf of other mesh points according to the proposed 802.11s amendment [1]. A set of Mesh STAs consist a Mesh Basic Service Set (MBSS) that may be used as a distribution system (DS). Each mesh STA is a member of exactly one MBSS.

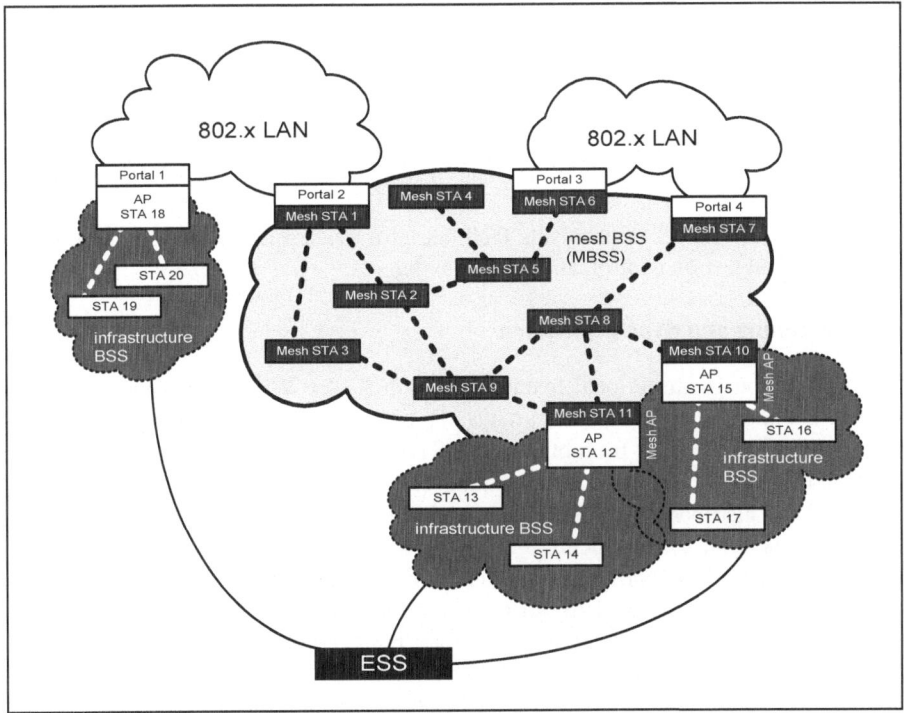

Fig. 1. The ESS 802.11s Network Architecture

Also, Mesh STAs that have additionally access point functionality are called *mesh access points (mesh APs)*. IEEE 802.11 stations that do not have mesh capabilities can connect to mesh APs in order to send data over the mesh network.

A Mesh STA that has a connection to a wired network and can bridge data between the mesh network and the wired network is called *Portal*. Therefore, in order to uniquely identify a Mesh STA in an IEEE 802.11s, a Mesh BSSID is assigned to each Mesh STA, similar to the use of Service Set IDentifier (SSID) to represent an ESS in legacy 802.11 networks [6].

An IEEE 802.11s network architecture is illustrated in Fig. 1.

3   Medium Access Control

The IEEE 802.11s employs the EDCA specified in IEEE 802.11e [7]. It also provides an optional Coordination Function (CF) that is called Mesh Deterministic Access (MDA) that allows supporting Mesh STAs that support MDA to access a certain period with lower contention, called MDA OPportunity (MDAOP). To use the MDA method for access, supporting Mesh STAs must be synchronized to each other. MDA capable Mesh STA negotiates with their neighbor Mesh STAs on the reservation of multiples of 32 μs time slots. Once accepted, the transmitter is referred to as the owner of the MDAOP.

During an MDAOP period, the owner of the MDAOP and the receiver follow a different procedure: The owner of MDAOP that has highest priority for medium access, attempts to use Carrier Sense Multiple Access with Collision Avoidance (CSMA/CA) with new backoff parameters to set up a Transmission Opportunity (TXOP), while the non owner of MDAOP has to defer its access by setting its Network Allocation Vector (NAV) to the end of the MDOAP or by using a carrier sensing scheme.

The proposed MAC mechanisms facilitate also synchronization, beacon collision avoidance, congestion control, and power saving.

3.1 Beaconing and Synchronization

Synchronization is an optional feature for Mesh STAs. With synchronization, each Mesh STA updates its timers with time stamp and offset information received in beacons and probe responses from other Mesh STAs. Thereby a common Time Synchronization Function (TSF) timer is maintained, called Mesh TSF timer.

In 802.11s also beaconing procedures are defined not only for synchronized Mesh STAs but also and for unsynchronized Mesh STAs. More specifically, the standard also specifies a beacon collision avoidance mechanism, called Mesh Beacon Collision Avoidance (MBCA) mechanism. MBCA mechanism is utilized to detect and mitigate the contiguous collisions among beacon frames transmitted from hidden nodes.

In contrast, unsynchronized Mesh STAs transmit their beacons as APs in BSS networks, each maintaining independent TSF timer and Target Beacon Transmission Times (TBTTs).

3.2 Intra-mesh Congestion Control

In an IEEE 802.11s network intra-Mesh congestion control is achieved by implementing the following three main mechanisms: Local congestion monitoring and congestion detection, congestion control signaling, and local rate control.

Although Mesh STAs may support multiple congestion control protocols, as stated in 802.11s (draft version 2.0), there is only one congestion control protocol active in a particular mesh network at a time. The default congestion control protocol specifies only congestion control signalling. Specific algorithms for local congestion monitoring and congestion detection, as wells as, local rate control algorithms are beyond the scope of the default congestion control protocol.

3.3 Power Management

All Mesh STAs have the capability to operate in power save mode. More specifically, two different power states are considered:

- *Awake:* the Mesh STA is able to transmit or receive frames and is fully powered.
- *Doze:* the Mesh STA is not able to transmit or receive and consumes very low power.

The manner in which an Mesh STA transitions between these two power states is determined by the Power Management mode of Mesh STA. These include:

- *Active mode:* the Mesh STA shall be in the Awake state all the time.
- *Power save mode:* the Mesh STA alternates between Awake and Doze states, as determined by the frame transmission and reception rules. The Mesh STA in Power Save mode can either operate in light sleep or in deep sleep mode.

4 Routing in WMNs

In the IEEE 802.11s, the default hybrid wireless mesh protocol (HWMP) combines the flexibility of reactive on-demand route discovery and the efficiency of proactive routing. Specifically, the reactive on-demand mode in HWMP is based on the Radio-Metric Ad hoc On-demand Distance Vector (RM-AODV) protocol, while the proactive mode is implemented by the tree-based routing. Such a combination in HWMP can achieve the optimal and efficient path selection. The draft standard (1.0) also defines an optional Radio Aware-Optimized Link State Routing (RAOLSR) that uses multipoint relays, a subset of nodes that flood a radio aware link metric, thereby, reducing control overhead of the routing protocol [5]. However, starting from draft 1.07, the optional routing protocol is removed from the 802.11s [6].

All modes of the HWMP operation utilize common processing rules and primitives. HWMP information elements are the Path Request (PREQ), Path Reply (PREP), Path Error (PERR) and Root Announcement (RANN). The metric cost of the links determines which paths HWMP builds. In order to propagate the metric information between Mesh STAs, a metric field is used in the PREQ, PREP and RANN elements.

The HWMP can support various radio metrics in the path selection, such as throughput, QoS, load balancing, power-aware, etc. The default metric is the airtime cost metric, which considers the PHY and MAC protocol overhead, frame payload, and the packet error rate to reflect the radio link condition. The airtime cost at each link c_a is given by

$$c_a = \left[O + \frac{B_t}{r} \right] \frac{1}{1 - e_f} \tag{1}$$

where O and B_t are constants for each 802.11 modulation type (see Table 1), and the parameters r and e_f are the data rate in Mb/s and the frame error rate for the test frame with size B_t respectively. The rate r represents the data rate at which the Mesh STA would transmit a frame of standard size B_t, based on current conditions and its estimation is dependent on local implementation of rate adaptation. The frame error rate e_f is the probability that when a frame of standard size B_t is transmitted at the current transmission bit rate r, the frame is corrupted due to transmission error; its estimation is a local implementation choice. Frame drops due to exceeding Time-To-Live (TTL) should not be included in this estimation as they are not correlated with link performance.

Table 1. Airtime Link Metric Constants

Parameter	Recommended Value	Description
O	varies depending on PHY	Channel access overhead, which includes frame headers, training sequences, access protocol frames, etc.
B_t	8192	Number of bits in test frame

5 Challenging Issues in WMNs

Although several standard drafts have been released by 802.11s, many issues still remain to be resolved. The following of this section discusses and points out challenges, especially at the MAC layer and routing layer (Table 2).

5.1 Medium Access Control

The medium access protocols, which are currently used in mesh networks, were initially designed for single-hop networks. Therefore, these protocols do not work well in IEEE 802.11s WMNs, since data transmission and reception at a node is not only affected by nodes within one hop, but within two or more hops away and also from the fact that a mesh STA MAC is usually equipped with multiple wireless interfaces built on either the same or different wireless access technologies. Therefore, there exist several major challenging issues that ought to be addressed, these include:

- **Multi-channel Operation:** Although a Mesh STA can have multi- radio interfaces, however, no multi-channel mechanism has been specified in 802.11s. In 802.11 draft (version 0.04), the concept of Common Channel Framework (CCF) was introduced in order to deal with multi-channel operation. However, because of many problems that were not resolved effectively, the proposal was removed from the draft [6]. Therefore, multi-channel operation is still an open issue.
- **Scalable MAC:** Scalability is an important factor for the performance of a mesh network. However, this issue has not been studied in depth in the 802.11 standard. In order to really achieve spectrum efficiency and improve the per-channel throughput, the scalable MAC protocol needs to consider the overall performance improvement in multiple channels [12]. Therefore, new distributed and collaborative MAC schemes must be proposed to ensure that network performance (e.g., throughput and even QoS parameters, such as delay and delay jitter) will not degrade as network size increases [12].
- **Network Integration:** In WMNs, mesh routers can operate in various wireless technologies, such as IEEE 802.11 and IEEE 802.15.4, and IEEE 802.16. Hence, in the MAC layer, advanced bridging functions should be designed. In this way, different wireless technologies can work together seamlessly [11], [12].

- **Adaptivity to Network Configuration Change:** In WMNs, new nodes can be joined and some nodes can be left from the network dynamically. Hence, the MAC layer and the associated channel assignment schemes need to be adaptive to these network configuration changes [11].
- **QoS:** Support for different QoS levels in a multiradio multi-channel architecture using IEEE 802.11e should be investigated [1].

5.2 Routing

Although the standard defines the HWMP as the default routing protocol, HWMP has several shortcomings. Scalability in this protocol is limited and it cannot support route optimization between two mesh STAs [6]. Also, more routing metrics are needed for multi-channel operation, as well as, interaction with MAC layer since HWMP is considered as a module at the MAC layer [6]. Therefore, there still exist several major challenging issues that ought to be addressed, these include:

- **Load balancing and QoS:** Most applications of WMNs are broadband services with heterogeneous QoS requirements. Therefore, routing algorithms are needed to provide QoS guaranteed paths or at least some support for QoS provisioning. Also, the proposed routing algorithms need to perform load balancing and to ensure that a router does not become a bottleneck node [2].
- **Integrated Routing/MAC Design:** In WMNs, the routing layer needs to work interactively with the MAC layer in order to maximize its performance. Integrating adaptive performance metrics from layer-2 into routing protocols or merging certain operations of MAC and routing protocols can be promising approaches [11].
- **Routing Metrics:** Although the airtime cost metric considers the PHY and MAC protocol overhead, frame payload, and the packet error rate to reflect the radio link condition and other factors, such as the power saving feature, the mobility, the multi-channel operation s may need to be devised in order to take into account the peculiarities of multi-channel multi-radio wireless mesh networks [11].

Table 2. 802.11s Challenging Issues

Layer	Issues
MAC	Multi-Channel Operation
	Scalable MAC
	Network Integration
	Scalability
	Adaptivity
	QoS
Routing	Load balancing & QoS
	Cross Layer Routing
	Routing Metrics

6 Conclusions

A promising solution for wireless environments in order to provide wireless Internet connectivity at lower cost than the classic Wireless Fidelity (WiFi) networks is the wireless mesh technology. A Wireless Mesh Network (WMN) consists of mesh routers and mesh clients. Mesh routers have minimal mobility and form the mesh backbone for mesh clients. Furthermore, in order to further improve the flexibility of mesh networking, a mesh router is usually equipped with multiple wireless interfaces built on either the same or different wireless access technologies.

In order to develop a mesh standard for local area wireless networks the IEEE 802.11 TGs has been established in 2003. Although a lot of progress has been made and a few new drafts have been released recently, there exist many issues that demand enhanced or even new solutions to 802.11s mesh networking. The paper aims to overview the latest IEEE 802.11s (Draft 2.02), especially at the MAC and routing layers, and to point out the challenges that these networks have to be overcome in these layers.

Acknowledgment

This work is part of the "Design and Development Models for QoS Provisioning in Wireless Broadband Networks" (03ED485) research project, implemented within the framework of the "Reinforcement Programme of Human Research Manpower" (PENED) and co-financed by National and Community Funds (20% from the Greek Ministry of Development-General Secretariat of Research and Technology and 80% from E.U.-European Social Fund).

References

1. Nandiraju, N., Nandiraju, D., Santhanam, L., He, B., Wang, J., Agrawal, D.P.: Wireless Mesh Networks: Current Challenges and Future Directions of Web-In-The-Sky. IEEE Wireless Communications 14(4), 79–89 (2007)
2. Akyildiz, I.F., Wang, X.: A survey on wireless mesh networks. IEEE Communications Magazine 9(43), 23–30 (2005)
3. Draft Standard for Information Technology - Telecommunications and Information Exchange Between Systems - LAN/MAN Specific Requirements - Part 11: Wireless Medium Access Control (MAC) and physical layer (PHY) specifications: Amendment: ESS Mesh Networking, IEEE Unapproved draft P802.11s/D2.02 (September 2008)
4. Draft Standard for Information Technology - Telecommunications and Information Exchange Between Systems - LAN/MAN Specific Requirements - Part 11: Wireless Medium Access Control (MAC) and physical layer (PHY) specifications: Amendment: ESS Mesh Networking, IEEE Unapproved draft P802.11s/D2.01 (July 2008)
5. Camp, J.D., Knightly, E.W.: The IEEE 802.11s Extended Service Set Mesh Networking Standard. IEEE Communications Magazine 46(8), 120–126 (2008)
6. Wang, X., Lim, A.O.: IEEE 802.11s wireless mesh networks: Framework and challenges. Ad Hoc Networks 6, 970–984 (2008)

7. IEEE P802.11e/D13.0, Amendment: Medium Access Control (MAC) Quality of Service (QoS) Enhancements (January 2005)
8. Hiertz, G.R., Max, S., Zhao, R., Denteneer, D., Berlemann, L.: Principles of IEEE 802.11s. In: 16th IEEE International Conference Computer Communications and Networks, Honolulu, Hawaii, USA, pp. 1002–1007 (2007)
9. MOTOMESH Duo by Motorola Company,
   `http://www.motorola.com/statichtml/MOTOMESH_Duo.html`
10. ORiNOCO Wi-Fi Mesh Series by Proxim Wireless Corporation,
   `http://www.orinocowireless.com/products/wifi_mesh/index.html`
11. Hossain, E., Leung, K.: Wireless Mesh Networks Architectures and Protocols. Springer, Heidelberg (2008)
12. Akyildiz, I.F., Wang, X., Wang, W.: A Survey on Wireless Mesh Networks. Computer Networks 47(4), 445–487 (2005)

IEEE 802.16 Packet Scheduling with Traffic Prioritization and Cross-Layer Optimization

João Monteiro[1,2], Susana Sargento[2], Álvaro Gomes[1], Francisco Fontes[1], and Pedro Neves[1,2]

[1] Portugal Telecom Inovação, Rua Eng. José Ferreira P. Basto 3810-106 Aveiro, Portugal
{it-j-monteiro,agomes,fontes,pedro-m-neves}@ptinovacao.pt
[2] University of Aveiro, Campus Universitário Santiago, 3810-193 Aveiro, Portugal
susana@ua.pt

Abstract. WiMAX is emerging as a broadband wireless access technology to satisfy end user expectations, containing a new set of advantages in terms of throughput, coverage and QoS support at the MAC level which allows convergence of several different types of applications and services. For that reason, the allocation of resources or scheduling becomes of greater importance. This paper focuses on a cross-layer scheduling optimization solution for IEEE 802.16. The relevant features of the proposed packet scheduling optimization scheme consist: of prioritization of users within the same traffic class, allowing for example to an operator, differentiated treatment among users, for instance distinguishing between premium or gold users and silver users; and also cross layer optimization which implies radio resource optimization and a more effective scheduler decision. Simulation scenarios are presented to demonstrate how the scheduling solution allocates resources through particular WiMAX MAC layer implementation in the NS-2 simulator. Results show that the new mechanism implementation results in an improvement to the simple Round Robin fashion present in the original simulation model, being able to increase differentiation between different classes and decrease packets delay, due to its cross-layer processing and traffic prioritization.

Keywords: QoS, WiMAX, Scheduling, NS-2, WiMAX Forum.

1 Introduction

Applications such as video and audio streaming, online gaming, video conferencing, Voice over IP (VoIP) and File Transfer Protocol (FTP), demand a wide range of QoS requirements such as bandwidth and delay. Existing wireless technologies that can satisfy the requirements of heterogeneous traffic are very costly to deploy in rural areas and "last mile" access. Worldwide Interoperability for Microwave Access (WiMAX) provides an affordable alternative for wireless broadband access supporting a multiplicity of applications. The IEEE 802.16 standard provides specification for the Medium Access Control (MAC) and Physical (PHY) layers for WiMAX. A critical part of the MAC layer specification is the scheduler, which resolves contention for bandwidth and determines the transmission order of users: it is imperative for

F. Granelli et al. (Eds.): MOBILIGHT 2009, LNICST 13, pp. 272–281, 2009.
© ICST Institute for Computer Sciences, Social-Informatics and Telecommunication Engineering 2009

a scheduler to satisfy QoS requirements of the users, maximizing system utilization and ensuring fairness among the users.

IEEE 802.16 [1][2] is a broadband wireless technology that already contains intrinsic QoS support, with the usage of Connection Identifiers (CID) to identify service flows with specific characteristics, the downlink and uplink classification and scheduling mechanisms. Nevertheless these mechanisms are not present in the standard [1][2] and were left for proprietary implementation by vendors.

In this paper, the QoS support for WiMAX is addressed, to provide service differentiation over WiMAX networks. It is proposed and evaluated a scheduling solution that uses prioritization and dynamic cross layer information for scheduling decisions in WiMAX networks. The base IEEE802.16 simulation model was implemented by a consortium under the WiMAX Forum [3][4][5] Application Working Group, specially involved in the realization of a WiMAX simulation model based in ns-2. At the moment, the model is only distributed to members and is under development. In order to validate the proposed scheduling solution, a set of QoS oriented scenarios have been simulated and the obtained results show that the implemented scheduler is able to efficiently differentiate between the traffic classes defined in the WiMAX model and achieve gains in throughput and delay, when compared to the base model.

The remainder of this paper is organized as follows. Section 2 provides an overview on WiMAX and section 3 describes the base simulation model proposed by WiMAX Forum. Section 4 details on the proposed scheduling scheme and section 5 discusses the simulated scenario and the obtained results, comparing the proposed algorithm in terms of QoS performance with the base WiMAX Forum simulation model. Finally, section 6 concludes this paper.

2   Overview of IEEE 802.16 Quality of Service

The physical channel defined in the IEEE 802.16 standard [1][2] operates either in PTP (point-to-point) or PTM (poin-to-multipoint) fashion, using a framed format. Each frame is divided in two subframes: the downlink subframe is used by the BS (Base Station) to send data and control information to the SSs (Subscriber Stations), and the uplink subframe is shared by all SSs for data transmission. In TDD (Time Division Duplexing) mode, uplink and downlink transmissions occur at different times since both subframes share the same frequency. Each TDD frame has a downlink subframe followed by an uplink subframe.

The 802.16 MAC protocol is connection-oriented. In this sense, an SS must register to the BS before it can start to send or receive data. During the registration process, an SS can negotiate the initial QoS requirements with the BS. These requirements can be changed later and new connections may be established.

Service flows (SF) in the WiMAX standard are used for establishing connections from SS to BS and vice-versa. Each SF is characterized by the set of QoS parameters that determine the QoS needed by the connection; for example, they can specify the maximum tolerated delay, required bandwidth, the way in which the SS can request the bandwidth, and the behavior of the scheduler.

The QoS requirements may be either per connection based or per SS based. For the purpose of supporting QoS at the MAC level, the BS must allocate slots using a specific algorithm based on the QoS requirements, bandwidth request sizes, or network parameters.

The basic approach for providing the QoS guarantees in the WiMAX network considers that the BS performs the scheduling for both the uplink and downlink directions; an algorithm at the BS has to then translate the QoS requirements of SSs into the appropriate number of slots. When the BS makes a scheduling decision, it informs all SSs about it by using the *UpLink* and *DownLink* management messages (UL-MAP and DL-MAP) in the beginning of each frame. These special messages define explicitly slots that are allocated to each SS in both the uplink and downlink directions. The algorithm to allocate the slots, the scheduling policy, is not defined in the WiMAX standards and is left open for proprietary implementations.

To support a wide variety of multimedia applications, the IEEE 802.16 standard defines five different scheduling classes, or traffic classes, each with different QoS requirements. Each connection between the SS and the BS is associated to one service flow. The Unsolicited Grant Service (UGS) receives fixed size data grants periodically. The real-time Polling Service (rtPS) receives unicast polls to allow the SSs to specify the size of the desired grant. QoS guarantees are given as bounded delay and assurance of minimum bandwidth. The extended real-time Polling Service (ertPS) uses a grant mechanism similar to the one for UGS connections. Moreover, periodic allocated grants can be used to send bandwidth requests to inform the required grant size. For the non-real-time Polling Service (nrtPS), the BS provides timely unicast request opportunities; besides that, the SS is also allowed to use contention request opportunities. Minimum bandwidth guarantees are also provided to nrtPS connections. The Best Effort service (BE) requests bandwidth through contention request opportunities as well as unicast request opportunities.

3   WMF 802.16 Model

This section briefly describes the ns-2 WiMAX forum release 2.1 of the 802.16 module, specifically addressing QoS and scheduling capabilities. The merge of the efforts to produce a scheduling model in ns-2 was taken from the independent development efforts supported by the *Application Working Group* (AWG) of the WiMAX Forum [7] and NIST [9]. This collaboration resulted in a release software module for OFDMA PHY [3][4][5]. The teams at *Rensselaer Polytechnic Institute* (RPI) and *Washington University in St. Louis* (WUSTL) are the primary development teams, among others, supported by the AWG. The authors of this paper are also involved in this group.

The model currently implemented is based on the IEEE 802.16 standard (802.16-2004) and the mobility extension 80216e-2005. A set of features are inherited and present in both models, such as WirelessMAN-OFDM with configurable modulation and TDD at the physical level; it also encompasses the standard management messages to execute network entry without authentication. At the MAC level, fragmentation and concatenation are supported as well. Nevertheless, this model features a series of new capabilities to the existing ones, not only with the introduction of an

OFMA physical layer, but also with the implementation of QoS and SFs, as the most important ones. A description of available features that were included in the WMF model are listed below [3]:

- OFDMA physical layer;
- Selectable fast fading models: ITU PED A, PED B, VEHIC A;
- Service Flow and QoS scheduling;
- ARQ (without ARQ blocks).

3.1 Packet Classification

In terms of packet classification, this model implements an approach using the destination MAC address located in the packet and the packet type to determine the proper CID. The data traffic transmission takes place through a general Data connection.

However, one of the most important enhancements in terms of QoS in the WMF model is the implementation of SFs. This module has the basic infrastructure for requesting and establishing SF with given QoS, although the actual establishment of connections and SF that are based on application requirements is not implemented. Each connection can be associated with a SF and corresponding QoS parameters. The list of flows is configurable in each SS. These provisioned flows are stored as static connections. They are established every time the SS attaches to a new BS. While the structure supports the definition of QoS flows, it is the scheduler that makes use of that information. Furthermore, no admission control mechanisms are provided. The model accepts all the flow requests from the mobile stations, hence congestion and packet loss might occur.

Some parameters used to configure the list of flows that must be setup after network entry are the following [5]:

- Direction - Downlink (DL) and Uplink (UL);
- Data Rate (bytes/s);
- Scheduling Type (BE/rtPS/nrtPS/UGS);
- Data size (bytes).

3.2 Scheduler Mechanism

This section describes the scheduler operation in the WiMAX Forum release of ns-2 module, including both scheduling mechanisms in the BS or SS.

The model presents two schedulers: one for the BS and one for the SS. The BS scheduler is responsible for filling up the downlink subframe. The SS scheduler is responsible for dividing the bandwidth allocated to it amongst its various connections. An interface is defined between the scheduler and the remaining code. The interface defines a set of input parameters and expects the map structure as an output.

The downlink interface returns the DL Map; the uplink interface returns the UL Map. To the downlink/uplink scheduler, a list of downlink/uplink connections is sent.

The DL scheduler is a round robin priority scheduler that allocates bandwidth to a connection when it has data to be sent. It performs round robins through various connections in the following order: UGS, rtPS, nrtPS, and BE. Prior to allocating bandwidth for data connections, it allocates bandwidth to basic, primary and secondary connections. These features and scheduling rules are similar in both models. The bandwidth allocation is performed in multiple of slots. Bandwidth in the uplink direction is allocated per SS. This means that, if a SS has multiple connections, the bandwidth allocated to it is represented by a single UL Map, that is, in a frame in which the allocated bandwidth is the aggregate bandwidth for all its connections. In the WMF model, for all other connections apart from UGS, the scheduler checks the bandwidth request (BWR) packet received from the SSs. For rtPS connections, the BS increments the allocated bandwidth by the amount of bandwidth required to send another BWR packet.

The UL scheduler needs to split this bandwidth amongst all its connections, as it serves the various connections in the same round robin fashion used in DL, and only proceeds to the next class if there is more bandwidth left in the uplink direction. For UGS, a fixed amount of bandwidth is allocated depending on the rate at which bandwidth has been reserved for the connection. Currently, an SS can have only one connection in the DL and UL direction: the scheduler checks which connection it is related to, and transmits data from that connection. For all connections, apart from UGS, a BWR packet is created and queued for transmission.

4 Priority-Based and Cross-Layer WiMAX Scheduler

In this section, a proposal for the WiMAX scheduler that is capable of allocating slots based on the QoS Service class, traffic priority or the WiMAX network and transmission parameters is described. To test the proposed solution, the QoS model for the IEEE 802.16d/e MAC layer in the NS-2 simulator developed by the WiMAX forum and detailed in the preceding section, was taken as base. In section 5, simulation scenarios and traffic are presented to demonstrate how the scheduling solution allocates resources in various cases. Simulation results will reveal an optimized scheduling solution that ensures the QoS differentiation of the different WiMAX service classes and sharing of the free resources in a fairly manner, taking into account the instant transmission conditions.

4.1 Description

The proposed algorithm depicted in Figure 1 represents an enhancement to the previous WiMAX Forum algorithm. This algorithm, called **Enhanced Round Robin** (eRR), by using the same approach as the simple round robin solution, introduces more elements in the decision making process of packet allocation in each frame . These elements are either used to distinguish traffic and applications, and to give transmission preference to terminals with best radio conditions and its connections. The dynamic decisions are based on the actual transmission channel conditions.

More specifically, not only the traffic is mapped using fixed priorities to different service classes but also different priorities are possible inside the same traffic class, making possible to distinguish with more granularity the kind of traffic to prioritize in terms of transmission. Also, apart from the static traffic prioritization, the scheduler performs cross layer information processing: it first prioritizes connections from terminals which present the highest received power signal strength, commonly called

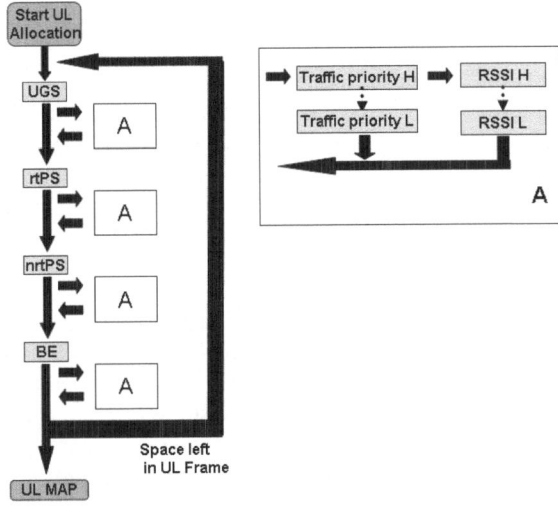

Fig. 1. Enhanced Round Robin Algorithm

RSSI (Received Signal Strength Indication), applying these rules in the case the connections are from the same traffic class and possess equal priority inside the respective class. According to this last rule, terminals with equal service class connections and traffic priority will be served in a certain order that privileges the ones that have better radio conditions, or are closer to the transmitting antenna. In practice, the algorithm initially performs the same round robin procedure as explained in the previous models, i.e. serving first connections in the following order: UGS, rtPS, nrtPS and BE. From the list of existing connections inside the same class, a priority is also established and used to allocate the corresponding packets in the given frame that is to be allocated at the present time. Assuming that more than one connection from the same class has equal traffic priority, the connections will be served taking into account the RSSI value for the given node, from the highest RSSI to the lowest, thus achieving some degree of radio resource optimization as well as guaranteeing that connections with better radio conditions are served minimizing also transmission errors.

This algorithm was implemented using the WMF IEEE802.16 Release 2.1 simulation model. The intended optimization considered only the uplink scheduler implemented in the model and modifications were mainly performed on the uplink frame allocation function *uplink_stage()* present in the *bsscheduler.cc*; the SF class was easily extensible to support the static traffic priority assignment using the configurable SF in the *tcl* simulation scripts. Apart from that, a structure is created each time an allocation is performed in which all the existing connections that are passed to the scheduler are stored with the instant value for their respective terminal's RSSI calculation. Finally, the information is processed in order to give the scheduler allocation decision based on the information that is present at the time.

5 Performance Evaluation

This section presents the results and performance comparison in order to evaluate the proposed model, in terms of efficiency, considering both throughput, delay and differentiation metrics. For this purpose, simulation scenarios were implemented to test QoS using a Point-to-multipoint (PMP) network topology,in which the number of SS's is increased We take WMF model as a basis for comparison with the new scheduler.

5.1 Simulation Parameters/Scenarios

The network topology considers differentiated traffic traversing the uplink direction from different hosts. A PMP wireless scenario is considered (Figure 2), in which single hosts are connected to SSs, with each host's traffic representing one connection flow per SS in the uplink direction. In order to test the different network topologies, assuring differentiation between the different service classes, we defined and implemented new traffic sources. As an example, BE traffic generator contains a variable packet size and interval to emulate FTP/web traffic, and an UGS traffic generator contains a constant transmission rate. rtPS traffic presents a variable packet size and interval, consistent with real time video transmission. Table 1. presents the different values adopted for these traffic generators.

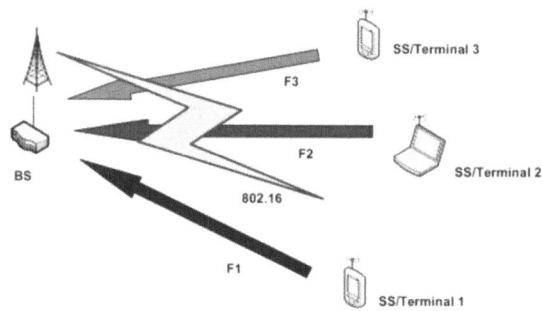

Fig. 2. PMP scenario

Table 1. Traffic generator parameters

Packet Type	Bitrate (Kb/s)	Packet Size (bytes)
BE	200	512 to 1024
UGS	200	300
rtPS	200	200 980

5.2 Simulation Results

In this scenario we have defined the relevant PHY layer simulation parameters. The most important used parameters are summarized in Table 2.

Table 2. Simulation parameters

Frequency	3.493 GHz
Bandwidth	20MHz
Frame duration	5 ms
Downlink ratio	0.3
Modulation	16 QAM
Cyclic prefix	0.25
Queue lenght	100 packets

Comparison results of both schedulers comprise throughput, delay, and traffic differentiation from an increasing number of SSs from 3 to 12 and one terminal attached to each SS, with only one of the respective traffic sources for UGS, rtPS or BE.

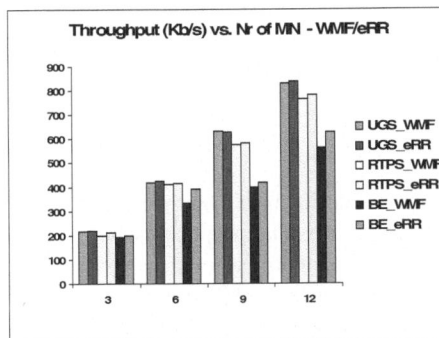

Fig. 3. Throughput WMF/eRR results   **Fig. 4.** Delay WMF/eRR results

Figure 3 shows the results with an increasing of terminals up to 12. It is visible the slight gain difference that can be achieved in throughput using the enhanced Round Robin solution as well as the observed earlier service class differentiation. In this particular case, the traffic priority was assumed to be equal among the same classes, being the decision parameters here in evidence only the service class and RSSI of respective terminal.

In relation to delay, as can be seen in Figure 4, there is also a slight decrease when using the proposed scheduling solution, meaning that in every case the scheduler either equals or outperforms slightly the existent Round Robin in the WMF model, expectable also since it is based on the same principle.

Figure 5 and Figure 6 illustrate the obtained differentiation inside the same service class and among different ones, using only traffic prioritization, RSSI based decision is not effective since all connections are of different classes and priorities. Results are presented for separate connections for rtPS and BE in terms of throughput and delay for the enhanced RR solution. The scenario consisted of terminals with connections with the rtPS and BE classes respectively, and different traffic priorities inside each

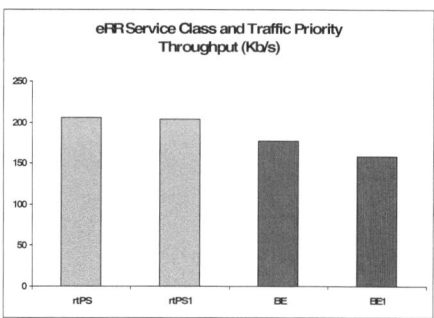 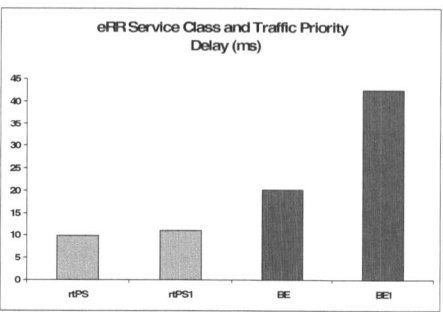

Fig. 5. Throughput values for distinct service classes w/ diverse traffic priority

Fig. 6. Delay values for distinct service classes w/ diverse traffic priority

service class, i.e rtPS1 has lower priority than rtPS connection and BE1 also in re spect to BE. Results show the intended differentiation in the scheduling decision as both classes are distinguished in terms of throughput and delay (better values for rtPS classes than BE ones) and traffic prioritization inside each particular class (better performance for rtPS and BE in relation to rtPS1 and BE1 respectively).

6 Conclusions

In this paper it was proposed and evaluated a novel solution based on the simple Round Robin scheduling model defined by the WiMAX Forum. The proposed scheme bases the scheduling decision on priority according to service class, a traffic priority inside the same class, and also a novel third principle using instant RSSI information to dis tinguish between similar connections. The enhancement proposed to the original scheduler present in the WMF model proved its efficiency in improving important QoS parameters and differentiation of traffic. Another important aspect to retain is the greater granularity, achieved with this enhanced scheduler, more close to the standards specification, in terms of service differentiation, bringing advantages not in capacity, but making possible for operator networks for instance to distinguish different kinds of users and its attributes as well as achieving radio resources optimization.

Acknowledgements

The authors would like to thank the opportunity to perform this work in the concept of Celtic LOOP project [8] and collaboration of the WiMAX Forum, Application Work ing Group [7].

References

[1] IEEE 802.16 Working Group, IEEE Standard for Local and Metropolitan Area Networks. Part 16: Air Interface for Fixed Broadband Wireless Access Systems, IEEE Std. 802.16-2004 (October 2004)

[2] IEEE 802.16 Working Group, IEEE Standard for Local and Metropolitan Area Networks. Part 16: Air Interface for Fixed Broadband Wireless Access Systems. Amendment 2: Physical and Medium Access Control Layer for Combined Fixed and Mobile Operation in Licensed Bands, IEEE Std. 802.16e (December 2005)

[3] WiMAX Forum, WiMAX End-to-End Network Systems Architecture Stage 2: Architecture Tenets, Reference Model and Reference Points, Release 1.1.0 (June 2007)

[4] WiMAX Forum, WiMAX End-to-End Network Systems Architecture Stage 3: Detailed Protocols and Procedures, Release 1.1.0 (June 2007)

[5] The Network Simulator NS-2 MAC+PHY Add-On for WiMAX (IEEE 802.16): ns2 release 2 Documentation, WiMAX Forum, August 13 (2007)

[6] The Network Simulator - ns-2 (2002), http://www.isi.edu/nsnam/ns/

[7] Application Working Group website,
http://www.wimaxforum.org/about/board/
Working_Group_Organizations/AWG

[8] LOOP project Website, http://www.theloopproject.com/

[9] Ns-2 NIST Website,
http://www.antd.nist.gov/seamlessandsecure.shtml

Towards Adaptable Networking: Defining the Protocol Optimization Architecture Requirements

Martin André[1] and Fumio Teraoka[2]

[1] National Institute of Information and Communications Technology
4-2-1 Nukui-Kitamachi, Koganei, Tokyo, 184-8795, Japan
andre@nict.go.jp
[2] Faculty of Science and Technology, Keio University
3-14-1 Hiyoshi, Kohoku-ku, Yokohama, Kanagawa, 223-8522, Japan
tera@ics.keio.ac.jp

Abstract. With the recent trend in computer networking that tends to make the networks more dynamic, appeared the need for network protocols to accommodate against changing conditions. In this context, it is important to rely on a clean architecture to share information between protocol entities and to perform optimizations, in order to achieve what we define as adaptable networking. In this paper, we try to identify the needs for a network protocol optimization architecture and describe the care that should be taken when considering such architecture. We perform a study based on observation of the implications of information sharing between network protocol entities and define some requirements for a successful adaptable network architecture. Then we show how these recommendations could be applied in the legacy layered model by presenting an example architecture that respects the identified principles. This work is expected to serve as a basis for any future adaptable network architecture.

Keywords: cross-layer, architecture, adaptable network, optimization.

1 Introduction

In the early years of computer networks, communication link with a neighbor was expected to be stable. The OSI model has established itself as the de facto standard for network communication, offering a clear separation between the network protocol layers. It made the architecture able to easily incorporate new links and nodes, and to accommodate new protocols, applications, and devices. Recently, this base statement has changed. With the apparition of wireless networks, network condition is varying in time and it is often impossible to rely on stable communication links. In addition, over the past few years, radically new network topologies and applications, some of them hardly fitting in the strict layers of OSI model, have appeared. The user has become mobile , expecting ubiquitous connectivity, and network ground shifted from static to dynamic. One of the new challenges is to adapt to such changing conditions.

F. Granelli et al. (Eds.): MOBILIGHT 2009, LNICST 13, pp. 282–291, 2009.

Consequently, the OSI model started to be frequently transgressed with the intention of making the protocol entities more aware of the environment. This violation of the strict layer boundary is often referred to as *cross-layer optimization*. However, the term "cross-layer" is frequently employed through the misuse of language to describe any technique of network protocol optimization. Indeed, optimizations are not only made possible thanks to communication between non-boundary layers but collaboration between distant nodes has to be envisaged too. Moreover, the term cross-layer implies a dependency on the legacy layered model and as a result excludes alternative network protocols models approaches. In the following of this paper, we will prefer the term *adaptation* that we define as the network protocols optimization thanks to information sharing.

A great amount of literature describes mechanisms that allow sharing of information in order to permit optimization in domains such as ad-hoc networks, Quality of Service (QoS), security, or mobility. The proposed mechanisms usually adopt a case by case approach that targets a single issue where a generic infrastructure that helps protocol entities to make the best decision depending on the environment and to accommodate against the changing condition would be necessary. Here also, quite some work has been made in that direction and several generic optimization architectures have been proposed, notably CLASS [1], ECLAIR [2], MobileMAN [3], and more recently Hydra [4].

The rest of this paper is organized as following. We will start by performing analysis of adaptable networking in Section 2, where we discuss the following subjects: extension of the scope of view, implementation of information coming from the network, communication model, and integration with the network protocols model. In Section 3 we present as an example a conceptual architecture that follows the previously identified precepts. Finally, Section 4 concludes this paper and proposes a direction for future work.

2   Adaptable Networking Study

In this section, we analyze the implications of information sharing and we try to identify the key design factors that are needed for an adaptable network architecture. This results in recommendations for a successful architecture.

2.1   Extension of the Scope of View

Supposing that a protocol works in an optimum manner at its own level, it can perform awfully at a higher level because it will, by the nature of the layering model, behave selfishly without taking into account other protocols' constraints. TCP, for example, achieves terrible performances on a link prone to errors. We introduce the concept of *scope of view* as the quantity of information visible by a protocol entity. The idea is to extend the scope of view of protocol entities in order to permit better decisions based on the environment. This is the very principle of adaptable networking.

The question we would like to answer is to what extent we should broaden the scope of view. We will examine the effect of the extension of scope of view on several important aspects such as complexity, reliability, latency, and performance either locally to a node or globally for a whole network. Fig. 1 depicts the effect of extending the scope of view over these properties, all represented vertically. The horizontal axis represents the four different identified levels of scope of view, with no information shared which reflects the strict conformance to the conventional OSI layered model, the information shared inside a node which is typically called cross-layer optimization, the information shared between nodes that consist in adding a local network view to a node, and finally the information shared between networks. The arrow shows the direction and the range where the effect increases, while the stripped and plain patterns respectively show that the effect is negative or positive.

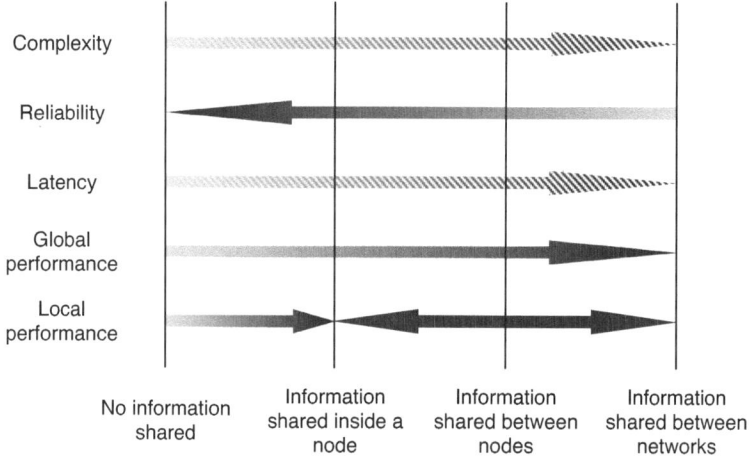

Fig. 1. Effect of extending the scope of view

The broader the scope, the higher the complexity of the architecture and consequently the higher the risk of stability issue. Indeed, "complex" and "unstable" are often synonyms. When more elements are involved, the risk of a failing entity increases too. In addition to complexity and possible failure risk, extending the scope of view also introduces overhead and thus latency. This is particularly visible when information begins to be shared among nodes, caused by the delay of messages transportation over the network. Latency can be problematic when adapting to a phenomenon that evolves rapidly. The delay introduced makes the decision not accurate and possibly makes the desired optimization under-efficient.

Despite these pernicious effects induced by the extension of the scope of view, allowing the sharing of information and the cooperation between protocol entities or nodes can be very beneficial to the global (at a network level) or the local (at a node level) performance. This is the whole point of adaptable networking.

Concerning the local performance however, the choice can be made, thanks to the information coming from the network, to voluntary decrease its performance in favor of the global performance. This choice will then depend on how priorities are set, either to optimize the performance of a single node or the one of a whole network eventually at the expense of some nodes constituting the network. There is no ideal scope of view that suits all constraints. Also, the choice of the scope of view level will rely on the decision that has to be made:

- some decisions have no influence outside of a protocol entity and don't need external information. This is the case of most protocol entities internal mechanisms.
- some decisions require only communication local to a node between protocol entities in order to guarantee optimal performance. For example, TCP may consider the type of data link before reducing the congestion window size after a packet loss is detected.
- some decisions require external information coming from the network in order to have a better vision and guarantee smart choice in full knowledge of the environment. Routing algorithms in a mesh networks scenario are concerned by this type of decision.

The optimization architecture has to deal with the challenge of determining the suitable scope of view for a given situation, i.e., finding a good consensus between advantages and drawbacks. The scope of view must be adapted to the situation and depends on the system requirements and on the decision that need to be made.

2.2 Network-Related Information

We've exposed that the scope of view is not something fixed and that several scenarios must be envisaged. The network view is a logical evolution to information sharing inside a node, and it is necessary under certain condition to achieve optimal results. The architecture should in consequence be able to consider information coming from the network as input for decision making.

A trivial approach would be to let the nodes exchange information directly, but this raises the issue of the security. How can one give credibility to a node? Indeed, the information coming from a distant node modifies the vision of the network environment and is susceptible to induce a decrease in the local performance. It is thus primordial to trust this information in the first step. Besides, a malicious node may inject false information leading to wrong decisions. The system must be prepared against potential attacks. In addition to the security and reliability problems, there is a concern about confidentiality. Some information may be sensitive and, in consequence, shouldn't be accessed by anyone. As a matter of fact, it is necessary to authenticate the nodes in order to establish a trusted network.

But authenticating the nodes isn't sufficient. Let us imagine the case where a node has been authenticated but which is spreading, intentionally or not, erroneous information. This node may be relaying outdated information for instance.

The optimization architecture needs to allow the validation of the information's relevancy by multiplying the source of information, for example by verifying to other nodes any information that can be double-checked. This could be the role of a central server to confront information announced by the nodes constituting the network and to detect inconsistency.

Finally, performance must also be taken into account. It is clearly more efficient in term of message exchange to gather nodes' information on a central server to achieve optimum network cooperation rather than to rely on direct communication between the nodes when unicast messages are exchanged. The latter one is expected to generate lots of messages for full cooperation between the nodes. However, multicast or broadcast message exchange between the nodes has the same order of complexity as communication with a central server. Fig. 2 represents the comparison between the central server and the distributed cooperation in unicast and shows how message exchange complexity is exponentially growing when the number of nodes is increasing.

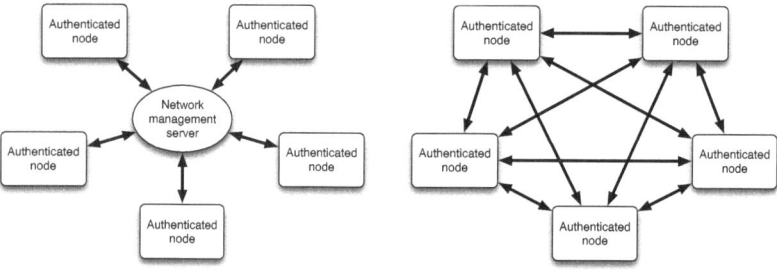

Fig. 2. Central server versus distributed cooperation

In conclusion, it is more appropriate for the implementation of the network view to introduce a new entity that acts as a trusted and reliable network management server, similarly to what is proposed in the IEEE 802.21 [5] developing standard aiming at providing a media independent handover mechanism. This choice has also the benefit to allow greater possibility for additional features. Direct communication between the nodes may not be completely left over, though, but it requires setting up an authentication mechanism.

2.3   Communication Model

There are conceptually three types of communication possible when talking about cooperation between functional units:

– Information data exchange;
– Event subscription and notification;
– Command requests and responses.

The IEEE 802.21 developing standard defines similar services in its Media Independent Handover Function (MIHF), respectively the Media Independent Information Service (MIIS), the Media Independent Event Service (MIES), and the Media Independent Command Service (MICS).

As shown in [6], introduction of dependencies between network protocol entities may lead to conflicting interactions and adaptation loops. The article also points out that the architecture is critical for adoption of technology, as it fixes for the most part the ease of development and the longevity. In other words, this is the role of the architecture to prevent potential conflicts. Considering the cooperation inside a node we can distinguish two types of communication between the protocol entities depending on whether they are subject to conflicts or not. On the one hand, information data exchange as well as event subscription and notification belong to the same category and are considered harmless. Command requests, on the other hand, are potentially harmful.

The communication for the first category of interactions, not likely to cause conflicts, should be performed directly between protocol entities to permit good reactivity for the exchange of information. This applies to information requests and event subscriptions and notifications. Concerning the communication for command requests, however, we should introduce a *supervision entity* that has a global view of the system requirements in terms of QoS, energy consumption, and security and that controls the potentially conflicting interactions.

Similarly to the cooperation inside a node, cooperation between the nodes requires a central authority having the knowledge of the whole network (see Sec. 2.2) and acting as a network management server. The envisaged type of communication is done through a publish and retrieve information manner between the nodes and the network management server, and commands are originated from the network management server to the nodes.

2.4   Integration with the Network Protocols Model

The layered model brings a lot of architectural concepts that are very efficient in today's network topology and which highly contributed to the Internet success. However, the model presents a certain rigidity which makes it difficult to accommodate evolutions, leading researchers to envisage alternatives. One consists in violating the strict layer boundary and is known as cross-layer. Others consist in non-layered approach to the design of network protocols. Among the new paradigms, the Role Based architecture (RBA, [7]) organizes communication in non hierarchical functional units called roles and defines the concept of network heap as a replacement of network stack. The SILO architecture [8] has an intermediate approach and proposes to build network silos (logical blocks stacks) that are assembled on-demand based on the application requirements. We need to think how the optimization architecture should fit in a network protocols model.

An adaptable network architecture has strong requirements that need to be considered with regards to the integration with a network protocols model. First, besides being essential for viability of a solution and strongly contributing to its adoption, maintainability has a substantial role to play. A good architecture has to be able to deal easily with evolution of the networks, in particular, it must take a special care about the case of a new protocol entity and the one of a new possible optimization.

Considering the case of a new protocol entity, it must integrate easily in the system, i.e., being readily able to communicate with existing protocol entities. For this purpose, the architecture needs to provide a standardized interface to communicate with protocol entities by means of a public Application Programming Interface (API). In addition, existing protocol entities should be able to interact with this new one with ideally no modifications. To this goal, we should introduce an *abstraction level* for the protocols' common capabilities that ensures forward compatibility between protocol entities.

By providing a standardized interface to communicate easily with a protocol entity, we also resolve the problems relating to the case of a possible new optimization. These are not restricted anymore to the limited vision of the architecture designer, and unforeseen interactions can easily be implemented. A public API gives assurance that the set of optimizations is extensible.

In conclusion, the optimization architecture should provide modularity to integrate smoothly in any existing or future network architecture, independently of the network protocols model. In addition, a great importance should be given to maintainability, by providing a public interface to communicate with the protocol entities and introducing an abstraction level for protocols' common capabilities.

3   Example Architecture

We will here present how the principles exposed in Section 2 can be put into practice in the legacy layered model. Fig. 3 represents an example architecture conforming to these principles.

3.1   Layer Abstraction

The first modification made to the legacy layered model consists in the addition of a level of abstraction to cope with protocols differences within the same layer and permit a generic use. It gives the assurance that the compatibility between protocol optimizations is maintained. In our example it takes the form of a standardized API to access the common layer's features and of an extension to the protocol entity (PE) called the *Abstract Entity* (AE), represented in light stripped pattern on Fig. 3. The AE is the interface used for all the communication and cooperation with the protocol entity. Protocol entities are not required to implement an AE, but this extension is a prerequisite in order to communicate in the optimization architecture.

RFC 5184 [9] presents an example of layer abstraction and proposes a standardized API for the link layer abstractions.

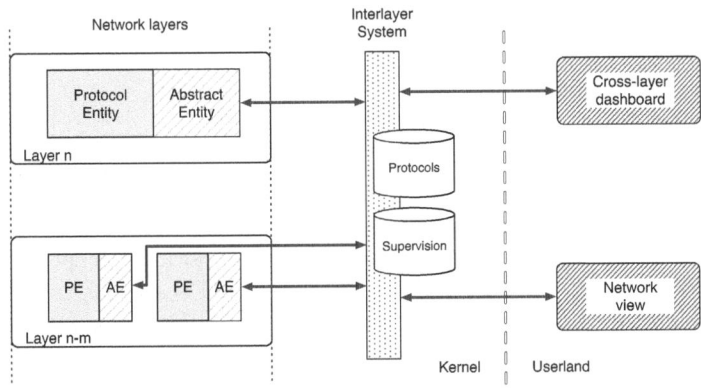

Fig. 3. An example architecture

3.2 Interlayer System

We introduce an entity called *interlayer system*, represented in dotted pattern on Fig. 3, that allows communication between all the protocol entities. It can be considered as the message delivery service, as every message exchange is done via the interlayer system. To this end, the interlayer system stores information about the protocol entities that are registered with it. In addition, it plays the role of the supervision entity that controls the potentially conflicting interactions thanks to a global view of the system requirements, as shown in Section 2.3. It thus stores all necessary information used to avoid interaction conflicts. Inappropriate controls can for example be rejected by the interlayer system either because it is in contradiction with global policy (QoS, energy, etc.) or it is conflicting with another control having higher priority.

3.3 Kernel-Userspace Bridge

The last part of this example architecture illustrate how a kernel-userspace bridge can be used to implement extensions, represented in dark stripped pattern on Fig. 3, that provide additional input for the decision making. Optimization will typically take place in the kernel space of the Operating System (OS) with no operation from the user. However, there are cases where external information input is necessary such as to provide to the nodes a certain knowledge of the network in order to do better optimizations, as shown in Section 2.2. The communication between the node and the network management server should be implemented in the user space of the OS as a daemon and connection with the interlayer system (Sec. 3.2) could be assured thanks to a kernel-userspace bridge.

Another example extension would be an optimization dashboard that gives the complete control of the optimization system to the user. It could be used to enable and disable certain optimization, set system-wide constraints determining the priorities, or even consult statistics. This extension could help researchers and developers to realize performance measurements of certain optimizations.

4   Conclusion

Our observations on the implications of information sharing for the network protocols resulted in recommendations usable for any adaptable network architecture that we will briefly summarize. There is no ideal scope of view for the network protocol entities, but rather an optimal scope depending on the constraints and it is the role of the architecture to find the best compromise. Network view, as part of the extension of the scope of view, is the natural evolution of cross-layer and makes a lot of sense in adaptable networking. In order to improve the potential for additional facilities, its implementation requires to place a new entity on the local network that provides a global view and acts as a management server. Direct communication between the nodes should still be allowed, but in any case, authentication of the nodes has to be provided. In the same manner, a logical entity that has the global knowledge of the constraints and which purpose is to regulate the communications inside the node is mandatory to assert the reliability of the architecture. Its role is to protect against inappropriate use and conflicting interactions. Concerning the integration of the optimization architecture in a network protocols model, designers should give priority to maintainability of their solution by providing a public interface to communicate with the protocol entities and introducing an abstraction level for protocols' common capabilities. Designers should also think as generic as possible in order to avoid dependence of the architecture over the design of the network protocols.

We presented an architecture that gives an example how the exposed principles could be put into practice in the layered model. The maintainability is achieved thanks to a layer abstraction and the use of a standard API. As for the reliability, a central entity called the interlayer system takes care of supervising all communications between the protocol entities. Finally, the example architecture presented in this paper can easily be extended by means of a kernel-userland interface, and one can imagine applications such as the implementation of the network view or a dashboard to control the system.

Our plans for future work will first be to confront existing solutions against the identified requirements and propose a new solution if none is found satisfactory. Some work remains to be done on the supervising entity. We will explore possible mechanisms that can be used to avoid and resolve the potential conflicts between the interactions. We will also investigate more in depth the communication methods between the nodes and the network management server. Finally, we have in the project to implement a solution to validate our concepts and prove the efficiency and the robustness of our architecture.

We expect this document to serve as a basis for any future work on adaptable network architecture.

References

1. Wang, Q., Abu-Rgheff, M.A.: Cross-layer signalling for next-generation wireless systems. In: Wireless Communications and Networking, vol. 2, pp. 1084–1089. IEEE Press, New York (2003)
2. Raisinghani, V.T., Iyer, S.: Cross-layer feedback architecture for mobile device protocol stacks. Communications Magazine 44(1), 85–92 (2006)
3. Borgia, E., Conti, M., Delmastro, F.: Mobileman: design, integration, and experimentation of cross-layer mobile multihop ad hoc networks. Communications Magazine 44(7), 80–85 (2006)
4. Choi, S.H., Perry, D.E., Nettles, S.M.: A Software Architecture for Cross-Layer Wireless Network Adaptations. In: 7th Working IEEE/IFIP Conference on Software Architecture, pp. 281–284 (2008)
5. IEEE Computer Society: Media Independent Handover Services. Draft Standard for Local and Metropolitan Area Networks. IEEE Computer Society (2008)
6. Kawadia, V., Kumar, P.R.: A cautionary perspective on cross-layer design. IEEE Wireless Communications 12(1), 3–11 (2005)
7. Braden, R., Faber, T., Handley, M.: From protocol stack to protocol heap: role-based architecture. SIGCOMM Comput. Commun. Rev. 33(1), 17–22 (2003)
8. Dutta, R., Rouskas, G.N., Baldine, I., Bragg, A., Stevenson, D.: The SILO Architecture for Services Integration, controL, and Optimization for the Future Internet. In: ICC 2007, IEEE International Conference on Communications, vol. 1, pp. 1899–1904. IEEE Press, New York (2007)
9. Teraoka, F., Gogo, K., Mitsuya, K., Shibui, R., Mitani, K.: Unified Layer 2 (L2) Abstractions for Layer 3 (L3)-Driven Fast Handover. IETF RFC 5184 (2008)

Uptake of Mobile ICT Health Services: Has the Time Come to become Commodity?

Pantelis A. Angelidis

University of Western Macedonia, Greece

Abstract. Personalized healthcare (pHealth) is a collective term aiming to reflect all modes of patient-centric healthcare delivery via advanced technology
means. Personalized health involves the utilization of micro and nanotechnology advances, molecular biology, implantable sensors, textile innovations and
mobile information & communication technology (mICT) to create individualized monitoring and treatment plans. pHealth proactively endorses the sense of
"one-to-one" communication to elevate healthcare delivery, optimize patient
services and ensure seamless from the patient point of view information exchange. Patient awareness, policy planning and technology progress are favoring phealth market penetration, while financing issues, political commitment,
and unavailability of technology infrastructures are fundamentally prohibiting
its expansion. This paper explores the drivers and barriers to the adoption of
phealth delivery schemes, including a discussion on interoperability issues. It
also presents case study results.

1 Introduction

Personalized healthcare (phealth) is a collective term aiming to reflect all modes of
patient-centric healthcare delivery via advanced technology means. Personalized
health involves the utilization of micro and nanotechnology advances, molecular
biology, implantable sensors, textile innovations and information & communication
technology (ICT) to create individualized monitoring and treatment plans. pHealth
proactively endorses the sense of "one-to-one" communication to elevate healthcare
delivery, optimize patient services and ensure seamless from the patient point of view
information exchange. Patient awareness, policy planning and technology progress
are favoring phealth market penetration, while financing issues, political commitment,
and unavailability of technology infrastructures are fundamentally prohibiting its
expansion.

Personalized healthcare will be achieved through a composite of scientific advances and new technology, and creative uses of information technology and human
thought in the practice of medicine. Scientific advances and discoveries, as well as
new technological capabilities, will be revolutionary. Innovation in the practice of
medicine will be evolutionary. The combination of revolutionary technologies and
evolutionary practices form information based medicine and will shape the future of
personalized healthcare [1].

F. Granelli et al (Eds.): MOBILIGHT 2009, LNICST 13, pp. 292–302, 2009.
© ICST Institute for Computer Sciences, Social-Informatics and Telecommunication Engineering 2009

Innovative computer and software technologies are deployed to provide vital patient data monitoring and connect clinicians with mobile patients via workstations, wireless devices and the Internet. Technology progresses to produce virtually invisible biosensors, implantable or integrated in the patients clothing or to small, portable devices, which enable continuous vital data transmission and allow the development of personalized treatment plans for the patient.

The concept of prevention prevails now against disease management and treatment plans. As patient-centric processes emerge, the citizens/patients undertake an active role in monitoring their health status, whereas e-wellness evolves to address the rising expectations of the e-health consumers, who are better informed, more demanding, and empowered. The empowered, worried-well, consumers require quality health services on the spot. The drivers are now connectivity, speed and personalization [2].

2 pHealth Service Models

2.1 pHealth Models of Care

Waves of technology incorporation and scientific discovers, have driven the sector from reliance on direct communication and physician experience, to a higher reliance on technology and community information. The expression "personalized healthcare" has become quite common as basic and clinical medical studies based on the decoding of the human genome continue to progress and social awareness of this trend increases. The idea here is to perform diagnosis and therapy after scientifically determining personal traits [1].

In the frame of the phealth models of care, individual physical characteristics are identified in order to depict the appropriate medical care protocol and associated risks for each person. Personal and family medical history joined with lifestyle and patient empowerment mentality foster the development of personalized pathways of care, which are enabled by research and technology implementation.

Bio-molecular information is explored in order to issue new medications; sensors are getting smaller, smarter, and implantable in order to ubiquitously monitor health state parameters; pervasive computing and data networks are being deployed for the timely information exchange; and the healthcare providers are challenged to keep up with new market and technology trends in order to meet increasing patient needs.

2.2 Compelling Drivers for Change

The deployment of personalized healthcare models prerequisites research orientation to fulfilling individual patient and carer needs.

The underlying factors for the emergence of new patterns of care include:

Changing trends: The reformulation of current service provision in order to meet patient demand for quality services to be provided anywhere – anytime is necessary, as population grows older and becomes more concerned about the quality and availability of health services

Adding varieties: The increasing demand for quality healthcare services forces the healthcare providers to adopt advanced tools and invest on human capital and infrastructure in order to be competitive in the new information society era. The traditional treatment plans in the traditional nursing areas are insufficient as life rhythms and technology advances lead the patient base to call for novel and sophisticated health products

Improving quality: The availability of information and knowledge accessibility created patient awareness, shaping the demand for service availability and efficacy, expertise deployed for development of medical plans and prevention rather than treatment.

Advancing technology: New technologies introduce highly innovative health products / services creating new needs in patient care and addressing "hidden" needs for specialized medical care.

2.3 Barriers to Emerging pHealth Adoption, Critical Success Factors and Dependencies

Although theoretically phealth applications effectively address patient needs, actual implementation is still at infancy stage and constrained by numerous factors, "unforeseen" by the research community that provides the new tools and methods for healthcare delivery.

Cost pressure from managed care entities and the lengthy approval processes are prime considerations in bringing a truly breakthrough product rapidly to the market [3]. The healthcare providers focus on healthcare cost containment in the short-run and neglect to examine cost efficacies created by the adoption of new technologies in the long-run, when the high initial investment costs are outranged by more effective patient and internal resources management and/or revenue generation from the provision of added value services.

Even when leadership with vision is overcoming the cost barriers and is budgeting resources so as to shift an organization to adopt new technologies, resistance to change is a common factor that scales down this kind of efforts. The commitment of the whole team (top management, healthcare workers, administrative personnel involved) is essential to the successful implementation of novel applications in the health market.

To this end, awareness and education of the healthcare professionals is also necessary in order to highlight the capabilities generated by the new tools and methods and diminish competitiveness between the traditional practices vs new technologies. Technophobia is often concealed behind reliability in familiar / traditional methods, leading thus to obstructionism behaviours. As a result, the sector in lagging behind in the adoption of novel products and services.

Selecting the right people to be involved is a critical success factor to every project implementation. Especially, when it comes to applying new methods or systems, recruiting people familiar to or interested in new technologies and practices and providing further motives for involvement is essential for the following reasons: a) The participants interested in the new applications are self-motivated and thus, willing to work towards the achievement of the project goals b) Provision of further motives (i.e. know-how transfer, scientific knowledge diffusion, self-esteem, etc) contributes to more active involvement, c) The project participants that are interested in carrying out the project also contribute to effective dissemination of the project outcomes.

Additionally, the commitment of the management of the organizations involved in the project is also a critical success factors as it a) ensures proper allocation of recourses for the effective project implementation, b) support throughout the whole process, c) provision of motives to the participants, d) strong collaboration and dissemination of the results.

3 Current Trends

3.1 The Vision

Specialized health-telematics enterprises aiming at the effective integration of mICT in the health sector and the optimization of the quality of the healthcare services, have started to spring in the recent years in Europe. These are mostly spin-off companies arising from research in academia and industry.

They aim at the commercial deployment of novel technologies generated in the laboratory environments; a particularly interesting area is the vital signs telemonitoring service concept, based on mobile networks and intelligent sensor devices, targeting citizens / patients on the move. This is aligned with the current international trend regarding the provision of healthcare services and in particular chronic disease management, bringing points of care closer to the patients and striving at the wellness of the person and prevention, instead of disease management and treatment plans. These services promote citizen health awareness and patient empowerment.

3.2 Market Frame

1) Service Concept

The service concept emerges, as points of care move closer to the patient and the citizen/patient undertakes a more active role in healthcare monitoring and prevention.

The system enables remote monitoring and transmission of the patients vital signs via wearable monitoring devices with mobile transmission capabilities over (locally) BT or Zigbee and (widely) GPRS/3G, WiFi and recently WiMAX. Such a system provides the possibility for doctor-patient ubiquitous communication and support, while the patient is at home, work, vacation (i.e. away from the traditional nursing areas). More than this, it triggers a patient-centric process, focusing on prevention rather than disease management and treatment and initiates patients' active involvement in healthcare.

The need to provide cost-effective healthcare services for continuous telemonitoring of vital signs to remote or on the move patients has been early identified, to bridge the gap in healthcare provision. This gap is created by the inability of healthcare providers to offer continuous monitoring, seamlessly to chronic patients and worried-well citizens.

Table 1. Stakeholder Groups

Stakeholder Groups	Mobinet Value
Private health-care providers	Expansion of current product /service portfolio Innovation; Leading position; Improved customer care; Provision of services to remote markets
Health care authorities	Long-term: savings on healthcare spending, social care; Short-term: provision of added value healthcare services; prevention of diseases
Public health-care providers	Long-term: savings on healthcare spending; social care; collaboration and networking; better allocation of resources; Short-term: provision of added value healthcare services; prevention of diseases
Insurance companies	Provision of advanced services to end customers Leading position; Improved customer care; Differentiation from competition
Healthcare associations	Networking; Exchange of scientific view; Ability to extract anonymous statistical data for reference and conduction of studies
Patients – Citizens	Improved care management; User friendly patient record management; Improved quality of life (for carers as well); Wellness & time savings
Athletes	Improved health status monitoring; Based on the above, risk minimization regarding health deterioration incidents in the field; Improved performance without jeopardizing health

2) Stakeholder Groups

3) Business Models

The following business models depict the overall business potential of the service:

1. Joint venture with a remote healthcare centre or a doctor in private practice: The local healthcare centre will be equipped with a portable device. A trained employee (medical auxiliary personnel, nurse, etc.) will be responsible for conducting the signal recording and transmit it to a specialized healthcare provider (private hospital, diagnostic centre) for the provision of diagnosis. This scenario concerns the provision of the service to patients in remote areas in cooperation with local healthcare professionals. For example the Greek topology (isolated islands, small villages away from the cities – hospitals – and not easily accessible by healthcare professionals) favors such a service.

2. Public Healthcare Providers: The rationale is the same as above (of the remote healthcare center). The public healthcare provider may as well provide the service in the manner described above. The savings are obvious for the patients, since the relevant price will be covered by the public insurance organizations. For the public healthcare provider introduction of the service results to provision of healthcare services in isolated areas, elevation of the provided services, and impact on the quality of life. The following scenarios have potential:

a. The Public Healthcare Provider operates the service: The remote healthcare centre will be equipped with portable devices. A trained employee (medical auxiliary personnel, nurse, etc.) will be responsible for conducting the signal recording and transmit it to a specialized healthcare provider (private/public hospital, diagnostic centre) for the provision of diagnosis.

b. Provision of the service to subscribed members: It has a potential, but since we are dealing here with public authorities the organisational issues are more complex. For example, with the new form of the Greek NHS, such scenarios are now possible depending on the decision of a local healthcare authority or public insurance company to provide phealth services.

3. Private Insurance Companies: This model addresses insurance organizations that provide healthcare services as well. The insurance premiums in these cases include provision of primary / secondary healthcare services, homecare, etc. The savings for the patient are more obvious under this model. The cost-effectiveness for the service is also more obvious for the insurance company. Resources previously allocated for homecare (i.e. personnel – nurses and doctors – etc) will be reduced.

4. Provision of the service to subscribed customers: A customer wishing to have an active role in monitoring his/her health status in order to enjoy an enhanced feeling of safety and an elevated quality of life. The newest, emerging business model in the market is about selling devices and service subscriptions directly to the general public and set up a more elaborate and modular service provision infrastructure to meet demand. This model foresees that monitoring devices are sold like mobile phones, together with subscriptions to medical call centres that can receive and diagnose a measurement, keep records of all transactions, even give access to the citizen's desired physician to view and diagnose the data and send recommendations through the service.

This is especially relevant in view of the new, "wellness" market that is the fastest growing area of health-related expenditure in recent years. Under such circumstances the target population includes everyone, and not only some chronic patient groups or other clinical conditions.

4 Case Studies

4.1 The Private-Hospital Project

The Private-Hospital project concerns the provision of the Mobinet telemonitoring services to post-surgery patients and patients with cardiovascular diseases by a well-know private hospital in Athens.

During the project pilot operation phase eight (8) patients were equipped with medical kits, each including three telemonitoring devices, an ECG recorder, a blood-pressure monitor and an oxymeter, as well as a mobile phone. Each patient participating in the pilot operation, following his discharge from the hospital, received a complete kit and training on the use of the devices and the mobile application for data transmission. The patients were then responsible for recording their vital data on a daily basis and sending it to the Mobinet web-center for review and consultation by the hospital specialized staff.

The project never proceeded beyond the pilot operation, as the local medical community perceived this novel service as a strong competitor that would reduce their clientele.

It is a complicated structure, where the cardiologists, external to the hospital, refer their patients for operation, but then fear that they will loose them as customers. The introduction of a personalized monitoring system fear that will push the patients closer to the hospital and that will eventually lead to them loosing them. Thus, they stood against it and the Hospital Medical Management made a quick decision to kill the project so as not to put into turmoil the cooperation with the external cardiologists.

Resistance to change is a common issue faced during the introduction of new methods in all industries. In the healthcare sector in particular, it is common that the medical community shows a negative attitude towards eHealth solutions. Although the project team worked towards the development of a network, where the cardiologists would be actively involved in the monitoring of their patients' health status following hospitalization, the project was eventually terminated.

4.2 RHA - Telemonitoring Pilot

The Regional Health Authority - telemonitoring pilot has been designed with the aim to facilitate General Practitioners in completing every-day tasks and providing quality primary healthcare services to citizens. It is being implemented by the health units that operate under the 3rd Regional Healthcare Authority in Greece, covering the Region of Central Macedonia so as to enhance access to specialized healthcare services in remote areas.

The pilot network implementation for the effective health monitoring in remote areas aims at the:

- Provision of advanced healthcare services, regardless of geographical limitations
- Preventive medicine
- Efficient human resources management (for the healthcare providers)
- Scientific personnel facilitation and diffusion of specialized knowledge

The project generates significant social benefits and enables healthcare professionals to allocate their time in an efficient and effective manner, as they are able to manage more patients, since telemonitoring allows the simultaneous monitoring of the health status of multiple patients. Patient management and also, data management for each patient is improved, facilitating medication management and the completion of administrative tasks for the healthcare professionals.

The RHA - telemonitoring pilot concerns interaction between GPs and experts. The 3rd RHM is responsible for the coordination and implementation of health care policies and services in the corresponding geographical region. In the frame of the project implementation, it equipped the health-centers with sets of telemonitoring devices in the five remote areas for the provision of health telematics services.

Early in the project design and planning, several health units expressed interest in participating to the provision of the telemonitoring services in rural areas. Still, those health units operate under different regional healthcare authorities that due to legal and organization complications restricted their participation to the project. As a result the project implementation team faced resistance to change from the selected participants and dealt with it by trying to provide motives, as the provision of anonymous medical data from the project database to the participants, so that the later will be able to present papers in scientific conferences and journals.

4.3 The Telecare Center

The Municipality of Trikala has designed a long-term strategic plan for the transformation of the local society, based on the opportunities created by the information society era. To this end, several e-services are planned and are being implemented. Part of this strategic plan is the establishment of a center offering advance health and social care services to the citizens of the region.

The telecare center constitutes a single entry point to health and social services. Medical intervention and social support is provided to all citizens, eliminating discrimination and other social isolation incidents. The citizens receive personalized health services and enjoy advanced community services at the same time. As a result, the telecare center constitutes an efficient channel for the provision of citizen-centric services, strengthening the role of the community public bodies in the society. Thus, the citizens feel safe and confident that the community takes their needs into serious consideration and respond to them.

The Municipality of Trikala has designed a broad strategic plan regarding the implementation of e-services in several fields of activity. To this end, the Municipality of Trikala implements different projects, guided by the common vision, that is the region to be pioneer in applying advance ICT technologies in everyday activities. As a result, the telecare center project is carefully designed and is being implemented by a committed team who puts effort in order to bring advanced technology closer to all citizens. The success of the project lies on its impact on the quality of lives of the end-users that is the citizens and their carers, while enhancing the social profile of the Municipality.

4.4 Rural Healthcare

A pilot study is to assess the performance of the development of a new telecare service for rural areas of Greece featuring a pilot telemetry network has recently reached its mid-life. The network was established in 2008 in 25 remote and isolated rural municipalities of Greece, 10 of them located in islands. The local primary health services were equipped with vital signs telemonitoring devices. At these points the family physicians record the vital signs of the patients with chronic diseases (cardiovascular and respiratory diseases). The data are transmitted through GPRS to a central webserver. Specialized physicians in Athens consult the recorded tests and provide advisory diagnosis to the local physicians.

A retrospective evaluation study was designed to evaluate the initial 6 months (of a 12 months duration planned) of full operational working period of the telecare network. Evaluation criteria measuring the adoption and the outcomes of the implementation of the specific telecare service were based on the recommendations of the WONCAon ICT to Improve Rural Health Care [4].

In total 777 different tele-consultations took place and 2206 logins in the online patients' health records database, with the level of adoption of the telecare services by the local health professionals in everyday practice to vary significantly.

What the study has concluded up to now can be summarized to that the introduction of telecare services for remote communities cannot automatically be a benefit for rural health workers and the communities that they serve. Ongoing support and commitment from all engaged partners is crucial in order to maximise the potential for successful and sustainable telecare services to rural communities.

5 Interoperability

pHealth interoperability brings a new challenge to healthcare in that interoperability quality needs to be delivered across many systems and devices from a broad range of implementers. This challenge is of a new dimension at a scale and in a market environment where the management of such processes among stakeholders is not yet in place. The aim of the efforts above is on the "what" needs to be done to deliver systems that will "plug and play" according to the specifications following the example of the IT industry.

Recently a number of coordinated efforts are attempting to overcome this barrier, i.e. lack of interoperability. On the one end the industry has formed open consortia and alliances, such as the IHE, COCIR and Continua, to create interoperability profiles for specific use cases based on international standards. On the other end the European Commicion has taken particular steps to encourage cross-border interoperability, the most prominent the Mandate 403/2007 to CEN/CENELEC/ETSI [5] and the Commission recommendation on cross-border interoperability [6].

A Quality Assurance process reinforcing interoperability may need to include some form of specification or labelling to allow for the easy identification by external parties to the implementation that the quality assurance was effectively and satisfactorily performed. Many such schemes involving or not third party testers have been used in the medical industry.

The most prominent labelling initiative in pHealth is the Continua Health Alliance [7], a group of technology, medical device and health and fitness industry players, committed to empowering consumers and patients world wide, to take an active role in their own care through the use of technology. They recently unveiled their first set of guidelines, based on proven connectivity standards that is hoped will help to increase assurance of interoperability between devices, enabling consumers to share information with care givers and service providers more easily. Manufacturers of products that meet these guidelines are permitted to use the Continua Health Alliance certification logo to promote their products. The logo will clearly identify certified products, making it easy for purchasers to choose products that work together seamlessly.

6 The Implementation Context Implications – Concluding Remarks

Recent research advances have made possible a viable solution regarding the provision of personalized health services, seamlessly from the patient point of view.

However, patient satisfaction is no longer an easy goal to achieve. Demographics and socio-economic forces endorse the healthcare industry transformation and modernization. Ease of use, system reliability, availability of the service are some of the critical factors that lead to a successful phealth scheme implementation. The attractiveness of the new systems or method reassures active user involvement, and much more importantly leads to user satisfaction and acceptance.

The most important asset in such phealth attempts is the broadly disciplined and highly skilled human capital, consisting of engineers, healthcare professional and marketers with long-term research experience in the health-telematics field.

Within the commercial context, research outcome should be perceived as the conversion of knowledge and ideas into a benefit, which is for commercial use, while promoting at the same time the public good.

Vidavo in particular aims to bridge the gap between research results and commercial deployment. In theory, the gap is created by the inability of the researchers to timely consider the actual user needs in order to develop products and services addressing those needs. Instead, they provide high technology, transformed in sophisticated products that target highly skilled users, creating new educational and training needs.

In practice, the gap between a successfully working prototype to a successful commercial deployment is widened by the users resistance to change, organizational complexities, financial implications, the sector's dependencies on old fashioned marketing policies, and regulatory vagueness. User attractiveness of the new systems (emerged as the outcome of research activity) is only one attribute that contributes to real implementation success and certainly is not sufficient.

As already discussed in previous paragraphs and depicted in the case studies described, uptaking of phealth is initially limited by the required investment costs. Once the financial obstacles are overcame, phealth uptake is constrained by the organizational complexities of the health sector, the reluctance to change of the health employees, combined with the technophobia of the medical community. Further concerns include the interoperability of the fragmented systems that prevail in the sector and the absence of a concrete legal / regulatory framework.

Still, phealth uptake can be effectively encouraged the strength and qualities of the partnership that attempts to initiate phealth services, the recruitment of the right people, system attractiveness, leading to strong demand, as well as political support and profound socio-economic benefits.

phealth implementation could prove to be a win-win situation for end users, healthcare providers and technology developers in the long-run. The short-term path to implementation still seems sterile, prolonging commercial exploitation life-cycle.

References

[1] Finkelstein, S.M., et al.: Home Telehealth Improves Clinical Outcomes at Lower Cost for Home Healthcare. Telemedicine and e-health 12(2) (2006)

[2] McKnight L - Tufts University – Medford, Massachusetts, Internet Business Models (2000)

[3] Andrews, R.R.: Products for Tomorrow, Medical Design Technology (April 2005)

[4] Meade, B.J., Dunbar, J.A.: A virtual clinic: telemetric assessment and monitoring for rural and remote areas. Rural and Remote Health 4 (online), 296 (July 2004)

[5] EUROPEAN COMMISSION, ENTERPRISE AND INDUSTRY DG, M/403 EN, Mandate to the European Standardisation Organisations CEN, CENELEC and ETSI in the field of Information and Communication Technologies, applied to the domain of eHealth, Brussels (March 6, 2007)

[6] EUROPEAN COMMISSION, RECOMMENDATION of 2nd, on cross-border interoperability of electronic health record systems, Brussels, COM (2008) 3282 final (July 2008)

[7] http://www.continuaalliance.org/home/

Impact of the Transmission Scheme on the Performance in Wireless LANs

Andreas Könsgen, Andreas Timm-Giel, Carmelita Görg[1], and Ronald Böhnke[2]

[1] Communication Networks
[2] Communications Engineering
Center for Computing Technologies (TZI)
University of Bremen
Otto-Hahn-Allee 1, 28359 Bremen, Germany

Abstract. In wireless LANs, different multi-user access methods such as TDMA, OFDMA and SDMA are available which can be used with or without channel knowledge at the transmitter and a single antenna (MISO) or multiple antennas (MIMO) at the receiver. A cross-layer scheduler is considered which can be configured with these different PHY methods as well as with knowledge about application requirements and channel conditions at the MAC layer. The scheduler computes priorities on the MAC layer that are handed over to the physical layer in order to keep quality-of-service constraints such as throughput and delay. In this paper, it is demonstrated that controlling the priorities by a QoS aware resource allocation method allows to meet the requirements by the applications under various channel conditions. MISO-SDMA has a relatively small performance penalty in comparison to MIMO-SDMA which gives the best result. For MIMO-TDMA and -OFDMA, channel knowledge at the PHY layer does not result in essential performance enhancement.

Keywords: Wireless LAN, cross-layer, MIMO.

1 Introduction

Wireless LANs have to meet increasing requirements nowadays and in the future: high data rates for each user, high spectral efficiency in the sense of a high total capacity and meeting several types of QoS requirements for different applications.

Up to now, most protocol stacks are designed according to the OSI model which defines seven layers from the physical layer up to the application layer, with an increasing degree of abstraction from the physical hardware. In legacy protocol stacks, these different protocol layers have been optimised independently of each other. This separation is in particular problematic for the design of the two lowest layers, which are the MAC and the PHY layer, because there are close mutual dependencies between these two layers. The QoS requirements have already to be considered by selecting the physical transmission method. Moreover, the actual channel conditions and the effects of these conditions for

F. Granelli et al. (Eds.): MOBILIGHT 2009, LNICST 13, pp. 303–314, 2009.
© ICST Institute for Computer Sciences, Social-Informatics and Telecommunication Engineering 2009

a QoS aware transmission have to be known when selecting a particular packet for the transmission.

To cope with these requirements, in the framework of the xLAYER project funded by the German Research Foundation (DFG), a cross-layer transmission system for wireless LANs is developed which is located inside the access point resp. base station and has full control of the channel access. The introduced transmission system extends the proposal of the IEEE 802.11n standards draft, where centralised channel access with assignment of user priorities is specified as Hybrid Coordinated Channel Access (HCCA). By means of this transmission system, a comparison of the scheduler performance in case of TDMA, OFDMA and SDMA was given in [6] where a statistical channel model [5] specified by the IEEE 802.11n Task Group was used. With the aim of a more precise modeling of the channel in case of indoor scenarios, a raytracing approach shall now be considered which allows a more realistic simulation of the signal propagation between the base station and the users. The model was investigated along with various physical transmission schemes in [1]. In this paper, the interaction between the newly introduced physical model and the higher-layer scheduler is highlighted. The QoS properties for different users are compared for different MAC scheduling methods along with the different approaches on the PHY layer. In this way, information also can be obtained on how much complexity on the PHY layer such as the requirement of channel knowledge and the number of antennas at the receiver side is required.

2   Cross-Layer Scheduler

The cross-layer scheduler deployed in the simulation includes two stages as shown in Fig. 1: the hardware-independent stage which is located in the MAC layer selects packets based on a certain scheduling strategy. For each user, a separate data flow with an own queue is maintained. The packets for each data flow are assigned a priority value according to the selected scheduling strategy. After the packets have been classified in this way, a list is handed over to the hardware-dependent stage inside the PHY layer. According to different transmission strategies, time slots, OFDM subcarriers or spatial transmission paths are assigned to the users according to the given priorities. For each user, the channel matrix is available for each OFDM subcarrier at regular sampling intervals. From the channel matrices and the user priorities, the allocation of the users to the channel resources and the resulting capacities are calculated. To do so, the physical layer scheduler maximises the weighted sum rate according to the priorities given by the MAC layer scheduler.

In each turn of the scheduler, at first, one packet is taken from each user provided that data is ready for transmission in the respective queue. For the user with the longest transmission time which results from the packet length and the available capacity, one packet is selected. The other users with shorter transmission times fill in the gap with further packets as far as possible. If finally the remaining gap is too small to transmit a complete packet, then only a part

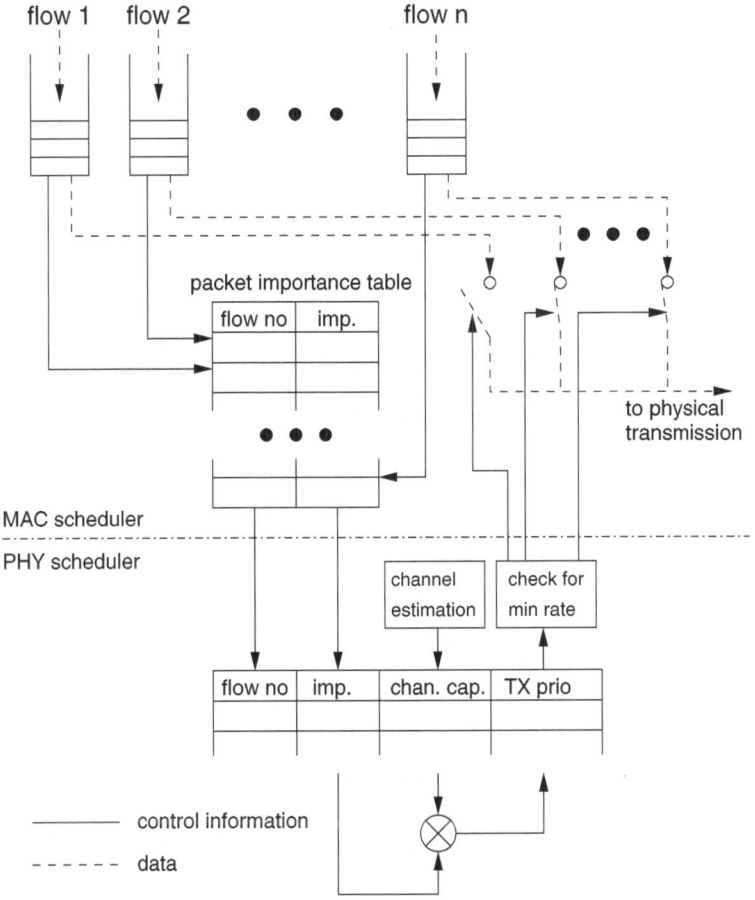

Fig. 1. Design of the parallelised cross-layer scheduler

of the packet is sent which fits into the gap and the remaining data is added to the next packet which is waiting in the queue. In this way, the available airtime is used whenever possible.

Two scheduling strategies are considered: in case of the modified Round Robin (RR) strategy, the MAC scheduler is unaware of queue states. In each turn of the scheduler, it assigns priorities in a linearly decreasing way as shown in table 1.

The quality-of-service (QoS) scheduler sketched in Fig. 2 compares the achieved throughput and the packet age against target values, where the target value $S_{u,\text{tar}}$ for the throughput is the amount of data transmitted by user u within a sliding time window; in order to control the delay, the packet expiry time $t_{u,\text{tar}}$ for the next packet that is at the top of the queue for user u is used. With $S_{u,\text{cur}}$ being the current throughput of a user and t_{cur} being the current system model time, The differences $S_{u,\text{tar}} - S_{u,\text{cur}}$ and $t_{u,\text{tar}} - t_{\text{cur}}$ are adaptively weighted and converted into a priority w_u for each user u as described in detail in [7].

Table 1. Priority assignment for modified Round Robin scheduling

	time slot						
	1	2	3	...	8	9	...
user 1	7	6	5	...	0	7	...
user 2	6	5	4	...	2	1	...
...				...			
user 8	0	7	6	...	4	3	...

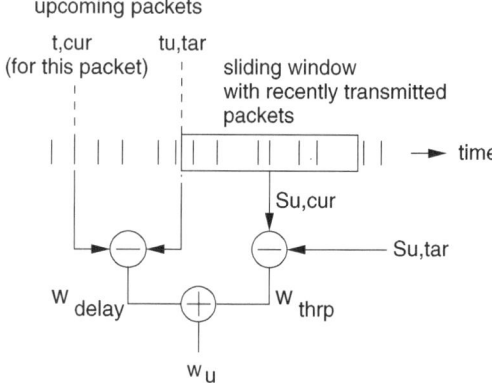

Fig. 2. Principle of the QoS aware scheduler

In this paper, only the downlink from the access point to the mobile stations is considered for the transmission of user data; in the uplink direction, only acknowledgement packets are transmitted.

3   Physical Transmission Methods

SISO (Single Input Single Output): In this case, the different locations of the users have little effect because both the access point and the mobile terminals only have one antenna. In this paper, this case is not furthermore considered.

MISO (Multiple Input Single Output): In the case of MISO, the access point exploits the different spatial locations between the users, however the mobile terminals do not. Users located at different positions however can be separated by the access point so that the sum rate over all users can be increased. The per-user rate is expected to remain unchanged.

If TDMA or OFDMA is used together with a MISO transmission, further cases can be distinguished regarding the usage of spatial diversity. *Space-Time Block Codes* (STBC) [8] improve the reliability of the transmission because they allow to combine the signals received at different antennas. In this way, it is likely that at least at one antenna a good signal is available. *Beamforming* (BF) [10] allows to focus the transmitted signal onto particular users during the transmission of a packet which uses the transmit power more efficiently. When combining MISO with SDMA, Dirty Paper Coding (DPC) can be deployed where users with higher priority are separated by "pre-subtracting" interference caused by signals for users with lower priority. For a sufficient number of users, the number of receive antennas becomes irrelevant [4] so that DPC then also can be deployed by using a single antenna at the user. In section 5 these anticipated effects of the different transmission methods are evaluated more closely by simulations.

MIMO (Multiple Input Multiple Output): In this case, both the access point and the mobile stations have more than one antenna. When combining with TDMA or OFDMA, successive interference cancellation (SIC) [3] is used if no channel knowledge is available at the AP or Singular Value Decomposition (SVD) in case that channel knowledge is perfect. In case of SDMA, the method to separate the users is DPC as already described in the section about MISO.

Considering an access point with N_T transmit antennas and mobile stations with N_R receive antennas each, the channel matrix $\mathbf{H}_u[m]$ of subcarrier m for user u has the size $N_R \times N_T$. The transmitter is described by a covariance matrix $\mathbf{\Phi}_{\mathbf{x}_u}[m]$ for each subcarrier m. Furthermore, the model considers an effective noise vector $\tilde{\mathbf{n}}_u[m]$ with the covariance matrix $\mathbf{\Phi}_{\tilde{\mathbf{n}}_u}[m]$ including additive white Gaussian noise and interference. The data rate $R_{u[m]}$ for user u and subcarrier m can then be calculated as [9]

$$R_u[m] = \log_2 \det(\mathbf{I} + \mathbf{\Phi}_{\tilde{\mathbf{n}}_u}^{-1}[m]\mathbf{H}_u[m]\mathbf{\Phi}_{\mathbf{x}_u}[m]\mathbf{H}_u^H[m]). \tag{1}$$

4   Simulation Setup

An indoor scenario is considered where $N_U = 8$ users are located in a room with the size $8\,\text{m} \times 6\,\text{m} \times 3\,\text{m}$ as shown in Fig. 4, where each of them is equipped with a mobile station which remains at a fixed place. Slight movements of the users are simulated by clusters of scatterers which slowly orbit around a center in front of the mobile stations. An access point with N_T transmit antennas is mounted at the ceiling. The access point keeps a separate queue for each user resp. data flow with a length of 50 packets. Each of the mobile stations has $N_R = 1$ receive antenna in case of MISO and $N_R = 4$ antennas in case of MIMO. An OFDM transmission system is used with $N_M = 32$ subcarriers working at a carrier frequency f_C of 5.2 GHz with a channel bandwidth of 40 MHz. Reflections on the walls and on the scatterers are simulated by a simplified raytracing model as specified in [2] which considers reflections up to second order. The signal-to-noise-ratio at the receiver averaged over all users, subcarriers and time samples is set to 20 dB.

An example for the channel characteristics is given in Fig. 3 which shows the absolute value of the channel matrices $\mathbf{H}$ as a function of the sample k and the subcarrier m for a particular user.

The channel coefficients which are the basis for the calculations of the physical-layer scheduler are updated every 4 ms.

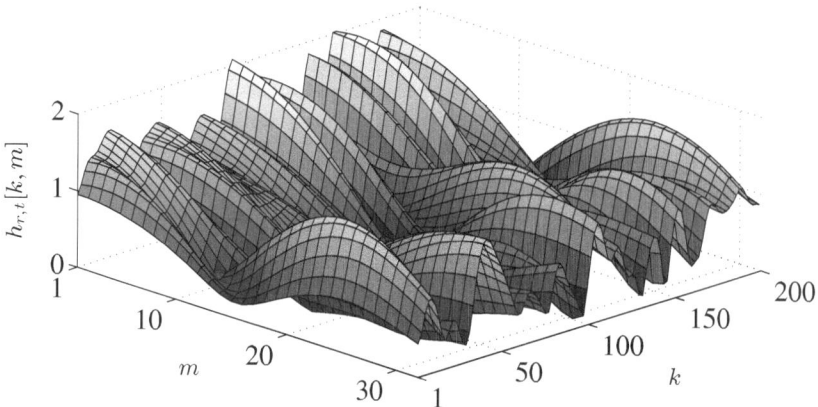

Fig. 3. Example for the characteristics of the channel model with m subcarriers for $k = 200$ time slots

The users 1, 2 and 3 have time-critical flows with delay constraints of 15 ms each; the other users have best-effort flows with a given data rate. The packet size is set to 40000 bytes for all users. The loads for the non-time-critical flows are set to values above the available channel capacities so that the system is always in saturation to allow a comparison between the different transmission schemes. The simulated model time is 10 seconds.

The simulator used for the investigations discussed in this paper is called WARP2; it implements the IEEE 802.11 protocol stack and has been extended with the two-stage MAC/PHY scheduler as described above.

5   Simulation Results

Fig. 5 shows the total throughput as well as the per-route throughputs which are achieved for the different transmission schemes where "dumb" resp. "smart" means that the transmission scheme works without resp. with channel knowledge. As expected, SDMA achieves the best performance, followed by MIMO-OFDMA and MIMO-TDMA. A notable fact is that the presence or absence of channel knowledge does not enhance the total throughput significantly in all cases.

The QoS scheduler keeps the requirements of the time-critical flows except for dumb MISO-TDMA and dumb MISO-OFDMA where a channel with a small total capacity has to be shared which results in failure of serving even the time-critical users. In this case, a call admission control would be needed to disconnect users whose requirements cannot be met, which will be future work. The remaining airtime is distributed among the non-time-critical flows. In most cases, the users are served in a fair manner in the sense that they get an amount of channel capacity which is proportional to their offered load. In certain situations, however, channel conditions prevent a fair distribution as it can for example be seen

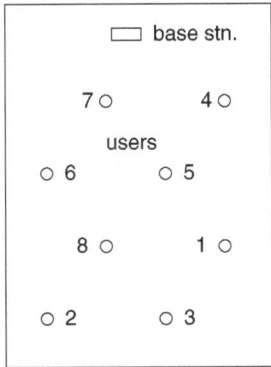

Fig. 4. Arrangement of the users inside the scenario

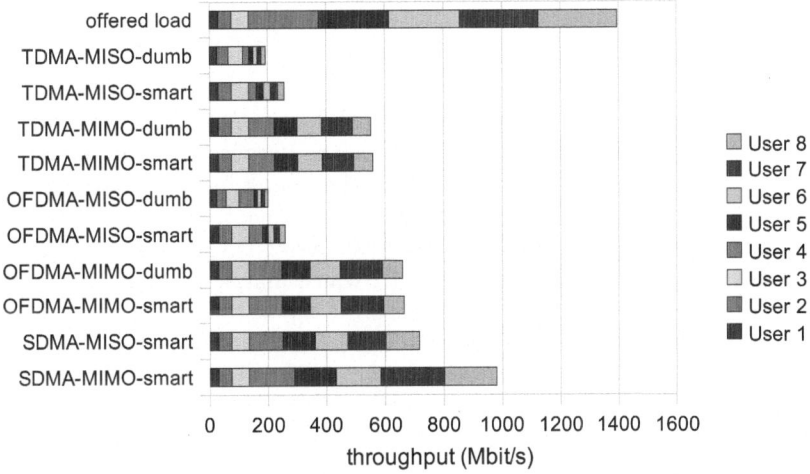

Fig. 5. Throughput for the different transmission methods, QoS scheduler

for user 8 who is served less than the other non-time-critical users in the TDMA
or OFDMA case. With SDMA, channel resources are sufficient so that the user
can be correctly served as well.

Except for the MIMO-SDMA scenarios, the achieved throughput for the non-
time-critical flows is significantly smaller than the offered traffic load. This means
that the system is in saturation; the respective queues are filled up to the maximum limit. Any packets which exceed the queueing capacity are dropped.

For MIMO-TDMA and MIMO-OFDMA the results for dumb and smart transmission are almost the same whereas there is a notable increase for MISO-
TDMA/-OFDMA. The reason is that in case of MIMO, the user gets a better average channel with less variation due to the antenna diversity which then

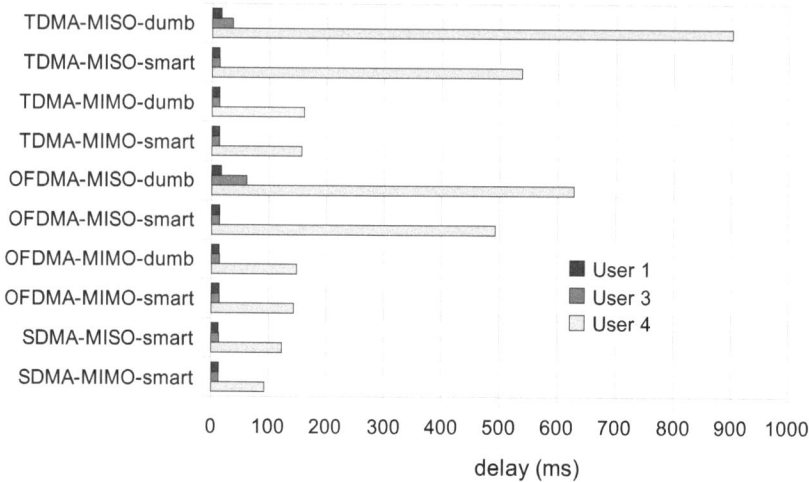

Fig. 6. Delay for the different transmission methods, QoS scheduler

cannot be enhanced much more by channel knowledge. Therefore, MIMO-TDMA/OFDMA is attractive for system designs because without the requirement for channel knowledge less feedback in the uplink from the receiver to the sender is needed. For maximum performance requirements, MIMO-SDMA is the method of choice, however also MISO-SDMA gives good results which are still higher than for MIMO-TDMA or -OFDMA which makes it attractive for small devices without space for multiple antennas.

The delays achieved by the above scenario are illustrated in Fig. 6. For overview reasons, the delay is shown only for three out of the eight users, two users with a time-critical flow and one user with a non-time-critical flow. The QoS requirements for the time-critical users are kept independent of the total capacity of the transmission system, except for dumb MISO-TDMA or -OFDMA where the total channel capacity is insufficient so that the users cannot be correctly served. The delays for the non-time-critical flows are high because the queues run full in this case. With increasing total capacity, the delays are reduced because due to the higher service rate, the dwell time of the packets inside the queue decreases.

An example for the behaviour of the queues during the time progress is given in Fig. 7 for the QoS scheduler in case of dumb MIMO-OFDMA. where the queue lengths are shown for three users for the first second of model time. User 1 and 3 get real-time service so that the queue lengths are kept at a maximum of 3 in order to avoid exceeding the maximum delay due to queueing. User 4 has a higher load and is a best-effort user, so that the offered load cannot be fully served and the queue are filled quickly up to the maximum length. The limited queue size also limits the delay of the queued packets, however packet loss will occur.

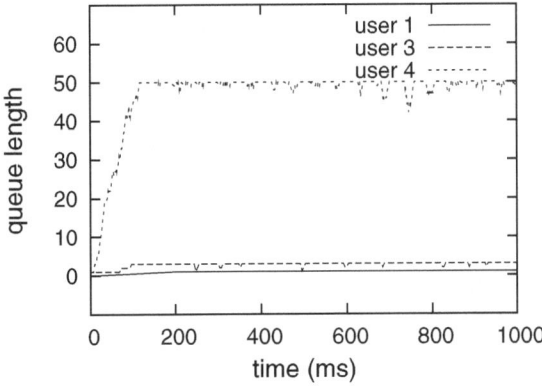

Fig. 7. Queue lengths for three users, QoS scheduler

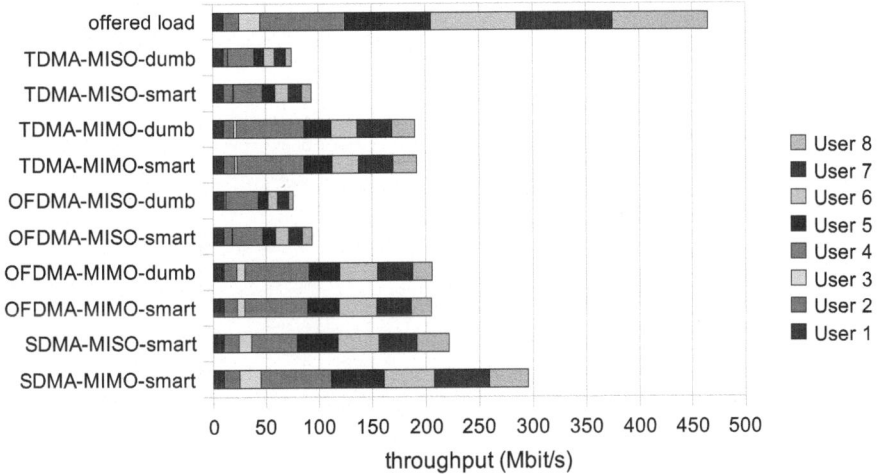

Fig. 8. Throughput for the different transmission methods, RR scheduler

For comparison, the same simulation setup has been run with RR as the MAC scheduling strategy. Fig. 8 shows the results for the achieved throughput. The total capacities in case of the different transmission schemes are similar to those achieved in the case of the QoS scheduler, however the allocation of troughput to individual users is largely different. The QoS requirements for the time-critical users cannot be kept except in the cases that the total available throughput is high. Despite of the regular assignment of priorities, even users with equal offered traffic load achieve significantly different throughputs due to different channel conditions: a user who is located at a distant position from the access point in average experiences a reduced channel quality in comparison to a user who is closer. User 4 is always assigned a disproportionately high capacity,

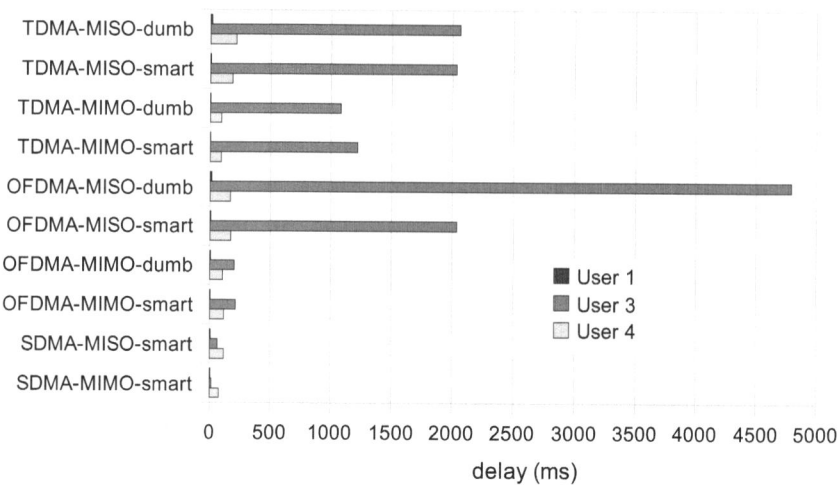

Fig. 9. Delay for the different transmission methods, RR scheduler

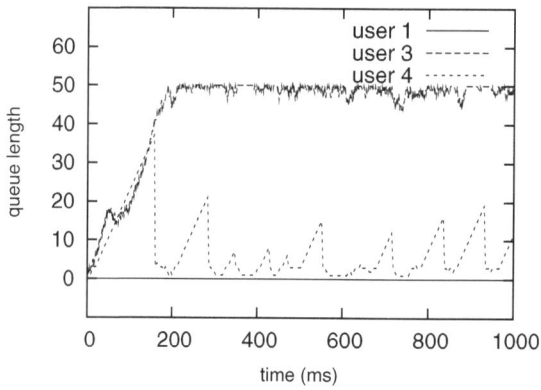

Fig. 10. Queue lengths for three users, RR scheduler

whereas the QoS scheduler discussed before reduces the service of user 4 to some extent in order to provide more capacity to other users. The delay results in Fig. 9 correspond to the throughput measurements. The QoS requirements for the time-critical users are are only met if the total capacity of the system is sufficiently high so that the system works in a best-effort manner.

Fig. 10 shows the temporal queueing behaviour of the users 1, 3 and 4. User 1 has a relatively small load so that the RR scheduler serves him frequently enough to keep the queue at zero size, i. e. no packet is backlogged, each packet is immediately prepared for transmission. User 3 has a load which is higher than the service rate, the queue is quickly growing. Since the user is time-critical, the packets expire so that the queue is flushed and the packets discarded in certain time intervals. User 4 has a higher load, but is not time-critical so that the packets are not subject to ageing. The queue rapidly grows to the maximum; any extra packets beyond the queue length are dropped.

6   Conclusion and Outlook

The performance of the cross-layer scheduler has been tested along with different MAC scheduling methods including and without knowledge about user requirements and queue states; different transmission methods with and without channel knowledge were investigated as well as transmission with single or multiple antennas at the receiver. With the QoS scheduler on the MAC layer, performance requirements can be kept for time-critical flows under a wide range of total system capacities. The remaining capacity is allocated to non-time-critical flows in a fair manner in the sense that each flow approximately gets the same portion of the available resources. The comparison with the RR scheduler shows that users with the same offered traffic load achieve different throughputs due to different long-term channel conditions; by the help of QoS scheduling, this problem can be balanced by assigning the priorities in a suitable way. MIMO-TDMA or MIMO-OFDMA transmissions show a relatively small speed penalty when transmissions with or without channel knowledge are compared so that effort of providing and processing channel state information can be saved. The speed penalty of MISO-SDMA vs. MIMO-SDMA is relatively small so that the scheme is suitable for small devices without space for multiple antennas.

References

1. Böhnke, R., Kammeyer, K.-D., Könsgen, A., Görg, C.: Smart MISO vs. Dumb MIMO for Cross-Layer Scheduling in Indoor Environments. In: Submitted to Proc. Second Int. Workshop on Cross-Layer Design (IWCLD), Mallorca, Spain (2009)
2. del Galdo, G.: Geometry-based Channel Modeling for Multi-User MIMO Systems and Applications. PhD thesis, Ilmenau University of Technology (2007)
3. Foschini, G.J.: Layered Space-Time Architecture for Wireless Communication in a Fading Environment when Using Multiple Antennas. Bell Labs Technical Journal 1(2) (1996)
4. Jindal, N., Goldsmith, A.: Dirty Paper Coding vs. TDMA for MIMO Broadcast Channels. IEEE Transactions on Information Theory 51(5) (May 2005)
5. Kermoal, J.P., Schumacher, L., Pedersen, K.I., Mogensen, P.E., Frederiksen, F.: A Stochastic MIMO Radio Channel Model with Experimental Validation. IEEE Journal on Selected Areas in Communications. Work supported by IST project I-METRA IST-2000-30148 20(6) (2002)
6. Könsgen, A., Herdt, W., Wang, H., Timm-Giel, A., Böhnke, R., Görg, C.: A Two-Stage QoS Aware Scheduler for Wireless LANs Based on MIMO-OFDMA-SDMA Transmission. In: Proc. First Int. Workshop on Cross-Layer Design (IWCLD), Jinan, China (2007)

7. Könsgen, A., Herdt, W., Wang, H., Timm-Giel, A., Görg, C.: An Enhanced Cross-Layer Two-Stage Scheduler for Wireless LANs. In: Proc. PIMRC, Athens, Greece (2007)
8. Tarokh, V., Jafarkhani, H., Calderbank, A.R.: Space-time block codes from orthogonal designs. IEEE Transactions on Information Theory 45(5) (July 1999)
9. Telatar, E.: Capacity of Multi-antenna Gaussian Channels. European Transactions on Telecommunications 10(6) (November-December 2000)
10. Van Veen, B.D., Buckley, K.M.: Beamforming: a versatile approach to spatial filtering. IEEE ASSP Magazine 5(2) (April 1988)

Increasing the Performance of OFDM-OQAM Communication Systems through Smart Antennas Processing

Nizar Zorba[1], Stephan Pfletschinger[2], and Faouzi Bader[2]

[1] University of Jordan
EE department
11942 Amman, Jordan
n.zorba@ju.edu.jo
[2] Centre Tecnòlogic de Telecomunicacions de Catalunya
Av. Canal Olímpic S/N
08860-Castelldefels, Barcelona, Spain
{faouzi.bader,stephan.pfletschinger}@cttc.es

Abstract. A novel filter bank based multicarrier (FBMC) transmission scheme is proposed, where the transmit antennas are employed to substantially reduce the inherent inter-carrier and inter-symbol interference. Since FBMC systems do not apply a guard interval, they can achieve higher spectral efficiencies than OFDM systems, although at the cost of additional inter-symbol interference (ISI). In this paper, we present a method which reduces the number of interference terms by employing a multiantenna precoding scheme based on spatial diversity, and the system can benefit from the multiuser gain, through an opportunistic scheduler at the transmitter side.

Keywords: OFDM, OFDM-OQAM, Multiple Antennas, Filterbank, MIMO.

1 Introduction

Very high communication rates in wireless systems can be achieved by multicarrier techniques, that can be further combined with the Multiple-Input-Multiple-Output (MIMO) technology to provide both efficiency and Quality of Service to the system. The channel in these broadband systems is typically frequency selective and one of the best multicarrier techniques that can be jointly used with MIMO is the Orthogonal Frequency Division Multiplexing (OFDM), that converts the frequency selective channel into a set of parallel frequency-flat channels. Such great characteristic made OFDM to be included in several communication standards as the IEEE 802.11n WLAN standard, while its OFDM Access (OFDMA) version is considered for the IEEE 802.16e WiMAX standard.

However, OFDM has a number of drawbacks that decrease its efficiency. One of these drawbacks is the need of a Cyclic Prefix (CP) to deal with the channel

F. Granelli et al. (Eds.): MOBILIGHT 2009, LNICST 13, pp. 315–324, 2009.
© ICST Institute for Computer Sciences, Social-Informatics and Telecommunication Engineering 2009

impulse response, which leads to an efficiency decrease of 10% - 20% that represents a huge amount of misuse in the invested resources [1]. It also requires a block processing to maintain orthogonality among all the carriers, which is a serious handicap for scalability, as it is impossible to introduce, in a block of carriers, one or several signals that are not synchronous with the rest of the block. Keeping in mind the heterogeneous nature of modern communications with users running different applications characterized by various rates, initialization times and QoS demands; OFDM can create a problem of synchronization over the system. Clearly, OFDM is already implemented in a lot of communication standards and it is attractive due to its low complexity, and it is now familiar to both the academia and industry; but to further increase the system efficiency, further research is developed in the communications arena to find alternative multicarrier schemes.

One of most promising proposals is the Filter Bank based Multicarrier (FBMC) transmission [2], that shows both enhanced performance and operational flexibility by exploiting the spectral efficiency of filter banks and the independence of the subchannels. While in OFDM, the subcarrier spectra have a strong overlap with adjacent subcarriers, in FBMC the transmission channel is divided into subchannels, providing a control over the allocation process, together with the scalability advantage. FBMC benefits from the OFDM advantages and combines them with the Offset Quadrature Amplitude Modulation (OFDM-OQAM), where no CP is needed, achieving higher spectral efficiency than the classical CP-OFDM as all the system resources are devoted to increase the whole system throughput.

As OFDM-OQAM (i.e. FBMC) does not use the CP, then the main complex task in this technique resides in the combat of the InterSymbol Interference (ISI) and the InterCarrier Interference (ICI), where these tasks are usually performed by the receiver through some complex operations, that have handicapped its implementation. The study of OFDM-OQAM was initially proposed more than 25 years ago [3], and its complexity was the main drawback behind the consideration of FBMC in realistic systems. But current processing capabilities at both the transmitter and the receiver make the objections to the FBMC approach to be unfounded. And recently, an increasing interest in FBMC has again emerged [2][4][5].

With the implementation of MIMO in almost all commercial standards, the system designer has an additional resource that can be employed to cancel the interference terms, and therefore to decrease the complexity related to FBMC schemes. Moreover, the availability of multiple users in the system is beneficial to enable the transmitter to select the user with the best channel conditions at each time, and by this way to increase its sum rate. This scheme is known as the Opportunistic scheduler [6], which has been commercially introduced in the UMTS-HSDPA standard.

Therefore, the objective of this paper is to propose a spatial diversity scheme through MIMO to cancel the ISI and ICI in the system, so that the implementation of the FBMC technique can be possible. In other words, MIMO will be in charge of the required interference mitigation in the system. Besides that, the system multiuser gain [6][7] is employed to increase the rate behaviour. To the best of the authors' knowledge, no such scheme is previously proposed in literature, so that the result of our work will be a communication technique that provides all the advantages of MIMO and FBMC with a complexity that enables for its consideration in practical systems.

2   System Model

We focus on the Downlink channel where V receivers, each one of them equipped with a single receiving antenna, are being served by a transmitter at the Base Station (BS) provided with n_t transmitting antennas. The case of $n_t = 2$ is considered along the paper for easiness in the results presentation and to align with all commercial implementations of the IEEE 802.11 *pre-n* and the proposals for all LTE systems, where its upgrade to any number n_t is straightforward. A wireless multiantenna channel $\mathbf{h}_{[1 \times n_t]} = [h_1(t) \ h_2(t)]$ is considered between the transmitter and each one of the users, where a quasi-static block fading model is assumed, which keeps constant through the coherence time, and independently changes between consecutive time intervals with independent and identically distributed (i.i.d.) complex Gaussian entries $\sim \mathcal{CN}(0, 1)$. Let $\mathbf{x}(t) = [x_1(t) \ x_2(t)]^T$ be the $n_t \times 1$ transmitted vector, while denote $r_v(t)$ as the received signal at the vth receiver as

$$r_v(t) = \mathbf{h}_v(t)\mathbf{x}(t) + z_v(t) = h_{1(v)}(t)x_1(t) + h_{2(v)}(t)x_2(t) + z_v(t) \qquad (1)$$

where $z_v(t)$ is an additive Gaussian complex noise component with zero mean and a variance of σ^2. The transmitted signal $\mathbf{x}(t)$ is a coded version of the i.i.d. data symbols $s_i(t)$ with $E\{|s_i|^2\} = 1$. For ease of notation, both the user and time indexes are dropped whenever possible.

3   Opportunistic Transmission

One of the main transmission techniques in multiuser scenarios is the opportunistic technique [6][7], where during the acquisition step, a known training sequence is transmitted for all the users in the system, and each one of the users calculates the received SNR, and feeds it back to the BS. The BS scheduler chooses the user with the largest SNR value for transmission to benefit from its current channel situation, and therefore improving the global system performance. This opportunistic strategy is proved to be optimal [6][7] as it obtains the maximum rate point.

4   OFDM-OQAM

In conventional OFDM systems each carrier from a total of M carriers is modulated using QAM, where a rectangular window is employed to shape each QAM

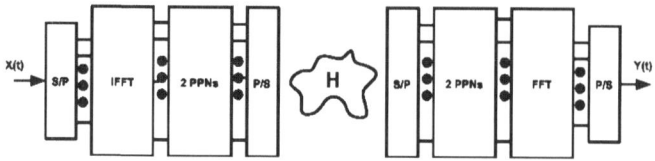

Fig. 1. System model for FBMC

symbol. To avoid ISI in the system a CP is used, so that thanks to the CP, no complex channel equalizers are needed at the receiver side, thus decreasing the system complexity, but at expenses of a lower system efficiency due to the 10% - 20% of resources that are employed for CP.

OFDM-OQAM is an alternative modulation that introduces an efficient pulse shaping in its modulation scheme through the use of accurately non-rectangular pulse shaping and thus, it generates less out-of-band radiation and provides better frequency localization. Therefore it can be employed without the need of a CP, so that it overcomes the loss of resources of conventional OFDM. But to employ non-rectangular pulse shaping, the OQAM is needed, where this modulation introduces through each of the carriers, a time offset between the real part and the imaginary part of the sent symbols. Removing the CP increases the system performance, but it requires for an alternative processing to remove ISI, which together with the accurate pulse shaping requirement, drives some complexity in the system that has been considered excessive for its implementation.

But thanks to current receivers processing capabilities, the complexity increase is affordable, so that along the complexity-performance tradeoff, the optimization goes to the performance side, which motivates the recent large interest in FBMC systems in realistic systems [2][4]. Remember that such employment of OFDM-OQAM also enables the system administrator to benefit from other advantages, mainly in terms of scalability and synchronization.

Fig. 1 shows a schematic of the FBMC system where two PolyPhase Networks (PPN), with a total of $N = 2 \times M$ subchannels, are included in the processing at both the transmitter and the receiver. The PPN is composed of a set of filters that are obtained by shifting the response of a low pass filter on the frequency axis; where the low pass filter is named the Prototype Filter in the research community [5][8]. As previously commented, the OQAM accomplishes a signal

Table 1. Example of OQAM symbols mapping to the PPNs

subchannel\time	t	t+1	t+2	...
SubChn.1 (PPN_1)	Re. part	Img. part	Re. part	...
SubChn.2 (PPN_2)	Img. part	Re. part	Img. part	...
SubChn.3 (PPN_1)	Re. part	Img. part	Re. part	...
...	...	...	...	...

separation between the real and the imaginary parts of the signal and performs an offset on them, so that 2 PPNs are employed in the transmission process, one for transmitting the real part of the symbol and the other one for the imaginary part, and performing a continuous switch between their roles, thus making the transmitted data along the PPNs to be as shown in Table 1. A main characteristic of OFDM-OQAM is that the data are transmitted as real numbers at twice the conventional Nyquist rate associated with the prototype filter. Therefore, the number of PPN subchannels (i.e branches) are twice that of OFDM carriers to separately account for both the real and imaginary parts of the symbol. The output of each two adjacent subchannels (over one time instant) are added and transmitted over a single OFDM carrier.

However, large values of ICI and ISI are generated in the system, and some processing at the transmitter and/or the receiver side must be accomplished to mitigate them. In addition, two neighbouring subchannels overlap, in order to fully exploit the available frequency spectrum. The consequence is an interference pattern between the subchannels as follows

Table 2. Example of an OFDM-OQAM interference pattern

	t-1	t	t+1
SubChn.1	Interference	Interference	Interference
SubChn.2	Interference	Desired signal	Interference
SubChn.3	Interference	Interference	Interference

As just commented, in the current State of the Art related to OFDM-OQAM (only with a single antenna), a symbol s is decomposed into its real part $d_1 = Re\{s\}$ and imaginary part $d_2 = Im\{s\}$, so that two adjacent subchannels are employed for its transmission in the same time instant, as follows:

Table 3. OFDM-OQAM setup in a single antenna scenario

	subchannel 1	subchannel 2
$antenna_1$	d_1 on PPN_1	d_2 on PPN_2

In the following time instant the order is reversed, so that the real part of the next symbol is transmitted in subchannel 2 and its imaginary part in the subchannel 1, to comply with the Offset philosophy.

To the best of the authors' knowledge, such interference mitigation is only proposed through some processing over different time samples [2], which beyond the introduced time delay, it does not show attractive results. In [2], an initial study of two transmitting antennas is performed, but the final result is still time dependant. In the current paper, we will propose some spatial interference mitigation that is jointly performed with a selection of the most appropriate user through the Opportunistic scheduler, as now shown in the next section.

5   Spatial Diversity in OFDM-OQAM

One of the drawbacks for the consideration of OFDM-OQAM in current commercial systems is the generated interference in the system due to the non-employment of the CP. On the other hand, the MIMO technology is already available in almost all OFDM-based wireless standards (e.g. IEEE 802.11n and IEEE 802.16e). Joining the two factors, notice that MIMO can be employed to carry out some interference mitigation in the system, which has its challenges when applied to the OFDM-OQAM interference pattern in Table 2, but such interference mitigation stands as one of the main milestones to make the OFDM-OQAM to be attractive, and to reduce its complexity as it avoids extra interference cancellation mechanisms at the receiver side, that use to be very complex.

Thanks to the consideration of MIMO, two simultaneous symbols can travel in the channel at the same time and through the same subchannel, so that a possible setup for the OFDM-OQAM transmission over the subchannels and on a single time instant is shown in Table 4, where the PPN order is switched over the two antennas.

Table 4. OFDM-OQAM setup in a two antennas scenario

	subchannel 1	subchannel 2
$antenna_1$	d_1 on PPN_1	d_2 on PPN_2
$antenna_2$	d_1 on PPN_2	d_2 on PPN_1

The two antennas are employed to provide the system with a Space Time scheme to enable interference mitigation at the receiver side through some signal processing. It is worth noting that with this approach, the second antenna is employed to transmit the same information as in the first antenna, implementing the same principle as the very well-known Alamouti scheme [9], but with some modification to enable its application to the OFDM-OQAM technology with all the challenges behind its consideration, mainly the large amount of generated interference in OFDM-OQAM.

Considering this setup, the received signal r_1 in the first subchannel states as

$$r_1 = (d_1 + jf_1)h_1 + (d_1 + jf_1)h_2 + z_1 \qquad (2)$$

where z_1 is the noise term received in the subchannel 1, and f_1 accounts for all the interference components [8] that arise from the filterbank usage at the first subchannel, where as shown in Table 2 this interference comes from the two adjacent subchannels and time instants. On the other hand, the received signal in the second subchannel is as

$$r_2 = (jd_2 + f_2)h_1 + (jd_2 + f_2)h_2 + z_2 \qquad (3)$$

with f_2 as the interference terms in the second subchannel.

The reader can wonder that if h_1 and h_2 are equal in magnitude and opposed in phase, then no signal will reach the receiver, but this is a hypothetical case with negligible probability. Even that, this case fails in the system outage consideration, exactly as the Alamouti scheme does [9].

5.1   Receiver Processing

The receiver now has two different arriving signals r_1 and r_2, one on each subchannel. Notice that two antennas operating on two subchannels (i.e. one single OFDM carrier) and on one time instant are employed to transmit a whole symbol (i.e. both its real and imaginary parts), then a full diversity rate [9] is obtained. The antennas are efficiently employed in the system with their diversity gain to help mitigating the generated OFDM-OQAM interference, thus decreasing the OFDM-OQAM complexity.

At the receiver side, some processing can be accomplished to obtain the following expression from r_1, as follows

$$y_1 = Re\{h_1^* r_1 + h_2 r_1^*\} = \left(|h_1|^2 + 2Re(h_1^* h_2) + |h_2|^2\right) d_1 + Re(h_1^* z_1) + Re(h_2 z_1^*) \quad (4)$$

where we can see that all the interference terms f_1 are removed thanks to the receiver processing. This is actually a great step for OFDM-OQAM as 8 interfering terms have disappeared. The price for that is some dependence on the channel phase due to the $2Re(h_1^* h_2)$ term, that can show positive and negative values depending on the instantaneous channel conditions of both h_1 and h_2. Notice that the information in d_1 is received with a great spatial antenna gain as it benefits from both h_1 and h_2. Moreover, the data component d_1 is received without any other data components, so that with the simple Matched Filter (MF) receiver, the data can be efficiently extracted. Obviously, this single equation is enough for the detection of d_1 but we still need another one for d_2. Remind that d_1 and d_2 are the real and imaginary parts of the same symbol, so that the symbol is correctly received only if its both parts are properly detected.

Thus we need for an additional equation for d_2, and applying a different processing for subchannel 2 at the receiver side, we get the following expression

$$y_2 = Im\{h_1^* r_2 + h_2^* r_2\} = \left(|h_1|^2 + 2Re(h_1^* h_2) + |h_2|^2\right) d_2 + Im(h_1^* z_2) + Im(h_2^* z_2) \quad (5)$$

where we also notice that there is not any interference term in the equation.

From the previous two equations, the detection of d_1 and d_2 seems to be solved as no more OFDM/OQAM interfering terms are shown in the equations. The problem that remains to be solved is the channel phase effect due to the $Re(h_1^* h_2)$ term. Notice that the channel phase effect can be positive or negative, where the receiver is interested in a positive value for the channels phase effect, so that the decoding process is improved.

To increase the performance of any wireless communication system, the Multiuser gain has to be taken into consideration [6], so that the system administrator can benefit from the channel conditions of the available users in the system to select the user with the best channel conditions. As OFDM/OQAM is targeted to high data rate systems, then it seems straightforward to tackle the opportunistic scheduling in its operation. This objective can be accomplished if we define the user with the best channel conditions as the one who shows a positive and high value for the $Re(h_1^*h_2)$ term, so that its selection enables the following condition

$$\left(|h_1|^2 + 2Re(h_1^*h_2) + |h_2|^2 \right) > \left(|h_1|^2 + |h_2|^2 \right) \tag{6}$$

to guarantee that the phase channel effect is always beneficial to the system performance. Therefore, the opportunistic scheduling is looking for scheduling the user showing

$$\max_{v=1:V} \left(|h_{1(v)}|^2 + 2Re(h_{1(v)}^*h_{2(v)}) + |h_{2(v)}|^2 \right) \tag{7}$$

which is later shown to offer higher multiuser gain than the standard Alamouti scheme. The latter can be also operated with an opportunistic scheduler, where the selected user will be the one showing

$$\max_{v=1:V} \left(|h_{1(v)}|^2 + |h_{2(v)}|^2 \right) \tag{8}$$

where the Alamouti scheme is shown [10] to highly benefit from the multiuser gain.

6   Simulations

The performance of the studied scheme is presented by Monte Carlo simulations, where the objective is to see the sum rate behaviour of the proposed scheme. We

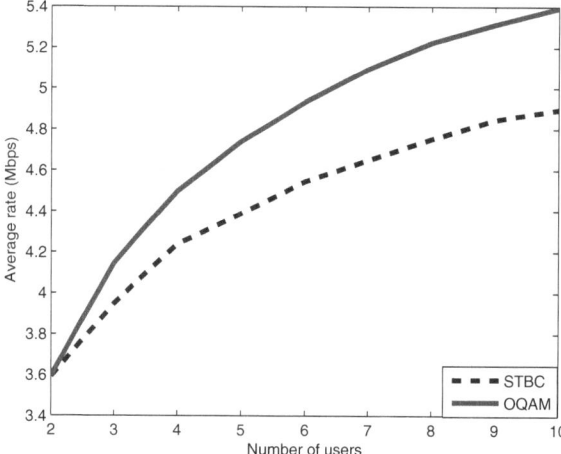

Fig. 2. Rate performance of Classical OFDM and OFDM-OQAM, both operated in the spatial diversity philosophy

consider a wireless scenario with $n_t = 2$ transmitting antennas, and a variable number of users each one equipped with a single-antenna. The transmitter runs a spatial diversity scheme over OFDM-OQAM, where a total transmitted power $P_t = 1$ is assumed with noise power $\sigma^2 = 1$. A total system bandwidth of $20MHz$ is considered in this scenario.

In Fig. 2, a scenario with a variable number of users is simulated, where the BS carries out the scheduling of the user with the best channel conditions following the selection algorithms in section 5.1. The comparison between the classical Alamouti scheme and our proposed multiuser OFDM/OQAM shows the better performance of or proposal, showing a higher benefit from the multiuser system capabilities. Remind that the OFDM-OQAM system operates without CP, then an additional 10%-20% gain is presented in the system and it is not included in the plots. Obviously, this gain comes at expenses of a higher complexity for the OFDM-OQAM strategy through its prototype filtering, but as we already commented along this paper, this complexity increase is more than affordable in current communication systems.

7   Conclusions

The paper proposed a spatial diversity scheme over OFDM-OQAM, where the generated interference is cancelled through some processing mainly at the receiver side of the communication process. The transmitter accomplishes a user scheduling to select the user with the best channel conditions to increase the systems performance. To the authors' knowledge, no previous proposals have been presented in the literature to deal with such scenario setup.

The obtained results show that OFDM-OQAM stands as a potential alternative to the classical OFDM, as its presents better rate behaviour thanks to the opportunistic scheduler to select the user with good phase component at each instant. Moreover, the proposed scheme does not employ the CP, which is a further increase in the system efficiency. Its advantages in terms of scalability and synchronization can be also attractive for the system. Therefore, OFDM-OQAM can be employed in certain scenario upon the requirements and restrictions for the system designer.

Acknowledgement

This work was partially supported by the European ICT-2008-211887 project PHYDYAS.

References

1. 802.16e-2005 IEEE Standard for Local and Metropolitan Area Networks Part 16, 3GPP (2006)
2. Bellanger, M.: Transmit Diversity in Multicarrier Transmission using OQAM Modulation. In: IEEE-ISWPC, Santorini (May 2008)

3. Hirosaki, B.: A Maximum Likelihood Receiver for an Orthogonally Multiplexed QAM System. In: IEEE-JSAC, September 1984, vol. (5) (1984)
4. ElTabach, M., Javaudin, J.P., Helard, M.: Spatial Data Multiplexing Over OFDM/OQAM Modulations. In: IEEE-ICC, Glasgow (June 2007)
5. Fusco, T., Tanda, M.: Blind Frequency-Offset Estimation for OFDM/OQAM Systems. IEEE Trans. on Signal Processing 55(5) (May 2007)
6. Viswanath, P., Tse, D.N., Laroia, R.: Opportunistic beamforming using dumb antennas. IEEE Trans. Inform. Theory 48, 1277–1294 (2002)
7. Knopp, R., Humblet, P.: Information capacity and power control in single-cell multiuser communications. In: IEEE-ICC, Seattle, USA (1995)
8. Bellanger, M.: Specification and Design of a Prototype Filter for Filter Bank Based Multicarrier Transmission. In: IEEE-ICASSP, Salt Lake City (May 2001)
9. Alamouti, S.: A Simple Transmit Diversity Technique for Wireless Communications. IEEE JSAC 16(8) (October 1998)
10. Akhtar, J., Gesbert, D.: Extending Orthogonal Block Codes With Partial Feedback. IEEE Transactions on Wireless Communications (November 2004)

Middleware Building Blocks for Architecting RFID Systems

Nikos Kefalakis[1], Nektarios Leontiadis[1], John Soldatos[1], and Didier Donsez[2]

[1] Athens Information Technology
19.5 Km Markopoulou Ave. 19002 Peania, Greece
`{nkef,nele,jsol}@ait.edu.gr`
[2] Université Joseph Fourier Grenoble
Avenue Centrale, Domaine Universitaire, 38041, Grenoble, France
`Didier.Donsez@imag.fr`

Abstract. RFID middleware is a cornerstone of non-trivial RFID deployments in complex heterogeneous environments. In this paper we present the principal middleware building blocks specified in the scope of the EPCglobal architecture. Alternative protocols and implementation frameworks for realizing these middleware blocks are also presented. At the same time we outline several middleware extensions to the EPCglobal architecture, towards meeting common requirements of automatic identification applications. Furthermore, we classify RFID applications into various categories based on their complexity, as well as based on their closed or open loop nature. Accordingly, we highlight the middleware blocks that are most important to each application category.

Keywords: RFID, Middleware, Architecture, EPCglobal, Business Event Generation.

1 Introduction

RFID middleware is gradually becoming the cornerstone of non-trivial RFID deployments in complex heterogeneous environments (e.g., logistics, supply chain management) comprising multiple readers, applications instances, legacy ICT systems, as well as sophisticated business processes and semantics. In such environments many distributed readers and antennas (e.g., in factories, warehouses, and distribution centers) capture RFID data, which must accordingly be conveyed to a variety of applications (e.g., enterprise resource planning (ERP) systems, warehouse management systems (WMS), corporate databases, process management systems).[1] Deployment and integration complexity are directly associated with the flexibility and versatility of the RFID middleware towards configuring and managing multiple heterogeneous devices, filtering and disseminating RFID data, translating low-level RFID data to high-level business semantics, as well as towards integrating RFID systems with legacy ICT systems and applications [5].

F. Granelli et al. (Eds.): MOBILIGHT 2009, LNICST 13, pp. 325–336, 2009.

The typical information flow within an RFID middleware system involves:

- Collecting RFID data from the physical readers, through reading the tagged items. At this level middleware implementations insulate higher layers from knowing what reader /models have been chosen. Moreover, they achieve virtualization of tags, which allows RFID applications to support different tag formats.
- Filtering the RFID sensor streams according to application needs, and accordingly emitting application level events. At this level middleware implementations insulate the higher layers from the physical design choices on how tags are sensed and accumulated, and how the time boundaries of events are triggered.
- Mapping the filtered readings to business semantics as required by the target applications and business processes. At this level middleware implementations insulate enterprise applications from understanding the details of how individual steps in a business process are carried out.

Software that implements any combination of these information flows can be conceived as an RFID middleware.

The above information flow is reflected in the EPCglobal architecture [2], where it is implemented based on the EPC-RP (Reader protocol), EPC-LLRP (Low-Level Reader Protocol), EPC-ALE (Application Level Events) and (EPC-IS) (Information Sharing) protocols and specifications [15]. Hence, the EPCglobal architecture specifies a middleware framework for a broad class of RFID applications. However it lacks some features that are extremely handy for many automatic identification applications. In this paper we present both the EPCglobal middleware layers, as well as additional middleware features, which are not completely covered by EPCglobal.

In this paper we introduce a middleware architecture (devised in the scope of the EC co-funded project ASPIRE [3]) which extends the EPCglobal architecture. To this end, we use the ASPIRE architecture to illustrate the various middleware layers and their possible implementations. Note that the proposed architectures have been devised in order to cover large scale open loop fully fledged RFID applications in the scope of inter-enterprise scenarios. Nevertheless, we are currently witnessing the proliferation of less complex closed loop applications, which can be implemented based on cut down versions of the proposed architectures. These applications require subsets of the presented middleware blocks as discussed in later sections.

The rest of this paper has the following structure: Section 2 discusses briefly the limitations of the EPCglobal architecture and introduces the ASPIRE architecture. This architecture is decomposed to the middleware building blocks dealing with readers and tags virtualization in Section 3, to filtering and collection blocks in Section 4 and Section 5 deals with the middleware blocks for addition of business context to RFID sensor streams. Section 6 classifies RFID applications into various categories and underlines the middleware building blocks that are relevant for each category. Finally, section 7 draws basic conclusions.

2 RFID Systems Architecture

The EPCglobal along with associated middleware implementations (see [4] for a comprehensive review) are subject to several limitations, some of which are inherent to the EPC architecture. Specifically, the most prominent of these limitations relate to the following areas [4]:

- Configurable Business Events Generation: Current middleware implementations do not provide support for configurable and automated translation of filtered data (i.e. ECReports) to business events (i.e. EPCIS Events). RFID developers are therefore still required to allocate programming effort in mapping ALE outputs to information sharing constructs. We strongly believe the configurable interpretation of RFID readings in a specific business context should be an essential functionality of any RFID middleware suite.

- Support and integration for sensor data: In addition to identifying objects many applications (e.g., cold chain management) need to detect and consume physical measurements (e.g., temperature, humidity, weight, acceleration (for shock-tracking), lighting). Hence, middleware frameworks must to provide the means to integrate sensors and accordingly make their data accessible by the applications. EPCglobal covers mainly the coding of things identifiers. While ALE reports can include (as extensions) physical measurements acquired by RFID sensor tags or sensors attached to the environment (e.g., RFID interrogator, container) at reading time, current middleware frameworks do not provide support for the consumption of these metrics. This is they do not cater for aligning the coding of these measurements with main international units, quantities standards and specifications (such as ISO 31-0, JSR 275, Open Geospatial Consortium GML, Google KML). Middleware frameworks must therefore provide support for adapting and using sensor readings in accordance to these coding schemes.

- Integration of Actuators: Experience with automatic identification applications manifests that there is often a need to quickly interact with the physical world based on a wide range of actuating functions such as locks, LEDs or mechanical controllers. Hence, RFID middleware frameworks need to be enhanced with actuator control frameworks.

- Reader Connectors and Virtualization: EPC-RP and EPC-LLRP prescribe reader protocol standards aiming at achieving vendor independence. In the current reader landscape however, there are still many readers that do not fully support these protocols. As a result there is still a need to provide an adaptation layer for non EPC-RP or EPC-LLRP compliant readers, similar to the HAL (Hardware Abstraction Layer) implementation of the Accada project for EPC-RP [1],[6]. Most important, a middleware suite should include a uniform interface for communicating with upstream EPC layers (e.g., ALE).

- End-to-End Management: Non-trivial RFID solutions are supported by highly heterogeneous infrastructures comprising multiple tags, readers, sensors, as well as a host of middleware components and servers. Managing such an infrastructure end-to-end is certainly asset towards facilitating the deployment and operation of RFID solutions. The EPC architecture and related middleware products emphasize on single reader management (e.g., based on the Reader Management Protocol) and do not support complete end-to-end management of the RFID solutions.

- Programmability and (Visual) Integrated Development Environments: Integrated development environments (IDEs) and visual tools are a key prerequisite to boosting RFID implementation. Most OSS RFID platforms do not provide complete integrated environments enabling visual development of RFID applications, which only few exceptions that are still in their infancy [7], [8]. In order for RFID deployment to go mainstream, complete IDEs enabling RFID consultants and business users to configure standards based solutions through minimal programming effort are urgently required.

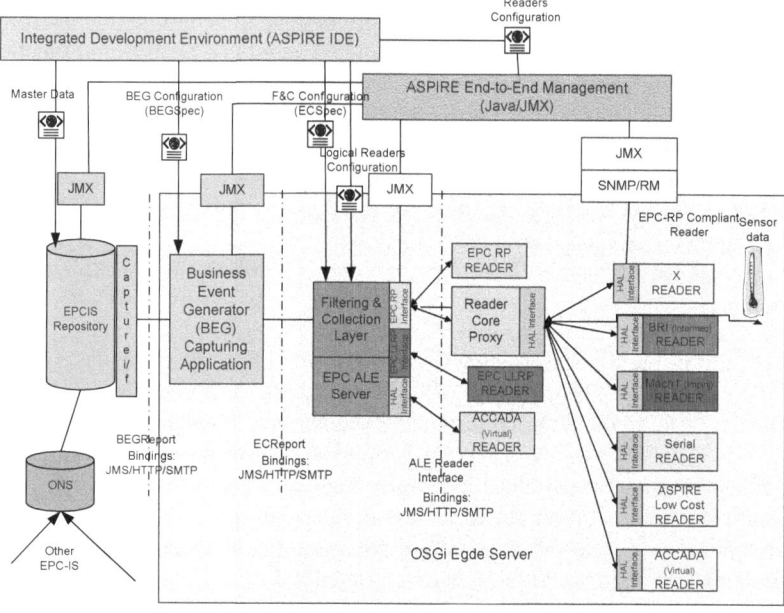

Fig. 1. ASPIRE Middleware Architecture

Driven by the above requirements and EPC limitations, the FP7 ASPIRE project has devised the middleware architecture depicted in Figure 1. It is based on the EPC architecture, but augments it with support for sensor data, end-to-end management, actuator control, as well as automated business context configuration functionality. These functionalities have been implemented in the scope of the AspireRfid [9] OSS project. Note that both the ASPIRE architecture and the AspireRfid project capitalize on lightweight container technologies, notably Open Services Gateway Interface (OSGi) (www.osgi.org) compliant for integrating and bundling the various middleware components comprising the architecture. In addition to being lightweight, an OSGi container constitutes a dynamic module system, which allows the deployment of various middleware blocks (described in later sections) as modules that can be flexibly (even at runtime) installed, started, stopped, activated, deactivated and updated. As a result, an OSGi based deployment facilitates the end-to-end management requirement, which is implemented based on JMX (Java Management Extensions) technology.

3   Readers and Tags Virtualization

3.1   Tags Virtualization

Tag virtualization capitalizes on a machine-readable version of the EPC Tag Data Standards specification [15]. This machine-readable version can be used for virtualizing underlying RFID tags through bridging and mapping different representations. Hence, a tag translation module is a (standalone or embedded) middleware component that enables the interpretation of machine readable version of the tag. The interpretation can be used in automated fashion. In addition the tag translation specification can be used for validating machine readable formats of the EPC tags. The TDT engine built in the scope of the AspireRfid [9] project supports the ISO15693, ISO14443, ISO15961, ISO15962, ISO15963, various GS1 formats (EAN/UPC, GS1 DataBar, GS1-128, ITF-14, GS1 DataMatrix, and Composite Component), as well as Bar Codes 1D and 2D: Note that barcode support is deemed particularly important given the vast number of legacy barcode applications, which need to be interoperable with emerging RFID applications.

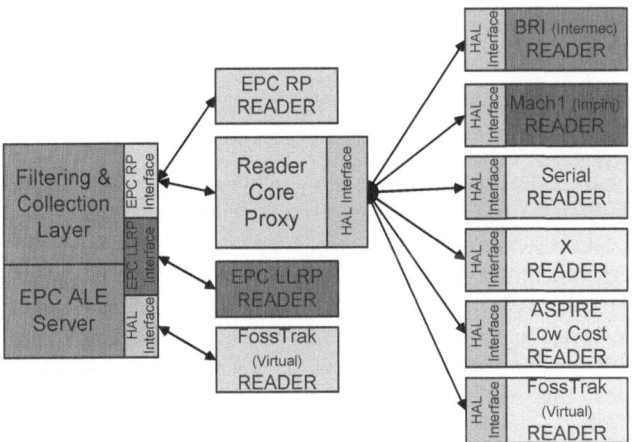

Fig. 2. Reader Virtualization Concept

3.2   Readers Virtualization

The role of the reader virtualization layer is to unify the way we interact with the miscellaneous hardware, by inserting a hardware abstraction layer and providing a fixed instruction set to the higher layers which require information from the hardware.

Specifications exist that satisfy the need for a norm at this level; namely the EPCglobal Reader Protocol (RP) and the EPCglobal Lower Level Reader Protocol (LLRP). These protocols define the standard bindings through which an application can send messages in a standardized format, as described in relevant standards [15].

Towards achieving reader virtualization a Hardware Abstraction Layer (HAL), ensuring a graceful mapping of the standardized messages to the low-level vendor specific reader communication primitives is specified. The methods of communication between the HAL and the hardware itself will vary, depending on the hardware vendor and it may require a serial connection, an Ethernet connection, etc. The protocols of communication may also vary from a raw TCP connection, to SSL and HTTP. The same will apply for the command and message encodings, which may be text, XML or binary.

The layers above the HAL exchange messages that conform to a well-defined format – XML or text – using a set of standard network interfaces – Serial, TCP and HTTP. Any combination of the aforementioned is allowed. Figure 2 depicts the reader virtualization concept.

4   Filtering and Collection (F&C)

RFID technology when used in a large scale deployment generates an enormous number of object reads. Many of those reads represent non-actionable "noise." To balance the cost and performance of this with the need for clear accountability and interoperability of the various parts, the design of the ASPIRE Architecture (Figure 1) seeks to:

- Drive as much filtering and counting of reads as low in the architecture as possible.
- Minimize the amount of "business logic" embedded in the Tags.

The Filtering and Collection Middleware by applying EPC ALE (Application Level Events) [15] is intended to facilitate these objectives by providing a flexible interface to a standard set of accumulation, filtering, and counting operations that produce "reports" in response to client "requests." The client will be responsible for interpreting and acting on the meaning of the report. The client of the ALE interface may be a traditional "enterprise application," or it may be new software designed expressly to carry out an EPC-enabled business process but which operates at a higher level than the "middleware" that implements the ALE interface. Section 6 later in this paper elaborates on different deployment configurations depending on the application scale and nature.

The ASPIRE filtering & collection middleware represents a single interface to the potentially large number of readers that make up an RFID system deployment. This allows applications to subscribe to a specific already defined specification, which is then used along with the Logical Reader definition to configure the corresponding reader devices using the EPC global reader protocol (RP) or low level reader protocol (LLRP) (Figure 2).

Once the readers capture relevant tag data they notify the middleware which combines the data arriving from different readers in a report that is sent according to a predetermined schedule to the subscribed applications. Since the middleware receives data from multiple readers, it provides specific filtering functionality depending on the already defined specifications. Redundant read events from different readers observing the same location are not included to the dispatched report, which accomplishes the reduction of filtering and delivers the level of required aggregation to the registered application(s) interpreting the captured RFID data.

The interfaces chosen to be used between the filtering & collection middleware and host applications is TCP/HTTP for the notification channel transferring XML reports and SOAP for the server operation programming (ECSpecifications definition/subscription, logical reader definition).

The primary data types associated with the ALE API (Application Programming Interface) are the ECSpec, which specifies how an event cycle is to be calculated, and the ECReports, which contains one or more reports generated from a single activation of an ECSpec. ECReports instances are both returned from the poll and immediate methods, and also sent to notification URIs when ECSpecs are subscribed to using the "subscribe" method of the specification.

An ECSpec describes an event cycle and one or more reports that are to be generated from it. It contains a list of logical Readers whose read cycles are to be included in the event cycle, a specification of how the boundaries of event cycles are to be determined, and a list of specifications each of which describes a report to be generated from this event cycle. There are two ways to cause event cycles to occur. A standing ECSpec may be posted using the define method. Subsequently, one or more clients may subscribe to that ECSpec using the subscribe method. The ECSpec will generate event cycles as long as there is at least one subscriber.

ECReports is the output from an event cycle. The essence of an ECReports instance is the list of ECReport instances, each corresponding to an ECSpec instance in the event cycle's ECSpec. In addition to the reports themselves, ECReports contains a number of "header" fields that provide useful information about the event cycle.

The ALE interface revolves around client requests and the corresponding reports that are produced. Requests can either be: (1) immediate, in which information is reported on a one-time basis at the time of the request; or (2) recurring, in which information is reported repeatedly whenever an event is detected or at a specified time interval. The results reported in response to a request can be directed back to the requesting client or to a "third party" specified by the requestor.

5 Business Context and Information Sharing

Adding business semantics to the low-level sensor streams is a key prerequisite to added-value deployments of RFID technology. For RFID to go mainstream companies must be offered tools and techniques for describing their RFID enabled business processes, without engaging in the low-level implementation details. To this end, a framework specification for describing business events must be provided. This framework should enable the description of business processes based on high-level semantics, which at the same time should be amendable by tools.

The EPC-IS framework [10] is standardized as an integral layer of the EPCglobal architecture. Its main function is to insulate enterprise applications from understanding the details of how individual steps in a business process are carried out at a detailed level. EPC-IS defines a data model for events associated with the lifetime of uniquely identified objects. As already outlined these events are industry and application agnostic. In this sense EPC-IS is a cross industry framework, which allows for industry specific vocabularies and extensions. Furthermore, the framework is a supplement to (and not a replacement for) existing enterprise information systems. Specifically, EPC-IS events are used to push/pull events to/from other enterprise systems such as ERP (Enterprise Resource Planning Systems), WMS (Warehouse Management Systems) and corporate databases.

EPC-IS can operate in the scope of a layered service-oriented architecture, through persisting supply chain events in a repository and accordingly sharing these events with internal and external applications. The sharing is accomplished through interfaces for capture and query of event data. EPCIS data (i.e. events) are represented as records of activity happening in real world. These events provide context ("What, where, when, why") and proliferate as more business is transacted. EPCIS events are interpreted based on descriptive information (so called "master data"), which provides context for the events such as descriptions of locations, products and business transactions. Note that "mater data" grow at different timescales comparing to EPCIS events. In particular, master data grow slowly as companies grow, not as more business is conducted.

EPCIS events described within the specification can be classified as follows (Figure 3):

- Object Events, which correspond to observations of a collection of EPCs during a specific business step at a specified Location & Time.
- Aggregation Events, which reflect a physical association of a set of EPCs with a parent EPC along with a business step at a Location & Time.
- Quantity Events, which correspond to statements about an object Class (not individual objects), including a quantity, a Location & Time.
- Transaction Events, which records objects associated with a wider business transaction.

Having these events at hand, consultants, researchers and engineers can use them to describe RFID enabled business processes [11]. The starting point is the documentation of the business requirements, comprising the archetypical use cases. Accordingly, it is important to break each use case into a series of discrete business steps. Each one of these steps needs to be modeled as an EPCIS event, according to the above mentioned core EPCIS event types. In rare cases a new type could be defined i.e. when existing types are not sufficient to describing a business step. Note that it is important to define any necessary extension fields, as well as the full range of vocabularies that populate each field. Furthermore, fixed lists of identifiers with standardized meanings for concepts like business step and disposition must be provided, along with rules for population of user-created identifiers like read point and business location.

A novel characteristic of the ASPIRE architecture (Figure 1) in terms of business context handling is the introduction and implementation of the "Business Event Generator" (BEG) middleware module. This middleware module undertakes the automated and configurable mapping of reports (stemming from the F&C module) to EPCIS events. This automation will greatly simplify the development of capturing applications (according to the EPCglobal architecture).

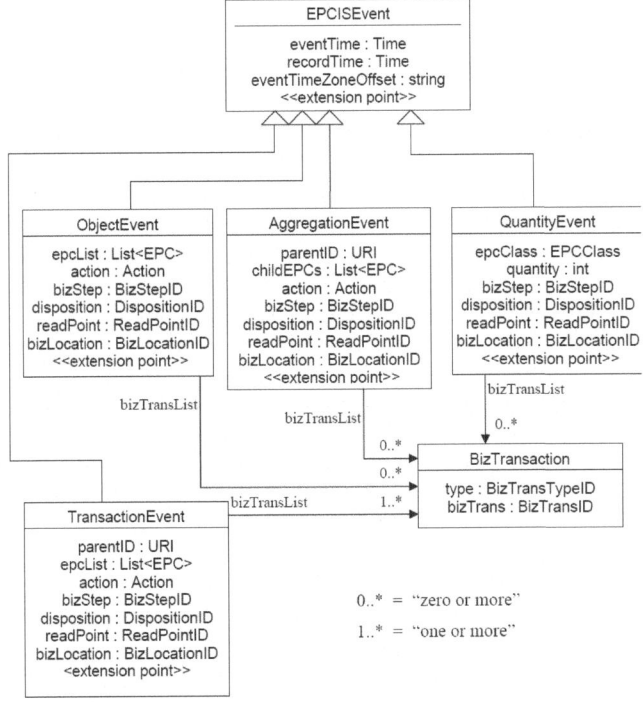

Fig. 3. Core EPCIS event types and the object-oriented relationship between them

6 Application Classification

The proliferating RFID applications, pilots and deployment vary in functionality and scale. The RFID vision was initially articulated through the specification of large scale "open loop" systems and deployments. Prominent examples of such systems are those developed and trialed by Wall-Mart and the U.S Department of Defense (DoD). A main characteristic of these systems is that they span different locations across multiple companies and/or organizations. Note that these trials manifested several problems, both technical (e.g., information sharing, interoperability and scalability at a large scale) and business ones (e.g., business model related issues). Large scale deployments can provide a crash test for protocols like EPCIS and ONS (Object Naming Service) [15].

Following these early complex and visionary deployments, the RFID community has gradually starting to dispel the hype (and its associated complexity). Hence, during the last couple of years we are witnessing a proliferating number of smaller scale solutions covering a wide range of asset tracking and inventory management scenarios, as well as other ROI (return-on-investment) generating case studies. These case studies focus on very specific business problems, which an AIS (automatic identification system) can solve even a single enterprise. A main characteristic of the smaller scale deployment is also the fact that tagging occurs at the case and pallet level tagging rather than item level. Furthermore, tracking, traceability and identification occur within a warehouse or a single supply chain. Note that these smaller scale solutions are also a reality for RFID vendors, which are gradually refocusing their strategies towards smaller-scale opportunities. Nevertheless, the vision still exists, since small applications (i.e. closed loop islands) could one day become integrated into larger scale open loop systems.

We believe that middleware developers and RFID consultants should prioritize middleware modules development based on the scale of the target application. Open loop solutions must pay emphasis on implementing the full range of middleware layers described in this paper. On the other hand smaller scale closed loop systems must prioritize the F&C, reading and tag virtualization building blocks. Moreover, for some very simple systems our experience shows that custom filters over a HAL for the target reader(s) could provide a rapid and acceptable solution. Table1 presents our view regarding the middleware building blocks that are required to implement each of the above application categories. This is based on our experience with RFID implementations and demonstrations across diverse deployments of varying scale (e.g., [9], [12], [13]).

Table 1. RFID application Classification and Middleware Building Blocks

Application Type/Middleware Block	HAL	EPC-RP, EPC-LLRP	F&C	Business Context
Simple	Yes	Recommended	Recommended	No
Simple Closed Loop	Yes	Yes	Recommended	No
Complex Closed Loop	Yes	Yes	Yes	Recommended
Open Loop	Yes	Yes	Yes	Yes

7   Conclusion

In this paper we have presented the middleware components and layers, which are commonly implemented in RFID applications. The EPCglobal architecture, as well as its extensions in the scope of the ASPIRE architecture provide a general middleware framework that can address the needs of many RFID applications. We argue however that the full range of middleware layers and building blocks are necessary only in the scope of large scale open loop middleware implementations. Simpler applications can leverage cut down versions of these architectures, towards economizing on performance overhead as well as implementation complexity and cost. Specifically, trivial applications can be implemented via customized filtering mechanisms on top of HAL layers or event the EPC-LLRP and EPC-RP protocols. Also, a wide range of closed loop intra-enterprise scenarios could be implemented without a need for sophisticated information sharing layer. Overall, we think that lightweight low-overhead implementations are essential for the smooth transition to fully fledged RFID deployments. This could reinforce a 'start small, think big' approach towards the Internet of Things (IoT) vision. The open source AspireRfid project of the OW2 (www.ow2.org) community provides distinct implementations of the various building blocks, in order to enable researchers and developers to gradually leverage the various middleware functionalities, as required by their target deployments.

Acknowledgments. Part of this work has been carried out in the scope of the ASPIRE project (FP7-215417). The authors acknowledge help and contributions from all partners of the project.

References

1. Floerkemeier, C., Roduner, C., Lampe, M.: RFID Application Development with the Accada Middleware Platform. IEEE Systems Journal 1(2), 82–94 (2007)
2. Architecture Review Committee, The EPCglobal Architecture Framework, EPCglobal (July 2005), http://www.epcglobalinc.org.
3. The ASPIRE FP7 Project, http://www.fp7-aspire.eu
4. Cezon, M., Vaudaux-Ruth, G., Laurens, L., Soldatos, J., et al.: Review of State-of-the-Art Middleware. ASPIRE Project Public Deliverable D2.1 (June 2008)
5. Sarma, S.: Integrating RFID. ACM Queue 2(7), 50–57 (2004)
6. The Accada RFID Middleware Project, http://www.accada.org
7. The Rifidi project, An open source IDE for RFID, http://www.rifidi.org
8. Sun's JCAPS (Java Composite Application Platform) for RFID, http://java.sun.com
9. AspireRfid project, http://wiki.aspire.objectweb.org/xwiki/bin/view/Main/WebHomewiki
10. EPC Information Services (EPCIS) Version 1.0, Specification, EPCglobal (April 2007), http://www.epcglobalinc.org
11. BEAWebLogic RFID Enterprise Server™, Understanding the Event, Master Data, and Data Exchange Services, Version 2.0, Revised: October 12 (2006)
12. Zarokostas, N., Dimitropoulos, P., Soldatos, J.: RFID Middleware Design for enhancing traceability in the Supply Chain Management. In: The Proc. of the 18th IEEE Personal Indoor and Mobile Radio Communications, Athens, Greece, September 3-7 (2007)
13. Rudametkin, W., Touseau, L., et al.: NFCMuseum: an Open-Source Middleware for Augmenting Museum Exhibits. In: IEEE International Conference on Pervasive Services (ICPS 2008), Sorrento, Italy, July 6-10 (2008) (Public demonstration)
14. Lampe, M., Floerkemeier, C.: High-Level System Support for Automatic-Identification Applications. In: Maass, W., Schoder, D., Stahl, F., Fischbach, K. (eds.) Proceedings of Workshop on Design of Smart Products, Furtwangen, Germany, March 2007, pp. 55–64 (2007)
15. EPCglobal standards, http://www.epcglobalinc.org/standards

Abbreviations

ALE – Application Level Events
BEG – Business Events Generator
EC –European Committee

ECReport – Event Cycle Report
ECSpec – Event Cycle Specification
EPC – Electronic Product Code
EPC-ALE - Electronic Product Code Application Level Events
EPC-IS - Electronic Product Code Information Service
EPC-LLRP - Electronic Product Code Low Level reader protocol
EPC-RP - Electronic Product Code Reader Protocol
ERP – Enterprise resource Planning
HAL – Hardware Abstraction Layer
ICT – Information and Communications Technologies
JMX – Java Management Extensions
ONS – Object Naming Service
OSGi – Open Service gateway Initiative
OSS – Open Source Software
RFID – Radio Frequency Identification
WMS – Warehouse management System

Multicost Energy-Aware Broadcasting in Wireless Networks with Distributed Considerations

Christos Papageorgiou*, Panagiotis Kokkinos, and Emmanouel Varvarigos

[1] University of Patras, Patras, Greece
[2] Research Academic Computer Technology Institute, Patras, Greece

Abstract. In this paper we propose an energy-aware broadcast algorithm for wireless networks. Our algorithm is based on the multicost approach and selects the set of nodes that by transmitting implement broadcasting in an optimally energy-efficient way. The energy-related parameters taken into account are the node transmission power and the node residual energy. The algorithm's complexity however is non-polynomial, and therefore, we propose a relaxation producing a near-optimal solution in polynomial time. We also consider a distributed information exchange scheme that can be coupled with the proposed algorithms and examine the overhead introduced by this integration. Using simulations we show that the proposed algorithms outperform other solutions in the literature in terms of energy efficiency. Moreover, it is shown that the near-optimal algorithm obtains most of the performance benefits of the optimal algorithm at a smaller computational overhead.

1 Introduction

Advances in battery lifetime during recent years have not kept in pace with the significant decline in computation and communication costs in ad hoc and sensor networks. Thus, considering the lack of any fixed infrastructure and the requirements for long operating lifetime, energy is a crucial resource limiting the performance and range of applicability of such networks. Furthermore, the cooperative nature of both ad hoc and sensor networks, makes broadcasting one of the most frequently performed primitive communication task. Being able to perform this communication task in an energy-efficient manner is an important priority for such networks.

In this paper we propose an optimal energy efficient broadcasting algorithm, called Optimal Total and Residual Energy Multicost Broadcast (abbreviated OTREMB) algorithm, for wireless networks consisting of nodes with preconfigured levels of transmission power. It is quite common that the nodes comprising wireless networks, either ad hoc or sensor, are not able to dynamically adjust

* C. Papageorgiou was supported by GSRT through PENED project 03EΔ207, funded 75% by the EC and 25% by the Greek State and the private sector.

F. Granelli et al. (Eds.): MOBILIGHT 2009, LNICST 13, pp. 337–346, 2009.

their transmission power, since their processing capabilities are inherently minimal. Our algorithm is optimal, in the sense that it can optimize any desired function of the total power consumed by the broadcasting task and the minimum of the current residual energies of the nodes, provided that the optimization function is monotonic in each of these parameters. The proposed algorithm takes into account these two energy-related parameters in selecting the optimal sequence of nodes for performing the broadcast, but it has non-polynomial complexity. We also present a relaxation of the optimal algorithm, to be referred to as the Near-Optimal Total and Residual Energy Multicost Broadcast (abbreviated NOTREMB) algorithm, that produces a near-optimal solution to the energy-efficient broadcasting problem in polynomial time. The proposed algorithms try to jointly maximize the network lifetime and minimize its energy consumption, by following the multicost routing approach [6]. Multicost routing has been verified to perform better than single-cost routing in terms of energy-efficiency for the case of unicast routing in wireless networks [16]. In this work we show that multicost routing schemes can also be used for energy-efficient broadcast communication.

The routing process (unicasting, multicasting or broadcasting) involves two levels: the information exchange level and the routing algorithmic level. The proposed algorithms focuses on the routing level and thus assumes that all the necessary information for the optimal broadcast schedule to be computed is instantly available at each node. This is also the approach followed by the majority of related works. As far as the information exchange level is concerned, we examine a distributed information exchange protocol and discuss the emerging tradeoffs regarding the algorithm's performance and the induced overhead.

In the performance results we evaluate our broadcast algorithms assuming instant and costless knowledge of the network information, obtaining in this way a performance upper bound of the proposed solutions, and ignoring the information exchange overhead. We compare the optimal and near-optimal algorithms to other representative algorithms for energy-efficient broadcasting. Our results show that the proposed algorithms outperform the other algorithms by making better use of the network energy reserves. Another important result is that the near-optimal algorithm performs comparably to the optimal algorithm, at a significantly lower computation cost.

The remainder of the paper is organized as follows. In Section 2 we discuss prior related work. In Sections 3 and 4 we present the optimal and near-optimal algorithms introduced in this paper for energy-efficient broadcasting. In Section 5 the simulations setting is outlined and the performance results are presented. Finally, in Section 6 we give the conclusions drawn from our work.

2   Related Work

Energy-efficiency in all types of communication tasks (unicast, multicast, broadcast) has been considered from the perspective of either minimizing the total energy consumption or maximizing the network lifetime. Most versions of both optimization problems are NP-hard [11,17,14]. Two surveys summarizing much of the related work in the field can be found in [5,1].

A major class of works in the field start with an empty solution which is gradually augmented to a broadcast tree. A seminal work presenting a series of basic energy-efficient broadcasting algorithms, like Minimum Spanning Tree, Shortest Path Tree and Broadcast Incremental Power (BIP), is [19]. The BIP algorithm maintains a single tree rooted at the source node, and new nodes are added to the tree, one by one, on a minimum incremental cost basis. In Broadcast Average Incremental Power (BAIP) algorithm [18] many new nodes can be added at the same step with the average incremental cost defined as the ratio of the minimum additional power required by a node in the current tree to reach these new nodes to the number of new nodes reached. The Greedy Perimeter Broadcast Efficiency (GPBE) algorithm [8] uses another greedy decision metric, defined as the number of newly covered nodes reached per unit transmission power. In [3], the Minimum Longest Edge (MLE) and the Minimum Weight Incremental Arborescence (MWIA) algorithms are presented. The MLE first computes a minimum spanning tree using as link costs the required transmission powers and then removes redundant transmissions. In MWIA, a broadcast tree is constructed using as criterion a weighted cost that combines the residual energy and the transmission power of each node. In [2], the Relative Neighborhood Graph (RNG) topology is used for broadcasting. In Local Minimum Spanning Tree (LMST) [13] each node builds a one-hop minimum spanning tree. A link is included in the final graph if it selected in the local MSTs of both its edge nodes. In [7] a localized version of the BIP algorithm is presented. All the aforementioned works assume adjustable node transmission power. One of the few papers that assumes preconfigured power levels for each node is [10], where two heuristics for the minimum energy broadcast problem are proposed.

Local search algorithms perform a walk on broadcast forwarding structures. The walk starts from an initial broadcast topology obtained by some algorithm and in each step, a local search algorithm moves to a new broadcast topology so that the necessary connectivity properties are maintained. The rule used at each step for selecting the next topology is energy-related and the algorithm terminates when no further improvement can be obtained. In [19], the Sweep heuristic algorithm was proposed to improve the performance of BIP by removing transmissions that are unnecessary, due to the wireless broadcast advantage. Iterative Maximum-Branch Minimization (IMBM) [12] starts with a trivial broadcast tree where the source transmits directly to all other nodes and at each step replaces the longest link with a two-hop path that consumes less energy. In [17], EWMA is proposed that modifies a minimum spanning tree by checking whether increasing a node's power so as to cover a child of one of its children, would lead to power savings. The r-Shrink heuristic [4] is applied to every transmitting node and shrinks its transmission radius so that less than r nodes hear each transmission. The LESS heuristic [9] permits a slight increase in the transmission power of a node so that multiple other nodes can stop transmitting or reduce their transmission power.

3   The Optimal Total and Residual Energy Multicost Broadcast Algorithm

The objective of the Optimal Total and Residual Energy Multicost Broadcasting (OTREMB) algorithm is to find, for a given source node, an optimal sequence of nodes for transmitting, so as to implement broadcasting in an energy-efficient way. In particular, it selects a transmission schedule that optimizes any desired function of the total power T consumed by the broadcasting task and the minimum R of the residual energies of the nodes, provided that the optimization function used is monotonic in each of these parameters, T and R. The OTREMB algorithm's operation consists of two phases, in accordance with the general multicost algorithm [6] on which it is based. In the first phase, the source node u calculates a set of candidate node transmission sequences $\mathcal{S}_u$, called set of non-dominated schedules, which can send to all nodes any packet originating at that source. In the second phase, the optimal sequence of nodes for broadcasting is selected based on the desired optimization function.

3.1   The Enumeration of the Candidate Broadcast Schedules

In the first phase of the OTREMB algorithm, every source node u maintains at each time a set of candidate broadcast schedules $\mathcal{S}_u$. A broadcast schedule $S \in \mathcal{S}_u$ is defined as $S = \{(u_1 = u, u_2, \ldots, u_h), V_S\}$, where $(u_1, u_2, \ldots, u_h)$ is the ordered sequence of nodes used for transmission and $V_S = (R_S, T_S, P_S)$ is the cost vector of the schedule, consisting of: the minimum residual energy R_S of the sequence of nodes $u_1, u_2, \ldots, u_h$, the total power consumption T_S caused when these nodes are used for transmission and the set P_S of network nodes covered when nodes $u_1, u_2, \ldots, u_h$ transmit a packet.

When node u_i transmits a packet at distance r_i, the energy expended is taken to be proportional to r_i^a, where a is a parameter that takes values between 2 and 4. Because of the broadcast nature of the medium and assuming omni-directional antennas, a packet being sent or forwarded by a node can be correctly received by any node within range r_i of the transmitting node u_i. Therefore, broadcast communication tasks in these networks correspond to finding a sequence of transmitting nodes, instead of a sequence of links as it is common in the wireline world. The assumption of omni-directional antennas is not necessary for the proposed algorithms to work, provided that we know the set of nodes $D(u_i)$ that can correctly decode a packet transmitted by node u_i; the performance results to be presented in Section 5, however, assume that omni-directional antennas are used.

Initially, each source node u has only one broadcast schedule $\{\emptyset, (\infty, 0, u)\}$, with no nodes, infinite node residual energy, zero total power consumption, while the set of covered nodes contains only the source. The candidate broadcast schedules from source node u are calculated as follows:

1. Each broadcast schedule $S = \{(u_1, u_2, \ldots, u_{i-1}), (R_S, T_S, P_S)\}$ in the set of non-dominated schedules $\mathcal{S}_u$ is extended, by adding to its sequence of

transmitting nodes a node $u_i \in P_S$ that can transmit to some node u_j not contained in P_S. If no such nodes u_i and u_j exist, we proceed to the final step.

Then the schedule S is used to obtain an extended schedule S' as follows:

- node u_i is added to the sequence $u_1, u_2, \ldots, u_{i-1}$ of transmitting nodes
- $R_{S'} = \min(R_i, R_S)$, where R_i is the residual energy of node u_i
- $T_{S'} = T_S + T_i$, where T_i is the (fixed) transmission power of node u_i
- the set of nodes $D(u_i)$ that are within transmission range from u_i are added to the set P_S.
- the extended schedule
  $S' = \{(u_1, \ldots, u_{i-1}, u_i), (\min(R_S, R_i), T_S + T_i, P_S \cup D(u_i))\}$ obtained in the way described above is added to the set $\mathcal{S}_u$ of candidate schedules.

2. Next, a *domination relation* between the various broadcast schedules of source node u is applied, and the schedules found to be dominated are discarded. In particular, a schedule S_1 is said to *dominate* a schedule S_2 when $T_1 < T_2$, $R_1 > R_2$ and $P_1 \supset P_2$. In other words schedule S_1 dominates schedule S_2 if it covers a superset of nodes than the one covered by S_2, using less total transmission power and with larger minimum residual energy on the nodes it uses. All the schedules found to be dominated by another schedule are discarded from the set $\mathcal{S}_u$.

3. The procedure is repeated, starting from the first step 1, for all broadcast schedules in $\mathcal{S}_u$ that meet the above conditions. If no schedule $S \in \mathcal{S}_u$ can be extended further, we go to the final step.

4. Among the schedules in $\mathcal{S}_u$ we form the subset of schedules S for which P_S includes all network nodes. This subset is called the *set of non-dominated schedules* for broadcasting from source node u, and is denoted by $\mathcal{S}_{u,\mathcal{B}}$.

3.2   The Selection of the Optimal Broadcast Schedule

In the second phase of the OTREMB algorithm, an optimization function $f(V_S)$ is applied to the cost vector V_S of every non-dominated schedule $S \in \mathcal{S}_{u,\mathcal{B}}$ of source node u, produced in the first phase. The optimization function combines the cost vector parameters to produce a scalar metric representing the cost of using the corresponding sequence of nodes for broadcasting. The schedule with the minimum cost is selected. In the performance results described in Section 5, the optimization function used is

$$f(S) = \frac{T_S}{R_S}, \text{for } S \in \mathcal{S}_{u,\mathcal{B}},$$

which favors, among the schedules that cover all nodes, those that consume less total energy T_S and whose residual energy R_S is larger. Other optimization functions and parameters (with or without weights) could also be used, depending on the interests of the network. The only requirement is that the optimization function has to be monotonic in each of its parameters.

Theorem 1. *If the optimization function $f(V_S)$ is monotonic in each of the parameters involved, the OTREMB algorithm finds the optimal broadcast schedule.*

Proof. Since $f(V_S)$ is monotonic in each of its parameters, the optimal schedule has to belong to the set of non-dominated schedules (a schedule S_1 that is dominated by a schedule S_2, meaning that it is worse than S_2 with respect to all the parameters, cannot optimize f). Therefore, it is enough to show that the set $\mathcal{S}_u$ computed in Steps 1-3 of OTREMB includes all the non-dominated schedules for broadcasting from node u.

We let $S = ((u_1, u_2, \ldots, u_h), (R_S, T_S, P_S))$ be a non-dominated schedule that has minimal number of transmissions h among the schedules not produced by OTREMB. Then for the schedule $S' = ((u_1, u_2, \ldots, u_{h-1}), (R_{S'}, T_{S'}, P_{S'}))$ we have that $R_S = \min(R_{S'}, R_h)$, $T_S = T_{S'} + T_h$, and $P_S = P_{S'} \cup D(u_h)$. The fact that S is non-dominated and was not produced by OTREMB, implies that S' was not produced by OTREMB either. Since S is a non-dominated schedule with minimal number of transmissions among those not produced by OTREMB, and S' was not produced by OTREMB and uses less transmissions, this means that S' is dominated. However, since S is non-dominated, this means that S' is also non-dominated (otherwise, the schedule S'' that dominates S', in the sense that it has $T_{S''} < T_{S'}$, $R_{S''} > R_{S'}$ and $P_{S''} \supset P_{S'}$, extended by the transmission from node u_h would dominate S), which is a contradiction.

4 The Near-Optimal Total and Residual Energy Multicost Broadcast Algorithm

The OTREMB algorithm finds the schedule that optimizes the desired optimization function $f(V_S)$, but it has non-polynomial complexity, since the number of non-dominated schedules generated by the first phase of the algorithm can be exponential. In order to obtain a polynomial time algorithm, we relax the domination condition so as to obtain a smaller number of candidate schedules. In particular, we define a *pseudo-domination* relation among schedules, according to which a schedule S_1 *pseudo-dominates* schedule S_2, if $T_1 < T_2$, $R_1 > R_2$, and $|P_1| > |P_2|$, where T_i, R_i, $|P_i|$ are the total transmission power, the residual energy of the broadcast nodes and the cardinality of the set of nodes covered by schedule S_i, $i = 1, 2$, respectively. When this pseudo-domination relationship is used in step 2 of the OTREMB algorithm, it results in more schedules being pruned (not considered further) and smaller algorithmic complexity. In fact, by weakening the definition of the domination relationship the complexity of the algorithm becomes polynomial (this can easily be shown by arguing that T_i, R_i and $|P_i|$ can take a finite number of values, namely, at most as many as the number of nodes). The decrease in time complexity, however, comes at the price of losing the optimality of the solution. We will refer to this this near-optimal variation of the OTREMB algorithm, as the Near-Optimal Total and Residual Energy Multicost Broadcast algorithm (abbreviated NOTREMB).

5   Information Exchange Protocols

The proposed algorithms require information that can be provided by information collection and dissemination mechanisms. An important categorization of the information protocols is whether they work in a centralized, decentralized or a distributed manner. In the centralized case, all the needed information is gathered by some central node that is accessible by all the other nodes of the network. In the decentralized scenario this information is gathered separately by each node. In the distributed case, only local information is available at each node and decisions are taken either based on this information, or by gathering on demand additional information, using an information protocol.

For our optimal and near-optimal broadcasting algorithms we can use a distributed information exchange mechanism. When a packet needs to be broadcasted, the source node broadcasts a control packet containing an empty schedule S. Every node u receiving this packet and not belonging to the set of covered nodes, updates the schedule S according to the way described in the first step of the enumeration of the broadcast sequences phase of the algorithms (Section 3.1). Then node u broadcasts the updated control packet to its neighbors. When all nodes are covered, the control packet is returned to the source node, using the information included in the sequence S. Before a newly received sequence S is added to the source node's set of schedules $\mathcal{S}_{u,\mathcal{B}}$, its domination relation with the other schedules in the set is checked. In the end, the optimization function is applied to the schedules in $\mathcal{S}_{u,\mathcal{B}}$ in order to select the best schedule. The algorithm's performance depends directly on the accuracy of the information at each node. The described information collection scheme can be used both in a periodic and an on-demand fashion. However, the tradeoff between the information accuracy and the induced overhead regarding network traffic and energy consumption needs to be considered.

In our performance results, we decided to focus only on the broadcast algorithm assuming instant and costless knowledge of the network information. Therefore, the results obtained there, can be viewed as an upper bound on the actual performance of the proposed solutions. This permits us to focus on the routing problem without having to deal with the implementation details of the information exchange protocol that could obscure the routing issues.

6   Performance Results

6.1   Simulation Setting

We implemented and evaluated the proposed algorithms, using the Network Simulator ns-2 [15]. We use a 4×4 two-dimensional grid network topology of 16 stationary nodes with distance of 50 meters between neighboring nodes. Each node's transmission radius is fixed at a value uniformly distributed between 50 and 100 meters. In our experiments the initial energy E_0 is taken to be equal for all nodes (5, 10 and 100 Joules). The proposed algorithms are compared against the fixed-power versions of the BIP [19], the MWIA [3] and the BAIP [18] algorithms.

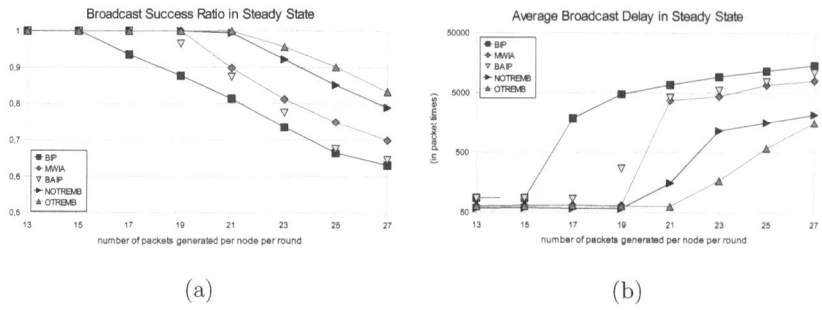

(a)   (b)

Fig. 1. The broadcast success ratio p and the average broadcast delay D in the steady-state of the algorithms evaluated, for a different number of broadcast packets N inserted in the network

We evaluate the proposed algorithms under the infinite time horizon model. In this model, the broadcasting strategies are evaluated assuming packets and energy are generated over an infinite time horizon, according to a round-based scenario. At the beginning of each round, the node energy reserves are restored to a certain level, and an equal number of packets N to be broadcasted is generated at every node. A round terminates when the residual energy of at least half of the network nodes falls below a certain safety limit. Packets that are not successfully broadcasted during a round, continue from the point they stopped (e.g., a node with residual energy levels below the safety limit) in the following round(s) until their broadcast is completed. The succession of rounds continues until the network reaches steady-state, or until it becomes inoperable (unstable). We use the following metrics:

- The average broadcast delay D of a packet is the time that elapses from the time instant it is generated at a source node, until the time it has reached all nodes of the network, possibly after several rounds.
- The broadcast success ratio p, defined as the ratio of the number of packets successfully broadcasted (reached all nodes of the network) over the total broadcast packets sent.

6.2   Simulation Results

Figure 1.a presents the broadcast success ratio p at steady state and the average broadcast delay D (in packet times), for a different number of broadcast packets N inserted at each node per round. We observe that even for relatively light traffic inserted in each round, the BIP, the MWIA and the BAIP algorithms are not able to successfully broadcast all the packets generated. The OTREMB and NOTREMB schemes have the maximum stability region (maximum broadcast throughput) and remain stable for up to $N = 21$ packets per node per round. By taking into account energy-related cost parameters and switching through multiple energy-efficient paths, both OTREMB and NOTREMB spread energy consumption more evenly and increase the volume of broadcast traffic that can be successfully served. In Figure 1.b, where the delay versus traffic is depicted, the load curves of the BIP, MWIA and BAIP algorithms are above those of the OTREMB and the NOTREMB algorithms. Since packets whose broadcast is not completed during a round fill the node queues and congest the network, the average delay of the BIP, MWIA and BAIP algorithms quickly becomes very large. Naturally, when the traffic load inserted increases beyond each scheme's maximum stable throughput, the delays will also become unbounded, and the success ratio p will start falling. The OTREMB and the NOTREMB algorithms have smaller average delay D and remain stable for higher loads than the other schemes considered. In both figures we observe that the NOTREMB algorithm performs comparably to the OTREMB algorithm.

7   Conclusions

We studied energy-aware broadcasting in wireless networks, and proposed an optimal (OTREMB) and a near-optimal (NOTREMB) algorithm, based on the multicost concept. A distributed information collection mechanism was also introduced in order to make the proposed algorithms suitable for distributed operation. Our results show that the proposed multicost algorithms outperform the other algorithms considered, consuming less energy and successfully broadcasting more packets to their destination. Moreover, NOTREMB has similar performance to that of the OTREMB while having considerably smaller execution time.

References

1. Athanassopoulos, S., Caragiannis, I., Kaklamanis, C., Kanellopoulos, P.: Experimental comparison of algorithms for energy-efficient multicasting in ad hoc networks. In: Nikolaidis, I., Barbeau, M., Kranakis, E. (eds.) ADHOC-NOW 2004. LNCS, vol. 3158, pp. 183–196. Springer, Heidelberg (2004)
2. Cartigny, J., Simplot, D., Stojmenovic, I.: Localized minimum-energy broadcasting in ad-hoc networks. INFOCOM 3, 2210–2217 (2003)
3. Cheng, M., Sun, J., Min, M., Li, Y., Wu, W.: Energy-efficient broadcast and multicast routing in multihop ad hoc wireless networks. Wireless Comm. and Mobile Comp. 6(2), 213–223 (2006)
4. Das, A., Marks, R., El-Sharkawi, M., Arabshahi, P., Gray, A.: r-shrink: A heuristic for improving minimum power broadcast trees in wireless networks. GLOBECOM 1, 523–527 (2003)
5. Guo, S., Yang, O.: Energy-aware multicasting in wireless ad hoc networks: A survey and discussion. ComCom 30(9), 2129–2148 (2007)
6. Gutierrez, F.J., Varvarigos, E., Vassiliadis, S.: Multi-cost routing in max-min fair share networks. Allerton Conference on Comm., Control and Comp. 2, 1294–1304 (2000)
7. Ingelrest, F., Simplot-Ryl, D.: Localized broadcast incremental power protocol for wireless ad hoc networks. Wireless Networks 14(3), 309–319 (2008)
8. Kang, I., Poovendran, R.: A novel power-efficient broadcast routing algorithm exploiting broadcast efficiency. VTC 5, 2926–2930 (2003)

9. Kang, I., Poovendran, R.: Broadcast with heterogeneous node capability. In: Global Telecommunications Conference, GLOBECOM, vol. 6, pp. 4114–4119 (2004)
10. Li, D., Jia, X., Liu, H.: Energy efficient broadcast routing in static ad hoc wireless networks. TMC 3(2), 144–151 (2004)
11. Li, D., Liu, Q., Hu, X., Jia, X.: Energy efficient multicast routing in ad hoc wireless networks. ComCom 30(18), 3746–3756 (2007)
12. Li, F., Nikolaidis, I.: On Minimum-Energy Broadcasting in All-Wireless Networks. In: LCN, pp. 193–202 (2001)
13. Li, N., Hou, J., Sha, L.: Design and Analysis of an MST-Based Topology Control Algorithm. In: INFOCOM, vol. 3, pp. 1702–1712 (2003)
14. Liang, W.: Constructing minimum-energy broadcast trees in wireless ad hoc networks. In: MobiHoc, pp. 112–122 (2002)
15. The Network Simulator NS-2, http://www.isi.edu/nsnam/ns/
16. Papageorgiou, C., Kokkinos, P., Varvarigos, E.: Multicost routing over an infinite time horizon in energy and capacity constrained wireless ad-hoc networks. In: Euro-Par, pp. 931–940 (2006)
17. Čagalj, M., Hubaux, J.-P., Enz, C.: Minimum-energy broadcast in all-wireless networks: Np-completeness and distribution issues. In: MobiCom, pp. 172–182 (2002)
18. Wan, P.-J., Calinescu, G., Li, X., Frieder, O.: Minimum-energy broadcast routing in static ad hoc wireless networks. INFOCOM 2, 1162–1171 (2001)
19. Wieselthier, J., Nguyen, G., Ephremides, A.: On the construction of energy-efficient broadcast and multicast trees in wireless networks. In: INFOCOM, pp. 585–594 (2000)

On the Performance of Intra-system Optimization of Virtual Manufacturing Communication Systems

Ainara Gonzalez, Angel Martin, Miguel Angel Urrutia, and Oscar Lazaro

Innovalia Association, CBT Comunicacion & Multimedia
Rodriguez Arias, 6, 605, 48008 Bilbao, Spain
maurrutia@cbt.es
{aigonzalez,amartin,olazaro}@innovalia.org

Abstract. New production environments will demand extensive exchange of communication information. Wireless communications will provide effective means to meet these demands. However, it is important that effective protocols are available to deliver the required QoS. Cross-layer algorithms are potential candidates that exhibit interesting features compared to monolithic traditional protocols. This paper analyses the performance of cross-layer enabled OLSR protocols compared to a hierarchical counterpart. The paper demonstrates that in the scenario defined cross-layer approach outperforms the intra-system optimization capabilities compared to a hierarchical OLSR enabled protocol.

Keywords: Cross-layer, OLSR, HLOSR, Virtual Manufacturing.

1 Introduction

The main objective is optimization to get efficient usage of the scarce radio resources. This will undoubtedly rely on cross-layer designs: Across-layer architecture encompasses an additional complexity relatively to a strictly-layered one, due to the fact that additional information besides the one that defines the basic service provided by the layer has to be exchanged. The need to exchange additional cross-layer information (CLI) leads to two fundamental questions:

- What information should be exchanged across protocol layers, and, how frequently should this exchange proceed?
- What are the adequate / efficient procedures to exchange this information?

This general question is currently being addressed by a large number of projects and authors. The benefits of cross-layer system design are mainly being applied in the area of mobile and wireless operators. In the recent years, the area of communications in the manufacturing is gaining importance. Traditional Ethernet and PROFIBUS factory systems are being enhanced to facilitate new means of automation. The role of wireless communication systems in terms of flexibility and self-configuration are attractive features that major manufacturers are trying to translate into business value to large companies.

F. Granelli et al. (Eds.): MOBILIGHT 2009, LNICST 13, pp. 347–356, 2009.
© ICST Institute for Computer Sciences, Social-Informatics and Telecommunication Engineering 2009

This paper is focused on the intra-system optimization scenario addressed by the European project LOOP. This project caters for the main data generation features related to future manufacturing environments and the communication needs and challenges. To this effect, the paper will initially discuss in Section 2 the concept of Virtual Part as a main source of production information. Next, section 3, will present the scenario and main challenges to be addressed by the communication environment. Then, the communication algorithms selected will be presented in Section 4 and the performance observed will be presented in Section 5. Finally, the main conclusions will be presented in Section 6.

2 Future Manufacturing Scenarios

As stated by Pat Byrne, president of Agilent Technologies' Electronic Measurements Group, the geographic diversification of manufacturing and R&D for many companies has created a challenge in maintaining quality and consistency. New products designed in one country may be prototyped in another and manufactured in yet another or even on another continent. The push to take advantage of the rich diversity of talent across the globe has increased our dependence upon robust measurement tools and techniques to ensure that the performance inherent in designs from the country of origin is maintained across the world at the end of the production line. Systems like the ones depicted in Fig. 1 do not fulfill the requirements set by such statements.

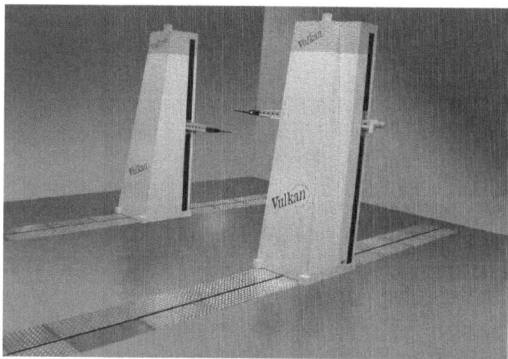

Fig. 1. Traditional Trimek Machine to Capture Dimensional Information of Manufactured Parts

To meet the challenges described above, it is necessary that a large amount of information is made available real-time, anywhere and anytime. The evolution, from a traditional system like the one depicted by Fig. 1, relies on the measurement of object dimensional features by extracting them from their corresponding Virtual Part (Fig. 2) and not from the physical object. A Virtual Part is 3D digital object that has a univocal relationship with its real counterpart, in terms of its dimensional, geometrical and surface characteristics. Hence, associated with the evolution of the manufacturing paradigm is associated the massive communication and management of information.

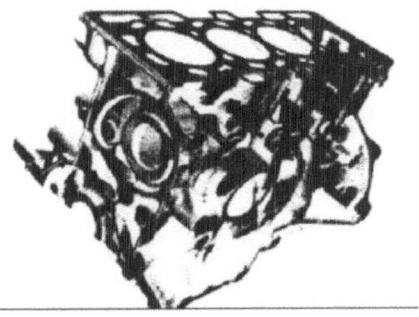

Fig. 2. Example of a Virtual Part

3 Communication Challenges

In order to identify the main communication challenges a scenario has been defined. The definition of a particular case is useful to specify the parameters of the scenario that we will simulate later. In our scenario we consider a car factory with two measurement areas in the same building with four production lines in each. The dimensions of these areas are 100x40m. Furthermore, in each production line there is a Trimek machine, which measures the pieces that are producing in the factory. Once a piece is measured, the machine sends its corresponding 3D Virtual Part to a virtual storage through a FTP connection.

All the information generated during the measurement process of the pieces must be stored in a virtual storage. Apart from the data, we will have to consider the transmission of video to carry out maintenance, calibration or repairing services in the production line

In the figure below we depict the defined use case, where each measurement area is equivalent to an ad-hoc network. There will be a Wi-Fi router in each ad-hoc network in order to have Internet access and it will work as cluster head in the hierarchical configuration. Hence, any component of one of the measurement areas will have the possibility to communicate with any component of the other measurement area.

In case the car factory opens a production plant in a different part of the plant or in a different country, an ad-hoc network connecting all the components of the new factory will be required. To that purpose the only device it should be configured it would be the wireless router. Once this wireless router has Internet access, the incorporated metrology gateways, nomadic workers and PCs would have connectivity with the rest of the components through Internet. Furthermore, these new components will be able to connect to Service of Technical Assistance (SAT) department of Trimek with guaranteed QoS.

The main challenges that the system will have to address are intermittent connectivity, extensibility, movement of users, and machines, high reliability, low cost and high throughput. For this reason an adhoc configuration has been selected for analysis.

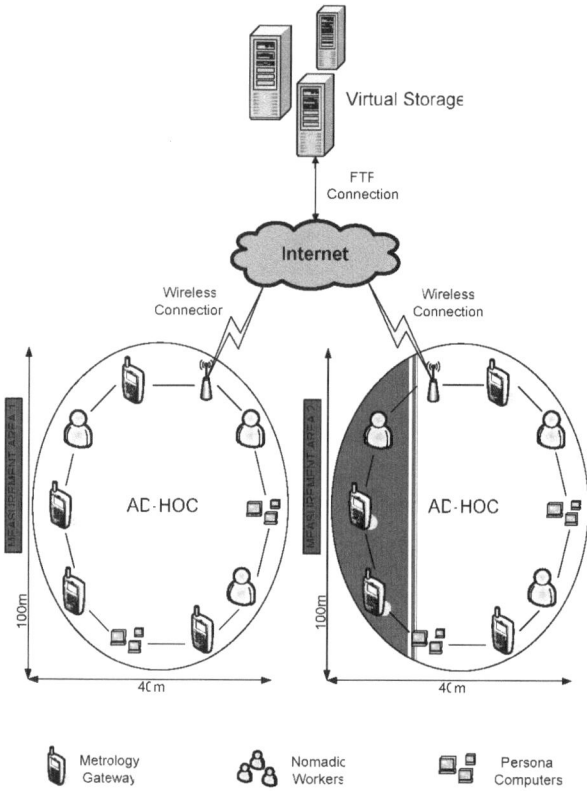

Fig. 3. Use Case Network Scheme

4   Intra-system Communication Optimisation

This section is devoted to analyze which are the most relevant aspects that can be derived in terms of network architecture and cross-layer protocol enhancements derived from the challenges presented in the previous Section.

4.1   Network Architecture

A network to serve the scenario above may be large in terms of both geographic expansion and the number of nodes. However, the penetration of the rate will vary along the deployment time and so would do the routing protocols selected for intra-system optimisation.

Therefore, a hierarchical architecture compared to a "flat" one, have been selected for investigation. The network hierarchy selected, as shown in Fig. 3, is based on a 2-tier hierarchy, which is a good tradeoff between network complexity and scalability.

As shown in the Fig. 3, the network is composed of a number of access networks connected through backbone nodes. The first tier is a backbone network composed of multi-hop connections with long distance wireless links connecting to several access networks. The backbone links are typically based on 802.11a links, and long distances between transmitters and receivers are achieved through directional antennas. The second tier is a mesh access network with short wireless links composed of a set of connected Mesh Routers (MRs) which serve as Access Points (APs) for end users. The connections between MR/APs and end users are typically based on 802.11b/g links. The backbone and access network itself is based on static topology however exhibits ad hoc features. In any case, the end users of this network can be either static (typically home users) or nomadic (typically visitors).

In brief, there are three categories of nodes in the proposed network architecture:

1. *Backbone nodes*: wireless devices used for backbone networks. Backbone nodes take part in routing.
2. *Mesh routers:* wireless devices used for mesh networking and serve as access points for end users. Mesh routers take part in routing.
3. *User equipments:* clients such as PCs, laptops, PDAs, wireless tablets etc. Users equipments are owned by either home users or visitors and do not take part in routing.

4.2 Network Characteristics

Inherited from ad hoc routing protocols, the routing strategies in Wireless Mesh Networks (WMNs) can also be classified as reactive, proactive or a hybrid of them. Although reactive protocols generate less overhead in general, they cannot provide instantaneous node and link status information since no messages are exchanged among mesh routers if there is no data traffic. This means that reactive routing protocols cannot provide real-time network availability information to system administrator, which is crucial from reliable service provisioning point of view. Therefore, the most representative proactive ad hoc routing protocol, Optimised Link State Routing (OLSR), has been selected as the baseline protocol for developing our routing strategy in LOOP networks.

Another reason for selecting OLSR is because of its legacy inter-network connection capability using Host and Node Association (HNA) messages. With this message, a gateway node is able to advertise its Internet *reachability* to all other nodes, so that they can access the Internet through the gateway. It is worth mentioning there that even though Radio-Aware OLSR has not been included in the newest version of the IEEE 802.11s mesh networking standard [1], the function of HNA has been integrated as part of their hybrid routing protocol.

However, the hop-count based OLSR specified in [2] is not able to fulfill the requirements for our targeted network. Therefore, a number of enhancements to the legacy OLSR protocol have been designed within the project, as presented in the following subsections.

4.3 OLSR Enhancement: Hierarchical Structure

There are two levels of hierarchy according to our network design where Level-1 hierarchy corresponds to connection among backbone network nodes, while Level-2 hierarchy corresponds to connection among mesh routers in access networks. An access sub-network which is connected to other access sub-networks is referred to as a cluster. A backbone node serves as the cluster head and advertises its reachability to other clusters periodically. The cluster heads are predefined, thus there is no need to develop an algorithm for cluster head selection. Each cluster-head uses HNA to advertise its reachability for both sides:

- **Inter-cluster.** HNA message advertises a cluster-head's connectivity of all nodes, including both mesh routers and Internet gateway nodes inside the same cluster, to other clusters. This message is sent to all other connected cluster heads using unicast packets (note this is different from the standard version of OLSR), or subnet-directed-broadcast packets. Both the mesh router and the gateways are advertised as connected subnets, specified by the netmask field in HNA.

- **Intra-cluster.** HNA message advertises a cluster-head's connectivity to other clusters, including also Internet gateways from another cluster. This message is sent to all mesh routers inside the same cluster. Both mesh routers and gateways from another cluster are advertised as connected subnets, and are specified by the netmask field in HNA.

- For both inter-cluster and intra-cluster HNA messages, an extended HNA format has been used, so that metric-based routing can be used for gateway selection of any mesh router. Various metrics, for instance, the airtime metric, can be used in our implementation [3]. Moreover, every mesh router in a cluster is advertised as a special type of "gateway", in which it acts as an AP for its clients, and therefore it generates HNA messages as well. The cluster-head, upon receiving this information, establishes an HNA Information Base, which is then used for building the inter-cluster HNA messages to be forwarded to other cluster heads.

4.4 OLSR Enhancement: Multi-homing with Load Balancing

HNA messages in OLSR allow gateway nodes to announce their network association (network address and netmask) with the Internet to other OLSR nodes. When multi-homed, the gateway which is closest to the end-user, in terms of the number of hops, is always chosen as the default gateway by the legacy OLSR. The other gateway will be used only if the default gateway is down, and the process of finding another gateway may take up to a few seconds.

With the implemented multi-homing enhancement, a node uses a metric-based policy to select the best gateway. These metrics include for example link and path capacity, traffic load and other QoS parameters, in addition to the number of hops.

Three types of load balancing have been considered in our network, namely load balancing among channels, paths and gateway nodes. Given that two or more channels co-exist between a pair of nodes, if one channel is close to congestion, another channel should be used. Similarly, if one path is over-loaded, the routing table calculation process will re-calculate a new path. This is triggered by including the traffic load information in a newly defined "LINKINFO" message, which has been implemented as a plug-in to OLSR.

4.5 OLSR Enhancement: Cross Layer Link Layer Notification

When a link break happens, the legacy OLSR will react to this change by exchanging "HELLO" and "TC" messages and this process may take up to a few seconds. With link layer notification, a new path, if existing, will be available immediately (e.g. in the order of milliseconds) after a link break. With this enhancement, we are able to provide the end-users with non-interrupted access.

The basis for this enhancement is to utilize link break information gathered at the MAC layer to impose OLSR routing table re-calculation. More specifically, the MAC layer detects the link break and sends an indication to the protocol layer. Upon receiving such an indication which is treated as a topology or neighbour change, OLSR shall conduct routing table re-calculation immediately.

5 Performance Evaluation

The first analysis carried out on the proposed enhancements is directed to understand which protocols are more effective in a factory configuration. In the scenario analysed we have only considered a single warehouse. The ns-2 simulator has been used.

The simulated scenario considers a 4000 square meter area representing the production lines. Each production line generates a fixed amount of data based on the pieces of work being measured – see Fig. 1. The pieces measured will vary in data size based on the digital information gathered. To our analysis we have considered such data sources as being Constant Bit Rate (CBR), since data will be produced at regular intervals – piece production interval – and the data provided will be always of the same size. Thus, a CBR source with varying bit rate is created and this background traffic is transmitted via ftp. The size of a cloud of points used in the simulation will be 200Mb more or less, and Trimek machines will digitalize a car door in 20ms. Thus, we will need a connection of 10Mbps for data. On the other hand, we must also consider the video of 2'8Mb that will be transmitted in each simulation.

The objective of the scenario is to evaluate the performance of a multimedia service that is carried out on top of this network. This service would permit that timely data for process configuration can be served on real-time. This information is used by production engineers to complement the information received in the form of virtual part.

The network is composed 20 nodes operating in the network. 6 nodes represent nomadic workers. 6 nodes are fixed terminals and the remaining 8 nodes are metrology gateways – routers. In the hierarchical case the nodes are evenly distributed in each cluster.

In the scenario considered, we are mostly interested to analyse the performance of both flat and hierarchical enhancements described in the previous Section. It is of great importance to understand which approach is more effective, namely cross-layer or hierarchical to deploy the correct solution based on the dimension addressed.

The objective of the analysis was to observe the performance of video communication over the network as the transmission of the virtual part was taking place. Such multimedia stream would be directed to experts in assisting the manufacturing decisions all over the plants that are normally very large. The video connection has a bit rate of 128 kbps and a QCIF format.

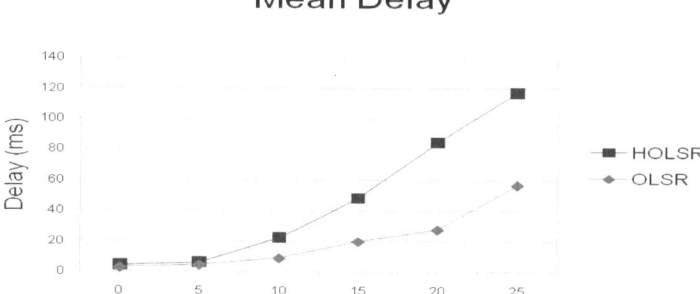

Fig. 4. Average Delay - Hierarchical OLSR vs enhanced-OLSR

To ensure the accuracy of the results 7 independent simulations have been carried out so that a confidence interval smaller than 1% of the results depicted is attained.

First we analyze, the performance of HLOSR and OLSR with increasing number of connections

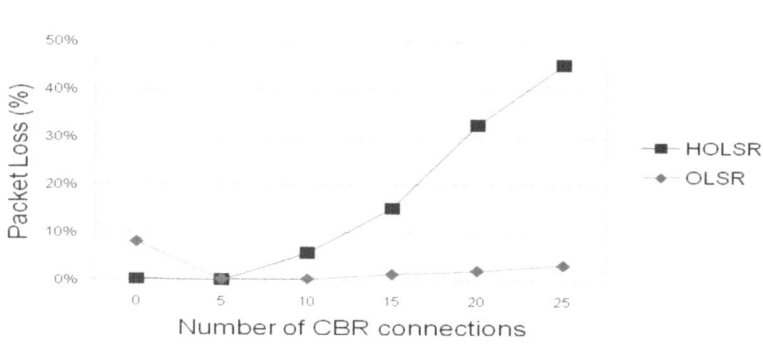

Fig. 5. Average Packet Loss - Hierarchical OLSR vs enhanced-OLSR

As it can be observed, from Fig. 4 and Fig. 5 the performance limiting factor is the Packet Loss. With the packet loss of 5% we can achieve a suitable Peak Signal to Noise Ratio (PSNR) on the video connections. Taking this value as reference we can conclude that a maximum number of 7 connections can be obtained with the hierarchical enhancement compared to the 25 supported with the cross-layer supported OLSR enhancement. This performance can be attributed to the fact that the scenario considered is relatively small in coverage area, so flat structures with a cross-layer support are more effective than hierarchical counterparts. This is related mainly to inter-cluster signaling and overheads created at the cluster heads, which do not prove effective over small areas.

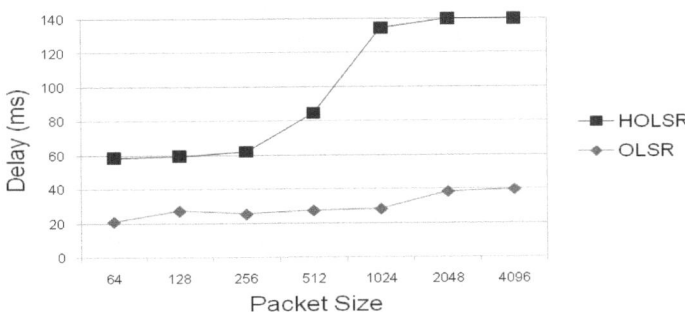

Fig. 6. Average Delay - Hierarchical OLSR vs. enhanced-OLSR

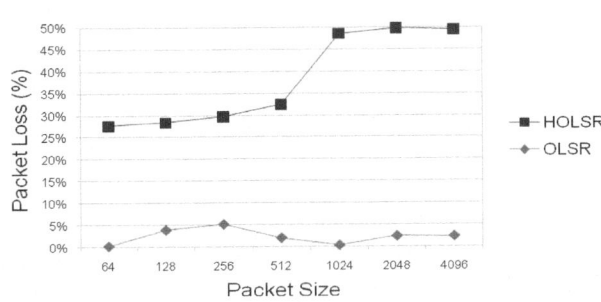

Fig. 7. Average Packet Loss - Hierarchical OLSR vs. enhanced-OLSR

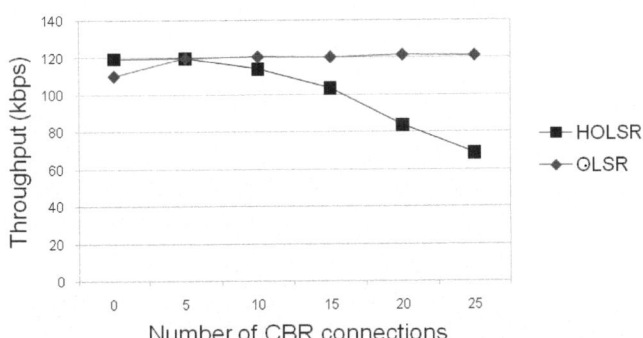

Fig. 8. Throughput - Hierarchical OLSR vs. enhanced-OLSR

The second analysis is carried out in terms of the packet size employed for transmission.

In terms of the packet size employed (Fig. 7), the performance is also very similar between both algorithms and the previous observations in terms of performance and intra-cluster heavy load traffic are reinforced.

To conclude the previous discussion, Fig. 8 shows that the average throughput obtained in the OLSR case is more favorable compared to the Hierarchical OLSR (HLOSR) one.

6 Performance Evaluation

This paper has presented the challenges posed by the future manufacturing scenarios and how cross-layer design could help to meet such demands. Two different strategies, namely cross-layer OLSR and HOLSR have been proposed to face the communication needs of such scenarios. To evaluate the performance of such algorithms a user scenario has been defined and the algorithms simulated. The results obtained suggest that due to the smaller area over which the communication network is deployed, intra-system optimization based on cross-layer approaches is more effective than the hierarchical counterparts.

References

1. IEEE 802.11 WG TGs, Draft Amendment to Standard IEEE 802.11TM: ESS Mesh Networking, P802.11sTM /D2.00 (May 2008)
2. Clausen, T., Jacquet, F.: Optimized Link State Routing Protocol (OLSR), IETF RFC 3626 (October 2003)
3. Aure, T., Li, F.Y.: An optimized path-selection using the airtime metric in OLSR networks: Implementation & Testing. In: Proceedings IEEE International Symposium on Wireless Communication Systems (ISWCS), Reykjavik, Iceland, October 21-24 (2008)

Performance Analysis of IEEE 802.15.4 Non-beacon Mode with Both Uplink and Downlink Traffic in Non-saturated Condition

Tae Ok Kim[1], Jin Soo Park[2], Kyung Jae Kim[1], and Bong Dae Choi[1,⋆]

[1] Department of Mathematics and Telecommunication Mathematics
Research Center, Korea University, Seoul, Korea
{violetgl,kimkjae,queue}@korea.ac.kr
[2] USN Service Division, KT, Seoul, Korea
vtjinsoo@paran.com

Abstract. We analyze the MAC performance of the IEEE 802.15.4 LR-WPAN with non-beacon mode and non-saturated condition in a star topology. Our approach is to model stochastic behavior of one device with both uplink and downlink traffic as a discrete time Markov chain. First, we propose an analytical model of a device with only downlink traffic. Then, by combining the model of a device with only uplink traffic in [3] and one with downlink traffic in this paper, we obtain the performance measures such as throughput, packet delay, energy consumption and packet loss probability of a device with both uplink and downlink traffic. Our results can be used to find the optimal number of devices so as to satisfy QoS (quality of service) on delay and loss probability.

Keywords: IEEE 802.15.4, Medium Access Control(MAC) protocol, CSMA/CA, Markov Chain, performance analysis.

1 Introduction

IEEE 802.15.4[1,2] is a standard toward low complexity, low power consumption and low data rate wireless data connectivity. Therefore IEEE 802.15.4 will play a key role as a MAC protocol at WSN(Wireless Sensor Network) where energy consumption is an important factor.

IEEE 802.15.4 low rate WPAN (LR-WPAN) allows two network topologies: star and peer-to-peer. In a star topology, every sensor device must communicate with a PAN coordinator, while in a peer-to-peer topology, all devices can communicate each other. In a star topology, network uses two types of network channel

---

⋆ This paper was presented in part at the IEEE International Symposium on Pervasive Computing and Ad Hoc Communications, May 2007. This research is supported by the MIC, under the ITRC support program supervised by the IITA, and supported by KT.

F. Granelli et al. (Eds.): MOBILIGHT 2009, LNICST 13, pp. 357–371, 2009.

access mechanism, non-beacon mode and beacon-enabled mode, depending on whether the network supports the transmission of beacons.

Diverse applications for wireless sensor network based on IEEE 802.15.4 have generated interests in analytical models of access mechanism based on CSMA/CA. Pollin et al.[6] and Park et al.[7] proposed analytic model for uplink traffic on IEEE 802.15.4 beacon-enabled mode under saturated condition where devices have always packets to send. In real environment, packets are generated in not too often, so that a device will have no packets to send or receive for most of time. Therefore we need to investigate non-saturated case where a device does not have packets to send or receive for some period of time. Misic et al.[8] analyzed performance of IEEE 802.15.4 with both uplink and downlink traffic in beacon mode under non-saturated condition by modeling of discrete-time Markov chains and the theory of M/G/1 queues. Also, Kim et al.[9] analyzed performance of uplink communication in non-beacon mode with unslotted CSMA/CA by busy cycle of M/G/1 queueing system.

This paper attempts to analyze the MAC performance of star-shaped IEEE 802.15.4 network in non-saturated condition running under non-beacon mode. Unlike beacon mode operation where nodes periodically wake up to listen to the beacon frame, nodes in non-beacon mode need to wake up only when they have packets to upload or when they request to download. This way, the nodes can save energy unless otherwise spent on listening to the beacon frame. For example, in practical situation like the forest fire monitoring system, the traffic is quite rarely generated and it is unnecessary that sensor nodes wake up frequently. The simulation results in Section 5 show that the non-beacon mode reduces the energy consumption significantly compared with the beacon mode without too much degrading downlink delay (See Fig. 5). This provides our motivation to investigate the non-beacon mode. In this paper, we assume that all nodes are synchronized, which can be realized by following method. The PAN coordinator broadcasts periodic signal (e.g. sine signal) for synchronization through an extra channel. When device has packet to transmit, the device synchronized by sensing the extra channel. In the environment that time clocks of all devices are synchronized by this approaches, we adopt the non-beacon mode with slotted CSMA/CA as MAC transmission procedure.

We model the stochastic behavior of a device with both uplink traffic and downlink traffic as a discrete-time Markov chain. Note that our Markov chain model of IEEE 802.15.4 is different from one of IEEE 802.11 [5], since no freezing of backoff counter operates during transmission of other devices in IEEE 802.15.4. We analyzed the performances of station with uplink traffic[3] and downlink traffic[4] individually under unsaturation condition. However, the practical situation where uplink and downlink traffic coexist gives us the motivation of this paper. So, first, we propose the analytical model[4] of a device with only downlink traffic with some modification. Then, by combining the uplink model and downlink model, we can make the analytical model of a device with both uplink and downlink traffics. Finally, we obtain the performance measures such as throughput, average packet delay, packet loss probability and energy consumption for the network with both uplink and downlink traffic.

This paper is organized as follows. In Section 2, we describe the MAC procedure for uplink and downlink in IEEE 802.15.4 non-beacon mode. In the Section 3, we propose the analytic model of a device with only downlink traffic under non-saturated condition and obtain performance measures from our analysis. In the Section 4, we make the analytic model of a device with both uplink and downlink traffics by combining the model of a device with uplink traffic in [3] and one with downlink traffic in Section 3 and obtain performance measures. Numerical results for performance measures of the network with both uplink and downlink traffic are presented in Section 5.

2 MAC Procedure for Uplink and Downlink in Non-beacon Mode

When applying the slotted CSMA/CA, the MAC sublayer is delayed for a random number j of backoff slots (the chosen random number of backoff slots is called backoff counter) in the range 0 to $2^{BE} - 1$. BE is the backoff exponent, which is related to how many attempts (the number of attempts is called backoff stages) a device has tried to access the channel. BE shall be initialized to the value of $macMinBE$ (this corresponds to the 0th backoff stage). The backoff counter j is decreased at the boundary of each backoff slot. In IEEE 802.15.4, the backoff counter is decremented regardless of the channel status contrary to that of IEEE 802.11 DCF. When the backoff counter reaches to zero, CCA is performed twice before transmitting a packet. CCA(clear channel assessment) is the procedure that a device listens to the channel to make sure the channel clear before attempting to transmit a packet. When the channel is found to be idle during two CCA periods, the device shall transmit its packet. When the channel is found to be busy during either the first CCA or the second CCA, backoff stage is increased by 1 and BE is increased by 1 until BE reaches $macMaxBE$ (let N be the backoff stage that BE reaches $macMaxBE$), and then a random backoff is tried again. If the transmitted packet suffers collision, it will restart from the beginning of procedure. If one of two CCAs fails at the Mth backoff stage where M is the maximum retransmission number, the packet is discarded.

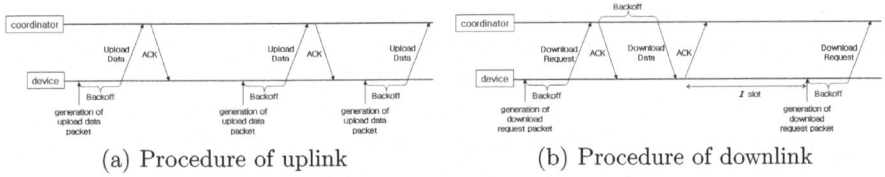

(a) Procedure of uplink   (b) Procedure of downlink

Fig. 1. Download sequences in a non-beacon network

The communication sequences for uplink and downlink in a non-beacon network are shown in Fig. 1. When a device has a data packet to transmit, it simply transmits its uplink data packet, using the slotted CSMA/CA, to the PAN coordinator. The coordinator sends an acknowledgment packet which notifies the successful reception of the uplink data packet.

When the coordinator has a data packet to send to a certain device, it stores the downlink data packet and waits until the device requests downlink transmission. A device makes a contact by transmitting a downlink request packet, using the slotted CSMA/CA, periodically. The coordinator sends an acknowledgment packet which notifies the successful reception of the downlink request packet and existence of downlink data packet. If data are pending, the coordinator transmits the downlink data packet, using the slotted CSMA/CA, to the device. The standard allows the coordinator to send a downlink data packet without using the CSMA/CA after the acknowledgment packet for downlink request packet. Analysis of this method is studied by Kim *et. al.*[?]. After receiving the downlink data packet, the device sends an acknowledgment packet which notifies the successful reception of the downlink data packet.

3   Analysis for a Device with Only Downlink Traffic

Let n sensor devices be associated with the PAN coordinator. For downlink in IEEE 802.15.4 non-beacon mode, a device sends a downlink request packet to the PAN coordinator periodically to check whether there is a downlink data packet at the PAN coordinator. We assume that a device generates a downlink request packet after fixed number I of slots from the moment of the completion of the previous downlink procedure (See Fig. 1). We also assume that it takes exponential random time with mean $\frac{1}{\lambda_d}$ that the PAN coordinator generates a downlink data packet destined to the tagged device after the previous packet is transmitted.

3.1   Mathematical Model

Let $s(t)$, $0 \le s(t) \le M$, be the backoff stage and $b(t)$ be the backoff counter. Let $(s(t), b(t))_r$ and $(s(t), b(t))_d$ denote the backoff stage and backoff counter for downlink request packet and downlink data packet, respectively. When the channel is idle at the first CCA, we define $b(t) = -1$. We assume that size of downlink request packet is fixed R in the unit of slots. Let $Tx_r[k]$, $1 \le k \le R$, represent the state of the kth slot of downlink request packet transmission. Let Rx_d represent the state of downlink data packet in transmission. We assume that the length of data packet measured in slots is geometrically distributed with mean $\frac{1}{1-P_{Rx,d}}$ where $P_{Rx,d}$ is the probability that current transmission continues at the next slot. Let the duration for both waiting and receiving ACK be A slots (Default value of A is equal to 2). Let $(-1, k)$, $1 \le k \le A$, represent the state of the kth slot of the duration for waiting and receiving ACK. Let $idle[k]$, $1 \le k \le I$, represent the state of the kth slot from the start of duration of fixed length I for generating downlink request packet. Define $Y(t)$ at t by :

$$Y(t) = \begin{cases} idle[k], & \text{when a device is in the } k\text{th slot of state before generating downlink request packet} \\ (s(t), b(t))_r, & \text{when a device is in the process of backoff for downlink request packet} \\ (s(t), -1)_r, & \text{when channel is idle at the first CCA for downlink request packet} \\ Tx_r[k], & \text{when a device is in the } k\text{th slot of a downlink request packet transmission} \\ (-1, k)_r, & \text{when a device is in the } k\text{th slot of waiting and receiving ACK} \\ & \text{for downlink request packet} \\ (s(t), b(t))_d, & \text{when PAN coordinator is in the process of backoff for downlink data packet} \\ (s(t), -1)_d, & \text{when channel is idle at the first CCA for downlink data packet} \\ Rx_d & \text{when a device receives a downlink data packet} \\ (-1, k)_d, & \text{when PAN coordinator is in the } k\text{th slot of waiting and receiving ACK} \\ & \text{for downlink data packet} \end{cases} \tag{1}$$

Then $Y(t)$ is a discrete Markov chain with one-step transition probabilities described in Fig. 3 for downlink procedure. Let $\pi_{(i,j)_r}$, $\pi_{(i,-1)_r}$, $\pi_{(-1,k)_r}$, $\pi_{Tx_r[k]}$, $\pi_{(i,j)_d}$, $\pi_{(i,-1)_d}$, $\pi_{(-1,k)_d}$, π_{Rx_d} and $\pi_{idle[k]}$ be the steady-state probability which can be obtained by solving the balance equations.

Next we will calculate the probability α of channel being busy at the first CCA, the probability β of channel being busy at the second CCA and the probability P_s of successful packet transmission. Note that these probabilities have the same values for downlink request packet and downlink data packet because these values are determined by the states of other $n-1$ devices. Since the probability of the channel being idle at the first CCA for the given device is equal to the probability that the all other $n-1$ devices are not in the states of $Tx_r[k]$, $(-1, k)_r$, Rx_d and $(-1, k)_d$. Therefore α is given by :

$$\alpha = 1 - (1 - \pi_d)^{n-1} , \tag{2}$$

where

$$\pi_d = \sum_{k=1}^{R} \pi_{Tx_r[k]} + \pi_{Rx_d} + \sum_{j=0}^{A} (\pi_{(-1,k)_r} + \pi_{(-1,k)_d}).$$

Note that in order to be eligible to sense the channel at the second CCA, the channel must be idle at the first CCA. So β is the probability that the channel is busy when the tagged device senses at the second CCA, given that the channel is idle at the first CCA, i.e,

$$\begin{aligned} 1 - \beta &= P\{\text{channel is idle at the second CCA} \mid \text{channel is idle at the first CCA}\} \\ &= \frac{P\{\text{channel is idle at the first CCA, channel is idle at the second CCA}\}}{P\{\text{channel is idle at the first CCA}\}} \\ &= \frac{\left\{1 - \pi_d - \sum_{i=0}^{M} (\pi_{(i,-1)_r} + \pi_{(i,-1)_d})\right\}^{n-1}}{1 - \alpha} \end{aligned} \tag{3}$$

The successful transmission probability, P_s, can be represented by :

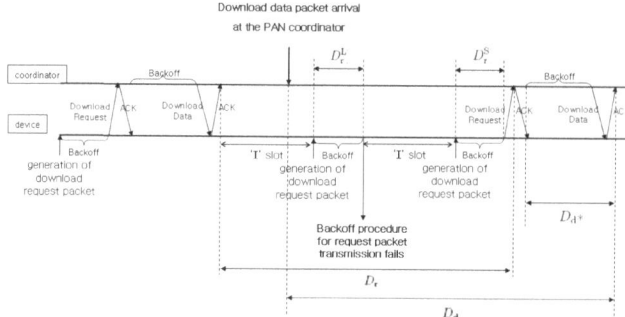

Fig. 2. Description of D_r, D_r*, D_d, and $D_r + I$

P_s = P{successful transmission | channel is idle at both the first CCA and the second CCA}

$$= \frac{\left\{1 - \pi_d - \sum_{i=0}^{M}\left(\pi_{(i,0)_r} + \pi_{(i,-1)_r} + \pi_{(i,0)_d} + \pi_{(i,-1)_d}\right)\right\}^{n-1}}{\left\{1 - \pi_d - \sum_{i=0}^{M}\left(\pi_{(i,-1)_r} + \pi_{(i,-1)_d}\right)\right\}^{n-1}} \quad (4)$$

Let e_d be the probability that there is a downlink data packet at the PAN coordinator when downlink request packet arrives at the PAN coordinator. This event occurs when downlink data packet arrives during the time duration, D_r, from the completion of one downlink procedure to the next arrival of downlink request packet at the PAN coordinator (See Fig. 2). The expected delay $E[D_r]$ are calculated by :

$$E[D_r] = \sum_{k=0}^{\infty}(P_{loss})^k(1 - P_{loss}) \cdot \{k(I \cdot \sigma + E[D_r^L]) + (I \cdot \sigma + E[D_r^S])\} + R \cdot \sigma \quad (5)$$

where P_{loss} is the probability of losing downlink request packet (given by (9)) and σ is the length of a slot. The expected delay $E[D_r^L]$ from the moment of generation of downlink request packet to the moment of discarding the packet and the expected delay $E[D_r^S]$ from the moment of generation of downlink request packet to the moment of beginning of downlink request packet transmission are calculated in Appendix 6. So, e_d is approximately calculated using $E[D_r]$.

$$e_d \approx 1 - e^{-\lambda_d \cdot E[D_r]} \quad (6)$$

To check the accuracy of the approximation (6), we simulated the system and it turns out that the approximation (6) is quite good (See Fig. 4).

Note that α, β, P_s and e_d in (2), (3), (4) and (6) express in terms of steady-state probability and vice versa. Therefore by solving nonlinear equation of (2), (3), (4), (6), balance equations of this Markov Chain and normalization condition, we obtain all necessary values such as steady-state probability, α and β.

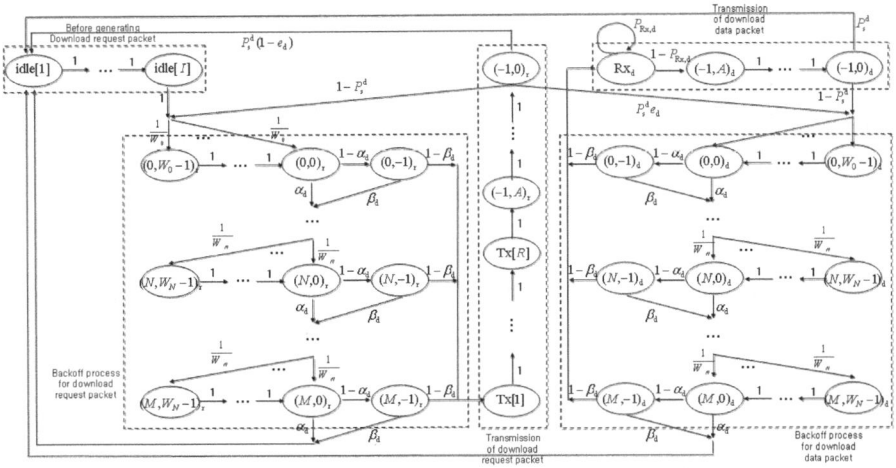

Fig. 3. Markov Chain for Downlink

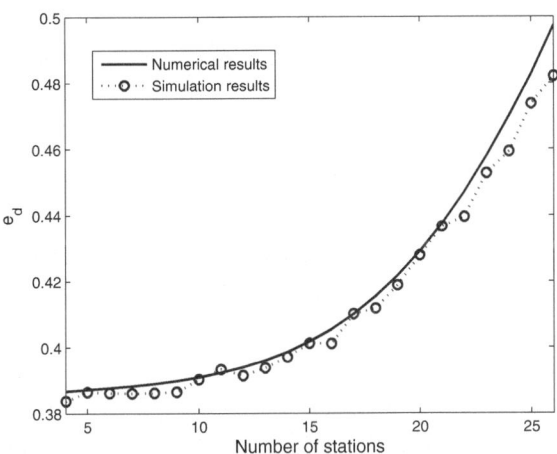

Fig. 4. Numerical and simulation results for e_d

3.2   Performance Measures

In this subsection, we obtain several performance measures such as throughput, delay, loss probability and energy consumption.

Throughput. The normalized system throughput S, defined as the fraction of time the channel is used to transmit downlink data packet successfully, is given as follows.

$$S = n \cdot \pi_{Rx_d} \cdot P_s \qquad (7)$$

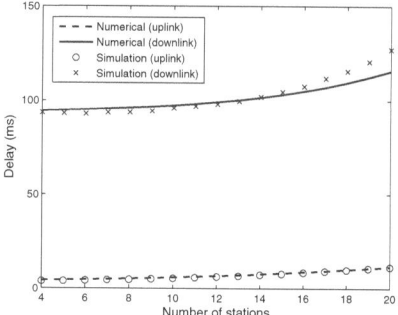

(a) Expected delay for uplink and downlink

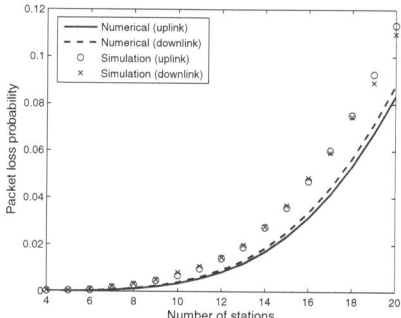

(b) Packet loss probability for uplink and downlink

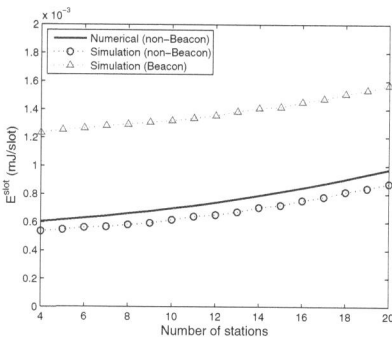

(c) Energy consumption for a device

Fig. 5. Numerical and simulation Results : Performance measures

Delay. The expected delay $E[D_d]$ from the moment of downlink data packet arrival at the PAN coordinator to service completion point is approximately calculated by :

$$E[D_{\mathrm{d}}] \approx E[D_{\mathrm{r}}] + A + E[D_{\mathrm{d}}*] - \frac{\int_0^{E[D_{\mathrm{r}}]} x \cdot \lambda_d e^{-\lambda_d x}\, dx}{1 - e^{-\lambda_d E[D_{\mathrm{r}}]}} \tag{8}$$

where $E[D_{\mathrm{d}}*]$ is the expected duration from the beginning of backoff procedure for downlink data packet transmission to the moment of service completion of the downlink data packet (given by (24) in Appendix 6). The last term in the right-hand side represents the average duration from the completion of previous downlink procedure to a arrival of next downlink data packet.

Packet Loss Probability. The probabilities of losing downlink request packet and downlink data packet are same and let it denoted by P_{loss}. Then we have

$$P_{\mathrm{loss}} = \sum_{v=0}^{M} \sum_{w=0}^{v} {}_v C_w \alpha^w \{(1-\alpha)\beta\}^{v-w}(1-\alpha)(1-\beta)(1-P_s)P_{\mathrm{loss}}$$
$$+ \sum_{w=0}^{M} {}_M C_w \alpha^w \{(1-\alpha)\beta\}^{M-w}\{\alpha + (1-\alpha)\beta\} \tag{9}$$

The general term in the first summation of (9) is the probability that the packet suffers loss after collision at the vth backoff stage in the first backoff procedure. Note that after collision the procedure starts from the 0th backoff stage again. The second term of (9) is the probability that the packet in the first backoff procedure suffers loss because channel is busy at the first CCA or the second CCA at the Mth backoff stage.

Energy Consumption. Since power is quite critical in a sensor network, energy consumption is the most important performance measure. To obtain the total lifetime of a battery, we need a concept of average energy consumption. Park et al. [7] and Pollin et al. [6] define the normalized energy consumption as the average energy consumption to transmit one slot amount of payload. Their definition has good explanation in saturation mode. However, in non-saturation mode, their definition mismatches with our intuition, as they [6] mentioned that the energy consumption increases as the arrival rate decreases, or equivalently idle period increases. See Fig. 9 in [6]. So, we calculate the average energy consumption E^{slot} per one slot(mJ/slot). Let E_{idle}, E_{Tx} and E_{Rx} be the energy consumption for idle slot, transmission slot and reception(or CCA) slot, respectively. Since energy consumption for reception slot and CCA slot are equal, we do not distinguish the valus. Let a^{idle}, a^{Tx} and a^{Rx} be the probabilities of slot being idle, being transmission, being reception(or CCA). Then,

$$a^{\mathrm{idle}} = \sum_{k=1}^{I} \pi_{idle[k]} + \sum_{i=0}^{M} \sum_{j=1}^{W_i-1} \pi_{(i,j)_{\mathrm{r}}}$$

$$a^{Tx} = \pi_{Tx_r[k]} + \left(\sum_{j=0}^{A} \pi_{(-1,k)_{\mathrm{d}}} \right) P_s$$

$$a^{Rx} = 1 - a^{\mathrm{idle}} - a^{Tx}$$

Note that a device consumes E_{Rx} per one slot when it waits the downlink data from PAN coordinator.

The average energy consumption E^{slot} per one slot is obtained as follows.

$$E^{slot} = a^{idle}E_{idle} + a^{Tx}E_{Tx} + a^{Rx}E_{Rx} \tag{10}$$

4   Analysis for a Device with Both Uplink and Downlink Traffic

We assume that uplink traffic has higher priority than downlink traffic in a sense that, if uplink data packet and downlink request packet attempt to transmit at the same slot, then uplink data packet is allowed to transmit and downlink request packet acts as if it suffers a collision. A device generates uplink data packets according to Poisson process with rate λ_u and the PAN coordinator generates downlink data packets destined to the tagged device according to Poisson process with rate λ_d, independently each other. We assume that a device and the PAN coordinator can accommodate only one uplink packet and only one downlink packet for the tagged device, respectively. A device generates a downlink request packet after fixed number I of slots from the completion of downlink procedure.

4.1   Mathematical Models

We combine the Markov chain model (called uplink model here) proposed in [3] for uplink traffic with different values α, β and P_s from ones in [3] and the Markov chain model (called downlink model here) proposed in Section 3 with different values α, β and P_s from ones in section 3. All traffics share a common channel and so they compete with each other to catch the common channel. Therefore, the probabilities α, β and P_s in each model (e.g. uplink model) relate not only with steady-state probability in its own model (e.g. uplink model) but also with steady-state probability in the other model (e.g. downlink model). For classification of index, we will use the subscript 'u', 'r' and 'd' for uplink data, downlink request, and downlink data, respectively. Note that, in the downlink model, the probability β_r for the downlink request packet are different from the probability β_d for downlink data packet because uplink traffic has higher priority than downlink traffic as mentioned in the beginning of this section.

The probability of the channel being idle at the first CCA for the tagged device is equal to the probability that any other transmission does not occur in the first CCA slot. Therefore α_u, α_r and α_d are given by

$$\alpha_u = 1 - (1 - \pi_u)^{n-1}(1 - \pi_d)^n \tag{11}$$

and

$$\alpha_r(= \alpha_d) = 1 - (1 - \pi_u)^n(1 - \pi_d)^{n-1} , \tag{12}$$

where

$$\pi_{\mathrm{u}} = \pi_{Tx_{\mathrm{u}}} + \sum_{j=0}^{A} \pi_{(-1,k)_{\mathrm{u}}} \quad \text{and}$$

$$\pi_{\mathrm{d}} = \pi_{Rx_{\mathrm{d}}} + \sum_{k=1}^{R} \pi_{Tx_{\mathrm{r}}[k]} + \sum_{j=0}^{A} \left(\pi_{(-1,k)_{\mathrm{d}}} + \pi_{(-1,k)_{\mathrm{r}}} \right).$$

Since β_{u} is the probability that the channel is busy when the tagged device senses its second CCA for uplink data transmission given that the channel was idle slot at the first CCA, β_{u} is given by :

$$\beta_{\mathrm{u}} = 1 - \frac{\left(1 - \pi_{\mathrm{u}} - \sum_{i=0}^{M} \pi_{(i,-1)_{\mathrm{u}}}\right)^{n-1} \left\{1 - \pi_{\mathrm{d}} - \sum_{i=0}^{M} \left(\pi_{(i,-1)_{\mathrm{r}}} + \pi_{(i,-1)_{\mathrm{d}}}\right)\right\}^{n}}{1 - \alpha_{\mathrm{u}}} \quad (13)$$

β_{r} is the probability that the tagged device fails in the second CCA for transmission of downlink request packet. The failure in the second CCA for transmission of downlink request packet comes from the following two cases. The first case is that the channel is busy at the second CCA for downlink request packet. The second case is that the the uplink data packet in the same device attempts transmission while the channel is idle at the second CCA for downlink request packet. Therefore, β_{r} is given by :

$$\beta_{\mathrm{r}} = 1 - \frac{\left\{1 - \pi_{\mathrm{u}} - \sum_{i=0}^{M} \left(\pi_{(i,0)_{\mathrm{u}}} + \pi_{(i,-1)_{\mathrm{u}}}\right)\right\} \left[\left(1 - \pi_{\mathrm{u}} - \sum_{i=0}^{M} \pi_{(i,-1)_{\mathrm{u}}}\right)\left\{1 - \pi_{\mathrm{d}} - \sum_{i=0}^{M} \left(\pi_{(i,-1)_{\mathrm{r}}} + \pi_{(i,-1)_{\mathrm{d}}}\right)\right\}\right]^{n-1}}{1 - \alpha_{\mathrm{r}}}$$

$$(14)$$

β_{d} is the probability that the channel is busy when the PAN coordinator senses it second CCA for downlink data transmission for tagged device. Therefore, β_{d} is given by :

$$\beta_{\mathrm{d}} = 1 - \frac{\left(1 - \pi_{\mathrm{u}} - \sum_{i=0}^{M} \pi_{(i,-1)_{\mathrm{u}}}\right)^{n} \left\{1 - \pi_{\mathrm{d}} - \sum_{i=0}^{M} \left(\pi_{(i,-1)_{\mathrm{r}}} + \pi_{(i,-1)_{\mathrm{d}}}\right)\right\}^{n-1}}{1 - \alpha_{\mathrm{d}}} \quad (15)$$

The successful transmission probability, i.e. $P_{\mathrm{s}}^{\mathrm{u}}$, $P_{\mathrm{s}}^{\mathrm{r}}$ and $P_{\mathrm{s}}^{\mathrm{d}}$, are all same and given by :

$$P_{\mathrm{s}}^{\mathrm{u}}(= P_{\mathrm{s}}^{\mathrm{r}} = P_{\mathrm{s}}^{\mathrm{d}}) = P\{\text{successful Tx} \mid \text{both the first CCA and the second CCA are succeed}\}$$
$$= \frac{\left\{1 - \pi_{\mathrm{u}} - \sum_{i=0}^{M} \left(\pi_{(i,0)_{\mathrm{u}}} + \pi_{(i,-1)_{\mathrm{u}}}\right)\right\}^{n-1} \left\{1 - \pi_{\mathrm{d}} - \sum_{i=0}^{M} \left(\pi_{(i,0)_{\mathrm{r}}} + \pi_{(i,-1)_{\mathrm{r}}} + \pi_{(i,0)_{\mathrm{d}}} + \pi_{(i,-1)_{\mathrm{d}}}\right)\right\}^{n-1}}{\left(1 - \pi_{\mathrm{u}} - \sum_{i=0}^{M} \pi_{(i,-1)_{\mathrm{u}}}\right)^{n-1} \left\{1 - \pi_{\mathrm{d}} - \sum_{i=0}^{M} \left(\pi_{(i,-1)_{\mathrm{r}}} + \pi_{(i,-1)_{\mathrm{d}}}\right)\right\}^{n-1}} \quad (16)$$

4.2   Performance Measures

In this subsection, we obtain several performance measures such as throughput, delay, loss probability and energy consumption.

Throughput. The normalized system throughput S, defined as the fraction of time the channel is used to transmit successfully for uplink data packet and downlink data packet, is given as follows.

$$S = n \cdot (\pi_{Tx_u} \cdot P_s^u + \pi_{Rx_d} \cdot P_s^d) \tag{17}$$

Delay. The average delay $E[D_u]$ from the moment of uplink data packet arrival at device to the moment of service completion point, can be obtained as the same form as the equation (6) in [3] by just replacing parameters α, β and P_s by α_u, β_u and P_s^u, respectively. Similarly, $E[D_r]$ is obtained as the same form as the equation (5) by just replacing parameters α, β and P_s by α_r, β_r and P_s^r in the equations for calculation of P_{loss}, $E[D_r^L]$ and $E[D_r^S]$, respectively. Then, $E[D_d]$ is calculated by the same form as the as the equation (8).

Packet Loss Probability. The uplink packet loss probability P_{loss}^u can be obtained as the same form with the equation (7) in [3] by just replacing parameters α, β and P_s by α_u, β_u and P_s^u, respectively. Let P_{loss}^r be the probability of losing downlink request packet and P_{loss}^d be the probability of losing downlink data packet. Then P_{loss}^r can be obtained as the same form with the equation (9) in 3.2 by just replacing parameters α, β and P_s by α_r, β_r and P_s^r, respectively. Moreover, P_{loss}^d can be obtained as the same form with the equation (9) in 3.2 by just replacing parameters α, β and P_s by α_d, β_d and P_s^d, respectively.

Energy Consumption. We calculate the average energy consumption E^{slot} per one slot(mJ/slot). Let a_u^{idle}, a_u^{Tx} and a_u^{Rx} be the probabilities of slot being idle, being transmission, being reception(or CCA) for uplink traffic, respectively. Then,

$$a_u^{idle} = 1 - \sum_{i=0}^{M}(\pi_{(i,0)_u} + \pi_{(i,-1)_u}) - \pi_{Tx_u} - \sum_{j=0}^{A}\pi_{(-1,j)_u}$$

$$a_u^{Tx} = \pi_{Tx_u}$$

Similarly, let a_d^{idle}, a_d^{Tx} and a_d^{Rx} be the probabilities of slot being idle, being transmission, being reception(or CCA) for downlink traffic. Then

$$a_d^{idle} = \sum_{k=1}^{I}\pi_{idle[k]} + \sum_{i=0}^{M}\sum_{j=1}^{W_i-1}\pi_{(i,j)_r}$$

$$a_d^{Tx} = \pi_{Tx_r[k]} + \left(\sum_{j=0}^{A}\pi_{(-1,k)_d}\right)P_s$$

Thus E^{slot} is calculated as follows.

$$E^{slot} = a_u^{idle}a_d^{idle}E_{idle} + (a_u^{Tx} + a_d^{Tx})E_{Tx} + \{1 - a_u^{idle}a_d^{idle} - (a_u^{Tx} + a_d^{Tx})\}E_{Rx} \tag{18}$$

5   Numerical Results and Simulation Results for Both Uplink and Downlink Traffic

In this section, numerical results for performance measures of the network with both uplink and downlink traffic are presented. For our numerical results, I is set to 500 backoff slots. The average length of a uplink data packet, $\frac{1}{1-P_{\mathrm{Tx,u}}}$, is set to 4 and the average length of a downlink data packet, $\frac{1}{1-P_{\mathrm{Rx,d}}}$, is set to 4. Note that $\sigma = 0.32$ms in case of 250 Mbps, 2.4 GHz. N and M are 2 and 4, respectively. W_0 is set to $2^3 = 8$ in our experiment. The energy consumptions at T_x, R_x, and CCA states are 0.0100224mJ, 0.0113472mJ and 0.0113472mJ, respectively, [7]. A device consumes 0.000056736mJ during idle state.

Fig. 5(a) depicts the expected delay $\mathrm{E}[D_{\mathrm{u}}]$ for uplink traffic and the expected delay $\mathrm{E}[D_{\mathrm{d}}]$ for downlink traffic. As the number of devices increases, $\mathrm{E}[D_{\mathrm{u}}]$ and $\mathrm{E}[D_{\mathrm{d}}]$ increase due to the exponential backoff by competitions of each other. Fig. 5(b) depicts the packet loss probability P_{loss}^u and P_{loss}^d for uplink traffic and downlink traffic, respectively. Also P_{loss}^u and P_{loss}^d increase as the number of devices increases. Fig. 5(c) depicts the average energy consumption E^{slot} per one backoff slot. Fig. 5 shows that the numerical results and simulation results for performance measures differ slightly. This may be caused by the analytical model where two approximations (6) and (8) are used. In Fig. 5(c), we also compare the energy consumption between the non-beacon mode and the beacon mode through simulation. For this comparison, we set the superframe duration by 96 backoff slots. From the simulation results, we find out that the non-beacon mode reduces the energy consumption about 50% compared with beacon mode, which the downlink delay is within $100msec$ (See Fig. 5(a)). Finally, our results are used for determining the optimal number of devices which can be accommodated in the system while supporting the required QoS on the expected packet delay and the packet loss probability. For instance, with the requirements of $\mathrm{E}[D_{\mathrm{u}}]\leq 20ms$, $\mathrm{E}[D_{\mathrm{d}}]\leq 100ms$, $P_{\mathrm{loss}}^u \leq 2\%$ and $P_{\mathrm{loss}}^d \leq 2\%$, the optimal number of devices in the network is from Fig. 5(a) and Fig. 5(b). With this case, we obtain from Fig. 5(c) that the average energy consumption E^{slot} per one backoff slot is 7.1×10^{-4}mJ/slot.

6   Appendix : Delay for Downlink

In this section, we obtain the expected durations $\mathrm{E}[D_{\mathrm{r}}^{\mathrm{L}}]$, $\mathrm{E}[D_{\mathrm{r}}^{\mathrm{S}}]$ and $\mathrm{E}[D_{\mathrm{d}}*]$ (See Fig. 2). $\mathrm{E}[D_{\mathrm{r}}^{\mathrm{L}}]$ is the expected time duration from the moment of generation of downlink request packet to the moment of discarding the packet, and $\mathrm{E}[D_{\mathrm{r}}^{\mathrm{S}}]$ is the expected time duration from the moment of generation of downlink request packet to the moment of beginning of downlink request packet transmission. To obtain $\mathrm{E}[D_{\mathrm{r}}^{\mathrm{L}}]$ and $\mathrm{E}[D_{\mathrm{r}}^{\mathrm{S}}]$, let P^{c} be the probability that a packet suffers collision in a backoff procedure. Then,

$$P^{\mathrm{c}} = \sum_{v=0}^{M} \sum_{r=0}^{v} {}_v\mathrm{C}_r \alpha^r \{(1-\alpha)\beta\}^{v-r}(1-\alpha)(1-\beta)(1-P_{\mathrm{s}}) \ . \tag{19}$$

Let $E[D_{\text{backoff}}^{\text{T}}]$ and $E[D_{\text{backoff}}^{\text{L}}]$ be the expected number of backoff slots that a packet experience until the moment of transmission attempt in a backoff procedure and the expected number of backoff slots that a packet experience until the moment of discarding in a backoff procedure, respectively. Then,

$$E[D_{\text{backoff}}^{\text{T}}] = \frac{\sum_{v=0}^{M} \sum_{r=0}^{v} {_v}C_r \alpha^r \{(1-\alpha)\beta\}^{v-r} (\sum_{i=0}^{v} \frac{W_i - 1}{2} + 2v - r + 2)}{\sum_{v=0}^{M} \sum_{r=0}^{v} {_v}C_r \alpha^r \{(1-\alpha)\beta\}^{v-r}}$$
(20)

$$E[D_{\text{backoff}}^{\text{L}}] = \frac{\sum_{r=0}^{M} {_M}C_r \alpha^r \{(1-\alpha)\beta\}^{M-r}}{\sum_{r=0}^{M+1} {_{M+1}}C_r \alpha^r \{(1-\alpha)\beta\}^{M+1-r}}$$
$$\times \{\alpha(\sum_{i=0}^{M} \frac{W_i - 1}{2} + 2M - r + 1 + (1-\alpha)\beta(\sum_{i=0}^{M} \frac{W_i - 1}{2} + 2M - r + 2)\}$$
(21)

Note that a downlink packet is discarded when the CCA fails at the Mth backoff stage. So, the expected duration $E[D_r^{\text{L}}]$ is given by :

$$E[D_r^{\text{L}}] = \sum_{k=0}^{\infty} (P^c)^k (1 - P^c) \left\{ k \left(D_{\text{backoff}}^{\text{T}} + R + A \right) + D_{\text{backoff}}^{\text{L}} \right\} \sigma .$$
(22)

The general term in (22) is the expected duration for the case that a packet is discarded after the kth collision. Similarly, the expected duration $E[D_r^{\text{S}}]$ is given by :

$$E[D_r^{\text{S}}] = \sum_{k=0}^{\infty} (P^c)^k (1 - P^c) \left\{ k \left(D_{\text{backoff}}^{\text{T}} + R + A \right) + D_{\text{backoff}}^{\text{T}} \right\} \sigma .$$
(23)

The general term in (23) is the expected duration for the case that a packet is successfully transmitted after the kth collision.

The expected delay $E[D_{\text{d}}*]$, from the beginning of backoff procedure for downlink data packet transmission to the moment of service completion of the downlink data packet, is given as follows.

$$E[D_{\text{d}}*] = \sum_{v=0}^{M} \sum_{r=0}^{v} {_v}C_r \alpha^r \{(1-\alpha)\beta\}^{v-r} (1-\alpha)(1-\beta)P_s \left(\sum_{i=0}^{v} \frac{W_i - 1}{2} + 2v - r + 2 + \frac{1}{1 - P_{\text{Rx,d}}} + A \right) \sigma$$
$$+ \sum_{v=0}^{M} \sum_{r=0}^{v} {_v}C_r \alpha^r \{(1-\alpha)\beta\}^{v-r} (1-\alpha)(1-\beta)(1-P_s)$$
$$\times \left\{ \left(\sum_{i=0}^{v} \frac{W_i - 1}{2} + 2v - r + 2 + \frac{1}{1 - P_{\text{Rx,d}}} + A \right) \sigma + E[D_{\text{d}}*] \right\}$$
$$+ \sum_{r=0}^{M} {_M}C_r \alpha^r \{(1-\alpha)\beta\}^{M-r}$$
$$\times \left\{ \alpha \left(\sum_{i=0}^{M} \frac{W_i - 1}{2} + 2M - r + 1 \right) + (1-\alpha)\beta \left(\sum_{i=0}^{M} \frac{W_i - 1}{2} + 2M - r + 2 \right) \right\} \sigma$$
(24)

The first summation and second summation of the equation (24) describe the cases of successfully transmission and collision in the first transmission attempt, respectively. The last summation of the equation (24) describes the case that a packet is discarded at the first backoff procdure because the channel is continuously sensed due to busy condition in CCA.

References

1. IEEE 802.15.4, Wireless LAN Medium Access Control (MAC) and Physical Layer (PHY) specifications for Low-Rate Wireless Personal Area Network (LR-WPANs) (2003)
2. IEEE 802.15.4, Wireless LAN Medium Access Control(MAC) and Physical Layer (PHY) specifications for Low-Rate Wireless Personal Area Network (LR-WPANs) (2006)
3. Kim, T.O., Kim, H., Lee, J., Park, J.S., Choi, B.D.: Performance Analysis of IEEE 802.15.4 with Non-Beacon-enabled CSMA/CA in Non-Saturated Condition. In: Sha, E., Han, S.-K., Xu, C.-Z., Kim, M.-H., Yang, L.T., Xiao, B. (eds.) EUC 2006. LNCS, vol. 4096, pp. 884–893. Springer, Heidelberg (2006)
4. Kim, T.O., Park, J., Choi, B.D.: Analytic Model of IEEE 802.15.4 with Download Traffic. In: Proceeding of The Second IEEE International Symposium on Pervasive Computing and Ad Hoc Communications (PCAC 2007) (May 2007)
5. Bianchi, G.: Performance Analysis of the IEEE 802.11 Distributed Coordination Function. IEEE Journal on Selected Areas In Communications 18(3) (March 2000)
6. Pollin, S., Ergen, M., Ergen, S.C., Bougard, B., der Perre, L.V., Catthoor, F., Moerman, I., Bahai, A., Varaiya, P.: Performance Analysis of the Slotted IEEE 802.15.4 Medium Access Layer. draft-jwl-tcp-fast-01.txt (2005)
7. Park, T., Kim, T., Choi, J.Y., Choi, S., Kwon, W.: Throughput and Energy Consumption Analysis of IEEE 802.15.4 Slotted CSMA/CA. Electronics Letters (2005)
8. Misic, J., Shafi, S., Misic, V.B.: Performance of a Beacon Enabled IEEE 802.15.4 Cluster with Downlink and Uplink Traffic. IEEE Transactions on Paraller and Distributed systems 17(4) (2006)
9. Kim, T.O., Park, J., Chong, H.J., Kim, K.J., Choi, B.D.: Performance Analysis of IEEE 802.15.4 Non-beacon Mode with the Unslotted CSMA/CA. IEEE Communications Letters 12(4) (April 2008)
10. Elson, J., Girod, L., Estrin, D.: Fine-Grained Network Time Synchronization using Reference Broadcasts. In: Proceedings of the Fifth Symposium on Operating Systems Design and Implementation (OSDI 2002), Boston (2002)

Efficient and Accurate WLAN Positioning with Weighted Graphs

René Hansen and Bent Thomsen

Department of Computer Science, Aalborg University
Fredrik Bajers Vej 7E, DK-9220 Aalborg Øst, Denmark
{rhansen,bt}@cs.aau.dk

Abstract. This paper concerns indoor location determination by using existing WLAN infrastructures and WLAN enabled mobile devices. The *location fingerprinting* technique performs localization by first constructing a radio map of signal strengths from nearby access points. The radio map is subsequently searched using a classification algorithm to determine a location estimate. This paper addresses two distinct challenges of location fingerprinting incurred by positioning moving users. Firstly, movement affects the positioning accuracy negatively due to increased signal strength fluctuations. Secondly, tracking moving users requires a low-latency overhead which translates into efficient computations to be done on a mobile device with limited capabilities. We present a technique to simultaneously improve the positioning accuracy and computational efficiency. The technique utilizes a weighted graph model of the indoor environment to improve positioning accuracy and computational efficiency by only considering the subset of locations in the radio map that are feasible to reach from a previously estimated position. The technique is general and can be used on top of any existing location system. Our results indicate that we are able to achieve similar dynamic localization accuracy to static localization. Effectively, we are able to counter the adverse effects of added signal fluctuations caused by movement. However, as some of our experiments testify, any location system is fundamentally constrained by the underlying environment. We give pointers to research which allows such problems to be detected early and thereby avoided before deploying a system.

1   Introduction

Indoor location-based services (LBS) hold promise for a multitude of valuable services to be built and has the potential of substantially leveraging the utility of existing outdoor solutions. Unfortunately, traditional "outdoor" technologies such as GPS and cellular networks do not work reliably and accurately enough in indoor environments. The GPS signal is not sufficiently strong to penetrate most buildings, while the accuracy of cellular positioning is limited by the denseness of GSM antennas in a particular area. A widely researched alternative is to make use of the existing 802.11 (WLAN) infrastructures which nowadays have become ubiquitous for providing wireless communication. Reusing WLAN for

F. Granelli et al. (Eds.): MOBILIGHT 2009, LNICST 13, pp. 372–386, 2009.
© ICST Institute for Computer Sciences, Social-Informatics and Telecommunication Engineering 2009

indoor positioning is appealing since it avoids the cost of specialized equipment used solely for positioning. Instead, systems can be implemented exclusively in software making indoor positioning a possibility in as large a population as the WLAN infrastructures themselves.

Positioning via WLAN can be done by using signal strength information from available access as the signal strengths vary with spatial changes. However, as radio waves are propagated in an indoor environment, they are also scattered, reflected, and attenuated by the obstacles they encounter. Signals may arrive at the receiver from several paths and when the multiple incoming waves combine at the receiver they can produce a distorted version of the desired waveform, a phenomenom known as *multipath interference* [3]. The end result is a highly unpredictable propagation patterns and the best results in the face of this situation has been achieved with the *location fingerprinting* technique which uses empirically measured signal strengths. The fingerprinting technique works in two phases: First, before a positioning system is deployed, signal strength information is recorded at a number of predefined locations throughout the area to be covered by the positioning system. The measured signal strength values and locations are saved in a database, also called a *radio map*. The process of collecting the signal strength measurements and building the radio map is commonly referred to as the *offline phase*. In the corresponding *online phase*, where the system is operational, a location estimate is obtained by first measuring the signal strengths and then searching the radio map for a closest match to the currently measured signal strength. Different methods can be used to arrive at a closest match and generally these are divided into two main categories: *deterministic methods* and *probabilistic methods*. Deterministic methods build a radio map by saving signal strength information as a scalar value, e.g., the mean signal strength, while probabilistic algorithms store the signal strength distributions from the access points. In the online phase, deterministic methods then estimate a location by comparing measurements by their value while probabilistic methods estimate a location by considering measurements as part of a random process [10,20]. The location fingerprinting technique has produced good positioning accuracy for *static localization*, i.e., location determination of a stationary user. However, users of an LBS service can in general be expected to the moving, for instance users who need navigation to a point of interest, and *dynamic localization*, or tracking a moving user, is more challenging for two reasons. Firstly, movement causes increased fluctuations in the signal strengths. The greater variance means the signal strengths in the radio map are less precise approximations. The result is more dispersed and erratic location estimates and hence the positioning accuracy deteriorates under movements. Secondly, movement necessitates frequent position updates to keep track of the users' current position. Performing localization on a central server and communicating the resulting locations to clients incurs a substantial communication overhead when the update interval is very short. A more scalable solution, which also preserves the privacy of the users, is to perform localization on the users' own devices. Reducing the computational overhead is necessary for constrained devices to meet the low latency requirements. Moreover, reducing the computational effort saves battery usage and thus helps prolong the lifetime of LBS services.

We present a technique based on a weighted graph model of the indoor environment that tries to improve both the positioning accuracy and computational efficiency of location fingerprinting. The underlying idea is to restrict attention - and hence the search space - to only those locations that are feasible to reach from a previously determined position. The intuition is to improve accuracy by filtering away even nearby locations that are impossible to reach due to obstructions or because they are too far away. Similarly, the search of the radio map is confined to a very small geographic region. We show that this region can be made as small as the positioning accuracy of the underlying algorithm permits and this is provably the smallest region possible.

We apply the technique in two real-world test-beds using a simple deterministic algorithm as the underlying base and only a bare minimum of training. This configuration is chosen in order to demonstrate the robustness of the graph technique. However, the graph technique is general and can be placed on top of any existing location system.

Our results indicate that we are able to achieve similar dynamic localization accuracy to static localization, i.e., the technique is resilient to the adverse effects caused by signal fluctuations. However, the results in one of our test beds also demonstrate that location systems are fundamentally limited by the underlying environment.

The rest of the paper is organized as follows: Section 2 discusses related research in 802.11 based localization with a focus on the research that addresses dynamic localization and computational efficiency. Moreover, the section discusses some of the fundamental constraints of 802.11 based localization. Section 3 introduces the weighted graph technique and gives the implementation details while Section 4 shows the results of the technique in two test beds. Finally, Section 5 concludes the paper.

2   Related Research

The *Radar* project [5], carried out at Microsoft Research, was the first to use commodity 802.11 hardware for indoor localization. Using a deterministic approach with a Nearest Neighbor algorithm this initial research effort was able to achieve a median accuracy of 2.94 meters. The Radar project has since become the defacto standard on which other location systems are measured. Bahl et al. [4] and Smailagic and Kogan [17] extended the Nearest Neighbor approach to produce better results by averaging k nearest neighbors on proximity to produce a location estimate. Recently, focus has shifted to probabilistic methods and several research efforts have achieved accuracies of 2 meters or better with at least 90% probability, i.e., locations estimates are within 2 meters of the actual location 90% of the time [2,11,12,16,18,20,22]. Since probabilistic algorithms draw information from signal strength distributions rather scalar values they can be trained to outperform their deterministic counterpart[19]. However, environmental factors and tunable parameters such as the building structure, number and placement of access points and number of samples often have a bigger impact than the choice of algorithm as witnessed in several experiments, e.g., [15,12,16,6].

The best-case positioning accuracies have been achieved with static localization when the tracked device remains at the same location. In this situation the signal strengths remain fairly stable and resemble the entries in the radio map. However, movement incurs increased signal strength fluctuation resulting in deteriorating positioning accuracy. Different approaches have been taken to modeling movement. Smagailagic and Kogan [17] propose a time-averaged location convergence technique to try and minimize the jumps of consecutive estimates. Bahl et al. [4] uses a Viterbi-like algorithm for tracking on top of a k-Nearest Neighbor algorithm. Ladd et al. [12] uses a sensor fusion technique together with spatial continuity assumptions to probabilistically reject outliers. The two techniques that are closest to ours are Xiang et al. [18] and Haeberlen et al. [7]. Xiang et al. [18] use a tracking-assistant algorithm implemented as a state machine. Here, each new state (location) is determined by the previous state and the allowed transitions which takes into account the topology of the environment. Haeberlen et al. [7] considers topology and movement in a similar fashion by using a transition graph to describe allowed movement.

Estimating a position by comparing with all entries in the radio map can become a very taxing operation, especially when the operation area increases or when performing localization on a resource-constrained device. Improving efficiency by reducing the search space of location fingerprinting has previously been addressed by using a notion of access point clustering [2,9,20,22]. An access point cluster defines a spatial region where a common set of access points can be heard.

Youssef et al. [22] use an *explicit* clustering technique where a number of clusters are defined by an administrator in the offline phase. Since access points may by intermittently missing, a subset q of the available access is used to define a cluster. This subset is comprised by the access points with the strongest signals as they are most likely to be present. In the online phase the current measurement is sorted in descending order and the q strongest access points are used to determine which cluster to search within to determine a location estimate. The *Locator* system [2] uses explicit clustering for a first-level clustering, and then proceeds with a further k-Means clustering during the online phase based on a previous position estimate and the physical closeness of locations. Similarly, the *Ariadne* system [9] uses k-Means clustering, but instead clusters a set of candidate positions with smaller mean square errors. The largest cluster is chosen and its center is picked as the location estimate.

Rather than explicitly clustering locations in the offline phase, the *Horus* system [20] uses an *implicit clustering* technique to defer clustering until the online phase. In the online phase the current measurement is once again sorted in descending order. Then, the algorithm proceeds incrementally: For the strongest access point, the probability of each location containing that access point is calculated. If the probability of the most probable location is significantly higher than the probability of the second most probable location according to a threshold parameter, this location is returned as the location estimate. Otherwise, the next access point in the sorted access point list is consulted. The process continues until a position estimate can be delivered.

A comparison of the two clustering approaches is performed by Youssef and Agrawala [21] which found that explicit clustering results in better positioning accuracy than implicit clustering but incurs a larger computational overhead.

3   Using Weighted Graphs for Efficient and Accurate Tracking

Weighted graphs are a fundamental data structure in computer science and have been applied in a host of different domains, including GIS mapping and road-network modeling. Similarly, weighted graphs are a natural formalism for modeling the topology of an indoor environment. As we will demonstrate weighted graphs can be further used to help provide accurate and efficient dynamic localization. We build a weighted graph model of the indoor environment by modeling fingerprinted locations as nodes with weighted edges to capture the distance between physically adjacent locations. Such a graph model can easily be built on top of an existing system or as part of the offline phase of a new location system.

In the online phase the weighted graph is utilized to provide accuracy and efficiency in the following way: Given a previously estimated position, the attention (and search space) is confined to *feasible* locations. These are the locations that are reachable within the update interval of consecutive location estimates by a user traveling at an assumed maximum speed. Thus, feasible locations can be found by performing a breadth-first search within the range ($max\_speed\_per\_second \times update\_interval$). With dynamic localization this range is not very large. Assuming an update interval of one second and a maximum pedestrian speed of three meters per second, the reachable range is 3 meters. In practice, this would in many cases correspond to only the neighboring nodes given the difficulty of distinguishing signal strengths at a finer granularity. As a result, the search space is very small and intuitively accuracy is improved by filtering away even nearby locations that cannot be reached.

Due to the noisy nature of the wireless channel this scheme is, of course, somewhat idealized and there is an inherent risk that a localization system could get "stuck" at a location. This would happen if the actual and estimated locations are not connected by a path, e.g., if they are separated by a wall with no doors connecting them. In order to handle this we introduce so-called *radius-nodes*. Radius-nodes are defined as the set of all non-connected nodes located within a specified euclidian distance of a node. Radius-nodes are used to account for estimation error to non-connected nodes of the actual location. As such the range of radius-nodes is determined by the precision of the underlying positioning system. As an example, if the system can deliver position estimates with 100% certainty within 10 meters, correspondingly the radius should be set to 10 meters in order to consider all relevant nodes. It follows that this is also the provably smallest geographical region of the radio map that needs to be searched since a smaller region would fail to consider all relevant nodes. Lemelson et al. give algorithms to estimate 802.11 positioning error which can be used to ascertain the range of radius-nodes[13]. Radius-nodes are implemented by augmenting each node with a list of its radius-nodes. Radius-nodes are found using a range query centered around the node in question. The range of radius-nodes can be set to different sizes in different parts of the building to reflect varying localization error.

In the online phase radius-nodes are used as a fall-back mechanism when the idealized scheme fails. The governing rule is to prefer connected nodes to a previously estimated position as this provides smooth tracking and avoids erratic jumps through obstructions. Only if there are strong indications that the currently selected track is wrong are these logical jumps allowed. To achieve this we operate with two distinct search spaces: A *primary search space* which consist of the reachable locations from a previously estimated location l and a *secondary search space* consisting of the radius-nodes of l. Both search spaces are searched and at any point a location estimate from the primary search space is preferred. However, if strong evidence suggest that a non-connected node is indeed more likely a "shift" can be made. A shift entails selecting the most probable non-connected node as the location estimate and adjusting the search space such that the primary search space now becomes centered around the new node. The notion of making a shift is illustrated in Figure 1.

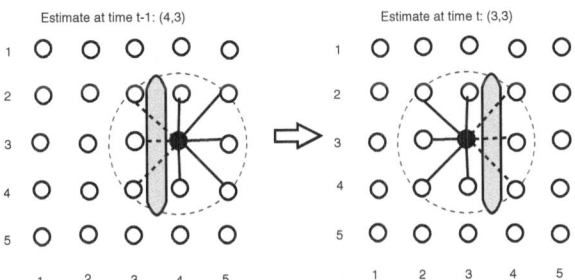

Fig. 1. A position estimate at time *t-1* (filled circle) has a primary search space (nodes connected by solid lines) and a secondary search space (nodes connected by dashed lines). At time t a radius-node is selected as the new position estimate. A "shift" is made by updating the primary and secondary search. The dashed circle indicated the range of radius nodes.

Evidence that a shift needs to be performed can come from a history of measurements. For example, if two or three consecutive estimates has been to radius-nodes this could indicate that a shift is needed. The range of radius-nodes, however, should be extended in accordance with the length of the history. So if a history-length of three seconds is used, the radius should be extended by $3 \times max\_speed\_per\_second$ since the user can move this distance further away

from the correct location before a shift is made. We advice using a history-length
of two seconds, and making a shift if two consecutive best estimates has been to
radius nodes. By using a short history-length the search space is kept small and
at the same time the system can quickly react to seemingly false assumptions.
Conversely, a longer history-length may yield even stronger indications but has
a correspondingly longer reaction time.

3.1   Algorithm and Data Structures

We now proceed to giving the details of the method that implements the graph
technique. The method has been implemented in C# and is presented in a combined C#/pseudo-code syntax in Listing 1 to improve readability. The method
is a template method where the individual steps can be customized to fit an existing system and application domain by adjusting the underlying parameters.

```
1  public void estimatePosition() {
2     Measurement meas;
3     Estimate est;
4     Estimate radiusEst;
5     List<Node> primSpace = graph.Nodes;
6     List<Node> secSpace;
7     History hist = new History();
8
9     while (ONLINE_RUNNING)
10    {
11       meas = MeasureSS(UPDATE_INTERVAL);
12       est = Compare(meas, primSpace);
13       nodes = FeasiblePositions(est);
14       secSpace = est.RadiusNodes;
15       radiusEst = Compare(meas, secSpace);
16       history.add(est, radiusEst);
17       if (history.correctionNeeded())
18          nodes = feasiblePositions(radiusEst);
19    }
20 }
```

Listing 1. Position estimation with radius-nodes

The *estimatePosition()* procedure is called when location estimation com-
mences. We start by describing the data structures involved. The *Measurement*
class encapsulates a signal strength measurement and the variable meas holds
the current measurement. The Estimate class represents a position estimate -
a node - along with the node's *score* which is a measure of how good the estimate is, e.g., highest probability or shortest distance. *Node* objects correspond
to nodes in the graph and the objects est and radiusEst represent the currently

best scoring node in the primary and secondary search space, respectively. The primary search space is represented in the `primSpace` list while the secondary search space is kept in the `secSpace` list. Finally, the `History` class represents a history of measurements (a queue of recent measurement), and a criteria for performing a correction. Performing a correction corresponds to making a shift as described in Section 3. The `History` allows for different queue lengths and criteria for correction to be set. It can be noted that in the code above the primary search initially consists of all nodes in the graph (line 5). In a large deployment, however, we would instead use one of the access point clustering techniques as described in Section 2 to reduce the initial search space.

The loop (lines 9-19) defines what happens during location determination: The signal strength is measured and saved into the `meas` object (line 11)[1]. The measurement is compared with the nodes in the primary search space (`primSpace`) using an underlying classification algorithm which results in in a position estimate (line 12). The primary search space is updated to be the set of reachable locations from the estimated node (line 13) and the secondary search space is updated to be the radius nodes of the estimated node (line 14). The signal strength is then compared with the radius nodes resulting in yet another estimate - to the best-scoring radius-node (line 15). Both estimates are added to the history (line 16) and if the history indicates that a correction is needed, a shift is made (line 17-18).

4   Experiments

The graph technique has been implemented in a prototypical LBS application running on the Windows mobile platform. This section discusses the setup and results from two separate experiments that were conducted with the application in order to test the robustness of the technique in different settings.

The experiments were performed using an HP Ipaq H5550 Pocket PC with an Intel XScale 400 Mhz processor. Experiments were conducted using both the in-built wireless Network Interface Card (NIC) as well as an SDIO NIC from Socket. An API from OpennetCF.org was used to query the NICs for the detected signal strengths.

In both experiments, the radio map was built by saving the average signal strength value of only ten measurements at each location. In the online phase a single measurement was used and the update interval was set to one second.

The system made use of a simple deterministic Nearest Neighbor algorithm similar to the one used in the original Radar project The choice of a bare minimum of signal strength samples coupled with a simple deterministic algorithm was made in order to "stress test" the robustness of the graph technique. A more advanced probabilistic algorithm coupled with more samples might have improved the accuracy further but this would also mean that the concept of radius-nodes would not be tested to its limits.

---

[1] OpennetCF provides an API for the .NET Compact Framework for querying signal strength information [1].

The experiments were performed by pressing a start/stop button which starts and stops the loop described in Listing 1. The results were recorded by logging all estimates produced during the test. This approach was taken, in contrast to clicking the actual position while the test was underway, as it gave a more realistic view of how the technique handles movement. Instead of stopping briefly to mark the current location, the test could be performed with a continuous walk. The assumed maximum speed were in both experiments set to three meters per second.

4.1   Aalborg University

The first test was performed in a wing of Aalborg University. A total of 16 access points could be heard throughout the building and at any one location the number of access points ranged from four to eight. The weighted graph was built by placing a node in offices and at locations where there were "forks" in the environment (the experiments were conducted at a holiday period which accounts for only some of the offices being included). The nodes were approximately three meters apart corresponding to the size of an office. The model consisted of 71 nodes and 75 edges and can be seen in Figure 2.

The figure also shows the radius that was chosen based on the observed static localization accuracy and chosen history-length. The worst-case static ac-

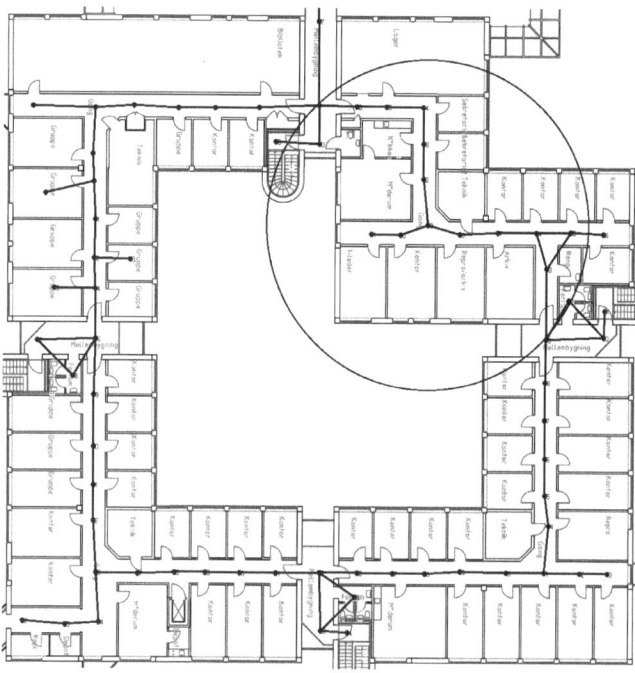

Fig. 2. Wing at Aalborg University modeled as a weighted graph. The size of the radius-nodes can be seen in the top-right circle.

curacy was nine meters and two consecutive best-scoring radius-estimates was used as the correction criteria which led to an additional three meters being added to the radius.

We completed eight walks around the building with both the graph technique and the original Nearest Neighbor algorithm. We used four start/stop points in the upper, lower, left, and right side of the building and tested by walking in both directions. The graph- and original technique were tested back to back. Immediately after completing a walk with the graph technique we took a tour with the original technique. This was done to avoid temporal differences in the comparison. The results of the tests can be seen in Figure 3.

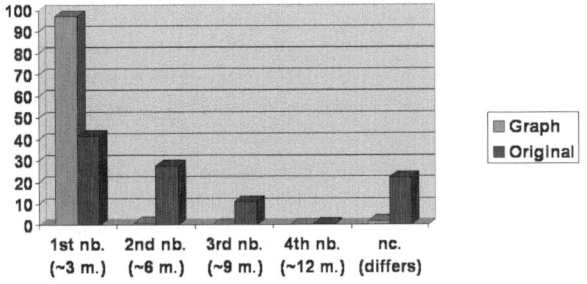

Fig. 3. Dynamic localization accuracy in Aalborg University

The first column represents location estimates that were at, or at most one node (approximately three meters) away from the actual location. Columns two through four represent location estimates that were two to four nodes away from the actual location, with the corresponding accuracy shown in parentheses, while the last column indicates those results that were at a non-connected node to the actual location. As can be seen the weighted graph technique massively outperforms the original Nearest Neighbor algorithm whose performance degrades significantly when moving. After the head-to-head comparisons a number of tests were run with the graph technique alone to test the robustness further. In these tests we played the role of an indecisive user who changes direction several times. This did not affect the accuracy adversely nor did using the either of the NICs. Since the activity level in the building was somewhat idle at the time of the experiments it cannot be ruled out that the accuracy would have been lower had there been more people and offices open. However, in the tests the three offices that were open were not selected as location estimates in the walks even though the doors were open. Instead, the few times the graph technique made a correction was when the operator passed by the midsections which can be found in the middle left, right, top, and bottom of the building. In most cases a wrong estimate here was resolved by the system "catching up" but sometimes the system chose to cut the corner.

Table 1. Computation time in Aalborg University

Card / Method	Graph	Radar
In-built	10-20 ms.	ca. 35 ms.
SDIO	30-40 ms.	ca. 60 ms.

The computational performance of the two techniques is summarized in Table 1.

Using the device's in-built NIC the computation time with the graph technique ranged from 10 to 20 milliseconds while the original Radar algorithm produced more stable results at around 35 milliseconds. The greater variance with the graph technique is due to a varying number of radius-nodes in different parts of the building. Using the Socket SDIO NIC added another 25 milliseconds to the computation time which is attributed to I/O operations. In comparison, we can see that the graph technique is approximately twice as fast as the Nearest Neighbor algorithm which searches the entire radio map. The graph technique searches less than half of the radio map, which can be seen by observing the radius in Figure 2, but adds a bit of extra programming logic which is the reason why the difference is not greater. However, since the computational complexity of the graph technique is constant the difference will increase further in even larger areas.

4.2   Randers Public Library

Experiments were also conducted in Randers Public Library. This section does not include a formal treatise of the results as the environment posed a pathological case for indoor positioning. Rather, we discuss the experiences from the experiments. The library had four access points installed but in certain parts of the building sometimes only two of the access points could be registered. A weighted graph model was created resulting in 142 nodes and 214 edges which can be seen in Figure 4.

As can be seen the weighted graph does not extend to the far right. This is because the building map was slightly outdated. As a result this section of the building was left out of the experiments. Upon building the radio map samples were taken at nine different locations in order to get a picture of the static localization accuracy. This revealed problems pertaining to delivering accurate positioning. In the left side of the building - at 60 out of the plotted 142 nodes it turned out - estimates jumped very far, as far as from one end of the building to the other. In this section it was observed that frequently only two of the access points could be registered so the problem was initially tried amended by using several samples in an effort to hear more access points. By using several samples we managed to pick up all four access points. However, this did not improve the accuracy. An analysis of the data in the radio map showed the signal strengths of three of the access points in this region to be highly overlapping. As Figure 5

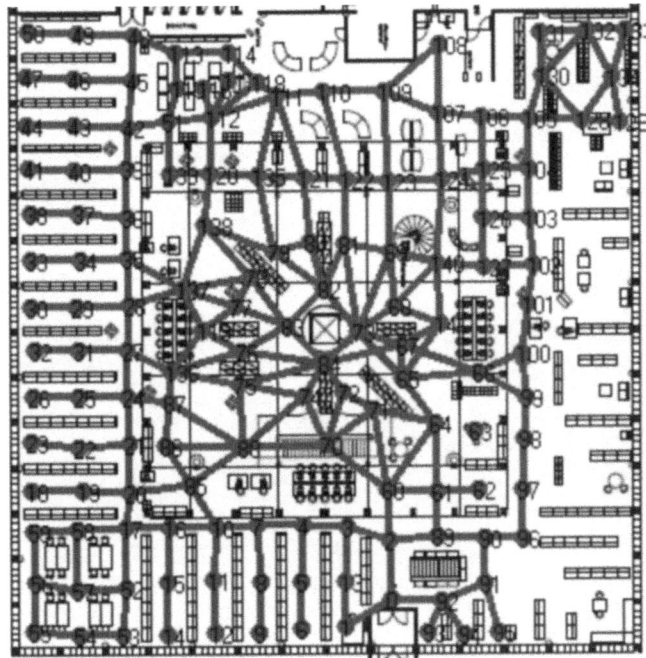

Fig. 4. Randers Public Library modeled as a weighted graph

shows especially two of the access points were affected showing a variance within only 10 dBm. Since it was not uncommon to see variances of 10 dBm simply by changing direction this made it impossible to perform sensible localization in this part of the building. In the middle- and right side of the building, however, the static localization accuracy was acceptable. The accuracy was lower than in Aalborg - more estimates were farther away from the actual location but the worst-case accuracy remained the same. As a result, the same setup as before was used.

The difference in dynamic localization accuracy was even more notable than in Aalborg University and the effect of movement was substantial as witnessed in Figure 6.

As can be seen, the graph technique initially does a very good job of confining the location but in the lower left corner it gets stuck whereafter it starts jumping back and forth while the user continues to the upper left corner. This illustrates the accuracy in a nutshell: While refraining from the left side of the building accuracy was substantially improved but any attempts to use the graph technique in the left side resulted in the system getting stuck. This was resolved only by changing the radius to the 40 meters locations could jump in the left side! Outside the left region the graph technique distinguished locations on opposite sides of bookshelves very well. In the mid-region where the obstructions were formed by low bookshelves (<1.5 meters high) and tables the system some times chose

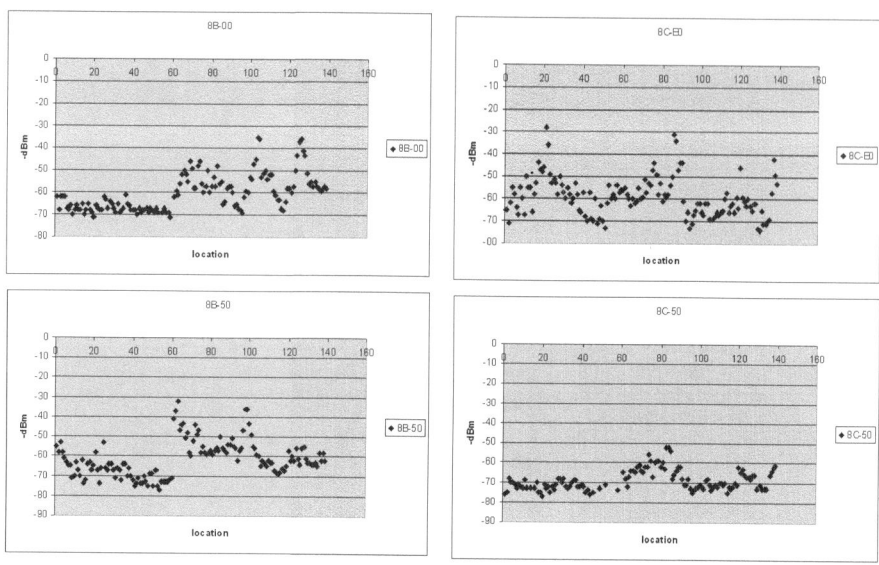

Fig. 5. Signal strengths from the four access points at Randers Public Library

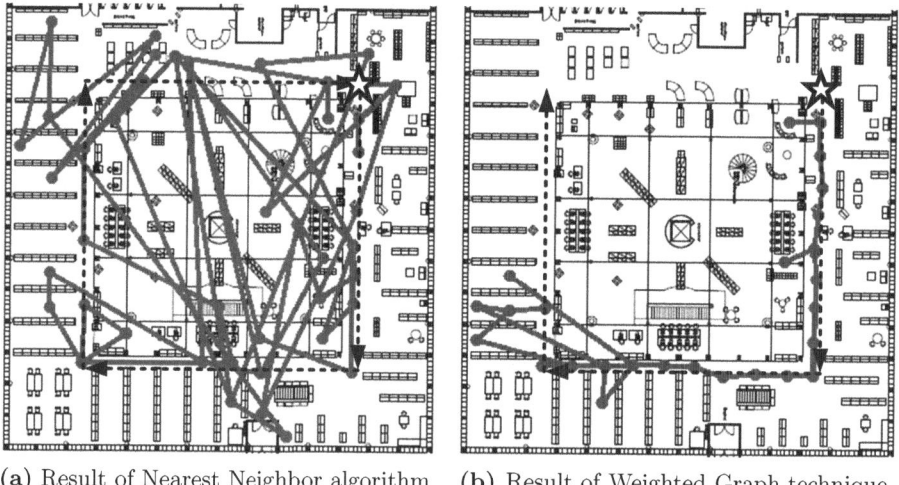

(a) Result of Nearest Neighbor algorithm   (b) Result of Weighted Graph technique

Fig. 6. Comparison of a walk in Randers Public Library. The star symbol indicates the starting point and the dashed line the actual route walked. The solid lines connect consecutive location estimates.

the wrong path around an obstruction but stayed within close proximity of the actual location in contrast to the Nearest Neighbor algorithm which exhibited very large estimation jumps when moving. The experiments in Randers Public Library are a testament that 802.11-based location systems are fundamentally limited by the available infrastructure. In some cases performance can only be improved to an acceptable level by restructuring or adding new access points. Although we were not free to modify the wireless infrastructure in this experiment, Lemelson et al. [13] presents a strategy for predicting the estimated location error based on the signal strength patterns in the radio map. The strategy clusters together locations with similar signal properties and the estimated error of a region is derived from the distance of the cluster of that region. This strategy allows potential trouble areas to be identified before deployment and appropriate countermeasures to be taken.

5   Conclusion

In this paper we presented a weighted graph technique for improving the computational efficiency and accuracy of dynamic localization (tracking of moving users) by using available 802.11 equipment. The weighted graph technique was used to impose topological constraints on the location system and improve the accuracy and efficiency of an underlying classification algorithm by only considering feasible location candidates based on the users' previous location and an assumed maximum traveling speed. The technique was found to provide a substantial improvement in dynamical localization accuracy by disregarding location estimation outliers caused by movement. At the same time the search space was confined to a very small geographic region which facilitates building scalable systems where the computation is performed on the users' own mobile devices. However, as experiments in one of the tests demonstrated the usability of an LBS service is fundamentally limited by the underlying wireless infrastructure that supports it. We provided pointers to research which allows such ailments to be discovered early and to be avoided before deploying an indoor location system.

References

1. Opennetcf.org, http://www.opennetcf.org/home.ocf
2. Agiwal, A., Khandpur, P., Saran, H.: Locator: location estimation system for wireless lans. In: WMASH 2004: Proceedings of the 2nd ACM international workshop on Wireless mobile applications and services on WLAN hotspots (2004)
3. Alexander, B.: 802.11 Wireless Network Site Surveying and Installation. CISCO Press (November 2004)
4. Bahl, P., Balachandran, A., Padmanabhan, V.: Enhancements to the RADAR User Location and Tracking System. Microsoft Research Technical Report (February 2000)
5. Bahl, P., Padmanabhan, V.N.: Radar an in-building RF-based user location and tracking system. In: INFOCOM 2000, Tel Aviv, Israel, pp. 775–784 (2000)
6. Elnahrawy, E., Li, X., Martin, R.P.: The limits of localization using signal strength: a comparative study. In: Sensor and Ad Hoc Communications and Networks, 2004. IEEE SECON 2004, pp. 406–414 (October 2004)

7. Haeberlen, A., Flannery, E., Ladd, A.M., Rudys, A., Wallach, D.S., Kavraki, L.E.: Practical robust localization over large-scale 802.11 wireless networks

8. Hightower, J., Borriello, G.: Location sensing techniques. Technical report. companion to "location systems for ubiquitous computing" appearing on pp. 57–66, august issue 2001 of ieee computer magazine, University of Washington, Computer Science and Engineering, Box 352350, Seattle, WA 98195 (July 2001)

9. Ji, Y., Biaz, S., Pandey, S., Agrawal, P.: Ariadne: A dynamic indoor signal map construction and localization system. In: MobiSys 2006 (2006)

10. Kjærgaard, M.B.: A taxonomy for radio location fingerprinting. In: Hightower, J., Schiele, B., Strang, T. (eds.) LoCA 2007. LNCS, vol. 4718, pp. 139–156. Springer, Heidelberg (2007)

11. Krumm, J., Horvitz, E.: Locadio: Inferring motion and location from wi-fi signal strengths. Mobiquitous, 4–13 (2004)

12. Ladd, A.M., Bekris, K.E., Rudys, A., Marceau, G., Kavraki, L.E., Wallach, D.S.: Robotics-based location sensing using wireless ethernet. In: MOBICOM 2002, September 23-26 (2002)

13. Lemelson, H., Kjaergaard, M.B., Hansen, R., King, T.: Error estimation for indoor 802.11 location fingerprinting

14. Battiti, M.B.R., Villani, A.: Statistical learning theory for location fingerprinting in wireless lans. Dit-02-086, University of Trento, Informatica e Telecomunicazioni (October 2002)

15. Roos, T., Myllymäki, P., Tirri, H., Misikangas, P., Sievänen, J.: A probabilistic approach to wlan user location estimation. International Journal of Wireless Information Networks 9(3) (July 2002)

16. Saha, S., Chaudhuri, K., Sanghi, D., Bhagwat, P.: Location determination of a mobile device using ieee 802.11b access point signals. In: IEEE Wireless Communications and Networking Conference (WCNC 2003), March 2003, pp. 1987–1992 (2003)

17. Smailagic, A., Kogan, D.: Location sensing and privacy in a context aware computing environment. IEEE Wireless Communication 9(5), 10–17 (2002)

18. Xiang, Z., Song, S., Chen, J., Wang, H., Huang, J., Gao, X.: A wireless lan-based indoor positioning technology. IBM J. Res. Dev. 48(5/6), 617–626 (2004)

19. Youssef, M., Agrawala, A.: On the optimality of wlan location determination systems. Technical Report UMIACS-TR 2003-29 and CS-TR 4459, University of Maryland, College Park, 2003 (2003)

20. Youssef, M., Agrawala, A.: The horus wlan location determination system. In: MobiSys 2005: Proceedings of the 3rd international conference on Mobile systems, applications, and services, pp. 205–218. ACM Press, New York (2005)

21. Youssef, M., Agrawala, A.: Location-Clustering Techniques for WLAN Location Determination Systems (2006)

22. Youssef, M.A., Agrawala, A., Shankar, A.U.: Wlan location determination via clustering and probability distributions. In: Proceedings of the First IEEE International Conference on Pervasive Computing and Communications (PerCom 2003) (2003)

Feasibility of a GNSS-Probe for Creating Digital Maps of High Accuracy and Integrity

Dimitris Vartziotis[1], Alkis Poulis[1], Alexandros Minogiannis[1], Panayiotis Siozos[1], Iraklis Goudas[1], Jaron Samson[2], and Michel Tossaint[2]

[1] NIKI Ltd, Ethnikis Antistatis 205, GR 45500, Katsika, Ioannina, Greece
{dimitris.vartziotis,alkis.poulis,alex.minogiannis,
panos.siozos,iraklis.goudas}@nikitec.gr
[2] European Space Agency-ESTEC, Keplerlaan 1, 2201 AZ Noordwijk, The Netherlands
{Jaron.Samson,Michel.Tossaint}@esa.int

Abstract. The "ROADSCANNER" project addresses the need for increased accuracy and integrity Digital Maps (DM) utilizing the latest developments in GNSS, in order to provide the required datasets for novel applications, such as navigation based Safety Applications, Advanced Driver Assistance Systems (ADAS) and Digital Automotive Simulations. The activity covered in the current paper is the feasibility study, preliminary tests, initial product design and development plan for an EGNOS enabled vehicle probe. The vehicle probe will be used for generating high accuracy, high integrity and ADAS compatible digital maps of roads, employing a multiple passes methodology supported by sophisticated refinement algorithms. Furthermore, the vehicle probe will be equipped with pavement scanning and other data fusion equipment, in order to produce 3D road surface models compatible with standards of road-tire simulation applications. The project was assigned to NIKI Ltd under the 1st Call for Ideas in the frame of the ESA - Greece Task Force.

Keywords: GNSS, EGNOS, Digital Maps, Accuracy, Integrity, ADAS.

1 Introduction

As the existing Global Navigation Satellite Systems (GNSS), such as GPS and GLONASS, have been available for well over three decades, their limited accuracy, reliability, availability and integrity prevent any advanced use in safety critical and precision demanding applications. As a consequence, the same holds true for publicly available GPS or GLONASS generated digital maps, since they were not compiled with the above mentioned precision and safety criteria in mind.

Today's available digital maps and GNSS positioning options are adequate mainly for turn-by-turn navigation uses. The ongoing research in advanced road passenger safety has already defined new precision requirements for enabling new enhanced applications of digital maps in safety applications and ADAS. Some critical examples, for which the automotive industry has already defined the preliminary requirements, are systems for Vehicle Control, Driver Warning and Exact Path Prediction.

F. Granelli et al. (Eds.): MOBILIGHT 2009, LNICST 13, pp. 387–396, 2009.

In addition, the novel use of digital 3D Road Surfaces in the vehicle engineering process has increased the market need for more precise datasets in advanced automotive simulation related to road-tire-vehicle interaction. The simulation of the vehicle mechanical system can be enhanced with the use of accurate input for the road geometry, in terms of its geometric characteristics, such as slope, elevation and curvature enhanced with metadata for their surface properties.

The required accuracy and integrity of digital maps for the forthcoming demanding applications can be achieved through the utilization of the latest sensor technology in combination with the more reliable, precise and publicly available European Navigation Systems such as EGNOS and later GALILEO. The ROADSCANNER methodology will refine existing digital maps with the use of proper algorithms in terms of precision and integrity. The combination of both technologies will lead to digital maps products/datasets certified and verified for use in critical applications.

2 Project Description

The "ROADSCANNER" project assessed the feasibility of a GNSS probe vehicle designed to generate enhanced digital maps (fig. 1). These digital maps will be useful in demanding and critical applications, such as safety related ADAS applications and automotive simulation.

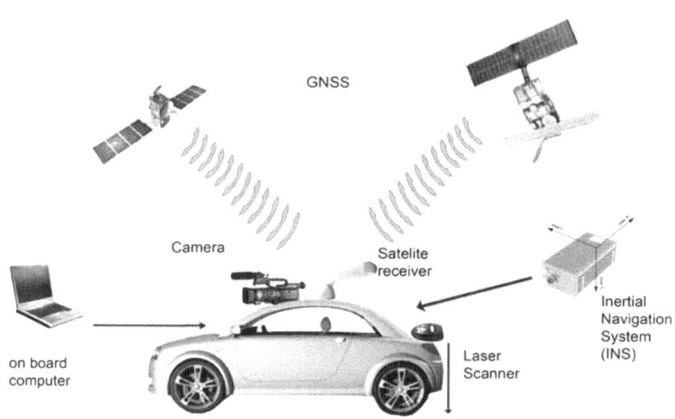

Fig. 1. The ROADSCANNER Preliminary Concept

This feasibility study initially examined the market potential of the probe vehicle and the enhanced digital maps. This market survey evaluated the digital maps and GNSS systems available today and the potential value of enhanced accuracy and integrity products. Furthermore, an additional state of the art technology market survey was conducted in order to identify existing components and possible solutions that can be adopted to build the vehicle probe and corresponding infrastructure required to achieve the ROADSCANNER's demanding objectives. After the state of the art evaluation, preliminary navigation equipment and map refinement algorithm tests took place at selected roads in Greece and Europe. Finally, the product specifications for the probe vehicle including all needed technologies, algorithms, software and hardware sub-components where defined, thus, concluding to a detailed development plan for the complete product.

3 Work Description

The Market Study has identified the needs for high accuracy and integrity digital maps and their applications to innovative ADAS pilots, digital road simulators, as well as current developments on similar vehicle probes. Digital maps and respective services are a profitable and expanding industry [1]. Furthermore, digital 3D roads are implemented in several simulation products for comfort simulation, tire simulation, vehicle durability, accident reconstruction and driving simulation.

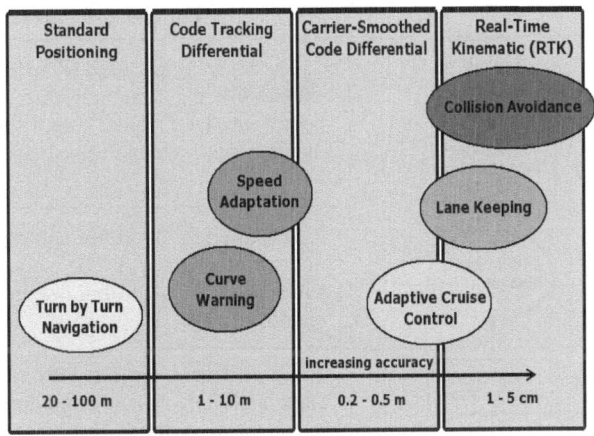

Fig. 2. Required navigation accuracy for various driver assistance systems [1]

However, the digital maps based ADAS applications are yet mainly a matter of research efforts especially due to their demanding requirements in terms of accuracy and integrity (Fig. 2), in transnational research projects like PReVENT [2], Cooperate Vehicle-Infrastructure System [3], NextMap [4], EuroRoadS [5], Gallant [6], etc.

With regard to existing vehicle probes, several types of commercial road and surface scanning vehicles are used in different kind of applications [7-13]. The equipment that is installed differs depending on the needs of each application like highway maintenance and survey, terrain mapping, city modelling, 3D reconstruction, standard road map production and geological survey.

The initial design of ROADSCANNER called for a modularized system comprising mainly of a positioning system, a road surface scanning system and the video camera system. To that end, a review was carried out for existing equipment fitting the requirements of the project, like GNSS receivers and services, accuracy enhancing services (e.g. DGPS), INS devices, road scanning devices, sensors integration and data flow management processes and finally on map refinement algorithms. The GNSS receivers research was focused on those that featured Dual frequency (L1/L2), GLONASS, SBAS (EGNOS) and a maximum number of channels. The road scanning devices presented three choices, namely, LIDAR devices, 3D laser scanning and photogrammetry.

3.1 Algorithm Implementation

A preliminary implementation of the algorithm was used in order to conduct a series of tests on captured data. The modular programming approach was chosen in order to facilitate successive component upgrades. Figure 3 describes the functionality of each module of the preliminary version used during the series of tests.

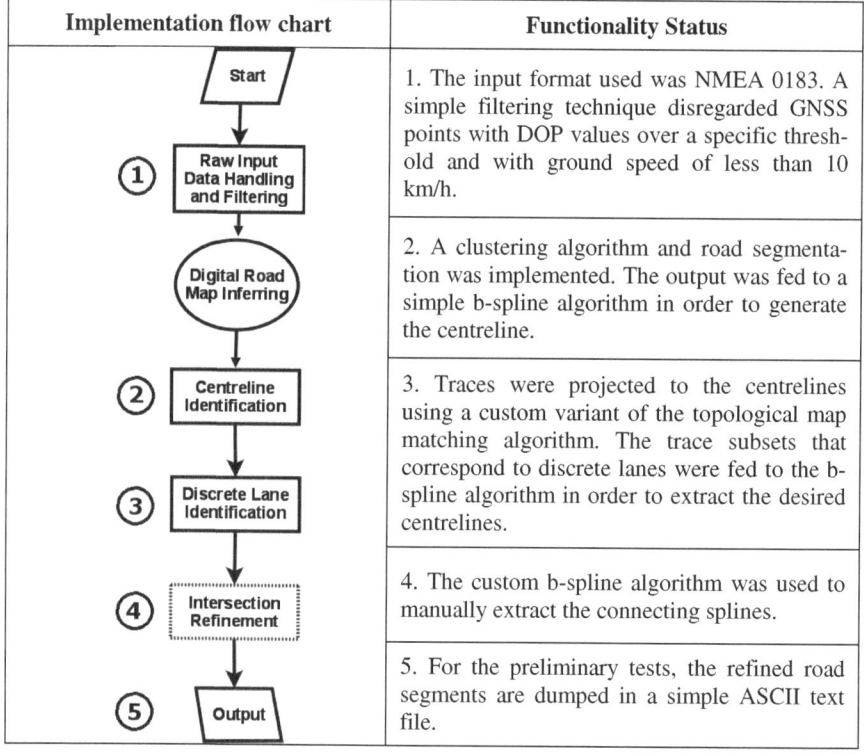

Implementation flow chart	Functionality Status
① Start → Raw Input Data Handling and Filtering	1. The input format used was NMEA 0183. A simple filtering technique disregarded GNSS points with DOP values over a specific threshold and with ground speed of less than 10 km/h.
Digital Road Map Inferring	2. A clustering algorithm and road segmentation was implemented. The output was fed to a simple b-spline algorithm in order to generate the centreline.
② Centreline Identification ③ Discrete Lane Identification	3. Traces were projected to the centrelines using a custom variant of the topological map matching algorithm. The trace subsets that correspond to discrete lanes were fed to the b-spline algorithm in order to extract the desired centrelines.
④ Intersection Refinement	4. The custom b-spline algorithm was used to manually extract the connecting splines.
⑤ Output	5. For the preliminary tests, the refined road segments are dumped in a simple ASCII text file.

Fig. 3. Preliminary algorithm implementation flow chart

3.2 Preliminary Tests

In order to validate the ROADSCANNER concept and in particular the map refinement algorithm employed, a series of tests was carried out. The tests were split in two parts that where carried out in Greece and in Germany. The choice of a European, apart from Greece, site was necessary for testing EGNOS, since there was no coverage in Greece at the time.

During the tests, a Septentrio PolarX2 receiver was used along with a Novatel GPS-702 antenna. PolarX2 is a general-purpose 48 channel receiver for high-end Original Equipment Manufacturer (OEM) applications. The receiver supports reception of L1 and L2 signals from up to 16 GPS satellites and additionally can track the L1 signals from up to 6 SBAS (EGNOS) satellites. Moreover, signals arriving from low elevation angles are attenuated. The RxControl software package developed by Septentrio was used to facilitate the data acquisition. RxControl is a Java-based graphical user interface to configure all types of the PolaRx2 receivers. With RxControl the activity of the receiver and log/post data both on site and remotely was monitored. Results were output mainly via industry-standard NMEA-0183 messages and secondly via Septentrio's binary format (SBF).

The algorithm performance is affected by the number of acquisitioned points due to the fitting procedures. Therefore, the receiver's output rate for the kinematics' tests had to be set at 10 Hz (receiver's maximum available) in order to achieve the highest performance. On the other hand, the receiver output rate for the static tests (where the EGNOS performance in Greece tested) was set at 1 Hz.

The tests in Greece were performed in selected routes around the city of Ioannina in the northwest area of the country, during February 2008. The test cases where as follows:

- Two way, single lane road segment
    - Length : approximately 4.9 km from start to end
    - Number of passes: 10 (both lanes)
    - Curvature was also evaluated in this test case

- Road segment with varying elevation
    - Length : approximately 10 km from start to end
    - Number of passes: 7 (both lanes)

- Intersection
    - Number of passes: 12 (equal to permitted traversals) x 2

- Roundabout
    - Number of passes: 8 (equal to permitted exits/entrances to roundabouts)

The tests in Germany were performed in the area of the city of Stuttgart, from the 4th to the 7th of April 2008. The hardware setup was the same with the one used for the tests in Greece. Similar to the tests in Greece, the following test cases were examined:

- Two way, single lane road segment
    - Length : approximately 3.3 km from start to end
    - Number of passes : 10 with standalone GPS (both lanes) 10 with EGNOS enabled (both lanes)
    - Curvature was also evaluated in this test case

- Road segment with varying elevation

  - Length : approximately 25km from start to end
  - Number of passes : 1 with standalone GPS (from start to finish), 1 with EGNOS enabled (from finish to start)

- Intersection

  - Number of passes: 12 (equal to permitted traversals) x 2

- Roundabout

  - Number of passes: 6 (equal to permitted exits/entrances to round-abouts)

The results of the road segment test confirmed that multiple passes can indeed improve the accuracy of a digital map. It is shown that the constant adding of new passes moved the centrelines away from the ones regarded as the initial base map. The first 5 passes exhibited higher Dilution of Precision (DOP) values, and as such are of lower quality in terms of relative accuracy. Figure 5 clearly shows that the addition of passes of better quality drives the centreline away from the initial result of low accuracy and towards a better and more correct result.

EGNOS was enabled during the tests in Germany. Figure 6 illustrates the DOP values and number of satellites used for both lanes. In comparison with the GPS standalone case (Fig. 6) a smaller number of satellites was used on all passes which led to higher DOP values. Overall Position DOP (PDOP) values average to 2.25, while Horizontal DOP (HDOP) to 1.16. By utilizing EGNOS, it is evident that a settlement between maximum relative accuracy and signal integrity has to be made.

Although EGNOS enabled tests were characterized by larger relative errors, the signal's validity was confirmed. Of course, in the final software version, reception of EGNOS could be improved by using SISNET technology. This can help avoiding situations where DOP values that do not correspond to reality (e.g. in case of satellite failure) lead to mistaken mapping results that cannot be monitored in any other way.

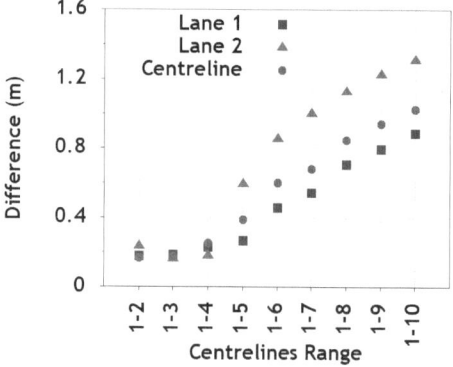

Fig. 4. Average Difference – Greece road segment test case

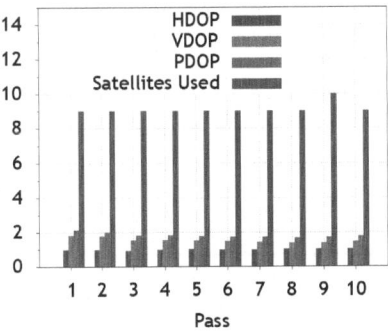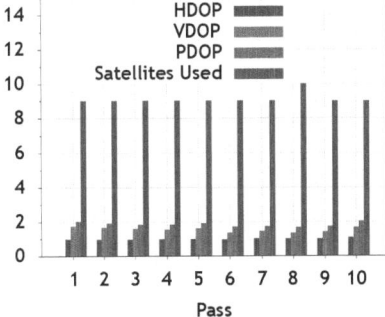

Fig. 5. DOPs & satellite usage for lane 1and lane 2 (Germany, using GPS signal)

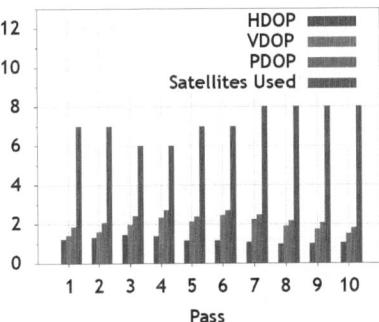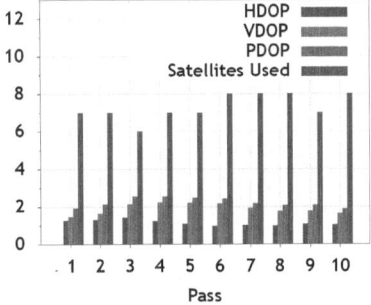

Fig. 6. DOPs & satellite usage for lane 1and lane 2 (Germany, using EGNOS signal)

3.3   Product Design and Development Plan

The ROADSCANNER's action flow of the client/system interaction procedures is the following:

- The client poses a request for creating a digital road map of high accuracy and integrity and/or a 3D digital road surface model for a specific area. In the first case either an existing digital map of the area is provided for refinement or a new one is requested to be created from scratch.
- The ROADSCANNER Service Centre receives the client's request and deploys the ROADSCANNER vehicle in the specified area for probing.
- After all measurements are made, accumulated data is transferred back to the Service Centre for processing.
- For the case of digital road maps, the consecutive passes of the probing vehicle from the designated area are combined (along with the existing digital map, if available) in order to produce a refined map with high accuracy and integrity characteristics.
- In case a 3D digital road surface model was requested, data from the probe vehicle is analyzed and combined in order to form a highly detailed 3D model of the road surface.

- The final product in both cases is bundled in standardized formats and delivered to the client.

The system functions are based on the principal that highly accurate digital road maps and models can be created by combining data from multiple probing passes over a road network.

Table 1. ROADSCANNER Requirements

Requirement	Coverage
Absolute accuracy (Digital map)	1 m
Absolute accuracy (3D road model)	1 m
Relative accuracy (Digital map)	0.2 – 0.5 m
Relative accuracy (3D road model)	0.2 – 0.5 m
Resolution (Digital map)	10 – 20 cm
Resolution (3D road model)	$\sim 1 \text{ cm}^2$

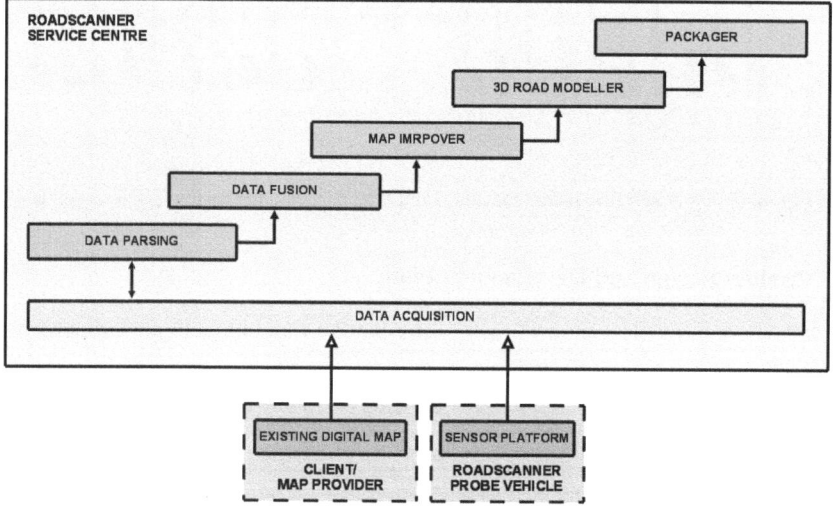

Fig. 7. ROADSCANNER system architecture

The ROADSCANNER product requirements are depicted on Table 1 and an architectural view on Fig 8. The functional specifications of the vehicle are presented in Table 2.

The vehicle probe includes three different platforms (fig. 9):

- The navigation platform
- The road scanning platform
- The computer and storage platform

Table 2. Functional specifications of the probe vehicle

Requirement	Value/Range
Operation speed width	30 - 80 km/h
Scanning width	3.75 m (Highway lane width)
Optimum number of road passes	10
Minimum road scanning resolution	1 point per cm^2
Minimum vehicle operational time	8 h

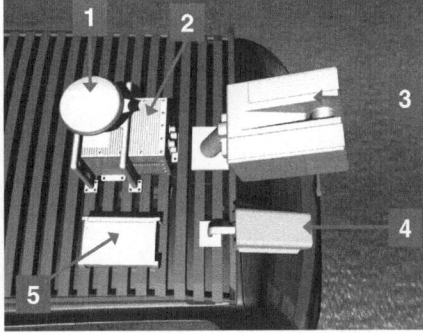

Fig. 8. Visualisation of the probe vehicle: 1) GNSS antenna, 2) INS unit, 3) LIDAR road scanner, 4) Video camera, 5) GNSS receiver

The navigation platform includes the GNSS device, the INS device and a distance meter. The road scanning platform includes the road scanning sensor and the video camera. Finally, the computer and storage platform will include two computer systems the communication system (Ethernet, USB, RS232 or RS422) and the storage unit (removable hard disks for storage). The navigation and the road scanning platform will be placed on a customised roof rack and only the computer and storage platform will be placed in a customised rack inside the vehicle.

In order to build the product a development plan was drawn including the final detailed product design, INS/GNSS integration, Data fusion and algorithmic software components, together with presentation and cooperation actions with the interested industrial stakeholders prior to vehicle construction.

4   Conclusions, Future Development

The main objective of the ROADSCANNER project was to assess the feasibility of a vehicle probe designed to produce digital maps of high accuracy and high integrity. The outcome of this project confirmed that the construction of the probe is feasible, due to following results of the respective tasks. The initial market analysis for existing and future applications, in need of enhanced digital maps and 3D road surface models, showed that there is a growing market trend for this kind of applications and intense research is focused on these subjects. Furthermore, the market study for the required equipment discovered that existing components can be used to construct the probe. Also, the processing and storage requirements are covered by existing technology.

The implementation of the proposed map refinement algorithm showed promising results, since it was able to reach a level of accuracy that can satisfy the needs of some ADAS applications. The algorithm was validated through the tests that took place in Greece and Germany. These tests, also displayed the importance of using EGNOS in terms of increased accuracy and data validity. Besides the construction of digital maps, the integrity feature of EGNOS was also reported to be crucial in safety applications. The increased accuracy introduced with the deployment of the GALI-LEO satellite constellation, will improve even more the obtained data in terms of accuracy. Using the output of the market study and the test results, the product design and the corresponding development plan followed.

References

1. European Commission, European Space Agency: Business in satellite navigation - An overview of market developments and emerging applications, Galileo Joint Undertaking (2003)
2. PReVENT Integrated Project, http://www.prevent-ip.org/
3. CVIS project (Cooperative Vehicle-Infrastructure Systems), http://www.cvisproject.org/
4. NextMap Project, http://ertico.webhouse.net/en/activities/activities/nextmap_website.htm/
5. EuroRoadS Project, http://www.euroroads.org/
6. Gallant Project (GaLileo for safety of Life Application of driver assistance in road Transport), http://www.crfproject-eu.org/
7. GSSI RoadScan, http://www.geophysical.com/RoadScan.htm
8. 3D Laser Mapping (3DLM), IGI Gmbh, http://www.streetmapper.net
9. The Stanford CityBlock Project, http://graphics.stanford.edu/projects/cityblock/
10. Mobile Mapping Van Brochure NA-8003-0107 (2007)
11. Newby, S., Mrstik, P.: Inertial Map the Kabul Road, GPS World (July 2005)
12. VISAT Mobile Mapping System, http://www.gisdevelopment.net/magazine/years/2004/nov/landbased.htm
13. RoadSTAR, Arsenal Research, http://www.arsenal.ac.at/roadstar/products_mob_roadstar_en.html

Extending Internet into Space – ESA DTN Testbed Implementation and Evaluation

Christos Samaras[1], Ioannis Komnios[1], Sotirios Diamantopoulos[1],
Efthymios Koutsogiannis[1], Vassilis Tsaoussidis[1,*]
Giorgos Papastergiou[2], and Nestor Peccia[3]

[1] Democritus University of Thrace
Department of Electrical and Computer Engineering,
12 Vas. Sofias Str., 67100 Xanthi, Greece
{csamaras,ikomnios,sdiaman,ekoutsog,vtsaousi}@ee.duth.gr
[2] Hellenic Aerospace Industry S.A.
P.O. Box 23, 32009 Schimatari, Greece
{papastergiou.georgios@haicorp.com}
[3] ESA/ESOC
Robert Bosch Strasse 5, D-64293 Darmstadt, Germany
{nestor.peccia@esa.int}

Abstract. As the number and complexity of space missions increases, space communications enter a new era, where internetworking gradually replaces or assists traditional telecommunication protocols. The Delay Tolerant Network (DTN) architecture has recently emerged as a communication system for challenged networks, originally designed for the Interplanetary Internet. In the context of our project with ESA called "Extending Internet into Space - ESA DTN Testbed Implementation and Evaluation" we intend to deploy a distributed, flexible and scalable DTN testbed for space communications. The testbed will provide the supportive infrastructure for the design and evaluation of space-suitable DTN protocols, architectures, and routing policies to allow efficient deep-space communications. Throughout the project, we will demonstrate the operational capabilities of the DTN protocols in space; design and evaluate novel transport protocols and architectures for reliable data transfer in space; and investigate routing algorithms that comply with ESA's policies and resource status.

Keywords: Delay-Tolerant Networking, Testbed, Deep-Space Communications.

1 Introduction

Currently, all space communications are static, inflexible, and involve prior scheduling of communication contacts. In this context, less sophistication was required from communication protocols; the link layer was the dominant layer for space communications; routing was never an issue; end-to-end reliability was frequently overlapping

---

* Technical/Scientific Leader.

F. Granelli et al. (Eds.): MOBILIGHT 2009, LNICST 13, pp. 397–404, 2009.

with reliability of a single hop; congestion and overflow were absent due to strict scheduling of communication activities and admission control; and the limited required sophistication was shifted to the application layer.

However, future missions become more complex and new communication architectures and protocols for backbone, access, and proximity networking need to be designed, validated and optimized. Two new major properties have changed the spectrum of potential architectural choices for space communications: (i) the multi-hop architecture, which is required to reach deep space and (ii) the increasing number of alternative communication paths that may be used to reach a single receiver. Along these two properties, the demand for interoperability among space agencies has also contributed towards the emerging field of Delay-Tolerant Networking (DTN) [1]. DTN architecture is essentially a communication system to provide data transfer services in challenged environments, featuring extreme operational characteristics such as high propagation delays or network partitions. DTN applicability spreads over a wide spectrum of networking environments. With respect to space, DTN is envisioned to support Internet-like services across interplanetary distances.

As DTN becomes a standard architecture included in the Consultative Committee for Space Data Systems (CCSDS) standardization procedures, a testing and verification infrastructure emerges along with a set of scenarios, operations and evaluation procedures. In this context, we are designing and building an appropriate DTN testbed to evaluate associated scenarios, mainly targeting Mars-to-Earth communications (for an example scenario, see Fig. 1). This testbed will allow for cost-effective evaluation and optimization of space communication designs for reliable and efficient data delivery, expose the corresponding constraints, and uncover potential tradeoffs. The testbed will verify the robustness of space communications network, which can potentially reduce mission risks and enhance space data transfer rates.

The remainder of the paper is organized as in the following. In Section 2, we elaborate on current space communications, which are mainly supported by CCSDS protocols, and discuss the DTN architecture. Our DTN testbed is presented in Section 3. We outline the goals and research directions of our project in Section 4. Finally, we conclude the paper in Section 5.

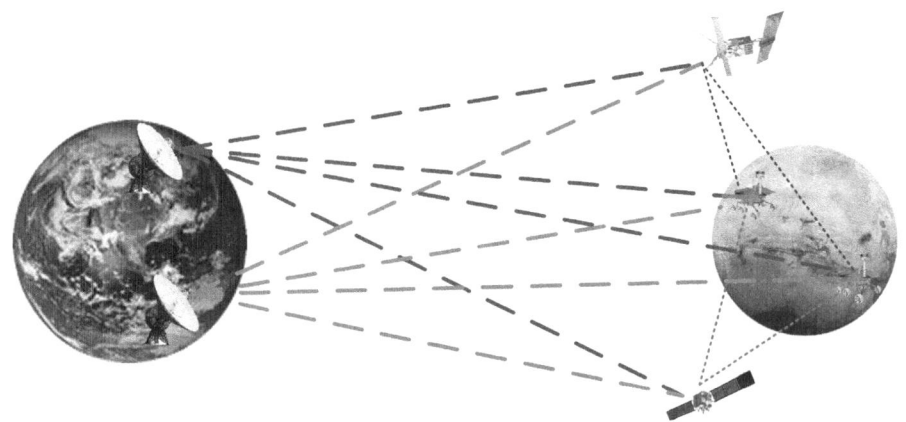

Fig. 1. Space Network Environment

2 Space Communications and DTN Architecture

2.1 Space Communications Today

Existing space communication infrastructure is built and managed around the operational targets of space missions and allows mainly for sending commands from mission control centers to spacecraft, and for receiving telemetry data from spacecraft to ground. In this context, the link layer is today the dominant layer for space communications. CCSDS has established a wide range of standards for space communications, including Telecommand and Telemetry Space Data Link Protocols [2]-[3], and Space Packet Protocol [4]. No provision is made for direct communication among space elements which would enable space internetworking using different layers at different points in end-to-end data flow.

Some technological advances have already been made towards the direction of end-to-end services. A recent achievement is, for example, the CCSDS File Delivery Protocol (CFDP) [5], which is designed for reliable file transfer across interplanetary links. CCSDS has also enabled onboard systems to have their own IP addresses. This is accomplished by either direct use of IP or an abbreviated form of IP that is the Network Protocol (NP) component of a four-layer stack of protocols, known as the Space Communication Protocol Standards (SCPS) [6]. Both of these capabilities allow dynamic routing through different paths in a connectionless fashion. However, it is inherent in the physics of space communications that both SCPS-NP and IP can not work well in areas of disrupted and long-delay communications that mainly characterize deep-space communications; IP is mainly bottlenecked by the conflicting nature of dynamic routing that requires frequent information exchange and the prohibiting nature of space communications that confine the concept of synchronization and accuracy.

Although the main target of CCSDS is to create a common protocol stack to be used by all agencies in the missions they plan, this is not what happens today, as the communication protocols used are mainly mission-specific. Inter-agency communication requires a common interoperable platform to allow for more natural communication that may replace a series of encapsulation and tunneling patches and help realize joint missions or multiple parallel missions of different space agencies. Furthermore, resource sharing and store-and-forward practices allow for enhanced connectivity and reduced mission costs. Typically, interoperability requires a level of convergence where different protocols can be deployed above or below that level. The experience from the Internet shows that interoperability is an issue along with cost and communication efficiency.

Another problem, present in space missions today, is the manual operation of communications to and from spacecraft. Indeed, no automation has been imported into the way space data transfers take place. Hence, as more and more assets are sent into space, scheduling of transmissions becomes too complex to be manually operated. That said, the need for deployment of an internetworking architecture in space is becoming more urgent than ever.

2.2 Delay-Tolerant Networking as a Candidate Architecture for Space

Delay/Disruption-Tolerant Networking has been proposed to overcome relevant problems that arise from current deep-space communication networks, such as absence of a common layer for interoperability among all space agencies; inefficiency in exploiting all communication opportunities; limited number of alternative communication paths that may be used to reach a single receiver; and the static and human-operated mission control. The architecture of DTN is based on, but not limited to that of the Interplanetary Internet [7]-[8]. DTN architecture embraces the concept of occasionally-connected networks that may suffer from frequent partitions and that may be comprised of more than one divergent set of protocols or protocol families. Nowadays, DTN has emerged as a recognized networking research area and has become an interesting idea for challenging environments within the terrestrial Internet as well. For example, DTNs are expected to provide connectivity to the edges of the current Internet infrastructure. DTN, as overlay architecture, provides Internet like services and is capable of extending store-and-forward architecture with permanent storage capabilities. Deep space missions are expected to sufficiently exploit an infrastructure that allows the efficient communication between in-space entities, such as explorer spacecraft, landed vehicles, and orbiters.

DTN employs an end-to-end message-oriented overlay, the so-called bundle layer, that exists at a layer above of the transport (or other) layer of the networks on which it is hosted and below applications. Devices implementing the bundle layer are called DTN nodes. A DTN-enabled application sends messages of arbitrary length that are transformed by the bundle layer into one or more protocol data units, called bundles. Bundles are thereupon forwarded by DTN nodes towards their destination.

Persistent storage (such as disk, flash memory, etc.) is employed to address discontinuous end-to-end connectivity, as it does not expect that network links are always available or reliable. The DTN architecture provides two features for enhancing delivery reliability: end-to-end acknowledgements and custody transfer. The latter mechanism permits delegating the responsibility for reliable transfer among different nodes in the network. When custody transfer is requested, the bundle layer employs a coarse-grained timeout and retransmission mechanism and an accompanying custodian-to-custodian acknowledgement signaling mechanism.

Therefore, custody transfer allows the source to assign retransmission responsibility and recover its retransmission-related resources relatively soon after sending a bundle. Not all nodes are required by the DTN architecture to accept custody transfers. Furthermore, a DTN node may have sufficient storage resources to sometimes act as a custodian, but may elect not to offer such services when congested or running low on power.

3   Testbed Architectural Design

Our testbed design emphasizes mainly on the properties of communication within deep space. We refer to our system as *DTN testbed* or simply *testbed*. During the initial deployment of the DTN testbed, our system will include modeled network links and real protocol implementations that operate at the network layer and above. Thus, no specialized hardware will be needed, all network parameters will be easily controlled, and the results will be reproducible. This will allow for initial cost-effective protocol evaluation and validation, and for smoother testbed extensions discussed later in the paper.

3.1 DTN Testbed Design Goals

The main design goals of the DTN testbed regarding its accuracy and efficiency are:

(i) *Dynamic control of network parameters.* The testbed should be able to emulate fundamental network parameters (such as bandwidth, packet error rate, propagation delay, and available connectivity), and adapt realistically and dynamically to changes in those parameters in real-time.

(ii) *Scalability.* Although current deep-space communications involve a limited number of communication nodes, the testbed should be able to scale well over a larger number of communication nodes to allow for emulation of any future deep-space communication scenarios, which will include several planetary surface networks and relay satellites.

(iii) *Transparency.* Network emulation should be transparent to upper layer protocols and applications, thus permitting their use and evaluation without need for modification.

(iv) *Flexibility.* The testbed should be flexible enough to: emulate any space communication topology; incorporate new protocols, applications and mechanisms; interoperate with other similar DTN testbeds; and provide a reusable infrastructure towards an actual hardware testbed.

3.2 Architecture

In this section we describe the functionality of basic testbed elements and analyze the interaction among them. In Fig. 2, we present the basic structure of the proposed testbed and the interconnection between testbed components.

The *Scenarios Description and Results Visualization System* consists mainly of two components: a tool for describing deep-space communication scenarios, and a tool for

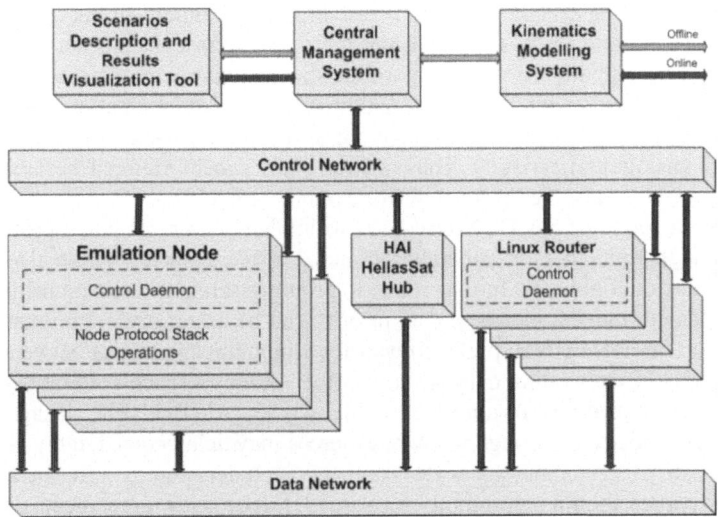

Fig. 2. DTN Testbed Architecture

visualizing emulation results. A *scenario description* contains information about the nodes in the network (e.g., landers, rovers, relay satellites, ground stations, and Mission Operation Centers), including initial position and movement (when applicable), communication protocols that will be used, link characteristics (e.g., frequency band, bandwidth, transmission power, etc.) and alike emulation parameters. The Results Visualization tool is used to present performance results graphically either in real-time (online) or after the emulation ends (offline), collected by all testbed elements through the control network. Performance results include file delivery time, retransmission overhead, status reports, etc. The Scenario Description tool passes scenario information to the Central Management System, while the latter forwards collected results to the Results Visualization tool.

The emulation system is divided into two discrete networks, namely a *Control Network* and a *Data Network*. The former is responsible for coordination of the emulation via control messages, while the latter moves data (such as output data) within the testbed. *Emulation nodes* represent distinct network nodes in the system, and a number of routers can be exploited to emulate multiple links, as is the case of a planetary surface network for example. The final component of the testbed is a real geostationary link through *Hellas Sat GEO Satellite* that will be integrated into the DTN testbed for testing purposes.

4   Goals and Research Directions

Core target of our DTN testbed is validation of the DTN architecture in space. Indeed, our project will serve as a means to reveal problems and deficiencies of current DTN specification [9], and propose relevant adaptations. DTN architecture will be tested in terms of: applicability to space networks; conformance with current space communication protocols and ESA's infrastructure; and performance.

Since manual routing as currently operated in space is costly and not scalable, more flexible routing schemes need to be established. Common IP-based routing protocols cannot operate in space, given that they rely upon constant network connectivity. A DTN-compatible routing algorithm should make communication decisions on the basis of locally available information such as: communication opportunities; expiration-time of messages to be delivered; performance history of communication paths, and analogous heuristics. Thus, static routes can be defined beforehand, but dynamic routes can also be discovered (for example, see [10]).

One major scope of our DTN testbed is to implement and evaluate protocols and mechanisms that enhance interoperability among space agencies, while also preserving their individual policies. In that context, a routing scheme based on priorities will be implemented. Indeed, in space, data priority can be associated with route priority. For instance, an agency may prefer to wait for communication links between its own assets, instead of forwarding data via some other agency's space systems, even if the latter contact opportunity becomes available sooner. Alternatively, an agency may decide that connectivity through its own resources may delay considerably, and go for a shortest path using another agency's resources. Such decision may fit to cases where high-priority data should be promptly forwarded. Inter-agency agreements can lay the ground for resource sharing between space missions, and DTN routing policies can exploit the offered contact opportunities.

The Delay-Tolerant Networking architecture essentially relies upon underlying network services adapted to the special networking conditions. In the context of deep-space communications, novel transport and application layer algorithms should be established for efficient and reliable data transfers. We have implemented and evaluated through ns-2 simulations [11] two protocols to address reliable data transports in space: Deep-Space Transport Protocol [12] and Delay-Tolerant Transport Protocol [13]. Deep-Space Transport Protocol (DS-TP) utilizes the hop-by-hop, store-and-forward message switching principle that governs today's space communications, and mitigates the need for congestion avoidance. DS-TP's novel, proactive retransmission scheduling allows for efficient and fast retransmission of corrupted packets due to high BERs or blackouts. Delay-Tolerant Transport Protocol (DTTP), like DS-TP, provides reliable data transfer over challenged network environments. It comprises a packet-oriented transfer approach over multi-hop and collaborative end-to-end paths, and acquires available bandwidth resources via its rate-based transmission behavior. DTTP can either operate as a standalone transport protocol (facing long delays and disruptions) or complement DTN architecture in space.

In the context of our project, we plan to further enhance DTTP and DS-TP functionality. Capabilities to be added include (but are not limited to) the following:

(i) *Proactive retransmission.* We will implement and evaluate retransmission mechanisms, that inject redundant data into the network to strengthen reliable transfers against packet losses. Alternatively, we will deploy packet-level erasure coding to allow for advanced recovering from packet losses. Coding rate can be dynamically adjusted, based on network measurements. We will investigate the trade-off between processing overhead, retransmission overhead, and recovering capability.

(ii) *Parallel data transfer.* Data transfer can be accomplished in parallel data paths, exploiting various communication opportunities. Sequence of application data is resumed at the receiver. Parallel data transfer can be implemented by preserving the original sequence number space. This feature requires explicit definition in the protocol header, so that the final destination can anticipate and merge data packets coming from different paths.

(iii) *Sending rate adaptivity.* In space settings, sending rate can be accurately characterized as temporarily constant. Sending rate adaptation can follow network events such as storage capacity exhaustion. These network events can be either perceived by senders through advanced mechanisms or explicitly signaled by receivers.

5 Conclusions

Currently, all space communications are static, inflexible, and manually preconfigured long before they actually take place. Network disconnections, potentially huge propagation delays, high link error-rates, and bandwidth asymmetries compose the space networking environment. Delay-Tolerant Networking is a candidate communication architecture in space, as it can stand long delays and network partitions. In this context, we construct a DTN testbed that integrates a variety of tools and protocols in order to emulate realistic deep-space communications scenarios with varying network elements, topologies, and operational parameters. The testbed will form a valuable platform for testing DTN architecture in space, and a research framework for evaluating new mechanisms and protocols.

References

1. Cerf, V., et al.: Delay-Tolerant Network Architecture, IETF RFC 4838, information (April 2007), http://www.ietf.org/rfc/rfc4838.txt
2. TC Space Data Link Protocol. Recommendation for Space Data System Standards. CCSDS 232.0-B-1. Blue Book. Issue 1. Washington, D.C. CCSDS (September 2003)
3. TM Space Data Link Protocol. Recommendation for Space Data System Standards. CCSDS 132.0-B-1. Blue Book. Issue 1. Washington, D.C. CCSDS (September 2003)
4. Space Packet Protocol. Recommendation for Space Data System Standards. CCSDS 133.0-B-1. Blue Book. Issue 1. Washington, D.C. CCSDS (September 2003)
5. CCSDS File Delivery Protocol. Recommendation for Space Data System Standards. CCSDS 727.0-B-2. Blue Book. Issue 2. Washington, D.C. CCSDS (October 2002)
6. Space Communications Protocol Specifications (SCPS), http://www.scps.org/
7. Burleigh, S., Hooke, A., Torgerson, L., Fall, K., Cerf, V., Durst, B., Scott, K.: Delay-Tolerant networking: an approach to InterPlaNetary Internet. IEEE Communications Magazine 41(6), 128–136 (2003)
8. Cerf, V., et al.: Delay-Tolerant Network Architecture: The Evolving Interplanetary Internet, IPNRG Internet Draft draft-irtf-ipnrg-arch-01.txt
9. Scott, K., Burleigh, S.: Bundle Protocol Specification, IETF RFC 5050, experimental (November 2007)
10. Burleigh, S.: Dynamic Routing for Delay-Tolerant Networking in Space, Flight Operations SpaceOps 2008, Conference (Hosted and organized by ESA and EUMETSAT in association with AIAA) (2008)
11. ns-2. The Network Simulator, http://www.isi.edu/nsnam/ns/
12. Psaras, I., Papastergiou, G., Tsaoussidis, V., Peccia, N.: DS-TP: Deep-Space Transport Protocol. In: IEEE Aerospace Conference 2008, Montana, USA (March 2008)
13. Samaras, C.V., Tsaoussidis, V.: DTTP: A Delay-Tolerant Transport Protocol for Space Internetworks. In: 2nd ERCIM Workshop on eMobility, Tampere, Finland (May 2008)

Conceptual Design of a Wireless Strain Monitoring System for Space Applications

Panagiotis Broutas, Stathis Kyriakis Bitzaros, Dimitrios Goustouridis,
Stavros Katsafouros, Dimitrios Tsoukalas, and Stavros Chatzandroulis

NCSR "Demokritos", Institute of Microelectronics
Athens, Greece
{panbrut,mpitz,gousto,sgk,dtsouk,stavros}@imel.demokritos.gr

Abstract. The conceptual design of the architecture of a wireless strain monitoring network suitable for space applications is presented. The system is a heterogeneous wireless network that consists of battery powered nodes and batteryless nodes that are able to harvest energy from an incident RF field. Battery powered nodes are based on the Zigbee standard. Both battery and batteryless nodes are envisioned to include sensors but some battery powered nodes could simply serve as relaying points to transfer data to the central computer. The structure of the batteryless nodes as well as remote powering and data transmission are analyzed.

Keywords: Wireless sensor networks, remote powering, structural health monitoring.

1 Introduction

Health monitoring of structures is a major concern in the space and aviation community, where the need for more sophisticated structural health monitoring (SHM) systems has been recognized [1]. The core of SHM technology is the development of self-sufficient systems that use built-in, distributed sensor/actuator networks not only to detect structural discrepancies and determine the extent of damage but also to monitor the effects of structural usage. SHM can provide early warnings of physical damage, which can be used to define remedial strategies before the damage compromises the spacecraft. Furthermore, it may be possible to quickly, routinely and remotely monitor the integrity of an air/spacecraft structure while in service.

Wireless SHM systems have many advantages over cabled systems because they are characterized by local computational ability, low cost of deployment and wireless networking functionality. In addition, massive interconnections between sensors and a central computer are avoided.

2 SHM System Architecture

The system under development is a heterogeneous wireless network of strain sensors which may be deployed to air or space vehicles. The system is envisaged to consist of

F. Granelli et al. (Eds.): MOBILIGHT 2009, LNICST 13, pp. 405–410, 2009.
© ICST Institute for Computer Sciences, Social-Informatics and Telecommunication Engineering 2009

both battery powered and batteryless nodes interoperating and communicating together and with a central computer which gathers, logs and evaluates data. Both battery and batteryless nodes may include sensors but some battery powered nodes could simply serve as relaying points to transfer data to the central computer. Batteryless nodes will harvest energy from incident RF waves.

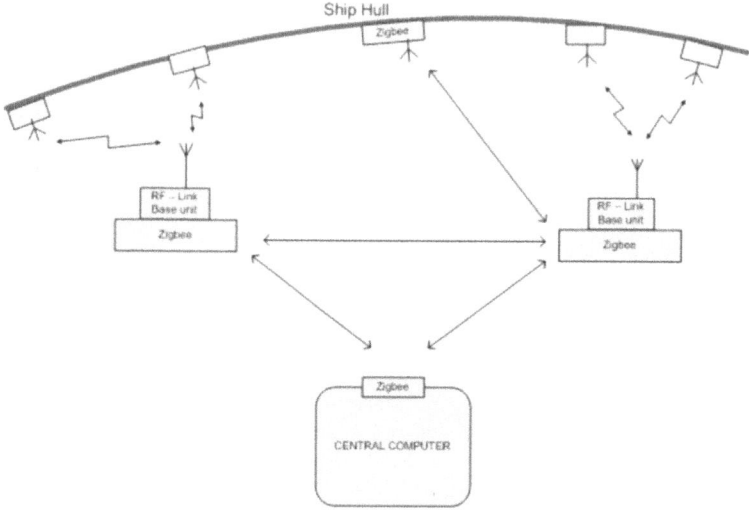

Fig. 1. Concept of the system which will demonstrate a short-range RF wireless system for Structural Health Monitoring

The general structure of the sensor network is depicted in Figure 1. Two different types of battery powered nodes will be built using Zigbee compliant transceivers [2]; nodes that will incorporate sensors and will have the capability to be used either as network coordinator or end device and nodes that will be connected to an application specific board incorporating the RF reader of the batteryless node and will be configured as an end device only.

In such a network the most crucial point is to collect and accumulate power from the reader RF field and then making it available to the batteryless sensor tag electronics. The block diagram of the batteryless sensor tag is depicted in Figure 2. It will consist of the tag antenna, an impedance matching circuit, the power harvester, a microcontroller (MCU), the sensor and an ASK modulator. In order to power the tag the reader will have to emit continuously until enough energy is accumulated to power the tag electronic circuitry, collect the sensor data and transmit them back to the base unit/reader.

The RF reader will comprise two basic subsystems: one for transmitting power to the remote sensor tag and one for reading backscattered data coming out of the tag (Figure 3). The reader functions will be controlled and coordinated by a dedicated low power microcontroller which will also be responsible for communicating with a corresponding Zigbee compliant node through its serial port. The subsystem which will be responsible for transmitting the power to the remote tag will comprise a frequency generator and an amplification unit, while the receiver will be a standard ASK receiver.

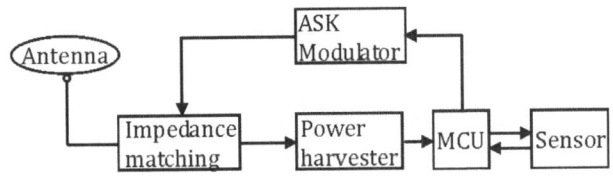

Fig. 2. Tag architecture

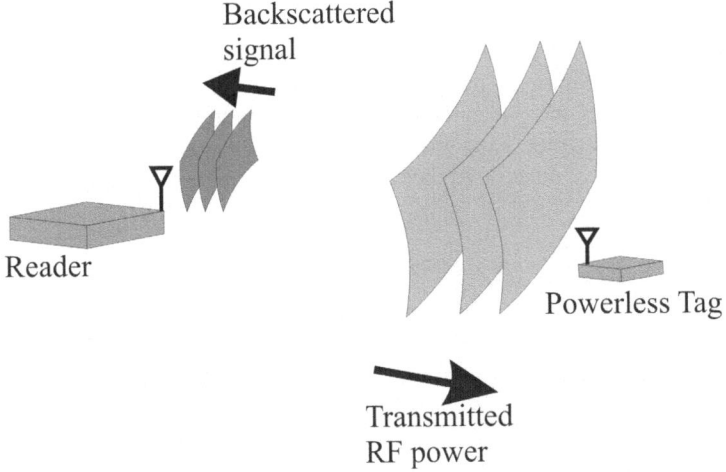

Fig. 3. The basic scheme of the passive link between the reader and the sensor tag is shown. The tag receives power from the remote reader and uses it to power its circuitry, perform the sensor measurement and modulate the backscattered signal with the sensor data.

2.1 Remote Powering

The balance between the available power at the tag and the power required to operate its electronic circuitry is the most important problem of the proposed RF wireless sensor network. According to Friis equation (1) tags receive only a portion P_{tag} of the power P_r emitted by the reader.

$$P_{tag} = A_{er}A_{etag} / (r^2\lambda^2 P_r) . \tag{1}$$

where lossless matched dipole antennas are assumed; A_{etag} is the effective aperture of the tag antenna at distance r and A_{er} is the effective aperture of the reader antenna, while the radiation wavelength is denoted by λ. In a practical circuit the voltage generated at the antenna terminals is of the order of only a few millivolts and has to be raised to such a level so as it would be possible to power practical circuits. This task implies that a voltage multiplication circuit follows the antenna and matching circuits (Figure 4). Insertion of the voltage multiplier in the tag circuitry comes however with an extra power loss, as the efficiency n of a six stage multiplier is approximately 30%.

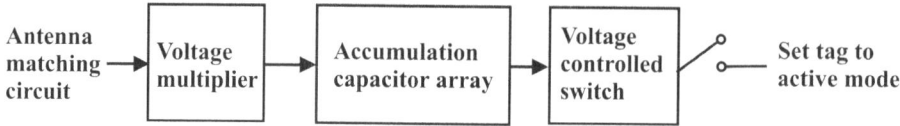

Fig. 4. Block diagram of the power harvesting unit architecture

Thus, the only way to power the tag circuitry is to accumulate enough energy over time by storing it in an accumulation capacitor array, and then making it available to the microcontroller to perform the required tasks.

The whole process will have to be controlled via a low power control circuit which has to be continuously on to perform its task. The power consumption of the control circuit equals the power consumption at standby mode. It should be noted that in order to charge the capacitors the minimum power available at the voltage multiplier input $P_{ant,min}$ must be higher than the power consumption in standby mode (formula (2)).

$$P_{ant,min} \geq P_{tag}/n . \tag{2}$$

As a result, the use of the capacitor array adds a limitation on the minimum amount of power that must be harvested by the antenna.

The time interval t_{charge} required to charge the capacitor array between operation cycles (the capacitor array has been fully charged once) is calculated by formula (3).

$$t_{charge} = t_{on} / (P_{load} - P_{standby}) . \tag{3}$$

Where t_{on} is the time interval required for measurement and data transmission of the remote tag and in the current implementation is equal to 125.1 msec. P_{load} is the power available at the terminals of the accumulation capacitor and is determined by the power emitted by a reader antenna and the power efficiency of the harvesting unit. The charging time versus distance for different levels of power emitted by the reader antenna is presented in Figure 5. From this figure it is evident that the charging time increases with distance and poses a severe limitation on the reader – sensor tag range that can be achieved. For example for 100mW of reader transmitted power less than 2m may be achieved even with long charging time.

2.2 Data Transmission and Backscattering Modulation

Tag is designed so as each tag of the network transmits data as soon as one measurement cycle is complete. Furthermore, in order to minimize the power required by the tag backscattering modulation is used. In addition, a simple anti-collision algorithm is employed, whereas in order to minimize the possibility of data collision each tag waits for a predefined time interval before transmission. The proposed transmission algorithm further reduces power consumption as the tags do not have to listen to the channel.

To calculate the minimum bits required to be transmitted by the tag we should take into account that at least two bytes must be initially transmitted in order to lock the

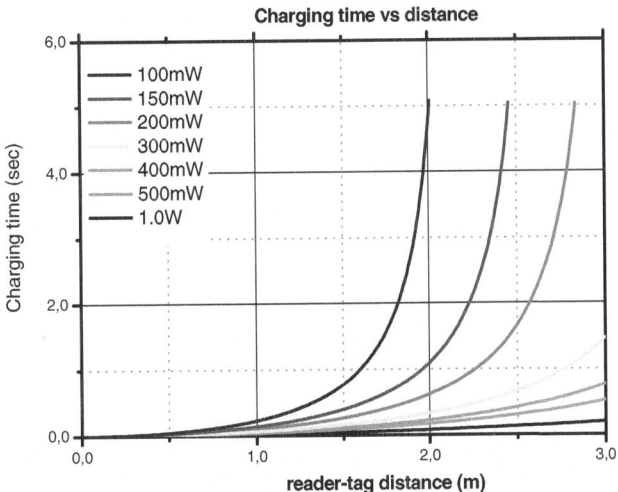

Fig. 5. Charging time versus distance for different reader antenna power (ERP). Perfectly matched dipole antennas have been considered for the reader and the tag. Operating frequency is 900MHz. The maximum power consumption of the tag is approximately $P_{tag} = 80\mu W$, while the estimated consumption in standby mode is about $P_{standby} = 12\mu W$.

PLL circuit of the reader. In order to minimize the data to be transmitted, the last 4 bits of this sequence of bits are also used as tag ID numbers. This way the system will be able to support up to 16 tags. The next two bytes will contain the actual sensor measurement. The tag transmission finally ends, with a checksum in order to ensure that the received data is not corrupt. The checksum algorithm is based on the addition of the assorted bits with the resulting byte appended to the transmitted data. Overall 5 bytes have to be transmitted in each reader-tag communication.

In backscattering modulation the tag modifies the amount of the incident radiation that it backscatters. According to [3] the actual backscattered power of the tag is approximately 1/3 of the absorbed power (that is -5dB). Thus, according to Friis equation the power P_{bsc} that is backscattered by the tag is given by formula (4), where P_r is the power transmitted by the reader, and G_r and G_t are the gain of the reader and the tag respectively. The wavelength is denoted by λ and the backscattering transmission loss by T_b.

$$P_{bcs} = P_rG_rG_tT_b(\lambda/(4\pi r))^2 . \qquad (4)$$

Using again the Friis equation the backscattered power received by the reader P_{rbsc} is given by formula (5).

$$P_{rbcs} = Pr(G_rG_t)^2T_b(\lambda/(4\pi r))^4 . \qquad (5)$$

If a reader that emits 1mW of power ERP, and dipole antennas of gain 2.2 for the reader and the tag are considered, then the values of the backscattered signal power for different distances are presented in Figure 6. In order to decipher the maximum range between reader and sensor tag that may be achieved one should take into

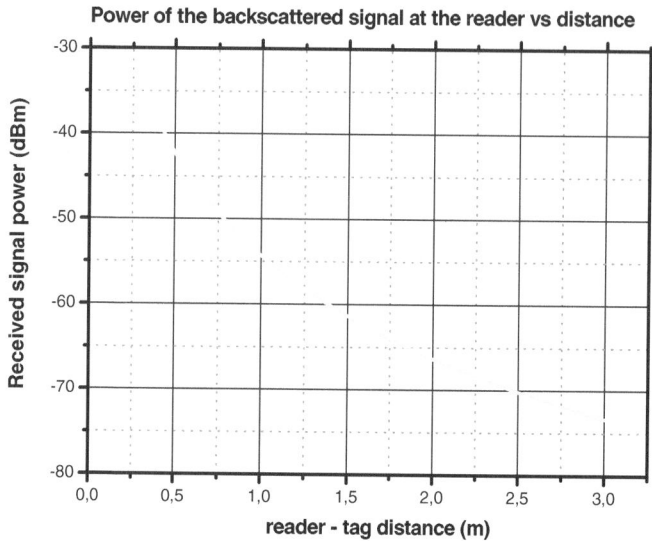

Fig. 6. Power of the backscattered signal received by the reader as a function of reader – tag distance. Operating frequency 900MHz. Dipole antennas have been considered for the reader and the tag (Reader transmitted power: 0dBm).

account that commercial readers are able to demodulate signals down to -80dBm, yields a range of more than 3m. Thus, in this implementation, the main limitation comes from the maximum distance at which the sensor tag may be powered, while the data transmission link does not impose any additional range limitation.

3   Conclusions

The basic considerations behind the design of a passive reader – sensor tag link for use in a wireless strain monitoring network suitable for space applications has been presented and major limitations explored. The system is a heterogeneous wireless network consisting of battery powered, Zigbee compliant, and batteryless nodes that are able to harvest energy from the incident RF field generated by reader. It has been found that it is possible to provide enough power to operate the sensor tag circuitry at a distance of up to 2m with 100mW transmitter power, while reading the sensor tag data does not pose any additional limitation when using backscattering.

References

1. Lynch, P., Kenneth, J.L.: A summary review of wireless sensors and sensor networks for structural health monitoring, Shock and Vibration Digest (2006)
2. Zigbee Alliance, http://www.zigbee.org/en/index.asp
3. Daniel, M.: Dobkin: The RF in RFID, passive UHF RFID in practice. Elsevier Inc., Amsterdam (2008)

Novel Metamaterials for Patch Antennas Applications

Theodore Zervos[1], Fotis Lazarakis[1], Antonis Alexandridis[1], Kostas Dangakis[1], Dimosthenis Stamopoulos[2], and Michalis Pissas[2]

[1] Institute of Informatics & Telecommunications,
National Centre for Scientific Research "Demokritos", Athens, Greece
{tzervos,flaz,aalex,kdang}@iit.demokritos.gr
[2] Institute of Materials Science,
National Centre for Scientific Research "Demokritos", Athens, Greece
{densta,mpissas}@ims.demokritos.gr

Abstract. In this paper we introduce the incorporation of magneto-electric materials into antenna design and the potential of controlling the behavior of the antenna by means of an external magnetic field. After an intensive study of magneto-electric material properties, a ferrimagnetic compound called Yttrium Iron Garnet (YIG) was found to be the best candidate for the novel antenna design. We provide a metamaterial patch antenna design where a part of the substrate is replaced by the YIG compound. After several design modifications the final model includes a circular-shaped YIG substrate just under the metallic patch and offers sufficient performance in terms of resonance, bandwidth and radiation efficiency. Additionally, in the presence of an external magnetic field the polarization becomes elliptical and the sense of the polarization (left or right) can be controlled through the direction of the magnetic field. That latter characteristic confirms the metamaterial-nature of the antenna.

Keywords: patch antenna, ferrimagnetic compound.

1 Introduction

Patch antennas technology is used for many years now. These antennas are low-profile robust planar structures and can achieve a wide range of radiation patterns. Moreover, they can be easily manufactured and thus they are considered as inexpensive solutions compared with other types of antennas. On the other hand, there are some limitations in patch antenna designs such as low gain, narrow bandwidth of operation and decreased radiation efficiency due to surface-wave losses. Recently, the use of artificially engineered structured materials, known as "metamaterials", has been investigated in order to overcome the above mentioned shortcomings.

Metamaterials refer to a large variety of complex structures that possess exceptional electromagnetic properties not readily available in nature. At microwave frequencies, metamaterials typically consist of periodic arrays of dielectric and/or conducting elements. The term metamaterials is quite general and is used for a wide range of modern concepts such as frequency selective surfaces (FSS), electromag-

F. Granelli et al. (Eds.): MOBILIGHT 2009, LNICST 13, pp. 411–419, 2009.

netic band-gap (EBG) materials, left-handed media (LHM), artificial magnetic conductors (AMC).

On the other hand, numerous recent studies in the area of materials science have been devoted in the development of preparation methods, in handling and understanding the physics and chemistry of nanomaterials, nanostructures and metamaterials with modulated physical and chemical properties.

The main goal of this work is to investigate the development of novel metamaterials produced by means of natural mechanisms and their application in patch antennas. After in depth studies, it was found that using a specific material as part of an antenna substrate, one can control the antenna characteristics by properly applying an external magnetic field. More specifically, we aim to design and implement a patch antenna with varying polarization - linear, left circular (LCP) and right circular (RCP) - according to an externally applied magnetic field.

In this paper, the design of a conventional patch antenna is firstly described. This antenna is used as a reference for our studies. Then, a brief description is given of the techniques and the materials that are available to implement the selected designs. Next, the design of the metamaterial-inspired patch antenna and the simulation results are presented. Finally, the paper is completed with the main conclusions of the presented work.

2 Conventional Antenna Design

The potential target of our research is an antenna suitable for satellite communication applications, thus we have selected as operating frequency band the Ku-band. More specifically the conventional patch antenna that will be used to evaluate the proposed metamaterial-inspired patch antenna was decided to operate at 14 GHz. The evaluation process will be based on the comparison of the characteristics and the performance metrics between the conventional (reference) and the metamaterial antenna. The conventional antenna should have a simple design for easy simulation and implementation. For this reason, a simple rectangular patch antenna has been adopted to be the reference conventional patch antenna design. This antenna has been modeled using an EM simulation software package [1]. The detailed design specifications and performance characteristics of the conventional patch antenna are given bellow:

A rectangular patch is used with sides 6.05 mm long, fed by a microstrip line which penetrates inside the patch via a notch in order to achieve the proper matching (Fig. 1 (a)). The ground plane dimensions are 16 x 14 mm and the substrate is Taconic TLY3, 1.524 mm thick dielectric plate with ε_r equal to 2.33. According to the simulation results, a deep resonance (<-25 dB) occurred in the frequency of operation (14 GHz) as can be seen in Fig. 1 (b). The bandwidth (measured at S_{11}<-10 dB) was calculated to be 785 MHz (5.6 %), while the radiation efficiency was around 95 %. The antenna gain on boresight was calculated to be 6.6 dBi. In Fig. 1 (c), the 3D absolute radiation pattern of the antenna is presented to be the anticipated half-plane pattern.

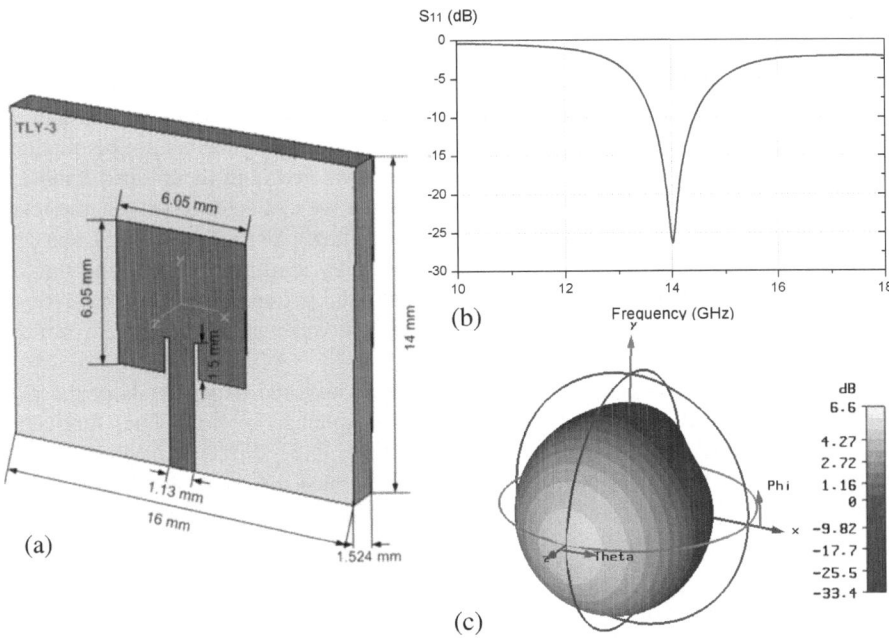

Fig. 1. The conventional patch antenna model (a), its reflection coefficient (b) and radiation diagram (c)

3 Materials with Modulated Properties

Before presenting the specific materials that we propose for implementation in the novel patch antennas and their highly unconventional dielectric, magnetic and crystal properties let us make a brief introduction on the basic underlying physics that could be very helpful to the non-experts.

Ferroelectricity is a term that is used analogously to ferromagnetism, in which a material exhibits a permanent magnetic moment. Thus a ferroelectric material exhibits a spontaneous electric polarization. The electric polarization can be modulated under the application of an external electric field. Thus ferroelectric materials can be used to make capacitors with tunable capacitance an opportunity that could be easily implemented in the design of planar patch antennas possibly enabling the control of the antenna characteristics under the application of an external electric field.

Ferroelectric materials exhibit nonlinear dielectric response and thereto demonstrate a spontaneous polarization in analogy to the spontaneous magnetization of the ferromagnetic materials. Similarly to their ferromagnetic counterparts, ferroelectric materials demonstrate their unconventional dielectric properties only below a certain phase transition temperature the so-called critical temperature.

Going a step farther, *multiferroics* are materials that have coupled dielectric, magnetic and crystal properties. The unusual coupling of these two order parameters enables the modulation of all dielectric and magnetic properties upon variation of a single external parameter that should naturally influence exclusively one of them.

Thus in such materials the dielectric properties may be varied upon application of an external magnetic field or similarly the magnetic properties can be varied upon application of an external electric field.

Such multiferroic materials can be employed as substrate and/or superstrates in patch antennas, possibly offering several novel characteristics that cannot be obtained by conventional design practice. It has been demonstrated both theoretically and experimentally that by using a ferromagnetic substrate we can achieve (a) tunable resonance frequency under the application of a dc magnetic field [2], (b) recurrent and tunable circularly polarized radiation [3]. Accordingly, aiming at controlling the antenna performance under the application of an external parameter such as magnetic or electric field we have focused our effort on these novel materials coming from the class of multiferroics.

The research objective is to investigate theoretically and experimentally the performance advantages of magnetic materials-based antennas over antennas fabricated using conventional substrates. To accomplish that, three different magnetoelectric compounds ($Al_{2-x}Fe_xO_3$, $Ga_{2-x}Fe_xO_3$ and $Y_3Fe_5O_{12}$) have been initially examined as components of the substrate of the novel patch antenna.

4 Metamaterial-Inspired Antenna

The use of ferromagnetic materials as elements of a patch antenna (more appropriate as a substrate) or as constituent elements in fabricating metamaterials, has not been sufficiently studied so far. Electromagnetic metamaterials consist of artificial arrangements of simple unit-cells/building-blocks that when arrayed produce effective media with tailored electromagnetic parameters, that is permittivity and permeability. For instance, such a metamaterial is realized by using periodic arrays of metallic split-ring resonators and wires with size and separation of the order of the radiation wavelength. Besides the success of these types of the metamaterials in improving, in some respect, the microwave devices' operation, significant Ohmic losses are encountered in their metallic parts preventing their use in high frequency applications.

An alternative approach in producing metamaterials that are not based on metallic elements could rely on using non-periodic or periodic arrangements of materials with high dielectric constant, multifunctional materials (magnetoelectric or multiferroic materials) and purely magnetic materials. In this context, we have decided to prepare, characterize, and implement (in patch antennas) these novel materials in simple and self assembled metamaterials form.

The modeling of the dielectric materials is well known and does not display any difficulties. In contrast, magnetic materials exhibit complicated behavior that can be accounted for under the consideration of numerous parameters. Accordingly, the implementation of magnetic properties in the modeling of patch antennas is a quite complicated task. The most important parameter that determines the behavior of magnetic materials in the microwave frequency range is the magnetic permeability tensor. Let us consider a ferromagnetic material under an applied static magnetic induction $B_{dc}=\mu_o H_{dc}$ parallel to the z-axis and an ac-magnetic induction $B_{ac}=\mu_o H_{ac}$. These dc and ac magnetic fields produce a dc- and ac-magnetic moment $M=M_{dc}+M_{ac}$. The equation of motion of the total magnetization is given by the Landau-Lifshitz equation [4]:

$$\frac{d(\mathbf{M}_{dc}+\mathbf{M}_{ac})}{dt} = -\gamma\left[(\mathbf{M}_{dc}+\mathbf{M}_{ac})\times(\mathbf{B}_{dc}+\mathbf{B}_{ac})+\frac{\alpha}{M_{dc}}\mathbf{M}_{dc}\times\frac{d(\mathbf{M}_{dc}+\mathbf{M}_{ac})}{dt}\right] \quad (1)$$

where $\gamma=ge/2m_e$ is the gyromagnetic ratio calculated from the Landé factor g in connection with the charge (e) and mass value (m_e) of an electron. The damping factor α is a dimensionless constant determined by the resonance line width ΔH through the relation $\alpha=\mu_0\gamma\Delta H/2\omega$. By solving this differential equation we can calculate the permeability tensor for the limiting case $H_{ac}<<H_{dc}$. A simple expression for the permeability tensor used in practice is given from the expression:

$$\mu_{ij} = \mu_0\begin{pmatrix} \mu_1(\omega) & \mu_2(\omega) & 0 \\ -\mu_2(\omega) & \mu_1(\omega) & 0 \\ 0 & 0 & 1 \end{pmatrix} \quad (2)$$

with:

$$\mu_1 = 1+\left[\omega_m(\omega_0+i\omega\alpha)\right]/\left[(\omega_0+i\omega\alpha)^2-\omega^2\right]$$
$$\mu_2 = i\omega\omega_m/\left[(\omega_0+i\omega\alpha)^2-\omega^2\right]$$

where $\omega_m=\gamma\mu_0 M_{dc}$ (here it is supposed that the external magnetic field saturates the magnetic material that is $M_{dc}=M_S$), and $\omega_0=\gamma\mu_0 H_{dc}$. The g, M_S and the permeability tensor are materials' parameters, while the external field, H_{dc}, is an external parameter. This plethora of parameters permits adequate flexibility in designing a patch antenna.

We have conducted a systematic and detailed study of some of the parameters of a simple patch antenna where the substrate consists of ferromagnetic/ferrimagnetic material, by varying the amplitude and direction of the external magnetic field (H_{dc}), the amplitude of the saturation moment (M_S), depending on the chemical composition of the material, and the width of the resonance (ΔH).

4.1 Initial Design

A basic characterization by means of magnetization (Superconducting Quantum Interference Device –SQUID–) and ferromagnetic resonance (FMR) measurements of the three candidate magnetoelectric compounds ($Al_{2-x}Fe_xO_3$, $Ga_{2-x}Fe_xO_3$ and $Y_3Fe_5O_{12}$) have been made. The critical temperature of the first two compounds is placed below T=300 K, making them poor candidates for room temperature applications. In contrast the $Y_3Fe_5O_{12}$ compound exhibits a critical temperature of T=550 K thus being appropriate for utilization in patch antennas operating at room temperature conditions. Further in this work the ferrimagnetic $Y_3Fe_5O_{12}$ compound, called Yttrium Iron Garnet (YIG), was used as a substrate in the proposed antenna design.

The initial design of the proposed antenna consists of the ferrimagnetic material that lies just under the radiating copper patch, and of a dielectric substrate around it as schematically presented in Fig. 2. More precisely, a rectangular block made of YIG having the same length and width with the metallic patch was placed at the centre of the antenna model and exactly under the radiating patch, replacing the dielectric (TLY3) substrate. The dielectric and magnetic properties of the YIG compound were determined the first from the literature and the latter by the magnetization and FMR measurements and they have properly set in the simulation software. These parameters are listed in Table 1. An external magnetic field having the maximum value of 5000 Oe was applied to the antenna having a direction towards +z, perpendicular to the substrate surface. According to the simulation results, the resonance frequency of the proposed antenna is 13 GHz while the bandwidth (measured at S11<-10 dB) is 306 MHz (2.35 %). The presence of the magnetic field has changed the antenna polarization from linear to elliptical with sense according to the direction of the applied field (towards +z is left while towards −z is right). The problem is that there is a large amount of losses resulting at low radiation efficiency and gain values comparing to the conventional antenna.

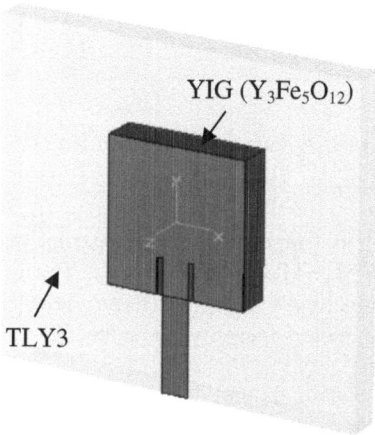

Fig. 2. Initial metamaterial-inspired antenna design

Table 1. Properties of the YIG compound

Dielectric properties		
Relative permittivity	$\varepsilon_r{}'$	10
Dielectric loss tangent	$\tan\delta$	0.0009
Magnetic properties		
Landé factor	g	2
Saturation magnetization	$4\pi Ms$	1700 Gauss
Resonance line width	ΔH	50 Oe

4.2 Final Model

A systematic study of the appropriate position, size and shape of the YIG part has been achieved using suitable simulation models. The aim of the study is the design of an antenna that would have polarization agility, which means it could change its polarization properties according to the value and the direction of an externally applied magnetic field. Additionally, the performance (e.g. radiation efficiency, gain) of the YIG-based patch antenna should be comparable to that of a conventional one. Firstly, rectangular YIG blocks with smaller size were tested in order to achieve higher efficiency and gain. Then, several different positions of the YIG block inside the substrate were investigated in order to find the one offering the highest overall performance. Also, cylindrical YIG blocks were simulated and compared to the rectangular ones. The similar performance of the metamaterial inspired antenna for the two substrate types in combination with the easier manufacturing of cylindrical YIG blocks, lead us towards the adoption of the cylindrical design to be the final patch antenna model.

The final design adopts a cylindrical YIG block, which has been placed at the centre of the dielectric substrate, underneath the rectangular patch [5]. The overall dimensions of the metallic patch have been slightly reduced to 5.54 x 5.54 (mm), resulting in a 8.5% reduction of patch area dimensions compared to the conventional patch antenna, in order to have resonance at 14 GHz. The dimensions set for the final design of the proposed patch antenna are illustrated in Fig. 3 (a). In Fig. 3 (b) the back side of the antenna model is presented, where it can be clearly seen how the YIG disc is being placed at the centre of the antenna exactly under the radiating patch.

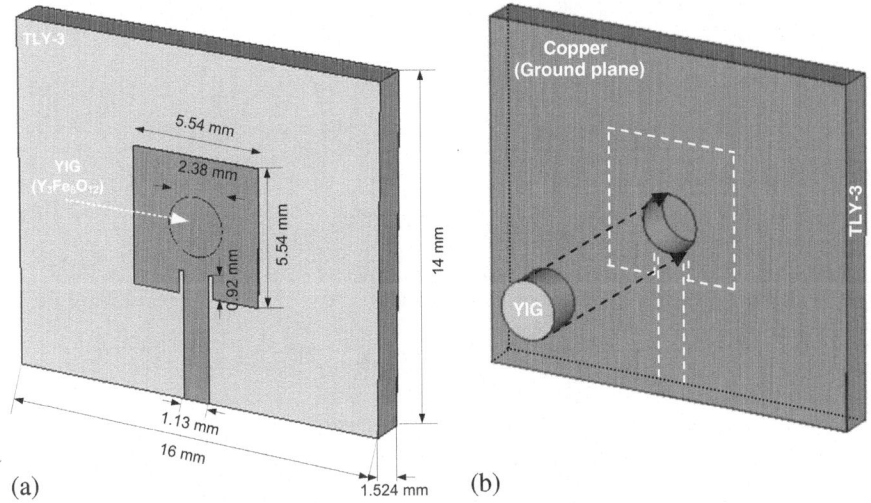

Fig. 3. Schematic representation of the proposed antenna model: (a) radiating element side view, (b) ground plane side view

5 Results

The reflection coefficient of the proposed antenna in direct comparison with that of the conventional patch is presented in Fig. 4. The bandwidth of the proposed antenna is 772 MHz (5.5 %) that is quite similar to that of the conventional one.

As mentioned above, the main characteristic of the proposed antenna is that the presence of an external magnetic field changes the polarization properties of the antenna. In this model, an external field of 5000 Oe has been applied towards the +z direction of the antenna coordination system, perpendicular to the substrate surface.

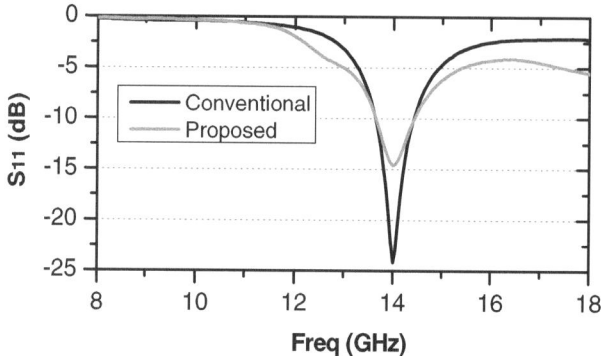

Fig. 4. The S_{11} parameter of the conventional and proposed antenna

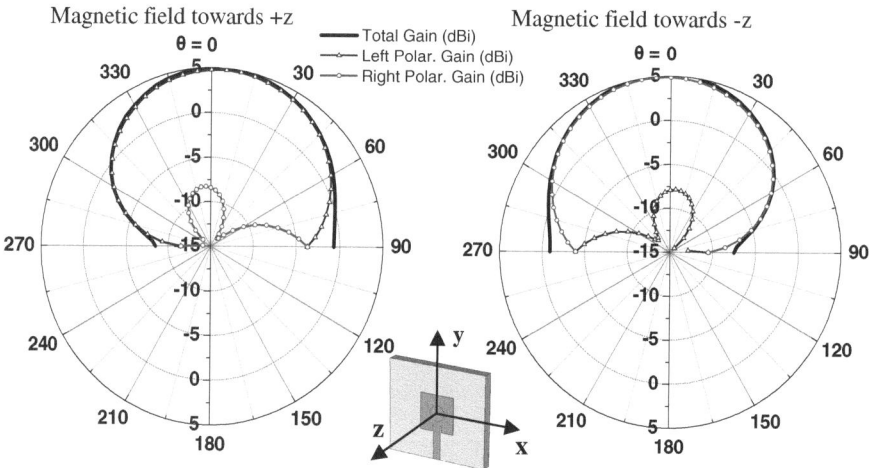

Fig. 5. Gain patterns in the xz plane indicating the change of the polarization sense

By applying this external dc magnetic field the polarization changes from linear (vertical) to left-handed elliptical. Alternatively, the same external field applied towards −z direction changes the polarization to right-handed elliptical one. As can be seen in Fig. 5, when the magnetic field is applied towards +z, the left polarized gain is very close to the total gain at the xz plane (nearly 0.4 dB difference at the boresight direction), while the right polarized gain is significantly smaller (almost -8 dBi). Similar conclusions can be extracted when the magnetic field is applied towards −z direction where now the right polarized gain is the dominant one and it is very close to the total gain.

Concerning radiation efficiency and polarization properties, Table 2 provides a concentrated view of the simulated results of the conventional patch and the proposed antenna. It is noted that the radiation efficiency of the YIG-based antenna is lower than that of the conventional, due to the relatively high permittivity of the YIG part that causes additional losses, but still with sufficient value (higher than 70 %).

Table 2. Comparison of the main antenna characteristics

	Conventional	Metamaterial inspired
Bandwidth (MHz)	785	772
Gain (dBi)	6.6	5.2
Erad (%)	95.3	71.4
Polarization	Vertical	Elliptical
Axial Ratio (dB)	200	4.1

Concerning polarization it has to be mentioned that it is almost circular since the axial ratio is 4.1 dB. All the above-mentioned characteristics of the YIG-based patch antenna are taken considering an external dc magnetic field of 5000 Oe applied perpendicular to its surface.

6 Conclusions

In this paper we introduced the utilization of a ferrimagnetic compound, namely Yttrium Iron Garnet (YIG) as a substrate in patch antennas in an effort to change and control the polarization under the application of an external dc magnetic field. We clearly demonstrated that the sense of the antenna polarization is strongly influenced by the YIG substrate since it changes at will in respect to the direction of the external magnetic field.

Further steps of the work include construction of the antenna and measurements of its antenna characteristics. At a first phase we will use a permanent magnet in order to prove the polarization agility of the proposed antenna and on a second step to investigate ways of altering the polarization of the induced external magnetic field (e.g. with the use of an electromagnet setup).

One of our main concerns is the potential application of the proposed antenna. A possible use of the proposed metamaterial inspired antenna could be as an array element. It could offer polarization diversity changing its polarization from left to right hand elliptical and vice versa according to communication needs.

Acknowledgments. This work was funded under ESA project, Contract No.: 20942/07/NL/ST/na.

References

1. CST Microwave Studio, CST GmbH-Computer Simulation Technology (2008)
2. Pozar, D.M., Sanchez, V.: Magnetic Tuning of a Microstrip Antenna on a Ferrite Substrate. Electronics Letters 24, 729–731 (1988)
3. Pozar, D.M.: Radiation and Scattering Characteristics of Microstrip Antennas on Normally Biased Ferrite Substrates. IEEE Trans. Antennas Propag. 40, 1084–1092 (1992)
4. Landau, L.D., Lifshitz, E.M.: Electrodynamics of Continuous Media, 2nd edn. Pergamon Pr., Oxford (1984)
5. Zervos, T., Stamopoulos, D., Lazarakis, F., Alexandridis, A.A., Pissas, M., Giannakopoulou, T., Dangakis, K.: Use of Multiferroic Materials in Patch Antenna Design. In: 3rd European Conference on Antennas and Propagation (2009)

On the Design of Direct Radiating Antenna Arrays with Reduced Number of Controls for Satellite Communications

Theodoros Kaifas, Katherine Siakavara, Dimitrios Babas, George Miaris,
Elias Vafiadis, and John N. Sahalos

RadioCommunications Laboratory, Department of Physics
Aristotle University of Thessaloniki, 541 24, Thessaloniki, Greece
tkaif@physics.auth.gr, skv@auth.gr, babas@auth.gr,
gmiar@physics.auth.gr, vafiadis@auth.gr, sahalos@auth.gr

Abstract. Our activity has to do with the design of Direct Radiating Arrays (DRA) for satellite communications. The objective is to have a reduced number of controls in order to minimize the manufacturing and operating complexity. The DRAs will create a set of simultaneously overlapped multi-beams in the frequency range of 20 GHz and will satisfy certain specifications (End of Coverage (EOC) gain, grating and side lobe levels). Radio-Communications Laboratory (RCL) shall consider the DRA design and shall mainly optimize the geometry of the array and develop the appropriate software tool. The design methods which are going to be used are the Fractal Technique and the Orthogonal Method(OM) in conjunction with the Orthogonal Perturbation Method (OPM). Some preliminary examples are presented and show the effectiveness of the design methods.

Keywords: Array Synthesis, Direct Radiating Arrays (DRA), Fractal Antennas, Orthogonal Method (OM).

1 Introduction

Satellite services in the near future are expected to be demanding in terms of data rate. In order to satisfy more advanced characteristics, the next generation of satellites will use new resources (Ka and Q/V bands). Most of the satellites offer an overlapped beam spot coverage with high gain performance to ensure high figures in terms of EIRP and G/T. This approach could relax the requirements for the earth stations. This is a fundamental aspect especially for mobile applications where it is important to reduce the size of the antenna terminals.

In order to overcome the problems, which are met in Focal Array Fed Reflectors, [1], such as the reflector deployment, the feed cluster/reflector alignment and the capability to reconfigure the beam spot coverage, innovative antenna configurations are investigated. Such an antenna configuration is the Direct Radiating Array (DRA), [2]. Unfortunately, a canonical approach for the DRA design foresees a very large

F. Granelli et al. (Eds.): MOBILIGHT 2009, LNICST 13, pp. 420–429, 2009.
© ICST Institute for Computer Sciences, Social-Informatics and Telecommunication Engineering 2009

number of controls. A drastic reduction of this number is possible by using appropriate sub-arrays or horn apertures with a dimension figure of several wavelengths. The position of the above can be optimized to create lattices that produce Sparse or Thinned arrays.

The satellite is planned to have a European coverage of 19 beams and the technical requirements are given in Table 1.

Table 1. Technical requirements of the DRA

Technical requirement	Value
Spots number	19
Spot diameter	$0.65°$
Inter - Spot distance (spot spacing)	$0.56°$
Tx band	19.7 – 20.2 GHz
Polarization f1/f4 beams	RHCP
Polarization f2/f3 beams	LHCP
Frequency f1/f2 beams	19.7 – 19.95 GHz
Frequency f3/f4 beams	19.95 – 20.2 GHz
Single Entry C/I	> 20dB
Aggregate C/I	> 14dB
D_{EOC}	> 43.8dBi
GL on earth (out of coverage)	< -20dB
GL out of the earth	< -10dB
Maximum Array Diameter	1.8m

The methods that are used to reach a suitable solution for the afore-mentioned problem are a) the Orthogonal, [3], in conjunction with Orthogonal Perturbation Method (OPM), [4], and b) the Fractal Array Technique, [5], [6].

In the first approach we perturb the element positions by combining an iterative technique with the orthogonal method, (OM). The final position of the elements of the array is found from the last iteration where the desired approximation of the pattern is obtained. It must be noticed that in the project frame, OPM is also going to be used for the optimization of the arrays designed by the other partners Consorzio Nazionale Interuniversitario per le Telecomunicazioni(CNIT).

In the fractal technique, a planar array of four elements is used as generator and the entire system is produced by four steps of development. Optimum performance is obtained by four means: a) the suitable choice of the scaling factor value, b) the thinning of the array, c) the array feeding via one step of quantization and d) the perturbation of the element positions. The number of control points is reduced by grouping the elements in uniformly excited blocks.

Using the synthesis techniques of CNIT and RCL, and after a final optimization, the most effective layout will be manufactured (Space Engineering) for the demonstrator.

In the following sections the description and some representative examples are given for both approaches.

2 Formulation

2.1 The Orthogonal Perturbation Method

An array is a linear system composed by N element radiators. The field of the array is a vector in a vector space consisting of all the possible fields that this system can produce. After defining a norm, one can use whatever set of N linearly independent and normalized fields to define this space. The total electric field is the summation of the element fields, which can be expressed in the following matrix form:

$$\overline{E}(\theta,\phi) = [\mathbf{W}]^t \cdot \left[\overline{F}\right] \tag{1}$$

where $[\ ]^t$ stands for transpose. The n-th element, $\overrightarrow{F}_n(\theta,\phi)$, of the vector $\left[\overline{F}\right]$ is the electric far field produced by the n-th radiating element. The n-th weight, W_n in [W] is the respective complex amplitude excitation.

In the orthogonal perturbation method we perturb the position of the elements. In this case we keep the excitation vector [W] constant while we permit in equation (1) the vector $\left[\overline{F}\right]$ to change. In the present study the element positions can be identified by up to two independent variables p_n, q_n. The field variation $\delta\overrightarrow{E}(\theta,\phi)$ of the array can be written as:

$$\delta\overrightarrow{E}(\theta,\phi) = \sum_{n=1}^{N} W_n \delta\overrightarrow{F}_n(\theta,\phi) = [\mathbf{W}]^t \cdot \left[\delta\overrightarrow{F}\right] \tag{2}$$

or:

$$\delta\overrightarrow{E}(\theta,\phi) = \sum_{n=1}^{2N} A_n \frac{\partial\overrightarrow{F}_n(\theta,\phi)}{\partial a_n} = [A]^t \cdot \left[\frac{\partial\overrightarrow{F}}{\partial a}\right] = [A]^t \cdot \left[\overrightarrow{G}\right] \tag{3}$$

where:

$$\frac{\partial\overrightarrow{F}_n(\theta,\phi)}{\partial a_n} = \frac{\partial\overrightarrow{F}_n(\theta,\phi)}{\partial p_n}; \quad A_n = W_n \, \delta p_n, \quad n \leq N$$

$$\frac{\partial\overrightarrow{F}_n(\theta,\phi)}{\partial a_n} = \frac{\partial\overrightarrow{F}_n(\theta,\phi)}{\partial q_n}; \quad A_n = W_{n-N} \, \delta q_{n-N}, \quad 2N \geq n > N \tag{4}$$

δp_n, δq_n are the variations of the independent variables p_n, q_n of the n-th element.

Following the OM in equation (3), [4], we reach:

$$\{\delta p_n\}_{\nu+1} = \{A_n\}_{\nu+1} / W_n, n \leq N$$
$$\{\delta q_{n-N}\}_{\nu+1} = \{A_n\}_{\nu+1} / W_{n-N}, 2N \geq n > N \tag{5}$$

The new element-coordinates follow:

$$\{p_n\}_{v+1} = \{p_n\}_v + \{\delta p_n\}_{v+1}$$
$$\{q_n\}_{v+1} = \{q_n\}_v + \{\delta q_n\}_{v+1}$$

(6)

Adjusting the method to the requirements and as they are given for the far field in a bound constraint form, the desired pattern is described accordingly. Suppose that for $(\theta,\varphi) \in \Omega$, (where Ω is a specific angular domain), $|\vec{E}| \leq B$, (where B stands for the bound value), is needed. Then the right part of (2) defines a multi-valued function C as follows:

$$C = \begin{cases} 0 & (\theta,\phi) \in \Omega \ and \ |\vec{E}| \leq B \\ \left\| \vec{E}^d - \vec{E} \right\|^2 & \text{otherwise} \end{cases}$$

(7)

Equivalently (2) can be left intact and the target pattern, in each iteration step, can be defined using the substitution:

$$\vec{E}^d \rightarrow \begin{cases} \vec{E} & (\theta,\phi) \in \Omega \ and \ |\vec{E}| \leq B \\ \vec{E}^d & \text{otherwise} \end{cases}$$

(8)

2.2 The Fractal Array Approach

The fractal technique is proposed for the synthesis of a DRA with planar configuration, which when operates as a phased array can produce nineteen pencil beams with the requirements given in Table 1.

The basic item of a fractally designed array is a generating sub-array and the entire radiating system can be formed via the recursive application of the generating sub-array under a specified scaling factor, which governs how large the array grows by each repetitive application of the generating array. Due to the nature of the fractal algorithm, the array factor of the nth stage is produced by the multiplication of all the factors of previous stages with the array factor of the nth stage. Moreover, at the nth stage the size of the array is equal to the one of the generating sub-array, multiplied by the nth power of the scaling factor and the element population is the (n+1)[th] power of the number of elements of generating array. The fractal procedure leads to a rapid growth of the antenna size, the rapid increase of the number of the elements and both of these parameters lead to large gain as it is preferable for the antenna under design.

The suppression of the grating maxima that come from the gradual increase of inter-element distances of the array can be obtained by suitable selection of the scaling factor. Furthermore the fractal arrangement of the DRA elements gives the potential to execute the beam scanning by grouping the elements in uniformly excited sub-arrays. As a consequence, the required phase shifters and drivers are less than one per element and the cost and complexity of the array is reduced.

For the implementation of the proposed DRA, a planar array (Fig. 1) of four elements positioned on the circumference of a circular ring of radius r, was selected as generator. The choice of the four elements was made under the criteria of high gain and low Side Lobe Level(SLL).

The general expression of the fractal array factor in the case of the ring generator case is

$$AF_F(\theta,\phi) = \prod_{p=1}^{N_F}\left[\sum_{n=1}^{N_g} I_n e^{j\delta^{p-1}\psi_n(\theta,\phi)}\right] \tag{9}$$

where

$\psi_n = kr\sin\theta\cos(\phi-\phi_n)+\alpha_n$, $k = 2\pi/\lambda$, N_g is the number of elements of the subbarray, r is the radius of the generator ring, p the stage of growth of the fractal development, δ the scaling or expansion factor of the array, I_n and a_n are the excitation current amplitude and current phase of the nth element located at $\phi = \phi_n$.

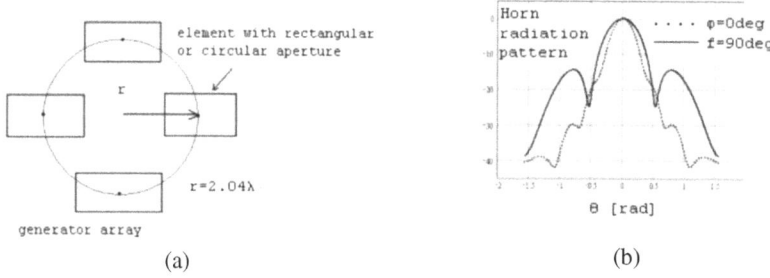

(a)   (b)

Fig. 1. a) The generating sub-array. b) Radiation pattern. Element type: Rectangular Horn , Directive gain ~17dB, Horn aperture about 3.98λx1.98λ, Waveguide: WR-51, a=12.9mm, b=6.48mm.

The selection of the values of r and δ was made under the requirements for the EOC directivity, the C/I ratio and the available area of occupation. Four steps of fractal development led to an array of 1024 elements and EOC directivity greater than 43.8dB, (Fig. 2a). To realize the feeding, the elements were grouped in blocks of 16, excited in phase and with equal amplitude. So, the number of control points is 64. With intense to ensure low complexity at the feeding network, all the blocks were initially considered as uniformly illuminated and solely a phase shift between adjacent blocks is necessary for the eighteen out of nineteen beams to be produced. The entire array fulfils the criteria of the directivity and occupation, for a large number of values of the pair (r, δ). The problem is that not all of these pairs guide the antenna to match the requirement of the C/I ratio. The selection of the proper pair can be made by an approximate process as follows. The array factor of Eq. (9), if we take into account that the elements of the DRA are uniformly excited, can be written as the multiplication of N_F array factors of uniform arrays of 2x2 elements. The suppression of the grating lobes could be obtained if we choose the values of r and δ such that the mth maximum of the factor of pth step coincides with the $(m-1)^{th}$ zero of the factor of the $(p-1)^{th}$ step.

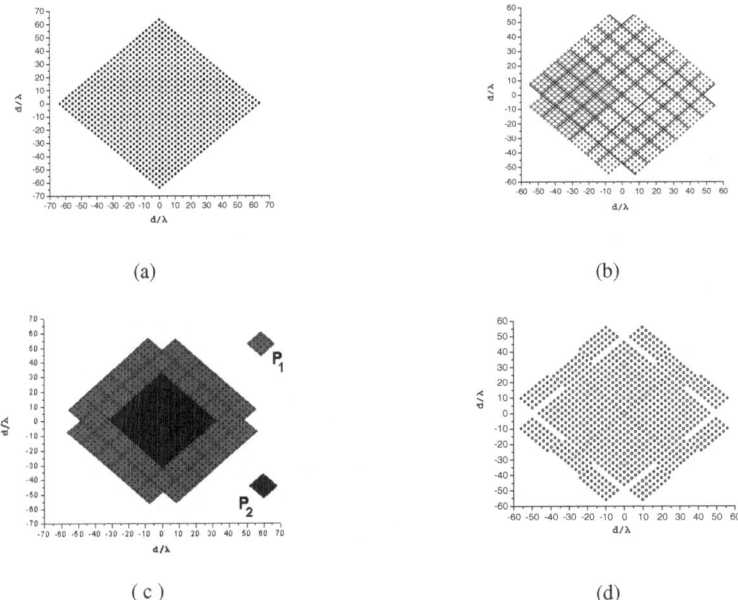

Fig. 2. a) The fractal fully populated array of 4th stage. Number of elements: 1024 b) The thinned array. Number of elements: 960, Number of control points: 60, Array size: 108.34λ x 108.34λ. c) Excitation of the blocks d) The proposed thinned array. Number of elements: 960, Number of control points: 60, Array size: 111.72λ x 111.72λ. Excitation: one step of quantization, P_2:P_1=2:1.

This evaluation is approximate as is valid solely for the central beam with maximum at $\theta=0°$ and $\varphi=0°$. In order the performance of the array to meet the requirements for all the nineteen beams and also to obtain better results, even for the central beam, some modifications to the geometry of the fully populated array as well as to the way of feeding were made:

a) The array was thinned by missing 16 elements from each one of its four vertices. So the number of elements of the proposed array is 960, grouped in 60 blocks of sixteen elements and the number of control points is reduced to 60 (Fig.2b)

b) Non-uniform excitation between the blocks is applied (Fig. 2c).

c) The value of δ was selected as $\delta=1.99$, that is a little smaller than 2 with intense the value of r to be $r=2.04\lambda$. With this pair of (r, δ) the size of the thinned array is 111.72λ x 111.72λ, whereas the value of r provides the potential to use, as radiating element, a horn antenna with aperture that gives directive gain ~17dB.

d) The positions of 16 blocks at the outline of the DRA were perturbed. So, the proposed array is that shown in Fig. 2d.

3 Results

3.1 Orthogonal Perturbation Method

In order to produce a sample result of the OPM we use a uniformly excited square array consisting of 121 (=11x11) circular apertures, (radius = 3λ), spaced 7.5λ apart. Each element is assumed to support a uniform aperture field. For the design of a broadside beam nearly 200 iterations of the OPM suffice to design an array that fulfils the requirements. The resulting array layout is given in Fig. 3. The broadside beam, accompanied by the bound constraints (red line) is given in Fig. 4a. Using one control point/ phase-shifter, per element the outermost beam is given in Fig. 4b.

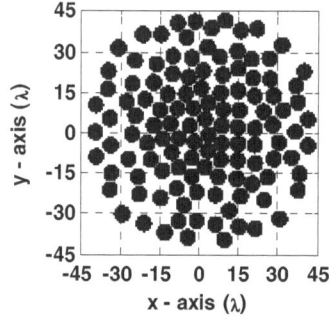

Fig. 3. The outcome of the OPM: Array Layout

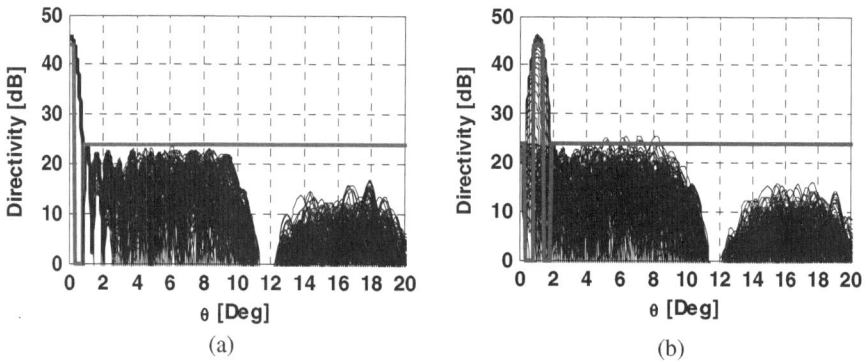

(a)   (b)

Fig. 4. The outcome of the OPM: (a) Broadside Beam, (b) Outermost beam

3.2 The Fractal DRA

Indicative results for the configuration of Fig. 2b and 2d are presented in figures 5 to 8. As source element was used a horn antenna with directivity 17.dB. The aperture horn was 3.97λ×1.97λ, it was fed by the waveguide: WR-51(a=12.9mm, b=6.48mm) and its radiation patterns on the two main planes are shown in Fig. 1b.

Figure 5 illustrates the radiation patterns of the central beam for $\varphi=0°$ - $90°$. The excitation of all the blocks is uniform, meanly P_2/P_1. It is shown that the antenna fails to fulfil the imposed requirements, as the C/I ratio is 10.97dB. Better results were received when one step of quantization was imposed. The results approached the requirements but are not judged satisfactory. In Table 2, analytical information about the operation of the DRA is presented.

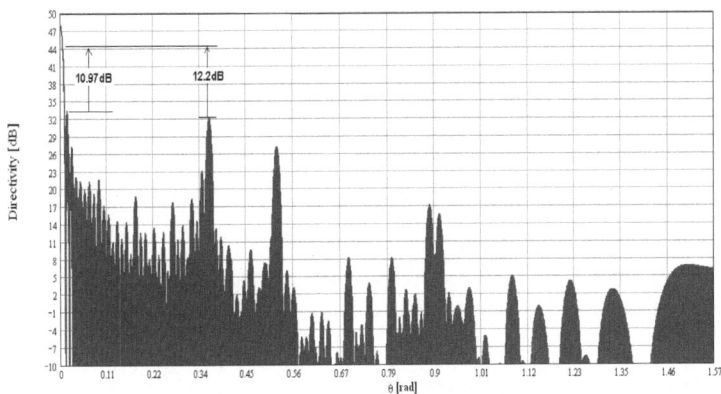

Fig. 5. Patterns, for $\varphi=0°$ - $90°$, of the central beam of the DRA of fig. 2d. Excitation: $P_2/P_1=1$.

Table 2. DRA records for various P_2/P_1 values

P_2/P_1	Max. directivity [dB]	EOC directive gain [dB]	C/I [dB]	Out of earth level [dB]
1	48.1	44.5	11	12.2
2	46.8	43.8	17.9	15.3
3	45.6	42.9	19.4	14.9
4	44.8	42.4	18.5	14.7

A small perturbation of 16 blocks at the perimeter, led the DRA operation to match the imposed requirements. In this case two steps of excitation with ratio $P_2/P_1 = 2$ are necessary. Analytical results are shown in figures 6 and 7. In Fig. 6 the radiation patterns of the central beam and two adjacent beams 1.12deg apart from the central are depicted. For all the three beams we have $D_{max} > 45dB$, C/I $> 23dB$ and $D_{EOC} > 43dB$. Respective results for the central beam and two adjacent beams $0.56°$ apart from the central are depicted in Fig. 7. In this case the records for beams C and D are: $D_{max}=45dB$, C/I$=20.2dB$ and $D_{EOC}=42.5dB$. The range in which the values of these parameters vary, for all the nineteen beams, are presented in summary in Table 3.

In figure 8 the patterns of the central beam for azimuth angles φ between $0°$ and -$90°$ are illustrated. It is shown that not only inside the coverage area but also out of it – on earth and out of earth - grating lobes with levels ~23dB and ~15.8dB, respectively, lower than that of the maximum radiation are ensured.

Table 3. Values of the performance parameters for all beams

Max. directivity [dB]	EOC directive gain [dB]	C/I[dB]	Out of earth level [dB]
45 - 47.9	42.5 – 45.6	20.2 – 24	<12.8

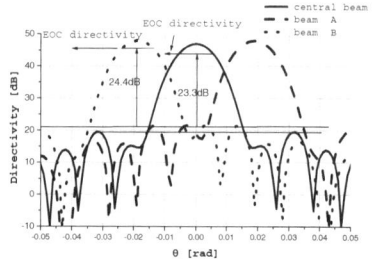

Fig. 6. The central beam and two adjacent-beams 1.12° apart from the central. <u>Beams A, B</u>: max. directivity 47.9dB, EOC directivity 45.6dB, C/I= 24.3dB <u>Central Beam</u> : max. directivity 46.9dB, EOC directivity 43.7dB, C/I= 23.4dB.

Fig. 7. The central beam and two adjacent-beams 0.56° apart from the central. <u>Beams C, D</u>: max. directivity 45.1dB, EOC directivity 42.5dB, C/I=20.2dB.

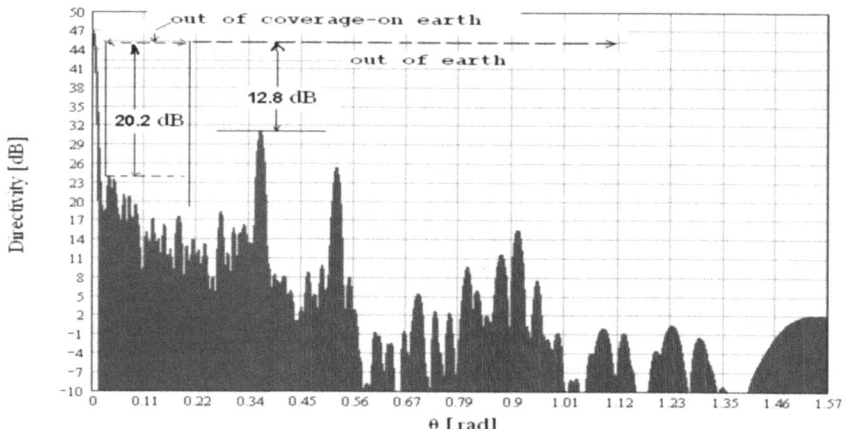

Fig. 8. Patterns, for φ= 0° - 90°, of the central beam of the DRA of fig. 2d. Excitation: P_2/P_1=2.

4 Conclusions

Two design methods for the synthesis of Direct Radiating Arrays (DRA) with a re-duced number of controls are given in the current work. The effort is part of the work included into the framework of a joint ESA program. The DRAs are to form a set of simultaneously overlapped multi-beams for satellite communications in the frequency range of 20 GHz and satisfy certain specifications (End of Coverage (EOC) gain, grating and side lobe levels).

The design methods are the Orthogonal Method (OM) in conjunction with a Perturbation (OPM) technique and the fractal technique. Some preliminary examples are presented and show the effectiveness of the design methods.

The OPM is enforced on a square array of uniformly excited circular element radiators. The desired pattern embodies the technical requirements of the broadside beam in the form of a bound constraints curve, (mask). The method is robust enough to produce an acceptable solution which can be further improved by using more than one type of elements and/ or quantized excitation. The ongoing effort is focused on those issues to further improve the method's performance.

The fractal technique is proved efficient for the design of a DRA of 960 elements which produce nineteen pencil beams for a satellite communication network. The inherent advantages of the fractal development permitted us to obtain, for all the beams, maximum and EOC directivity higher than 45dB and 42.5 dB respectively. C/I ratio larger than 20.2dB was attained by proper selection of the scaling factor, the perturbation of the positions of some blocks and applying one quantization step 2:1 of feeding. Moreover the fractal arrangement of the elements of the array and the ability to thin the array offered the potential to group the elements in blocks of uniformly excited elements. So, the number of the control points was reduced to 60, that is sixteen times smaller than it would be if one per element feeding was implied.

References

1. Mangenot, C., Lepeltier, P., Cazaux, J.L., Maurel, J.: Ka-band fed array focal reflector receive antenna design and development using MEMS switches. In: JINA Conference 2002, invited presentation, November 12-14, Nice, vol. II, pp. 337–345 (2002)
2. Mailloux, R.J.: Phased Array Antenna Handbook. Artech House, Boston (2005)
3. Sahalos, J.N.: Othogonal Methods for Array Synthesis: Theory & the ORAMA Computer Tool. John Wiley & Sons Ltd, West Sussex (2006)
4. Kaifas, T.N., Sahalos, J.N.: On the Geometry Synthesis of Arrays With a Given Excitation by the Orthogonal Method. IEEE Trans. Antennas and Propag. 56(12), 3680–3688 (2008)
5. Baliarda, C.P., Pous, R.: Fractal Design of Multiband and Low Sidelobes Array. IEEE Trans. Antennas Propagat. 44, 730–739 (1996)
6. Werner, D.H., Haupt, R.L., Werner, P.L.: Fractal Antenna Engineering: The Theory and Design of Fractal Antenna Arrays. IEEE Antennas and Propagation Magazine 41, 37–58 (1999)

Author Index